Kister Distillation Symposium 2017 (Full Papers)

Topical Conference at the 2017 AIChE Spring Meeting and
13th Global Congress on Process Safety

San Antonio, Texas, USA
26 – 30 March 2017

ISBN: 978-1-5108-4280-9

Printed from e-media with permission by:

Curran Associates, Inc.
57 Morehouse Lane
Red Hook, NY 12571

Some format issues inherent in the e-media version may also appear in this print version.

Printed by Curran Associates, Inc. (2017)

For permission requests, please contact AIChE
at the address below.

AIChE
120 Wall Street, FL 23
New York, NY 10005-4020

Phone: (800) 242-4363
Fax: (203) 775-5177

www.aiche.org

Additional copies of this publication are available from:

Curran Associates, Inc.
57 Morehouse Lane
Red Hook, NY 12571 USA
Phone: 845-758-0400
Fax: 845-758-2633
Email: curran@proceedings.com
Web: www.proceedings.com

TABLE OF CONTENTS

DISTILLATION 2017

IN THE FOOTSTEPS OF NEIL YEOMAN: PROVIDING INSPIRATION AND MENTORING TO NEW ENGINEERS

Our leading session honors the immense legacy of Neil Yeoman, AIChE Fellow and a great contributor to AIChE and distillation technology, who made it a life mission to mentor and inspire young engineers into top professionals. It is natural that the highlight of our symposium this year is providing inspiration to the up and coming young engineers.

Searching the web under *mentoring*, we found a study by Bob Audrey and Paul Cohen, published in their book "*Working Wisdom: Timeless Skills and Vanguard Strategies for Learning*" (1995). They found that the five most commonly used techniques among mentors are:

1. *Accompanying*, which is caringly taking part, side by side, in the learning and thinking process of the young engineers;

2. *Sowing*, which is preparing the learner for a change that is not presently understood or even acceptable to the learner, but will make sense in the future;

3. *Catalyzing,* which plunges the learner into a change, forcing the learner to think and seek his or her own path;

4. *Showing,* using your own example; and

5. *Harvesting,* creating awareness of what has been learned by experience.

China's great sage, Confucius, made the famous statement that "*learning without thinking is useless, while thinking without learning is dangerous*". Looks like he was 2500 years ahead of Audrey and Cohen, highlighting the two key values: *learning and thinking*. We hope that our symposium will provide the inspiration and mentoring to our up and coming engineers to learn, think, and achieve wonderful progress in our beloved field of distillation.

Our 2017 Distillation Symposium features improvements and innovations in troubleshooting, debottlenecking, simulation, and optimization that will help us pass our experiences to those who will fill up our shoes in the future. We have papers reporting cutting edge technology in distillation process control, dividing wall columns, FLNG applications, and in fundamental modeling of equipment and processes. There are papers reporting successful application of high performance trays including the ConSep, HiFi Plus, and Spray Trays. There are new breakthroughs on topics that we all thought we knew a lot about like mist elimination, dry tray pressure drop, tray operating windows, and the hidden theory behind the successful O'Connell plot for tray efficiency prediction.

We have papers on topics that have given us grief but were infrequently addressed in the literature. Instability due to multiple steady states in condensers, reboilers, and moving valve trays, makes headlines in several of our papers. Feed and reflux distributors, often tricky, are addressed by superb troubleshooting papers. Tray vibrations, dealing with VLE uncertainties, highly challenging refinery fractionator debottlenecks, pumparound optimization, and wash sections of vacuum towers are discussed in outstanding papers. Last but not least, exciting troubleshooting techniques such as gamma scans, thermal scans, judicious testing, and event timing analysis are featured in many great papers.

A special treasure for young as well as experienced engineers is our 3-hour tutorial session on Tower Inspections. The question "What should I look for when inspecting a tower", seldom answered at university, was answered by one of the industry's greats, Doug Bouck, who spent decades in the school of hard knocks.

It is our hope that the ideas and learning described in our symposium will become sharper pencils in the hands of practicing engineers, that will solve today's and avoid tomorrow's problems, and pave the way to efficient, high-capacity and trouble-free distillation.

Henry Kister
Chair
Distillation 2017

DRY TRAY PRESSURE DROP OF SIEVE TRAYS REVISTED

By

Daniel R. Summers, P.E.
SULZER CHEMTECH USA, Inc.
&
Dr. Tony J. Cai
Fractionation Research, Inc.

Presented at the AIChE Spring National Meeting
Topical 8:
Kister Distillation Symposium

Paper 4c
March 27, 2017
San Antonio, TX

Abstract

The sieve tray has been in the market place as a distillation device for many decades. It has been used extensively on distillation tray equipment worldwide as a highly efficient vapor/liquid-contacting device. Many people have examined dry tray pressure drop data for sieve tray decks including both authors. The primary author has published a recent paper correlating the data into an equation.[1] However, he had left the "door open" for additional refinement of this correlation for those applications that might have high tray thickness to hole diameter ratios (e.g. tiny hole applications in aluminum as had been practiced in air separations). The secondary author made a recent presentation[2] on this very subject employing Computational Fluid Dynamics (CFD) for very thick trays. This current paper will combine both papers into a single definitive correlation for the dry tray pressure drop of sieve trays. An improved correlation with the proper functionality for the dry tray pressure drop of sieve trays is presented that incorporates the effects of tray thickness, sieve hole diameter and open area and is compared to experimental data.

Discussion

As was stated in the primary author's original paper[1], "Dry tray pressure drop is an extremely important hydraulic parameter that aids the tray designer in many different ways. Obviously it is one of the major contributing parameters in the overall tray pressure drop. But more importantly, its magnitude can tell an experienced designer whether he or she is near flood, is at turndown, has a stable tray, or may be approaching spray fluidization. Finally, it can be used to understand the overall behavior/performance of a tray. Dry tray pressure drop is a fundamental building block of most other hydraulic parameters and its accuracy (or inaccuracy) has far reaching consequences."

The dry tray pressure drop hydraulic parameter is quite a simple concept. For fixed opening devices, the dry tray pressure drop can be reduced from Bernoulli's principles and takes on the form of the following equation:

$$\Delta \mathbf{P_{DRY}} = \mathbf{12}\ \boldsymbol{\rho_V}\ (\mathbf{V_H}/\mathbf{Cv})^2 / (\mathbf{2}\ \mathbf{g_C}\ \boldsymbol{\rho_W}) \qquad \textbf{(Eq. 1)}$$

Where,

ΔP_{DRY} = dry tray pressure drop, inches of Water
V_H = Hole Velocity, ft/sec
ρ_V = Vapor Density, lb/ft^3
C_V = Hole (Orifice) discharge Coefficient
ρ_W = Water Density = 62.4 lb/ft^3
g_C = Acceleration of Gravity = 32.2 ft/sec^2

If one combines all the constants from Equation 1 into a single constant Cp, the equation becomes very simple:

$$\Delta \mathbf{P_{DRY}} = \mathbf{Cp}\ (\mathbf{V_H^2})\ \boldsymbol{\rho_V} \qquad \textbf{(Eq. 2)}$$

Where,

$$Cp = 1 / (334.76\ C_V^2) \qquad \text{(Eq. 3)}$$

The previous work defined a new correlation to handle the complexities of Cp with respect to open area, tray thickness and hole diameter. However, two data points forced the author to make the correlation have a maxima at a tray thickness to hole diameter ratio of about 0.35, see Figure 1. This maxima was never discovered or observed by anyone previously dealing with this topic.

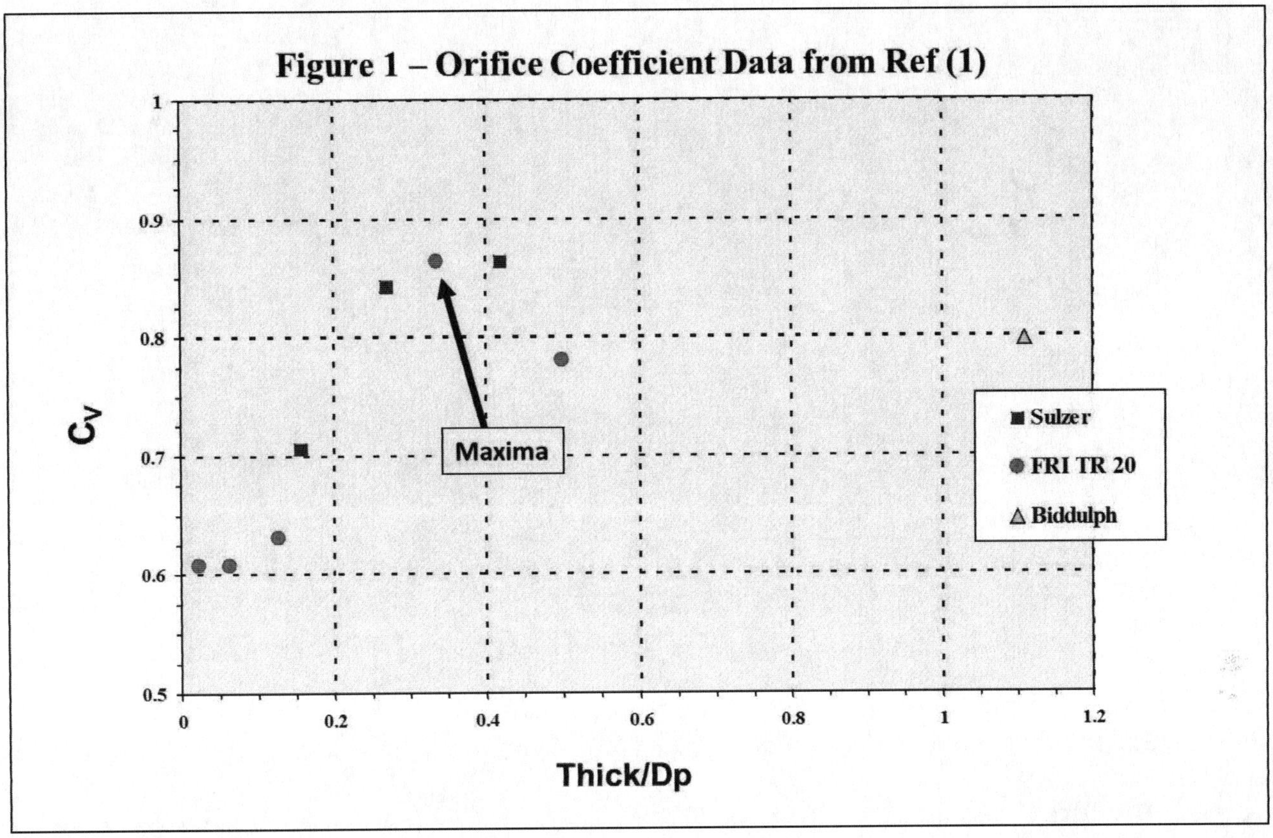

When Dr. Tony Cai presented his recent paper[2], it became obvious that this maxima was incorrect. Figure 2 shows the data from both papers of Summers and Cai. One can readily see that a different correlation (with no maxima) needed to be developed.

First however, the two "extraneous" data points from the work of Summers needed to be dispelled. The first is the rightmost FRI data point shown in Figure 1. This is a 1959 data point from Run #36 of a FRI Topical Report[4] and cannot be reproduced today. Since then, FRI has made a number of dry tray pressure drop measurements in a simulator with thicker trays[3] and as a result the authors can no longer support this data point. This work[3] is proprietary to FRI until 30 years after the report is written. The other data point is from Biddulph[5] and the authors discovered that this point is for smooth openings and was operated at very low Reynold's numbers. Again this data point cannot be supported. Therefore Figure 2 below has purposefully omitted these 2 data points.

The new curve is now a typical Sigmoid Function[6] that has the classic "S" shape. Selected dry tray pressure drop data used in the development of the new dry tray pressure drop correlation[2,3,4,7] is summarized in Appendix A. FRI has allowed us to use one of the data points from reference 3 that helps to "fill in the gap" at a thickness to hole diameter ratio of 0.6.

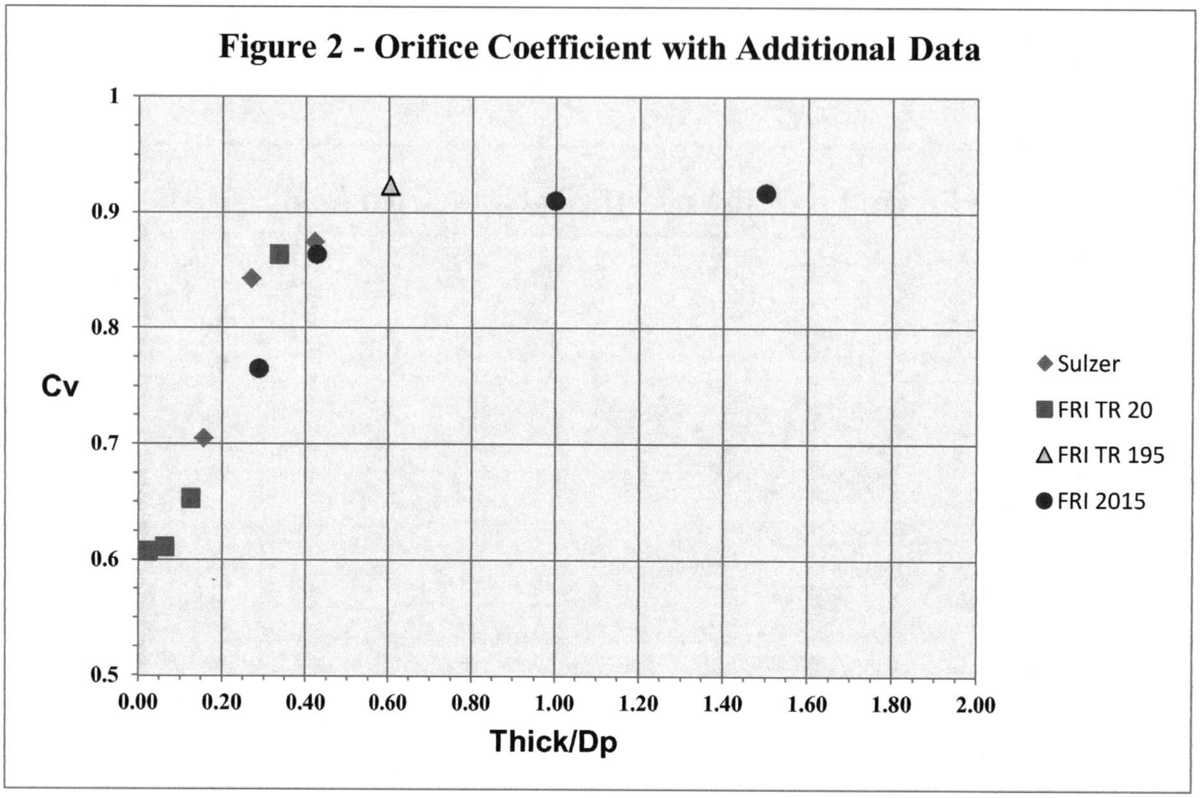

As was used in 2009, the open area effect on the orifice coefficient is reflected in the term "K" as defined here and first established by Kolodzie and Van Winkle[8];

$$\mathbf{K = C_V\,(Dp/Pitch)^{-0.10}} \qquad \textbf{(Eq. 4)}$$

The current authors found no data to support the need for any revision to this equation. However, the "K" equation itself needed to be updated. The new form of the Sigmoidal "K" equation is;

$$\mathbf{K = 0.997 - \{0.34 / [1.0 + (4.925\ TT/Dp)^{3.582}]\}} \qquad \textbf{(Eq. 5)}$$

Where,

Dp = Hole Diameter, inches
TT = Tray Thickness, inches
Pitch = Center to center distance between holes, inches

This equation, when compared to the data collection, is shown here in Figure 3. This equation is only good for Hole Reynolds numbers greater than 4000. For most applications this will be true. Keep in mind however that when the Hole Reynolds number is low, the dry tray pressure drop is also very low and the accuracy of its value becomes insignificant in comparison to the total tray pressure drop.

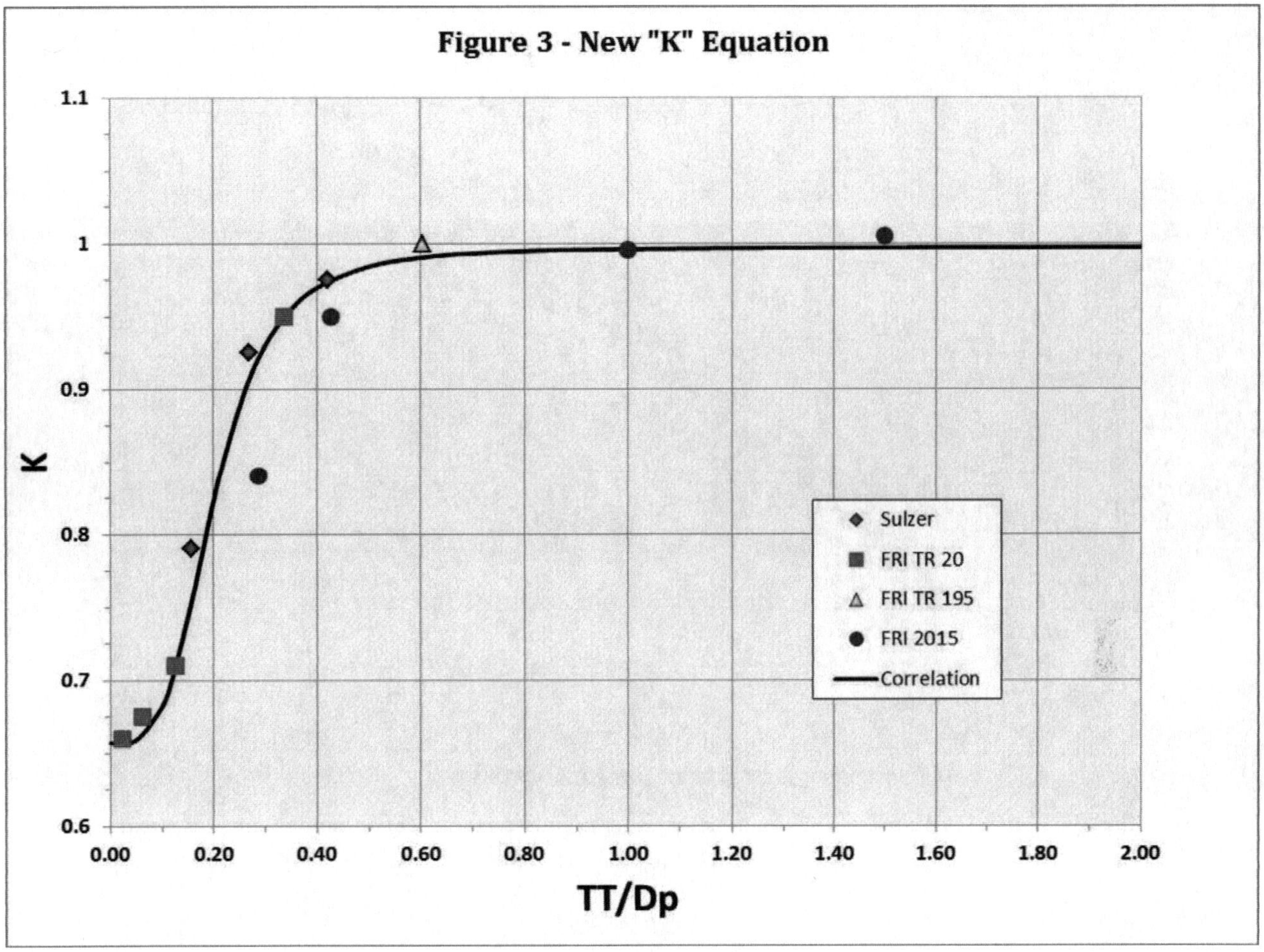

Applying the improved K equation to the above mentioned data, and limiting that data to open areas that are 20% of the active area or less, Figure 4 was generated. This figure clearly shows that the improved correlation is accurate by keeping the data points within a reasonable error band.

In conclusion, an improved and definitive sieve tray dry tray pressure drop equation is generated that matches, with mean absolute relative error 10.6%, data from four different sources. This equation can be used for all commercial sieve tray applications provided the holes are punched from the top in the downwards direction.

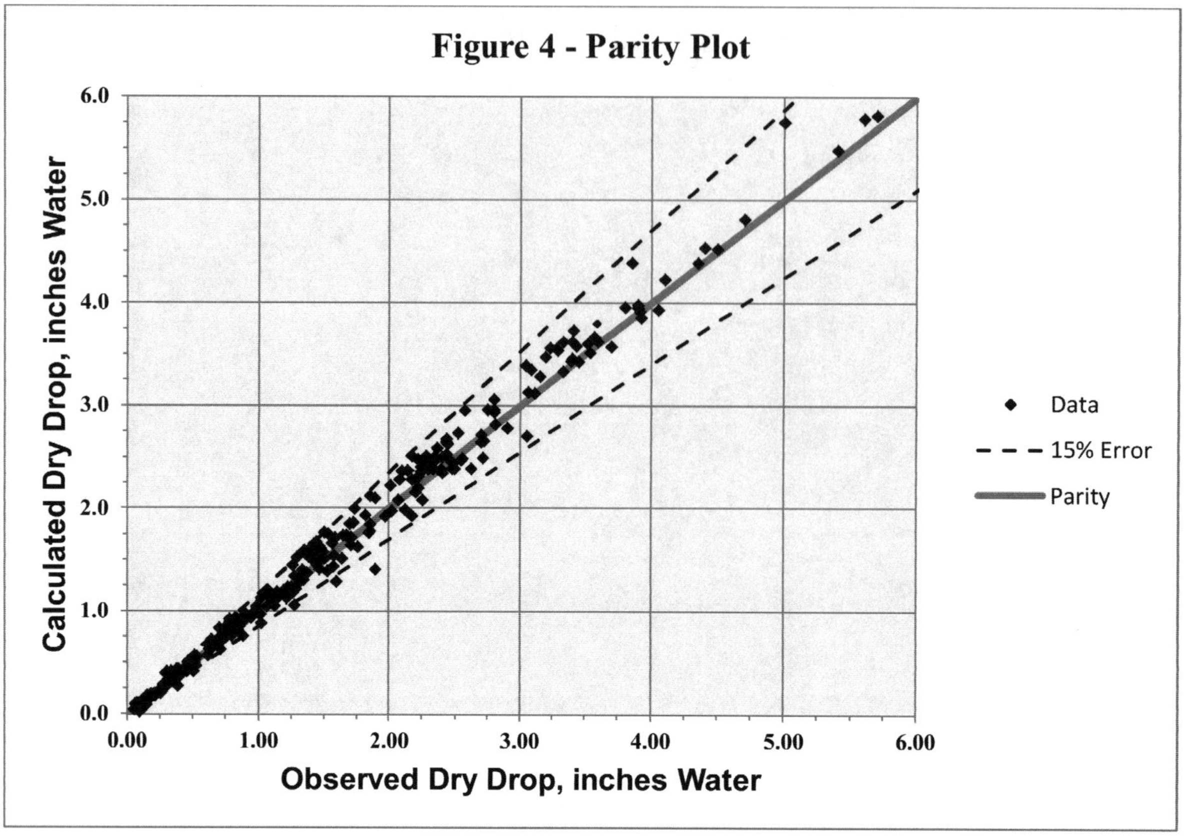

References

(1) Summers, D.R., "Dry Tray Pressure Drop of Sieve Trays", Chemical Engineering, June 2009, pp 36-39.

(2) Li, X. P., Whiteley, J. R., Resetarits, M. R., Cai, Tony, "CFD Simulation of Thick Sieve Tray Dry Tray Pressure Drops", 2015 AIChE Spring Meeting *Kister Distillation Symposium*, paper 206a.

(3) Cai, T. J., Resetarits, M. R., "Thick Sieve Tray Dry Pressure Drops", Fractionation Research Inc., Topical Report 195 (2013)

(4) FRI Topical Report No. 20, D. K. Clark and J. D. Miller (October 15, 1959)

(5) Biddulph, M. W., and Thomas, C. P., "Effect of Liquid Surface Tension on Small Hole Distillation Sieve Tray Pressure Drop", AIChE Journal, Volume 41, No. 4 pp 819-827 (1995)

(6) von Seggern, D., *CRC Standard Curves and Surfaces with Mathematics, 2nd ed.* Boca Raton, FL: CRC Press (2007)

(7) Sulzer Chemtech Laboratory Data 1993-1998, proprietary information

(8) Kolodzie, P. A. and M. Van Winkle, "Discharge Coefficients through Perforated Plates", AIChE Journal, Volume 3 pp. 305-312 (1957)

Appendix A – Dry Tray Pressure Drop Data from Sulzer and FRI

ρ_{V}, lb/ft^3	Dp, in.	TT, in.	Pitch, in.	V_H, ft/s	ΔP_{DRY} Data In. H_2O	ΔP_{DRY} Correlation In. H_2O
0.070859	0.5	0.0781	1.3010	15.179	0.09	0.1043
0.070989	0.5	0.0781	1.3010	30.303	0.33	0.4163
0.07084	0.5	0.0781	1.3010	45.550	0.82	0.9386
0.070622	0.5	0.0781	1.3010	60.921	1.44	1.6737
0.069759	0.5	0.0781	1.3010	77.094	2.43	2.6476
0.069968	0.5	0.0781	1.3010	92.236	3.59	3.8011
0.076136	0.5	0.0781	1.3010	14.127	0.09	0.0970
0.076783	0.5	0.0781	1.3010	28.017	0.32	0.3849
0.07699	0.5	0.0781	1.3010	41.912	0.82	0.8636
0.077233	0.5	0.0781	1.3010	55.707	1.38	1.5305
0.077078	0.5	0.0781	1.3010	69.773	2.33	2.3962
0.07693	0.5	0.0781	1.3010	83.889	3.4	3.4571
0.070308	0.5	0.0781	1.3010	15.298	0.07	0.1051
0.071529	0.5	0.0781	1.3010	30.074	0.3	0.4131
0.071823	0.5	0.0781	1.3010	44.927	0.78	0.9257
0.074305	0.5	0.0781	1.3010	57.902	1.35	1.5908
0.074558	0.5	0.0781	1.3010	72.132	2.22	2.4772
0.074619	0.5	0.0781	1.3010	86.487	3.23	3.5642
0.076096	0.5	0.0781	1.3010	14.135	0.08	0.0971
0.076573	0.5	0.0781	1.3010	28.093	0.3	0.3859
0.077052	0.5	0.0781	1.3010	41.878	0.77	0.8629
0.077354	0.5	0.0781	1.3010	55.620	1.29	1.5281
0.077842	0.5	0.0781	1.3010	69.089	2.14	2.3727
0.078361	0.5	0.0781	1.3010	82.357	3.04	3.3940
0.067875	0.5	0.0781	1.3010	15.847	0.11	0.1088
0.067918	0.5	0.0781	1.3010	31.673	0.38	0.4351
0.067881	0.5	0.0781	1.3010	47.536	0.95	0.9795
0.067878	0.5	0.0781	1.3010	63.384	1.64	1.7414
0.067901	0.5	0.0781	1.3010	79.203	2.7	2.7200
0.067858	0.5	0.0781	1.3010	95.104	3.91	3.9193
0.068194	0.5	0.0781	1.3010	15.773	0.13	0.1083
0.068045	0.5	0.0781	1.3010	31.614	0.37	0.4343
0.067834	0.5	0.0781	1.3010	47.569	0.9	0.9802
0.06756	0.5	0.0781	1.3010	63.683	1.53	1.7496
0.06746	0.5	0.0781	1.3010	79.721	2.52	2.7378
0.067151	0.5	0.0781	1.3010	96.106	3.8	3.9606
0.074667	0.5	0.0781	1.3010	14.405	0.1	0.0989
0.074776	0.5	0.0781	1.3010	28.769	0.35	0.3952
0.075167	0.5	0.0781	1.3010	42.928	0.86	0.8846
0.075382	0.5	0.0781	1.3010	57.075	1.48	1.5681
0.075597	0.5	0.0781	1.3010	71.140	2.48	2.4431
0.075623	0.5	0.0781	1.3010	85.339	3.54	3.5169
0.072153	0.5	0.0781	1.3010	14.907	0.12	0.1024
0.072656	0.5	0.0781	1.3010	29.608	0.36	0.4067
0.072982	0.5	0.0781	1.3010	44.214	0.87	0.9110
0.073729	0.5	0.0781	1.3010	58.354	1.47	1.6032
0.07385	0.5	0.0781	1.3010	72.823	2.43	2.5009

0.073844	0.5	0.0781	1.3010	87.395	3.52	3.6016
0.072677	0.5	0.0781	1.2996	14.769	0.12	0.1012
0.072947	0.5	0.0781	1.2996	29.428	0.36	0.4033
0.072933	0.5	0.0781	1.2996	44.150	0.86	0.9076
0.072808	0.5	0.0781	1.2996	58.968	1.47	1.6163
0.072733	0.5	0.0781	1.2996	73.786	2.43	2.5281
0.072227	0.5	0.0781	1.2996	89.163	3.57	3.6660
0.072094	0.5	0.0781	1.2996	14.888	0.1	0.1020
0.072339	0.5	0.0781	1.2996	29.675	0.33	0.4067
0.072748	0.5	0.0781	1.2996	44.262	0.83	0.9099
0.0731	0.5	0.0781	1.2996	58.732	1.39	1.6099
0.073258	0.5	0.0781	1.2996	73.257	2.28	2.5100
0.073149	0.5	0.0781	1.2996	88.039	3.33	3.6197
0.074214	0.5	0.0781	1.2996	14.463	0.12	0.0991
0.074313	0.5	0.0781	1.2996	28.887	0.34	0.3959
0.074351	0.5	0.0781	1.2996	43.308	0.84	0.8903
0.07433	0.5	0.0781	1.2996	57.760	1.41	1.5832
0.074198	0.5	0.0781	1.2996	72.329	2.35	2.4782
0.074018	0.5	0.0781	1.2996	87.006	3.43	3.5772
0.071129	0.5	0.0781	1.2996	15.090	0.1	0.1034
0.071075	0.5	0.0781	1.2996	30.203	0.33	0.4139
0.071	0.5	0.0781	1.2996	45.352	0.83	0.9323
0.07104	0.5	0.0781	1.2996	60.435	1.42	1.6565
0.071067	0.5	0.0781	1.2996	75.515	2.36	2.5874
0.07095	0.5	0.0781	1.2996	90.768	3.41	3.7319
0.073254	0.5	0.0781	1.2996	14.652	0.11	0.1004
0.073445	0.5	0.0781	1.2996	29.228	0.33	0.4006
0.073482	0.5	0.0781	1.2996	43.820	0.81	0.9008
0.073517	0.5	0.0781	1.2996	58.399	1.4	1.6007
0.073533	0.5	0.0781	1.2996	72.983	2.34	2.5006
0.073167	0.5	0.0781	1.2996	88.018	3.4	3.6189
0.073499	0.5	0.0781	1.2996	14.603	0.06	0.1001
0.073692	0.5	0.0781	1.2996	29.130	0.29	0.3992
0.074011	0.5	0.0781	1.2996	43.507	0.78	0.8944
0.074205	0.5	0.0781	1.2996	57.858	1.34	1.5859
0.074808	0.5	0.0781	1.2996	71.739	2.27	2.4580
0.074766	0.5	0.0781	1.2996	86.135	3.29	3.5415
0.073247	0.5	0.0781	1.2996	14.654	0.07	0.1004
0.073609	0.5	0.0781	1.2996	29.163	0.3	0.3997
0.073938	0.5	0.0781	1.2996	43.550	0.79	0.8953
0.074208	0.5	0.0781	1.2996	57.855	1.38	1.5858
0.07442	0.5	0.0781	1.2996	72.113	2.29	2.4708
0.074493	0.5	0.0781	1.2996	86.451	3.28	3.5544
0.075741	0.5	0.0781	1.2996	14.171	0.13	0.0971
0.076118	0.5	0.0781	1.2996	28.202	0.35	0.3865
0.076477	0.5	0.0781	1.2996	42.104	0.8	0.8656
0.076455	0.5	0.0781	1.2996	56.155	1.38	1.5392
0.076349	0.5	0.0781	1.2996	70.291	2.24	2.4084
0.076278	0.5	0.0781	1.2996	84.428	3.19	3.4713
0.072399	0.5	0.0781	1.2796	14.371	0.14	0.0952
0.072423	0.5	0.0781	1.2796	28.732	0.37	0.3805

0.072433	0.5	0.0781	1.2796	43.093	0.85	0.8561
0.072486	0.5	0.0781	1.2796	57.415	1.45	1.5208
0.072378	0.5	0.0781	1.2796	71.875	2.39	2.3798
0.072366	0.5	0.0781	1.2796	86.265	3.45	3.4274
0.071806	0.5	0.0781	1.2796	14.490	0.12	0.0959
0.071803	0.5	0.0781	1.2796	28.980	0.34	0.3838
0.071977	0.5	0.0781	1.2796	43.366	0.82	0.8615
0.072121	0.5	0.0781	1.2796	57.705	1.39	1.5285
0.072258	0.5	0.0781	1.2796	71.995	2.33	2.3837
0.072246	0.5	0.0781	1.2796	86.408	3.39	3.4331
0.072934	0.1875	0.0781	0.4569	12.935	0.08	0.0460
0.073089	0.1875	0.0781	0.4569	25.815	0.21	0.1837
0.073297	0.1875	0.0781	0.4569	38.612	0.42	0.4121
0.073403	0.1875	0.0781	0.4569	51.409	0.69	0.7316
0.07342	0.1875	0.0781	0.4569	64.246	1.15	1.1428
0.073472	0.1875	0.0781	0.4569	77.040	1.7	1.6445
0.073904	0.1875	0.0781	0.4569	12.765	0.06	0.0454
0.074128	0.1875	0.0781	0.4569	25.453	0.17	0.1811
0.074302	0.1875	0.0781	0.4569	38.090	0.42	0.4065
0.074351	0.1875	0.0781	0.4569	50.753	0.72	0.7222
0.074319	0.1875	0.0781	0.4569	63.469	1.21	1.1290
0.074238	0.1875	0.0781	0.4569	76.246	1.75	1.6275
0.068905	0.1875	0.0781	0.4569	13.691	0.06	0.0487
0.069007	0.1875	0.0781	0.4569	27.342	0.18	0.1945
0.069067	0.1875	0.0781	0.4569	40.977	0.43	0.4373
0.069016	0.1875	0.0781	0.4569	54.676	0.75	0.7781
0.068839	0.1875	0.0781	0.4569	68.521	1.28	1.2189
0.068647	0.1875	0.0781	0.4569	82.455	1.84	1.7601
0.074798	0.1875	0.0781	0.4523	12.361	0.08	0.0430
0.075325	0.1875	0.0781	0.4523	24.550	0.18	0.1709
0.075902	0.1875	0.0781	0.4523	36.545	0.4	0.3815
0.07613	0.1875	0.0781	0.4523	48.580	0.68	0.6762
0.076433	0.1875	0.0781	0.4523	60.484	1.12	1.0524
0.076441	0.1875	0.0781	0.4523	72.574	1.63	1.5152
0.075381	0.1875	0.0781	0.3902	9.128	0.09	0.0229
0.075487	0.1875	0.0781	0.3902	18.230	0.14	0.0917
0.075578	0.1875	0.0781	0.3902	27.312	0.25	0.2060
0.075523	0.1875	0.0781	0.3902	36.443	0.39	0.3665
0.075611	0.1875	0.0781	0.3902	45.501	0.61	0.5720
0.075483	0.1875	0.0781	0.3902	54.693	0.84	0.8251
0.071909	0.5	0.1345	1.2817	12.382	0.1	0.0484
0.072411	0.5	0.1345	1.2817	24.592	0.21	0.1923
0.0725	0.5	0.1345	1.2817	36.843	0.45	0.4322
0.072434	0.5	0.1345	1.2817	49.169	0.74	0.7691
0.072376	0.5	0.1345	1.2817	61.511	1.19	1.2027
0.072032	0.5	0.1345	1.2817	74.165	1.67	1.7402
0.075189	0.5	0.1345	1.2817	11.842	0.09	0.0463
0.075566	0.5	0.1345	1.2817	23.566	0.2	0.1843
0.075772	0.5	0.1345	1.2817	35.252	0.42	0.4136
0.07578	0.5	0.1345	1.2817	46.998	0.7	0.7352
0.075627	0.5	0.1345	1.2817	58.866	1.11	1.1510

0.075529	0.5	0.1345	1.2817	70.731	1.56	1.6596
0.075201	0.5	0.1345	1.2702	11.840	0.05	0.0462
0.075598	0.5	0.1345	1.2702	23.556	0.15	0.1839
0.076161	0.5	0.1345	1.2702	35.072	0.38	0.4107
0.076597	0.5	0.1345	1.2702	46.497	0.64	0.7260
0.076736	0.5	0.1345	1.2702	58.016	1.02	1.1323
0.0767	0.5	0.1345	1.2702	69.651	1.46	1.6313
0.075	2.625	0.063	6.5000	32.000	0.63	0.6369
0.075	2.625	0.063	6.5000	45.000	1.29	1.2594
0.075	2.625	0.063	6.5000	59.500	2.22	2.2018
0.075	2.625	0.063	6.5000	70.800	3.11	3.1175
0.075	2.625	0.063	6.5000	80.000	3.9	3.9803
0.075	2.625	0.063	6.5000	85.400	4.4	4.5358
0.075	1	0.063	2.3750	29.000	0.46	0.5110
0.075	1	0.063	2.3750	39.000	0.84	0.9242
0.075	1	0.063	2.3750	53.100	1.58	1.7133
0.075	1	0.063	2.3750	61.700	2.22	2.3133
0.075	1	0.063	2.3750	69.500	2.8	2.9351
0.075	1	0.063	2.3750	73.500	3.15	3.2827
0.075	1	0.063	2.3750	15.500	0.13	0.1460
0.075	1	0.063	2.3750	21.700	0.27	0.2861
0.075	1	0.063	2.3750	26.700	0.39	0.4332
0.075	1	0.063	2.3750	30.600	0.51	0.5690
0.075	1	0.063	2.3750	34.000	0.63	0.7024
0.075	1	0.063	3.0000	27.200	0.46	0.4711
0.075	1	0.063	3.0000	38.300	0.88	0.9340
0.075	1	0.063	3.0000	46.900	1.33	1.4005
0.075	1	0.063	3.0000	54.100	1.72	1.8636
0.075	1	0.063	3.0000	59.900	2.08	2.2846
0.075	1	0.063	3.0000	68.200	2.75	2.9615
0.075	1	0.063	3.0000	83.000	3.85	4.3864
0.075	1	0.063	3.0000	95.100	5	5.7585
0.075	1	0.063	3.0000	115.100	7.35	8.4353
0.075	1	0.063	3.0000	123.800	8.6	9.7586
0.075	1	0.063	3.0000	130.000	9.4	10.7606
0.075	1	0.063	3.0000	137.300	10.5	12.0030
0.075	1	0.063	3.0000	48.800	1.5	1.5163
0.075	1	0.063	3.0000	68.100	2.57	2.9529
0.075	1	0.063	3.0000	83.000	4.35	4.3864
0.075	1	0.063	3.0000	95.400	5.6	5.7949
0.075	1	0.063	3.0000	105.200	6.9	7.0466
0.075	1	0.063	3.0000	113.800	8.2	8.2458
0.075	1	0.063	3.0000	122.900	9.35	9.6173
0.075	1	0.063	3.0000	45.300	1.34	1.3066
0.075	1	0.063	3.0000	62.500	2.55	2.4872
0.075	1	0.063	3.0000	84.300	4.5	4.5248
0.075	1	0.063	3.0000	99.000	6.1	6.2405
0.075	1	0.063	3.0000	109.000	7.5	7.5649
0.075	1	0.063	3.0000	117.000	8.7	8.7161
0.075	0.5	0.063	1.1875	22.900	0.38	0.2778
0.075	0.5	0.063	1.1875	39.200	0.76	0.8139

0.075	0.5	0.063	1.1875	52.500	1.44	1.4598
0.075	0.5	0.063	1.1875	61.100	2.02	1.9773
0.075	0.5	0.063	1.1875	68.300	2.54	2.4707
0.075	0.5	0.063	1.1875	72.500	2.9	2.7840
0.075	0.5	0.063	1.3125	35.200	0.62	0.6695
0.075	0.5	0.063	1.3125	49.900	1.27	1.3455
0.075	0.5	0.063	1.3125	66.400	2.29	2.3824
0.075	0.5	0.063	1.3125	76.100	3.06	3.1293
0.075	0.5	0.063	1.3125	84.500	3.92	3.8583
0.075	0.5	0.063	1.3125	88.500	4.1	4.2322
0.075	0.5	0.063	1.5000	111.000	6.8	6.8379
0.075	0.5	0.063	1.5000	99.400	5.4	5.4834
0.075	0.5	0.063	1.5000	84.400	3.9	3.9533
0.075	0.5	0.063	1.5000	64.100	2.2	2.2803
0.075	0.5	0.063	1.5000	46.600	1.07	1.2052
0.075	0.5	0.063	1.1875	39.800	0.7	0.8390
0.075	0.5	0.063	1.1875	55.400	1.41	1.6256
0.075	0.5	0.063	1.1875	66.800	2.1	2.3634
0.075	0.5	0.063	1.1875	71.100	2.43	2.6775
0.075	0.5	0.063	1.1875	76.000	2.8	3.0592
0.075	0.5	0.063	1.1875	79.500	3.08	3.3475
0.075	0.5	0.063	2.0000	81.800	4.05	3.9334
0.075	0.5	0.063	2.0000	115.200	7.9	7.8014
0.075	0.5	0.063	2.0000	140.000	11.5	11.5218
0.075	0.5	0.063	2.0000	161.100	14.5	15.2565
0.075	0.5	0.063	2.0000	179.900	18.5	19.0251
0.075	0.5	0.063	2.0000	177.600	19.6	18.5418
0.075	0.5	0.063	2.0000	168.300	17.1	16.6507
0.075	0.5	0.063	2.0000	144.900	12.8	12.3425
0.075	0.5	0.063	2.0000	108.300	7	6.8948
0.075	0.5	0.063	2.0000	78.000	3.7	3.5765
0.075	0.5	0.063	2.0000	35.600	0.75	0.7450
0.075	0.5	0.063	2.0000	45.900	1.25	1.2385
0.075	0.5	0.063	2.0000	54.100	1.7	1.7205
0.075	0.5	0.063	2.0000	60.600	2.2	2.1588
0.075	0.5	0.063	2.0000	67.100	2.7	2.6467
0.075	0.5	0.063	2.0000	78.500	3.6	3.6225
0.075	0.5	0.063	2.0000	90.500	4.7	4.8146
0.075	0.5	0.063	2.0000	99.500	5.7	5.8198
0.075	0.1875	0.063	0.4375	26.700	0.24	0.2101
0.075	0.1875	0.063	0.4375	37.700	0.5	0.4189
0.075	0.1875	0.063	0.4375	50.800	0.88	0.7606
0.075	0.1875	0.063	0.4375	60.000	1.27	1.0610
0.075	0.1875	0.063	0.4375	66.100	1.59	1.2877
0.075	0.1875	0.063	0.4375	69.100	1.89	1.4072
0.075	0.1875	0.063	0.4688	36.800	0.41	0.4047
0.075	0.1875	0.063	0.4688	51.900	0.79	0.8049
0.075	0.1875	0.063	0.4688	63.100	1.14	1.1898
0.075	0.1875	0.063	0.4688	72.700	1.49	1.5793
0.075	0.1875	0.063	0.4688	80.400	1.81	1.9316
0.075	0.1875	0.063	0.4688	87.400	2.17	2.2826

0.075	0.1875	0.063	0.5000	37.200	0.45	0.4189
0.075	0.1875	0.063	0.5000	50.000	0.79	0.7567
0.075	0.1875	0.063	0.5000	68.400	1.46	1.4162
0.075	0.1875	0.063	0.5000	81.000	2.12	1.9860
0.075	0.1875	0.063	0.5000	88.800	2.62	2.3869
0.075	0.1875	0.063	0.5000	94.600	3.05	2.7089
0.075	0.5	0.25	1.5000	64.400	1.58	1.1953
0.075	0.5	0.25	1.5000	78.600	2.18	1.7806
0.075	0.5	0.25	1.5000	90.700	2.91	2.3710
0.075	0.5	0.25	1.5000	100.300	3.65	2.8995
0.075	0.5	0.25	1.5000	108.600	4.3	3.3992
0.075	0.5	0.25	1.5000	123.000	5.7	4.3604
0.075	0.5	0.25	1.5000	121.000	5.85	4.2198
0.075	0.5	0.25	1.5000	113.000	5	3.6802
0.075	0.5	0.25	1.5000	101.200	3.9	2.9518
0.075	0.5	0.25	1.5000	86.000	2.82	2.1317
0.075	0.5	0.25	1.5000	64.300	1.55	1.1916
0.075	0.5	0.25	1.5000	45.000	0.77	0.5836
0.075	0.2	0.063	0.4375	170.400	10.8	8.6351
0.075	0.2	0.063	0.4375	162.000	10	7.8047
0.075	0.2	0.063	0.4375	152.000	8.7	6.8709

Multiple Steady States in 2-Pass Hydraulics of Moving Valve Trays

By

Matthew Olsson, Fluor, Houston, Texas, USA

and

Henry Z. Kister, Fluor, Aliso Viejo, CA, USA

Presented at the Distillation Topical Conference, AIChE Spring Meeting, San Antonio, Texas, March 26-30, 2017

Summary

Our study reveals that multiple steady state vapor/liquid distributions may be experienced in 2-pass moving valve trays at turndown. Due to the symmetry of 2-pass trays, a perfect split of both vapor and liquid between the passes is always one possible distribution. The others involve having uneven vapor distribution. Some of these steady states effect wide variations in L/V ratios from panel to panel. This maldistribution is likely to lead to major efficiency reductions.

The maldistribution described in this paper is not due to poor distribution of vapor or liquid to the trayed section from an inlet, and it is also not caused by differences in tray geometry between one side and the other, though we have the capability to evaluate these situations.

The cause of the maldistribution is that at turndown, the valves close, and while they are closing, the dry pressure drop remains constant regardless of the vapor rate. This makes it easy for the vapor to swing from one pass to another, forming a number of alternative steady states.

Using our model, we investigated the effect of uniform weeping on the steady state multiplicity. While shifting the steady state some, uniform weeping does not largely impact the steady states. We also looked at the effect of weir loads on the steady state multiplicity. The multiplicity appears to exist both at the lower and higher weir loads. Our findings are that this is a vapor-driven phenomenon with little effect on or influence by liquid unless the L/V ratio is extremely high.

One strategy that appears to alleviate the effects of the multiplicity issue is using valves of different weight. This is a common practice used by valve tray manufacturers. Our study shows that this practice can significantly lessen the severity of maldistribution at turndown, but can also cause a wider range of loadings to experience some maldistribution.

Background

Moving valve trays are known to provide a turndown of 4-5:1 from the maximum operating loads (1,2). However there have been reports (3) and experiences in which turndown of multi-pass trays, including 2-pass trays, fell short of the expected turndown, with indication that maldistribution played a role (3).

In an attempt to gain understanding of the loss in turndown, we applied the Fluor Multi-pass Maldistribution Model (MMM) (4) to some hypothetical tray designs to explore the mechanisms detected. This paper describes our investigation.

Model Tray Design

The tower modeled has 8 ft internal diameter, and contains six 2-pass trays at 24" tray spacing. The top tray, Tray 6, is fed by an internal feed pipe and center false downcomer. The bottom tray, Tray 1, has a sealed center downcomer.

Even trays have side downcomers, and odd trays have center downcomers. The downcomers are straight, with the same width at the top and bottom. The side downcomer width is 11 inches, and the center downcomer width is 10 inches. The clearance between the bottom of each downcomer and the tray deck is 1 inch for both side and center downcomers. 1.5-inch inlet weirs are used to seal the downcomers, and the outlet weirs are 2 inches tall.

The tray deck is 14-gauge. Round moving valves are used, all with 16-gauge thickness and 7/16" leg length. The open slot area is 11% of the bubbling area.

To link with a system that has been studied extensively and for which abundant data and reliable flood and efficiency data and predictions are available, we chose to perform the calculations for the separation of cyclohexane from normal heptane at 24 psia and total reflux. Abundant data for this system in single-pass sieve and valve trays have been published and are available in the literature (5-7).

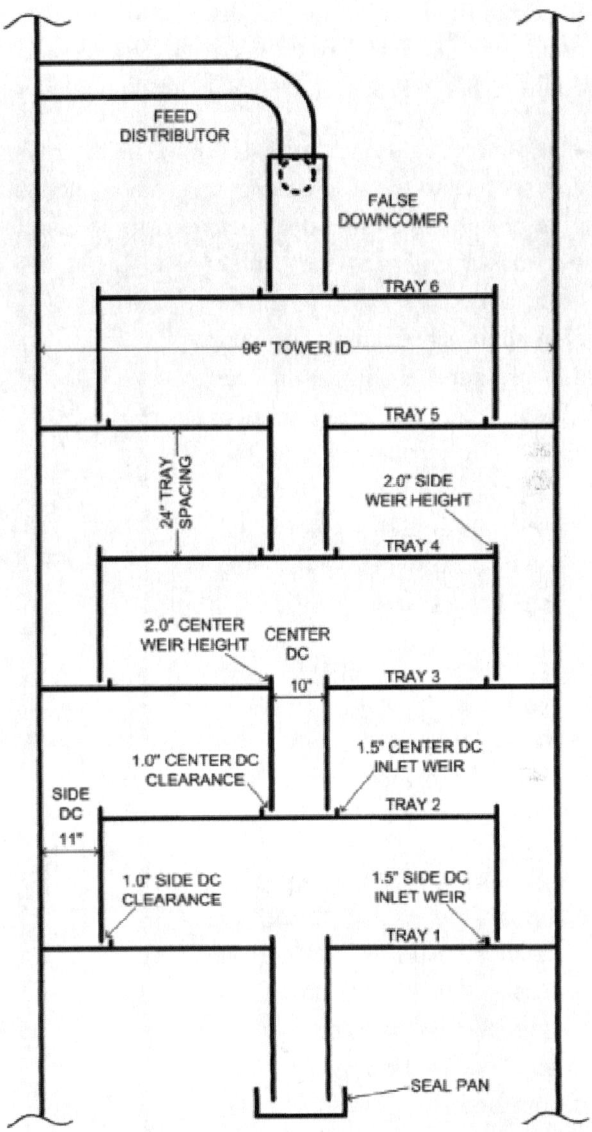

Figure 1. Key Tray Dimensions for Modeled Column.

Modeling Using the Multi-pass Maldistribution Model (MMM)

Our multi-pass maldistribution model is used to evaluate vapor-liquid distributions on multi-pass trays and to find hydraulically balanced distributions. Two equalities are required for a hydraulically balanced distribution for 2-pass trays, and both have to do with shared downcomers. For 2-pass trays, only the center downcomer is shared, and in the modeled column, Trays 1, 3, and 5 have center downcomers as illustrated in Figure 2.

The first equality is pressure drop. On trays for which liquid flows from the center downcomer to the side downcomers, the center downcomer divides the vapor space, and the pressure on either side does not need to be equal. On trays for which liquid flows from the side downcomers to the center downcomer, vapor is continuous above the tray, and the pressure equalizes in the shared space i.e. every two trays. For a balanced solution, the total pressure drop (through both trays) on either side of the shared downcomer must be equal.

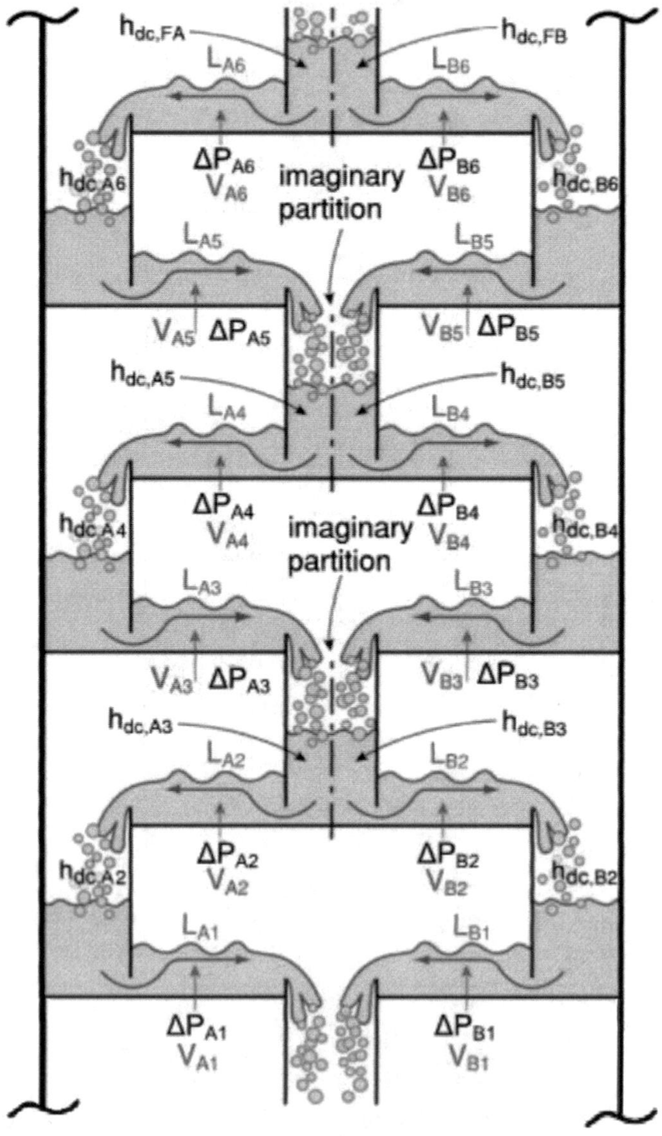

Figure 2. Key terms used in the multi-pass maldistribution model (MMM) equations.

The second equality is downcomer backup. As the center downcomer is shared, the calculated downcomer backup for either side must be equal.

The model used for this paper is an extension of our model to calculate maldistribution on 4-pass trays, presented previously (4). The 4-pass model was intended for tray design and rating, dividing the tower in half and hydraulically balancing the asymmetrical passes around the off-center downcomer to find a single solution. The tray design could be refined to match the L/V ratio on either side of the off-center downcomer.

The 4-pass model has only two

trays which are hydraulically linked, and the liquid rate to and from the center or side downcomer did not change. To obtain thorough insight into valve tray maldistribution, we developed a 2-pass maldistribution model that has 6 trays, hydraulically linked top to bottom, with a liquid feed (false downcomer) to the top tray and a vapor feed to the bottom tray. Total vapor and liquid loadings are constant throughout the tray section for both the 4-pass model and the new 2-pass maldistribution model.

With the 4-pass model, when the vapor loadings were such that one or more tray panels were operating in the region where valves open and close, the maldistribution would be much worse than if all valves were open or all were closed. We recommended avoiding this transition operating region throughout the tray section at design rates.

With 2-pass moving valve trays at turndown, the trays are operating in this transition region. The 2-pass maldistribution tool was designed to search for multiple hydraulically balanced solutions in the transition region. It was also designed to study the effects of feed maldistribution on the tray section, and it has features to calculate weeping and use multiple valve weights on a tray deck.

For the hydraulic calculations, we used the hydraulic equations described in our earlier work (4). For pressure drop calculation, we used the Klein method (8) and the Glitsch method (9), as we have had a lot of success with both. For downcomer backup, we used the classic downcomer backup equation in Perry's Handbook with the Klein method, and the Glitsch downcomer backup equation with the Glitsch method.

The vapor and liquid distributions were closely analyzed at 18%, 22%, 29%, 36% and 40% of the estimated jet flood point.

Steady State Multiplicity

2-pass trays are symmetrical, so a perfectly balanced solution with a 50%/50% split of vapor and the same for liquid is always a hydraulically balanced solution.

However, our MMM analysis found that for 2-pass moving valve trays, at some low loadings there is possibly more than one hydraulically balanced vapor and liquid distribution, a phenomenon we refer to as steady state multiplicity. Such multiplicity shows up when using the Klein pressure drop method (8), but not when using the Glitsch pressure drop method (9). The difference in the behavior of these methods at turndown is due to the combination of the valve tray dry pressure drop characteristic and the aeration factor.

For a moving valve tray, as the gas flow rate increases, the dry pressure drop increases to the point that the valves begin to open, the closed balance point (Figure 3). Then, the dry pressure drop stays constant (for the Klein method) from the gas flow rate when the first valve opens to the rate when all valves are open, the open balance point. Dry pressure drop again increases with the gas flow rate once all valves are open.

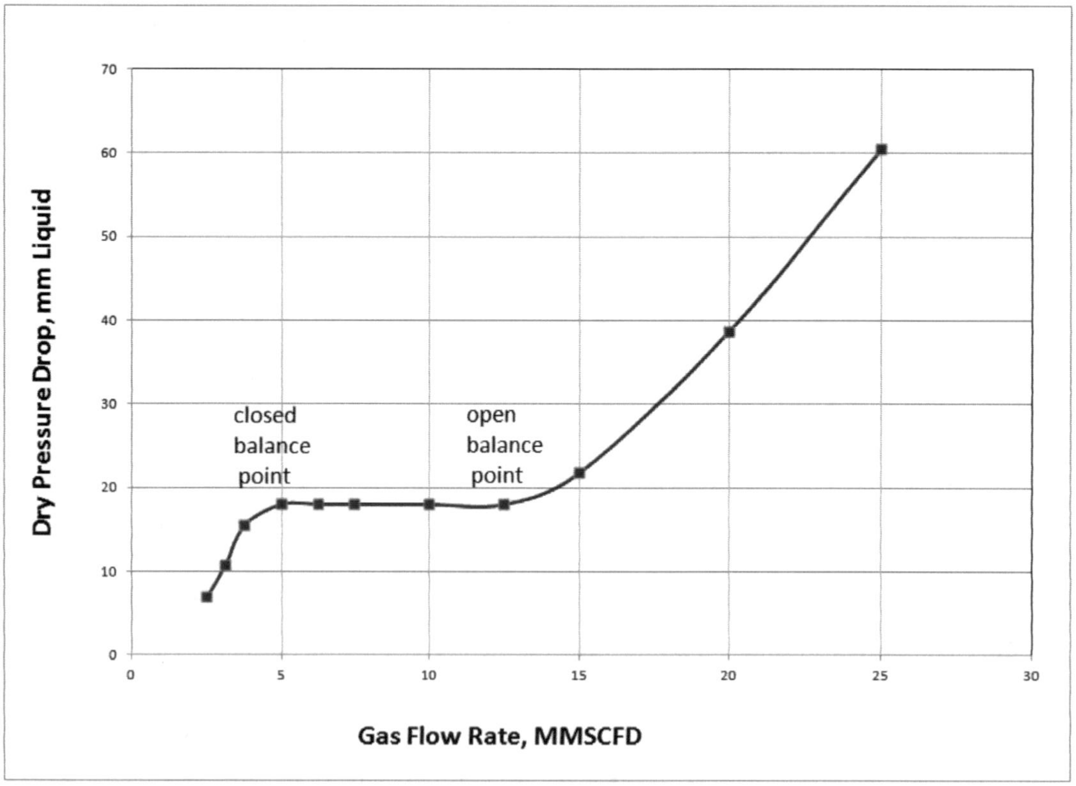

Figure 3. Typical Valve Tray Pressure Drop Characteristic

Multiple Solutions with the Klein Pressure Drop Method

Figure 4 shows the dry pressure drop, pressure drop through liquid, and total pressure drop through Trays 4 and 5 on either side (A and B) of the center downcomer from Tray 5. The loading is 29% of the estimated jet flood point, and the liquid is evenly distributed between passes A and B for both trays. The x-axis is the fraction of vapor to pass A, with the remainder to pass B.

The dashed lines represent the calculated dry pressure drop through both trays for either pass. The pass A dry pressure drop resembles the dry pressure drop characteristic in Figure 3, and pass B is a mirror image. The dotted lines represent the pressure drop through the tray liquid for either pass. The liquid rate remains the same, but per the Klein method, the pressure drop decreases with increasing vapor load as the liquid becomes more aerated.

The solid lines at the top represent the total pressure drop through the trays, the sum of the dry pressure drop and the pressure drop through the liquid. As with the dry pressure drop, the total pressure drop increases with increasing vapor flow rate below the closed balance point and above the open balance point. However, at the rates when the valves open and close, the total pressure drop decreases with increasing vapor flow rate. This is because the dry pressure drop is constant for this region while the pressure drop through the liquid decreases with increasing aeration.

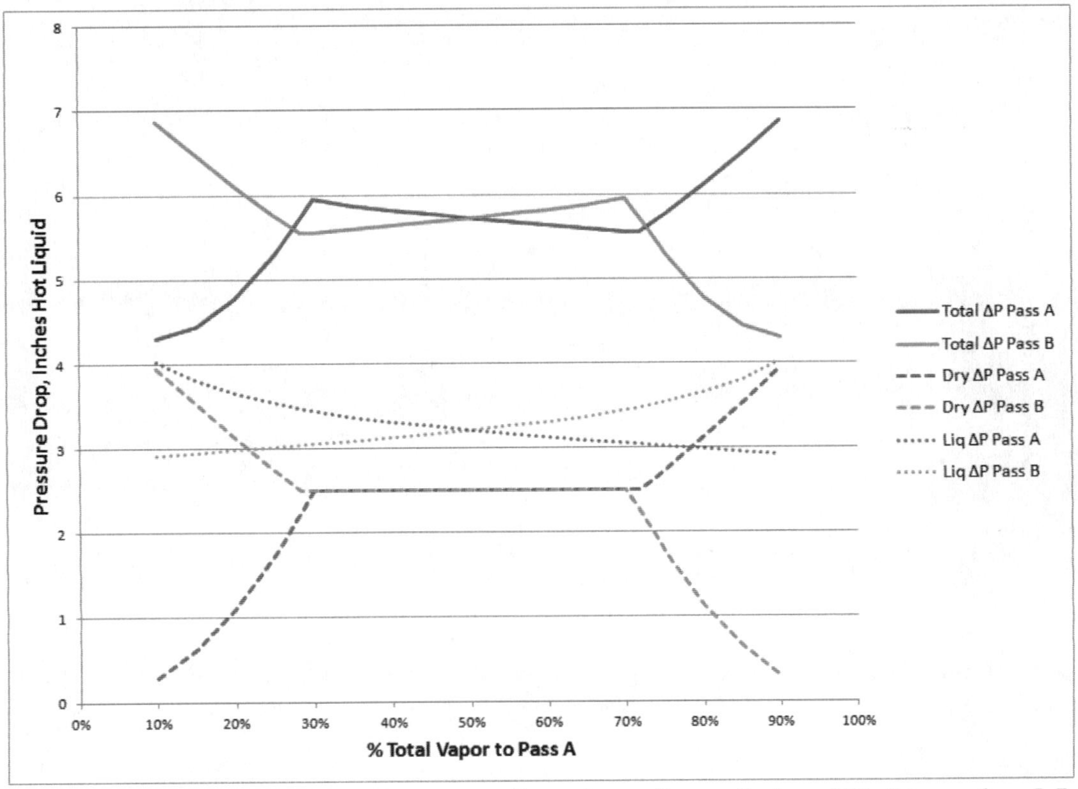

Figure 4. Dry, liquid, and total pressure drop through panels A and B of trays 4 and 5. Klein pressure drop equation, 29% jet flood.

The resulting shape of the total pressure drop curve includes a local maximum at the closed balance point and a local minimum at the open balance point. The total pressure drop curve of pass A intersects that of pass B three times. Each intersection constitutes a potential hydraulically balanced solution, where the pressure is equalized below Tray 4 and again above Tray 5, and the total pressure drop on either side of the center downcomer is equal.

To be a full solution, the downcomer backup in the center downcomer from Tray 5 also needs to be equal for each side. Figure 5 plots the downcomer backup from Tray 5 along with the total pressure drop through Trays 4 and 5 from Figure 4. Their shapes are similar, because the total pressure drop through Tray 5 is a major component of the downcomer backup. The other components (weir height, head over weir, head loss under downcomer, hydraulic gradient) are equal, because the liquid is split evenly between passes A and B for this diagram. The closeness of values of the total pressure drops to the downcomer backups is deceiving, as the plotted pressure drop is the total for two trays (4 and 5), while the downcomer backup is for only one downcomer (from tray 5), so the values of the downcomer backup per tray are of the order of about twice the total pressure drop per tray.

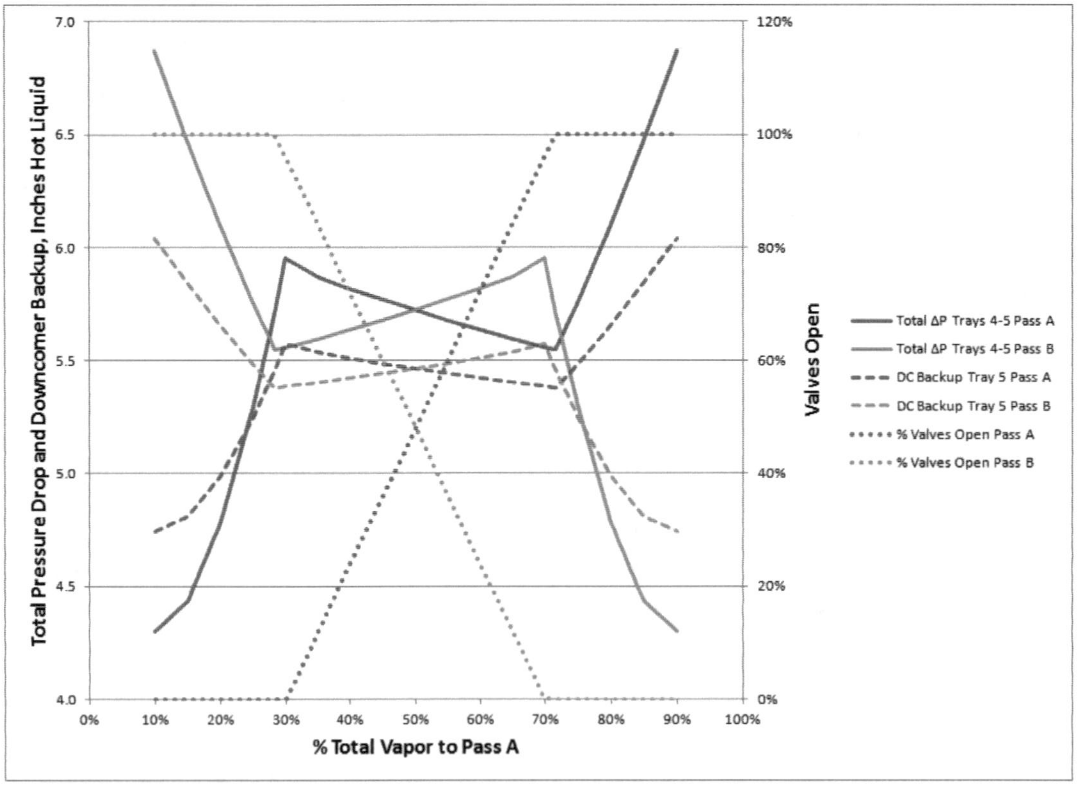

Figure 5. Total pressure drops and downcomer backups through panels A and B of trays 4 and 5. Klein pressure drop equation, 29% jet flood. At this load, multiple solutions exist.

Of course, the solution with both liquid and vapor split evenly between the passes (50% vapor to panel A, 50% liquid to panel A) is a solution on this diagram, as both the pressure drop curves and downcomer backup curves intersect at this point. At about 27% vapor to panel A, and for its mirror image at about 73%, the pressure drop curves and downcomer backup curves each intersect, at nearly the same vapor rate. To line these points up, only a slight liquid maldistribution is needed. For the 27% vapor to panel A, the solution is found with about 50.5% liquid to panel A. For 73% vapor to panel A, the solution is found with about 49.5% liquid to panel A. Each solution other than the even split has this slight liquid maldistribution. Based on our model, the only ways to significantly maldistribute the liquid are either to have a physical difference between the two sides or to have uneven weeping.

Variation of the Multiple Solutions with Percent Jet Flood

Figure 6 is similar to Figure 5, but based on 18% of the estimated jet flood. When the vapor and liquid are evenly distributed, less than 1% of the valves on each panel are open. The only regions present on Figure 6 are the increase in pressure drop with vapor rate when the valves are closed, the local maximum at the closed balance point (at about 50% of the vapor to Panel A or B), and a portion of the region where the valves are opening. Again, multiple solutions are found at the points where all valves are closed on one panel or the other, but, because of the closeness of the evenly split intersection to the closed balance point, the multiple solutions are not much different from each other, so the maldistribution is low. For only slightly less vapor than 18% flood, the evenly split intersection would be the only hydraulically balanced solution because it is in the region where pressure drop increases with vapor rate as all valves are closed.

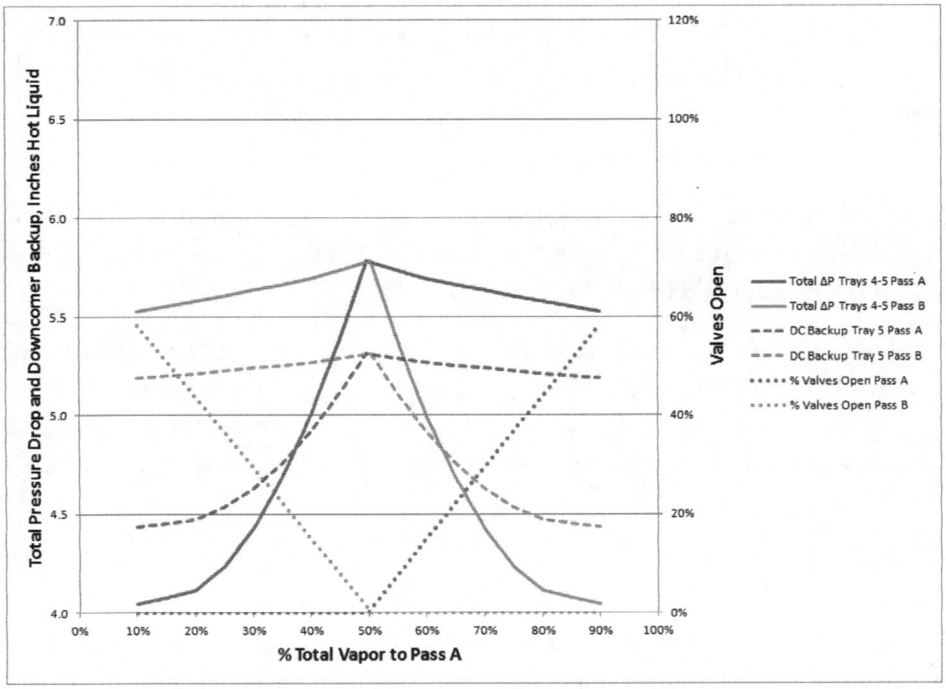

Figure 6. Total pressure drops and downcomer backups through panels A and B of trays 4 and 5. Klein pressure drop equation, 18% jet flood. At this load, multiple solutions still exist, but very close to each other, hardly visible.

For 22% jet flood (Figure 7), over 20% of the valves on each panel open at the evenly split intersection. As with 18% of jet flood, the other solutions are near the points where all valves are closed on one panel or the other. Because of the increased vapor load, these points are farther apart than in Figure 6, so the potential maldistribution at these loadings is greater than for 18% of jet flood.

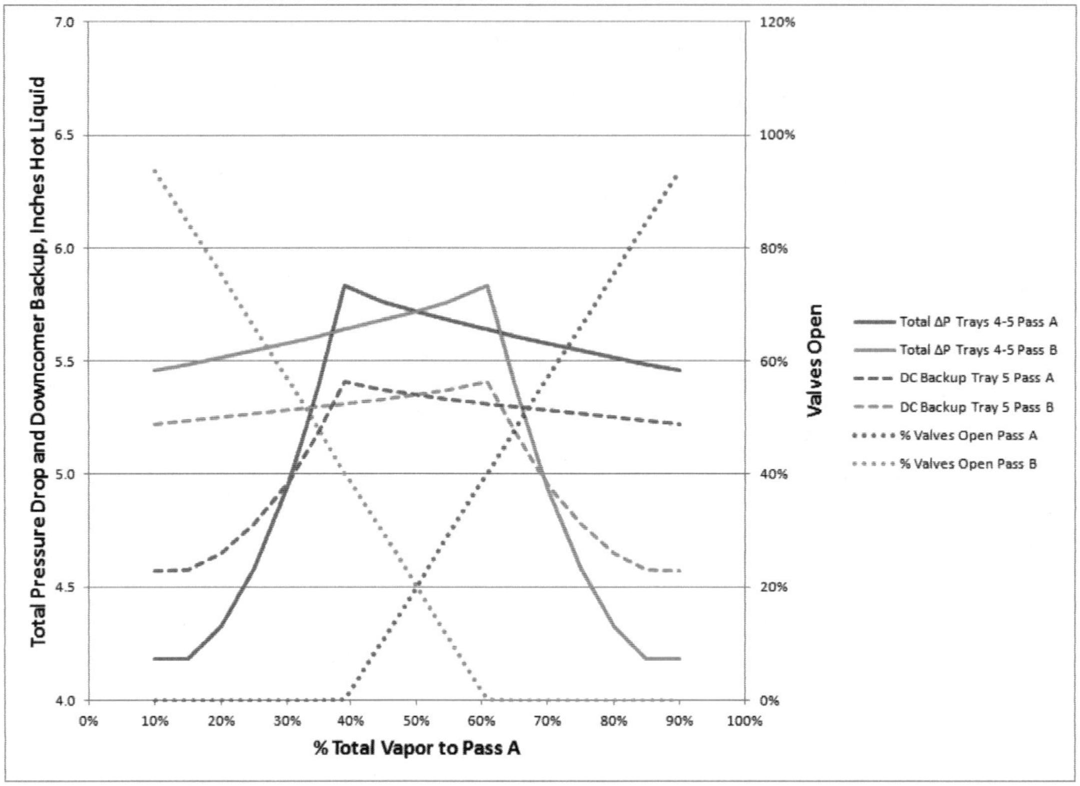

Figure 7. Total pressure drops, and downcomer backups, through panels A and B of trays 4 and 5. Klein pressure drop equation, 22% jet flood. At this load, multiple solutions exist, but closer together than in Figure 5.

For 29% jet flood (Figure 5), nearly 50% of the valves are open at the evenly split intersection, so this is in the center of the pressure drop curve. The other solutions are found at points where all valves are open on one panel and closed on the other. These solutions at 29% of jet flood (27% and 73%) are farther apart than for 22% of jet flood (37% and 63%), so this case has the highest worst-case maldistribution.

For 36% jet flood (Figure 8), about 75% of the valves are open at the evenly split intersection, so this point is closer to the open balance point. The other solutions are at points where all valves are open on one panel and some on the other. Because of the proximity to the open balance point, the multiple solutions are closer together than for 29% jet flood, so it has lower worst-case maldistribution. If the vapor loadings were increased further to the open balance point, the evenly split intersection would be the only hydraulically balanced solution.

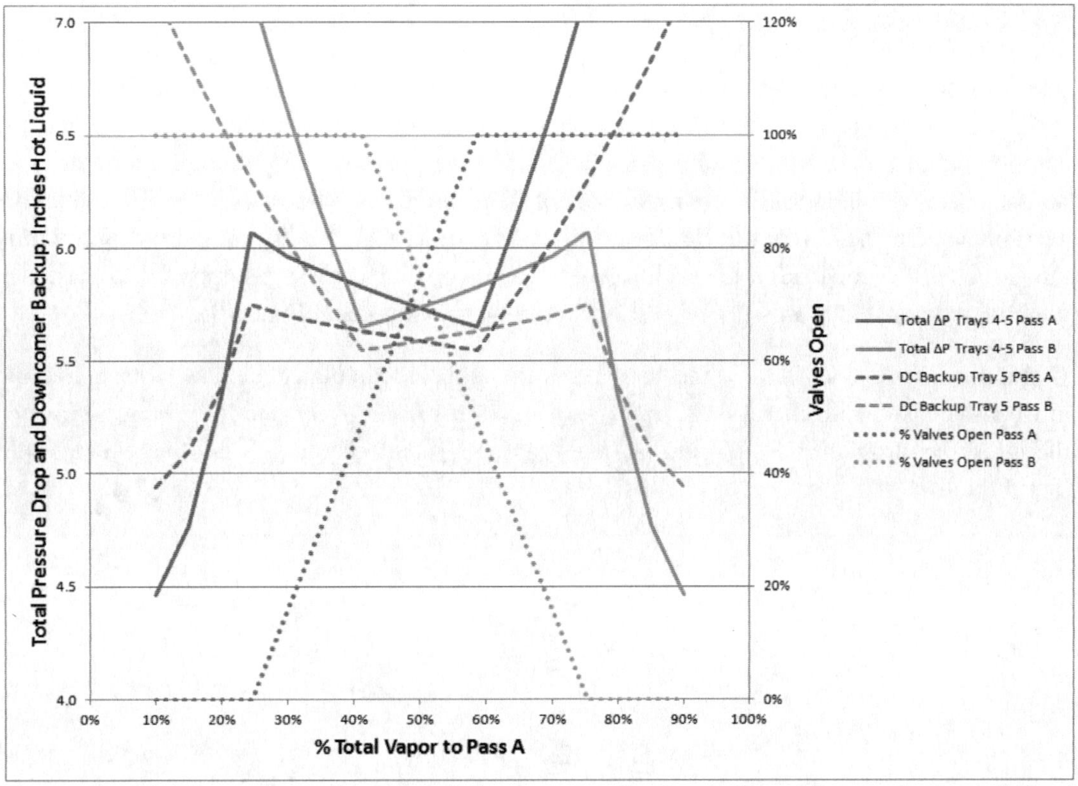

Figure 8. Total pressure drops and downcomer backups through panels A and B of trays 4 and 5. Klein pressure drop equation, 36% jet flood. At this load, multiple solutions exist, and again, closer together than in Figure 5.

The common thread for each of these loads is that at the evenly split intersection, both panels had some valves open but not all, and this is necessary for multiple solutions to exist. Also, the further away from either the open or closed balance point the evenly split intersection falls, the greater the potential maldistribution of the other solutions.

One difference between these loads is how the total pressure drop and downcomer backup for the alternate solutions compare with those for the evenly split solution. For 18% and 22% of jet flood, the maldistributed solutions actually have lower total pressure drop and downcomer backup than the evenly split solution. For 29% jet flood, the pressure drop and downcomer backup for the alternate solutions are slightly lower than for the evenly split solution, as well. For loadings only slightly higher than 29% jet flood, the three solutions would be equal, so switching between balanced solutions may be favored. For 36% jet flood, pressure drop and downcomer backup for the alternate solutions are higher than at the evenly split solution, and this should favor the evenly split solution. Comparing these findings, the implication may be that when greater than 50% of valves are open, even distribution is favored, but when less than 50% of valves are open, maldistribution is favored.

No Multiple Solutions for the Glitsch Pressure Drop Method

The Glitsch pressure drop method does not show multiplicity (Figure 9) because of two differences with the Klein method. The main difference is that for the Glitsch method the aeration factor for the tray liquid is constant, 0.4, so the pressure drop through the liquid does not decrease with increasing vapor load (at constant liquid) as it does with the Klein method. The other difference is that according to the Glitsch method, the dry pressure drop slightly increases with vapor load in the region where valves open and close, as opposed to the Klein method, where the dry pressure drop stays flat in that region (as in Figure 3).

Combining these two factors, the total pressure drop as calculated by the Glitsch method always increases with increasing vapor load. This results in only one intersection between the pressure drop curves of pass A and pass B, which is always at the evenly split distribution.

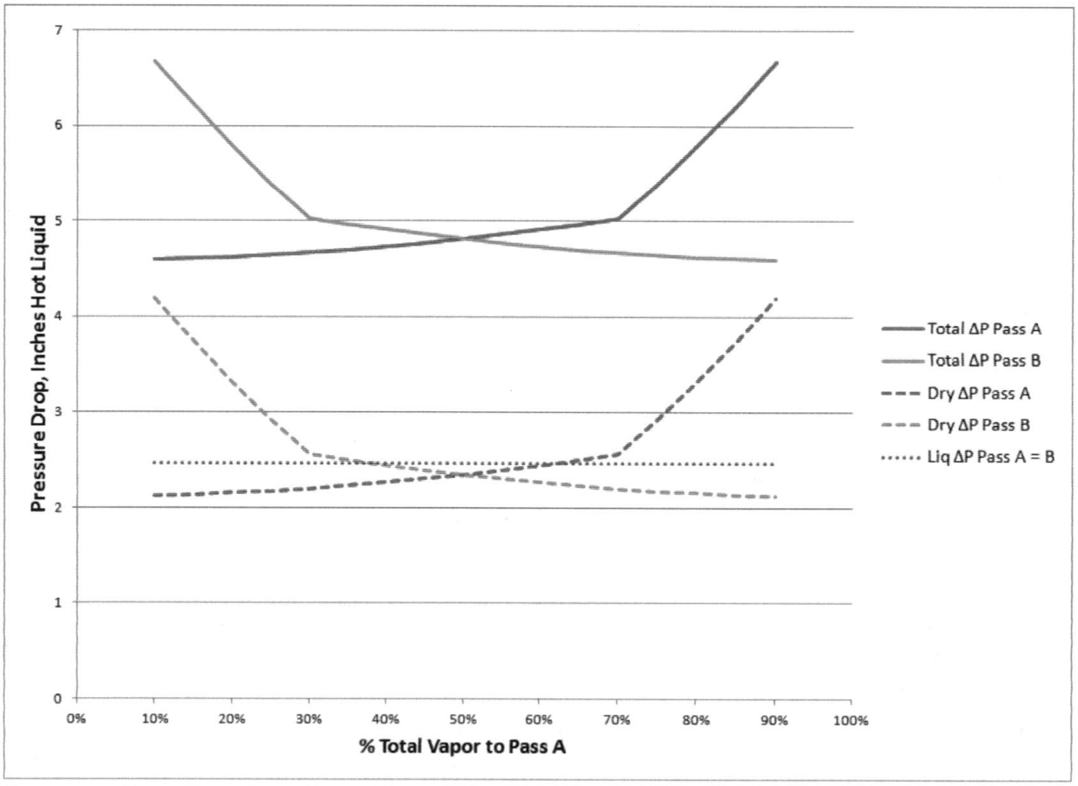

Figure 9. Dry, liquid, and total pressure drop through panels A and B of trays 4 and 5. Glitsch pressure drop equation, 29% jet flood.

Which Correlation is Better Here: Klein or Glitsch?

This brings the question: Which correlation do we believe? The key to the answer is to appreciate that both correlations focus on the region in most important to designers, i.e., when the valves are open and pressure drop is high. This is the region where dry pressure drop dominates. In this region both correlations predict much the same and their trends are robust and reliable.

Neither correlation was intended to address distribution in multi-pass trays at turndown. At turndown, measured pressure drops are low and subject to greater errors, so it is likely that the pressure drop trends are inaccurate. Further, due to the flatness of the pressure drop curve in this region, instability is likely, and gas can flip-flop from one panel to another, similar to the instability described by Brierley et al. (10).

An argument supporting the suitability of the Klein correlation for this analysis is that there is considerable evidence to suggest that due to increased aeration the liquid pressure drop on a tray diminishes as the vapor load is raised. This is featured in the Bolles and Fair correlation for sieve trays (11), the Colwell clear liquid height and froth density correlation (12), and even by a control issue reported by Buckley et al. (13). The Klein method's approximation of a liquid head that diminishes with an increasing vapor flow rate appears much more realistic than the Glitsch method's approximation that the aeration factor for the tray liquid is constant, 0.4, regardless of the vapor load. We therefore proceed on the basis of the Klein pressure drop correlation.

Distribution Criteria

Tray Stability

For low vapor loads, a criterion termed *tray stability factor*, η, was proposed by Summers et al. (3). It is a variation of the Froude number, comparing the upward inertial force of vapor to the downward gravitational force of liquid.

$$\eta = (\Delta P_{DRY}/H_{CL})^{0.5}$$

ΔP_{DRY} = calculated dry tray pressure drop, inches of liquid

H_{CL} = hydrostatic head, inches of liquid = total pressure drop – ΔP_{DRY}

Summers found that stability factors exceeding 0.6 limit weeping to below 30% and render trays stable. Note that Summers' method (3) derived the hydrostatic head H_{CL} values from total pressure drop measurements by subtracting from them calculated dry pressure drops as defined above. In the absence of measured total pressure drop data, Summers (14) recommends using the Colwell clear liquid height correlation (12) for calculating stability factors, and reports excellent experience with it. The values for the stability factors for our model runs are plotted on Figure 10 below. Figure 10 shows that the stability factors for the modeled trays were above 1.05 down to 20% of jet flood, suggesting no instability, less than 30% weeping, and minimal adverse effects due to weeping.

Summers et al. (3) also proposed an additional crude stability criterion for 2- and multi-pass trays to determine whether the total vapor loading is sufficient to make it likely that vapor will flow through each tray pass. The criterion is whether:

$$\Delta P_{ACTIVE} > \Delta P_{INACTIVE}$$

For this test, liquid would be distributed evenly among the tray passes. The inactive pass would have no vapor loading, and the full vapor loading would be divided among the remaining passes. In the case of 2-pass trays, all vapor would travel through one pass, with the liquid split evenly. The pressure drop through the inactive side is the sum of the weir height and the head over the weir, using the Francis Weir Formula (2). The pressure drop through the active side is the sum of the dry pressure drop and hydrostatic head calculated for the aerated liquid from Colwell (12). An equivalent expression for the multi-pass test is:

$$\Delta P_{ACTIVE}/\Delta P_{INACTIVE} > 1$$

We applied this test to the loadings used in our model and added the results to Figure 10. As long as the $\Delta P_{ACTIVE}/\Delta P_{INACTIVE}$ ratio is >1, this Summers et al. criterion expects no additional issues due to maldistribution between the passes. Figure 10 shows that this ratio exceeds unity for loadings above 25% of jet flood and is below unity at lower loadings.

Overall, of the two Summers et al. stability criteria, one (multi-pass instability) is consistent with our findings of 2-pass valve tray maldistribution at turndown, but only below 25% jet flood. The other (tray instability) does not anticipate this maldistribution. Neither anticipates the modeled maldistribution in the range of 25% to 40% of flood as is discussed below.

Bolles Distribution Ratios

Maldistribution varies the liquid to vapor (L/V) ratio from pass to pass. Pinching, and therefore poor separation, occurs in the low L/V regions, while the high L/V ratios in other regions only slightly offset the separation loss. Overall, observed efficiencies diminish, sometimes grossly. Bolles (15) defined the distribution ratio Φ as the ratio of the maximum pass L/V to the minimum pass L/V, and recommended keeping this distribution ratio below 1.2 for good tray efficiency. Summers (16) tightened Bolles' criterion to 1.1.

Distribution ratios can readily be calculated from our MMM model, since our model calculates the L and V for each pass for a given hydraulically balanced solution. Figure 10 shows the distribution ratios for the "worst-case maldistribution" from the MMM for each percent of jet flood based on the Klein pressure drop method. The term "the worst-case maldistribution" is used because we found up to three hydraulically balanced solutions loadings up to the vapor rate where all valves are open. One of the three is an evenly split distribution (distribution ratio of 1). The other two (mirror images in Figures 5 to 8) both have the same high "worst-case maldistribution" distribution ratios.

Figure 10 shows that for the tests between 20% of flood and 40% of flood, the distribution ratios were well above the maximum of 1.2 specified by Bolles, reaching a huge ratio of about 2.6 at 30% of flood. We have seen significant losses in tray efficiency in multi-pass trays when the distribution ratios exceeded 1.3, so 2.6 is huge.

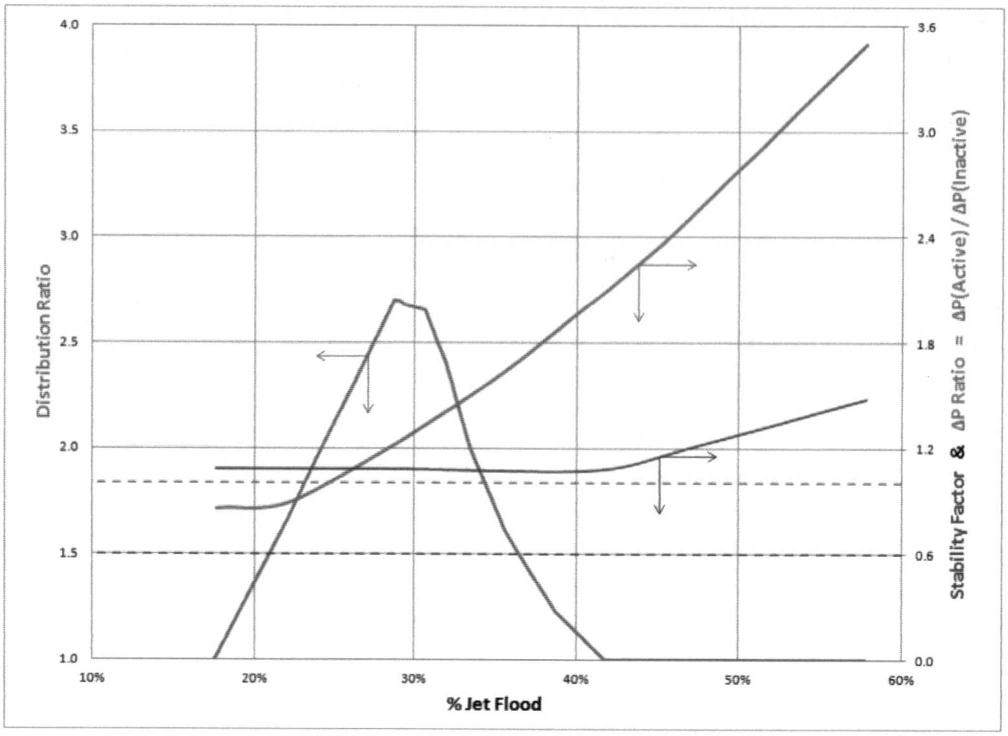

Figure 10. Applying the Summers et al. hydraulic metrics to explore instability: Stability Factor (red curve) and ΔP Ratio (green curve), compared with Bolles Distribution Ratio for Worst-Case Maldistribution calculated from our MMM model (blue curve).

What affects the Steady State Multiplicity?

Multiple Valve Weights

Tray manufacturers to often use two different valve weights on each tray panel, typically with 50% light gauge valves and 50% heavy gauge valves (17), to improve efficiency at turndown. The different weights are evenly spread across the tray (typically one row of lights, one row of heavies, rows perpendicular to the liquid flow) with the light valves forming the front row facing the incoming liquid. We explored the efficacy of this strategy to alleviate steady state multiplicity.

The dry pressure drop characteristic of a two-weight valve tray is similar to that of a single weight valve tray, except that there are two flat regions instead of the one in Figure 3, one for each valve weight. On Figure 11, for 29% flood, as the gas flow rate increases, the dry pressure drop increases to the point that the light valves (16-gauge) begin to open, the *light closed balance point*. Then, the dry pressure drop stays constant to the rate when all light valves are open, the *light open balance point*. Dry pressure drop again increases with the gas flow rate once all light valves are open, up to the point that the heavy valves (14-gauge) begin to open, the *heavy closed balance point*. Then, there is a second region in which the dry pressure drop is constant, up to the gas rate that all valves are open, the *heavy open balance point*. Above this point, dry pressure drop increases with increasing gas flow rate.

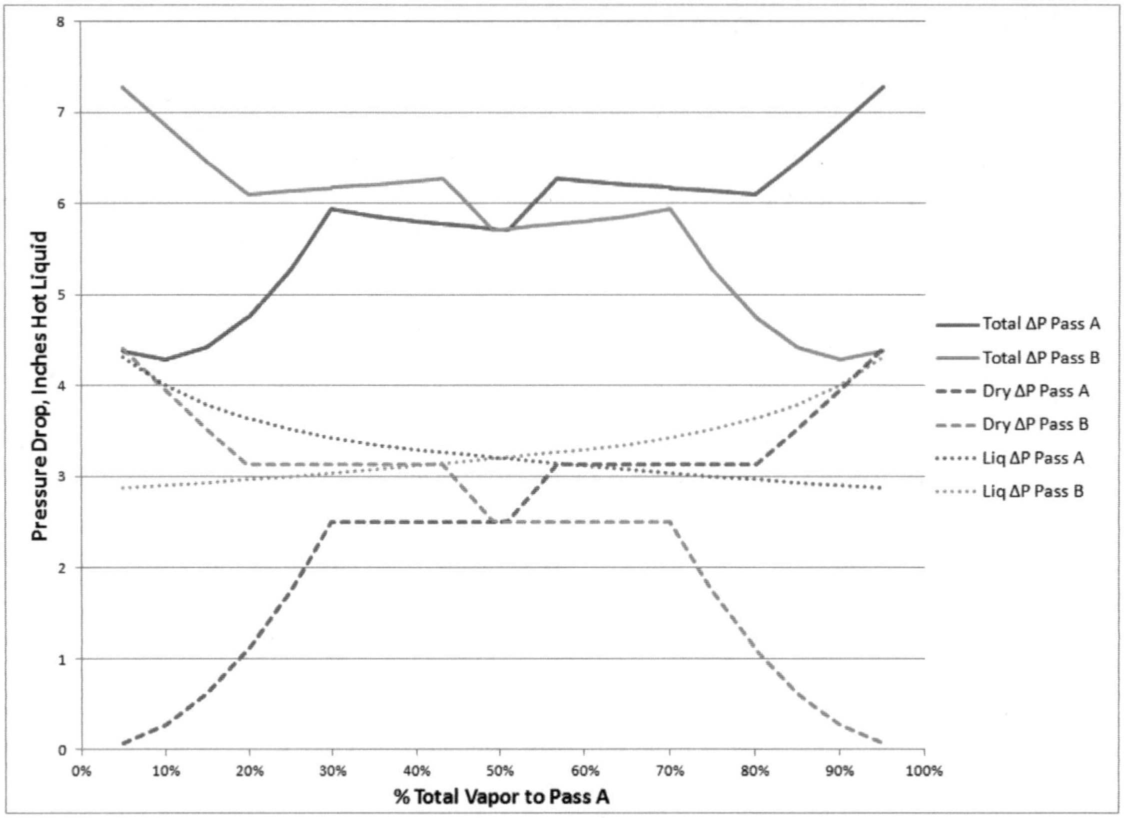

Figure 11. Dry, liquid, and total pressure drop through panels A and B of trays similar to trays 4 and 5, but with two weights of valves, 14-gauge and 16-gauge. Klein pressure drop equation, 29% jet flood.

Figure 11 shows that for 29% jet flood, a two-weight design would nearly eliminate the potential maldistribution seen in Figure 4. However, Figure 13 shows that at 40% of jet flood loadings, which had very little potential maldistribution with a single valve weight (Figure 12a), a two-weight valve design could experience significantly more maldistribution (Figure 12b).

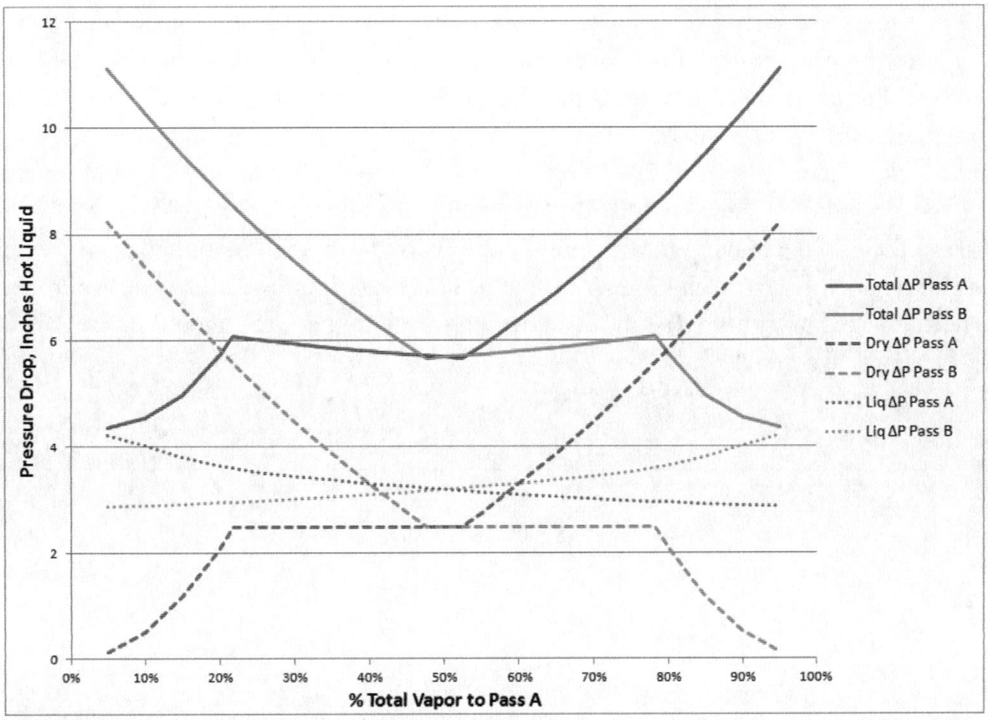

Figure 12a. Dry, liquid, and total pressure drop through panels A and B of trays 4 and 5. Klein pressure drop equation, 40% jet flood.

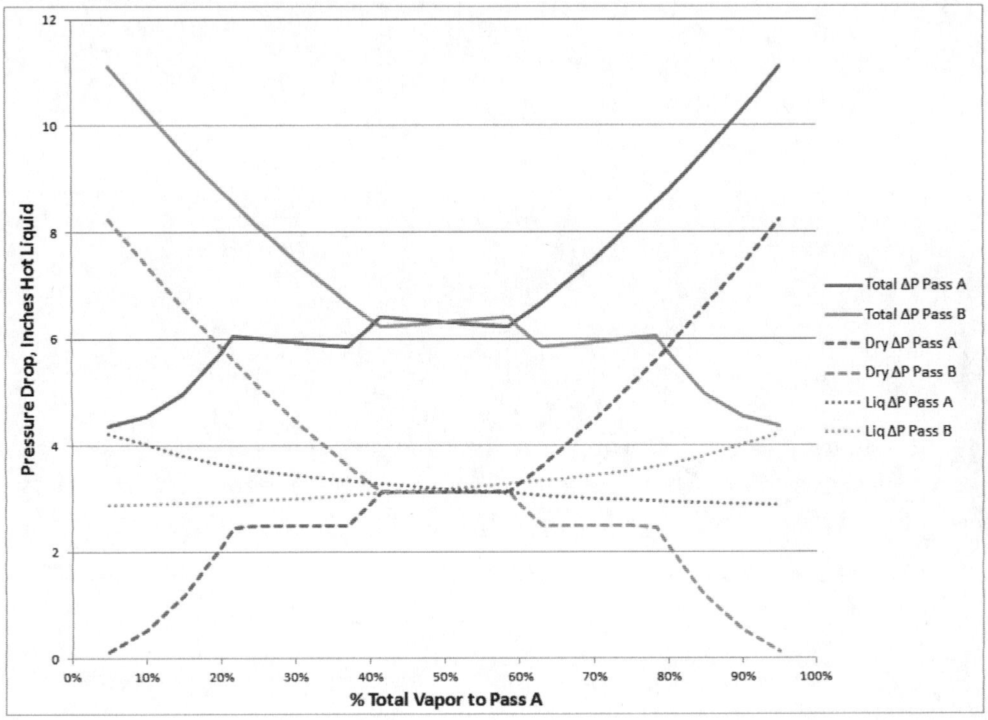

Figure 12b. Dry, liquid, and total pressure drop through panels A and B of trays similar to trays 4 and 5, but with two weights of valves, 14-gauge and 16-gauge. Klein pressure drop equation, 40% jet flood.

Figure 13 compares the distribution ratios for the worst-case maldistribution for the single valve weight to those of similar trays containing 50% light (16-gauge) and 50% heavy (14-gauge) valves. For the two valve weights the distribution ratios are much smaller, indicating a tremendous improvement. The turndown also improved, with the major maldistribution peak moving from about 30% to 23% of jet flood. On the debit side, the opening of the heavy valves generated a new maldistribution peak centered around 40% of jet flood. However, at this peak the maximum distribution ratio was 1.4, much less than the maximum in the single weight valve case, and in half of that peak the distribution ratios were within the Bolles criterion of 1.2 maximum. So the efficiency loss due to that peak should be much less than with the single-weight valves.

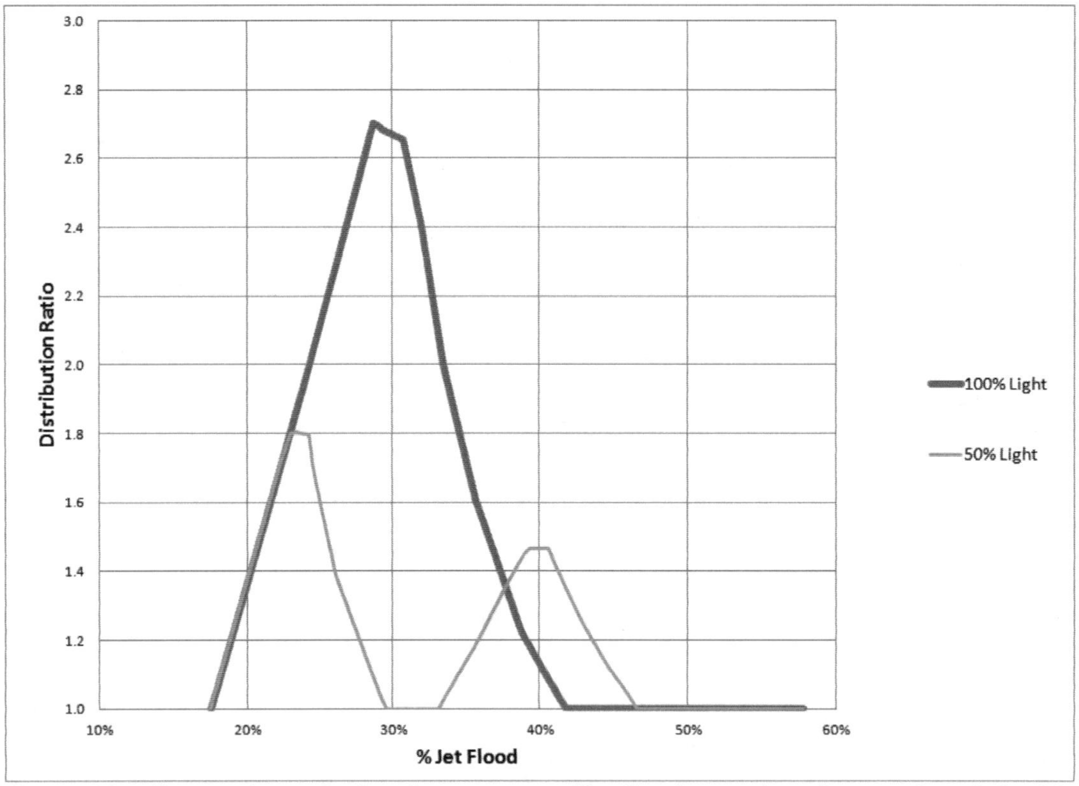

Figure 13. Comparing calculated worst-case distribution ratios of trays 4 and 5 with uniform valve weights (blue) to the calculated worst-case distribution ratios with two weights of valves, 14-gauge and 16-gauge (green).

Heavy Valve Gauge

To explore the effect of valve thickness, we ran the model again, pairing the 16-gauge valves with 12-gauge valves (Figure 14).

Comparing Figure 13 to 14, the distribution ratios are quite similar, and both are much lower than for the single weight valves as described earlier. Compared to the 14-gauge/16-gauge valves, the 12-gauge valves open at higher vapor rates than the 14-gauge valves. The maldistribution peaks associated with the heavy valves closing are shifted away from those of the light gauge valves. This adversely affects distribution closer to design rates, but it increases the range of good distribution between the peaks, where light valves are open and heavy valves are closed. There is no difference between the figures at vapor rates where only light valves are open.

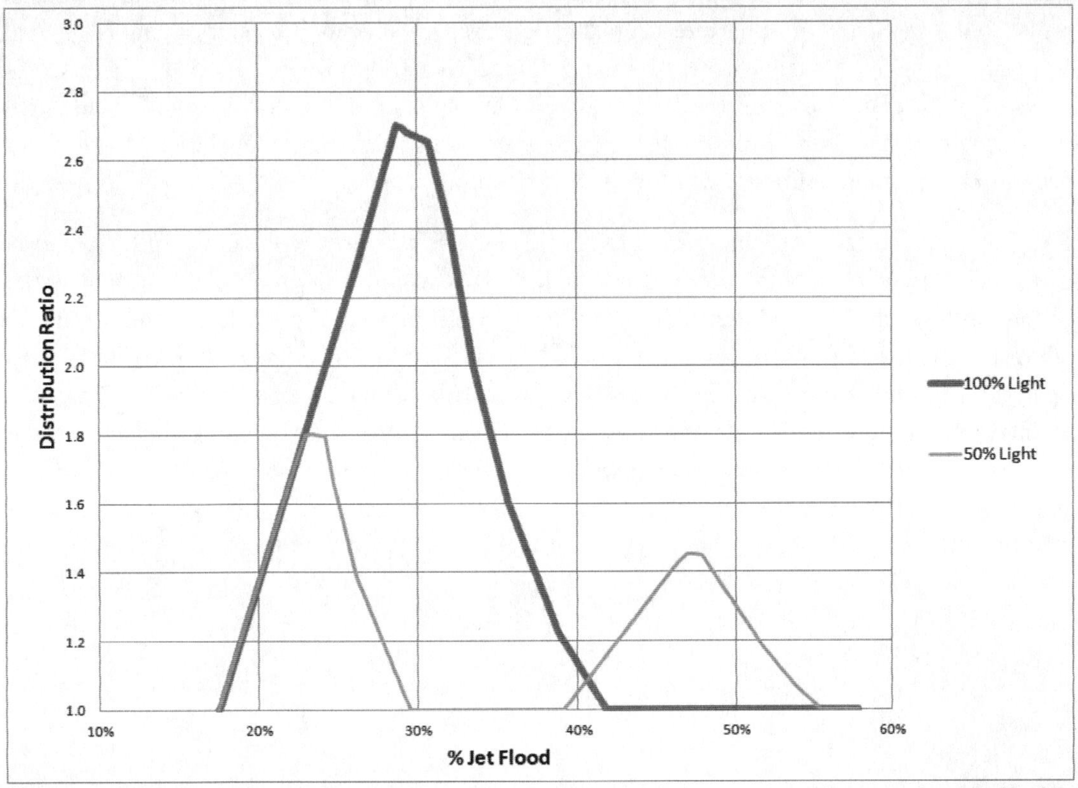

Figure 14. Comparing calculated worst-case distribution ratios of trays 4 and 5 with uniform valve weights (blue) to the calculated worst-case distribution ratios with two weights of valves, 12-gauge and 16-gauge (green).

Multiple Valve Weight Strategy

Figures 13 to 14, show that employing a multiple valve weight strategy tremendously reduces the distribution ratios for the worst-case maldistribution, and therefore has a beneficial effect at turndown. However, some loss of efficiency may be encountered at higher loads due to the opening of the heavier valves which is encountered at higher percent of jet flood value. Since it is also common that loadings vary significantly throughout a trayed section, while the tray design does not, at any point in time different trays in the tower may operate at different points on the same diagram. Whatever the case, the valve weight strategy should be reviewed for consistency with the predicted tower operation.

Vapor Tunnels

One strategy that is sometimes employed for multi-pass trays to allow for vapor to redistribute based on the pressure drop through each tray is to use vapor tunnels (4, 16). Vapor tunnels are passages through a shared downcomer which allow vapor to pass based on pressure differential between the tray passes on either side of the downcomer. These may also equalize the clear liquid height on the tray being fed from the shared downcomer, depending on their location vertically in the downcomer.

For this 2-pass trays case, vapor tunnels through the center downcomer would equalize the pressure on either side of the center downcomer. This would decouple the vapor distribution of the tray with the center downcomer from the distribution to the tray below. This would not address the cause of maldistribution at turndown, which is that the vapor rate is insufficient to open all the valves on the tray decks, resulting in no dry pressure drop difference between the evenly split solution and the maldistributed one. The difference would be that with vapor tunnels, every tray would have its own vapor distribution, rather than every two trays, and by this reasoning vapor tunnels should not significantly change the performance of a trayed section.

Weir Loads

Since our analysis was based on the cyclohexane-normal heptane system at 24 psia and total reflux, the weir loads were low. For any of the loadings between the closed balance point and the open balance point, the weir loading was less than about 2 gpm/inch for the side downcomers and less than about 1.5 gpm/inch for the center downcomers.

To determine if the multiplicity would persist at higher liquid loadings, we evaluated the turndown behavior of an alternate case. The alternate case has the same internal diameter, downcomer dimensions, and valve type as those in our model. The difference is that the weir loading was raised to 8 gpm/inch at 80% of flood. The L/V ratio for this case was 2.4, much higher than the L/V ratio of 1.0 for the total reflux case. An L/V ratio of 2.4 is of the order used in a stripping section of a fractionator, where 2-pass trays are frequently used.

The increase in liquid rate and L/V ratio did not increase the influence of the liquid on maldistribution. The maldistribution chart for the alternate case, Figure 15, looks the same as the one for the lower weir loads in Figure 13. The peaks appear to have shifted to the right. This is because for a given vapor rate, the increased liquid rate increased the percent jet flood. The vapor rates at which valves opened and closed did not change from Figure 13.

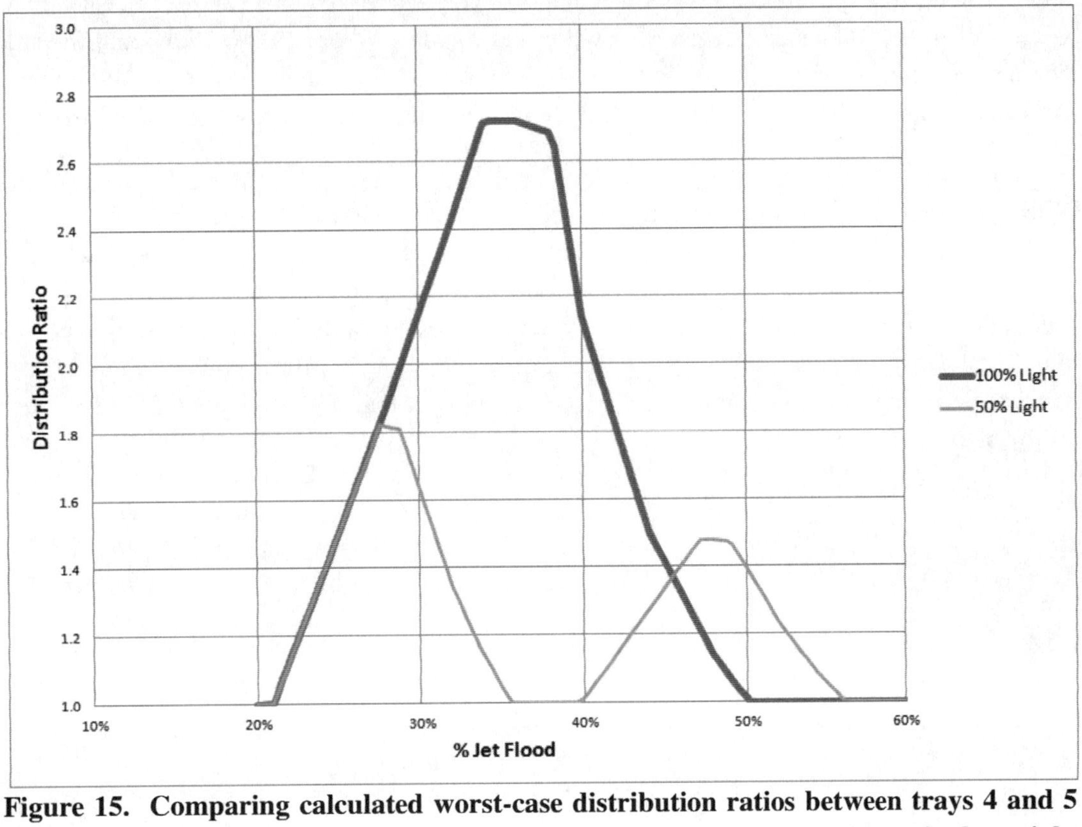

Figure 15. Comparing calculated worst-case distribution ratios between trays 4 and 5 for alternate loadings with higher weir loading than the total reflux, single weight 16-gauge valves (blue) vs. two weights of valves, 14-gauge and 16-gauge (green).

Unique Multiplicity at Extremely High L/V Ratios

For the alternate case in Figure 15, the L/V ratio was increased such that the weir loading would be 8 gpm/in at 80% of flood, and there was not a significant change in behavior from the total reflux case. Indeed, this is true for even higher weir loadings as well. For these cases, using multiple valve weights (16-gauge and 14-gauge) would be a good solution to limit maldistribution at turndown.

However, in some cases, the L/V ratio can be increased further, well above the normal range, such that the limiting factors in tray design are downcomer choke flood and downcomer backup flood rather than jet flood. We evaluated a second alternate case with a much higher L/V ratio to determine whether such liquid-dominated loadings would experience any unique behavior.

For the second alternate case, we used an L/V ratio of 8.6, much higher than the first alternate case (2.4) or the total reflux case (1.0). Unlike for the previous alternate case, the tray geometry needed altering for this high L/V case. The downcomers employed for the lower L/V ratios were far too small for the 8.6 ratio. To achieve a comparable design, the diameter of the tower was increased from 96" to 108", with all the additional area added to the downcomers, maintaining straight downcomers. The active and the valve open slot areas were unchanged. The side downcomer width was increased from 11" to 17.3", and the side downcomer clearance was increased from 1" to 4". The center downcomer width was increased from 10" to 17.75", and the center downcomer clearance was increased from 1" to 3".

With this design and 8.6 L/V ratio, at the vapor rate for which about half the valves are open (29% jet flood for total reflux loadings and modeled column trays), the weir loading is 9.3 gpm/in, and the trays are at 39.3% jet flood. As the loadings are increased to 80% jet flood, the weir loading is 19 gpm/in, the downcomer inlet velocity is just below 0.51 ft/s, and the center downcomer backup is 12.9" clear liquid. These values push against the maximum good practice limits (2), but do not violate them, giving at least a reasonable design for much of the operating range.

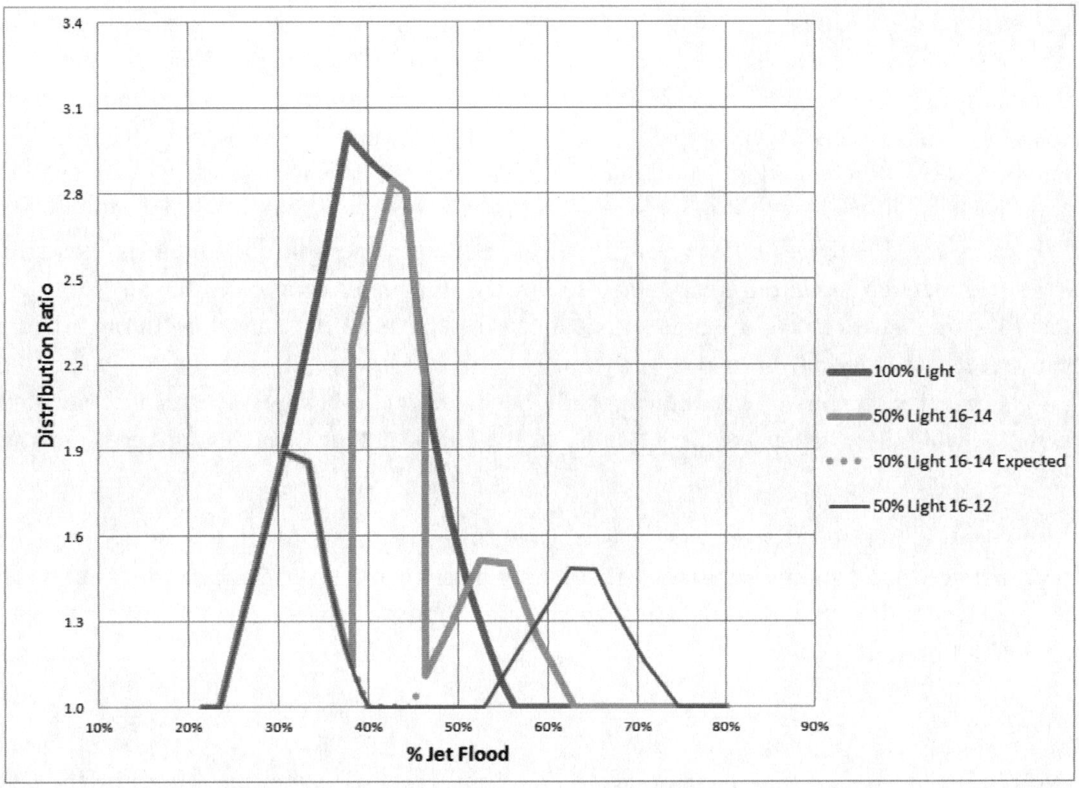

Figure 16. Comparing calculated worst-case distribution ratios between trays 4 and 5 of the larger diameter tower with high L/V loadings, single weight 16-gauge valves (blue) vs. two weights of valves, 14-gauge and 16-gauge (green) or 12-gauge and 16-gauge (red).

The distribution ratios for the 16-gauge single valve weight design and for two multiple valve weight designs (16-gauge/14-gauge & 16-gauge/12-gauge) are shown in Figure 16. For the single valve weight design, as well as for the 16-gauge/12-gauge design, the distribution ratio curves are similar to those for the lower L/V ratios (compare to Figure 14). In contrast, the design using 16-gauge and 14-gauge valves appears to no longer reduce peak maldistribution as effectively as it does for the lower L/V cases.

The portion of the distribution ratio curve for the 16-gauge/14-gauge design which deviates from the other two-weight curves is between 38% flood and 46% flood. The conventional solutions are still possible and shown as the dotted line, but for this range there are more maldistributed solutions. For the portion that the curve shares with the peak of the single-weight curve, all light and heavy valves are open on panel A, and all valves are closed on panel B.

For each of the designs discussed in this paper, the liquid is not maldistributed more than 0.5-1%, and that is also the case for the 8.6 L/V ratio. The difference in pressure drop through liquid for panel A and panel B is then due to the difference in the aeration factor, based on the amount of vapor maldistribution. For lower L/V ratios, the difference in pressure drop through liquid at turndown is very small, as on Figure 13, and is only significant enough to lead to an alternate solution for a single-weight design. With an 8.6 L/V ratio, the difference in pressure drop through liquid becomes much more pronounced. With the relatively small difference between the dry pressure drop at which the 16-gauge and the 14-gauge valves open, the more pronounced difference in pressure drop through liquid can exceed the dry pressure drop difference between the opening of the 16-gauge and 14-gauge valve weights. This allows for the behavior where heavy valves can be open on panel A, while light valves are closed on panel B. This is unique behavior for very high L/V ratios.

In order to eliminate this behavior, it is possible to increase the difference in dry pressure drop between the valve weights such that it is greater than the difference in pressure drop through liquid (due to aeration). This can be done by using 16-gauge and 12-gauge valves, shown in Figure 16.

Weeping

We estimated for each pass the amount of liquid that weeps through the valves on the tray deck and bypasses the downcomer. To do this, we used the Colwell and O'Bara modification (18) of the Lockett and Banik correlation for sieve trays (19), found in *Distillation Design* (20). The correlation uses the Colwell clear liquid height (12), which is a function of the weir load. The weeping liquid was removed from the weir load, and an iterative solution needed to be performed for each tray pass.

As this is a sieve tray correlation, it uses the full open area. This produced results with over 80% of tray liquid weeping for 18% flood and over 40% of tray liquid weeping for 22% flood. These results are unrealistic for moving valve trays because the valves can close, lowering the open area and increasing the vapor velocity, two factors which reduce weeping.

To correct for this, the open area was adjusted based on the number of valves open on the tray. For the Klein correlation, the closed balance point and open balance point are calculated in terms of the vapor velocity through the fully open hole area, with the closed balance point at a lower velocity and the open balance point at a higher velocity. The number of open valves is found by interpolating between these two points. The open area for weep is then adjusted based on the number of valves open. Closed valves are not completely closed, however. As an approximation, the velocity ratio between the open balance point and the closed balance point was used as the area ratio between an open valve and a closed valve.

With the area correction in place, the weeping was limited to less than 5% of tray liquid for all cases, making it mostly irrelevant to the distribution behavior discussed in this paper. It is possible that this is an overcorrection and that more weeping is possible at turndown. This would especially be true in the region where valves are opening and closing with no effect on the dry pressure drop, as some valves could be open for vapor while others are open but mostly weeping. This path was not explored as our model does not have a method to find solutions with this behavior, other than arbitrarily.

Uneven Weeping

We also explored the effects of uneven weeping on vapor and liquid distribution. We specified that for one side of the tower, 20% of the tray liquid would weep, while for the other side, 0% would weep. As the same amount of liquid weeps on each tray, the weeping liquid hydraulically does not exist on the tray, over the weir, in the downcomer, or under the downcomer escape, so it has no hydraulic influence anywhere. With this in mind, the conclusion was not surprising. This 1/5 of the tray liquid which weeps was 1/9 of the overall liquid. 4/9 of the overall liquid flowed across the tray and into the downcomer, with the other 4/9 on the opposite side. The vapor distribution was unaffected. The calculated clear liquid height, total pressure drop, and downcomer backup were all slightly lowered, but there was no hydraulic difference between the weeping side and the non-weeping side.

Final Thoughts

For 2-pass trays, the maldistribution occurs at turndown due to the multiplicity mechanism described in this paper that produces stable but maldistributed hydraulic steady states in which the vapor becomes highly maldistributed in the regions where valves open and close. We expect similar steady state multiplicities at turndown in multi-pass trays (with 3, 4, 6, 8 passes). Worse, these multi-pass trays do not have pressure equalization across the entire tray, as each vapor space is shared by a maximum of two passes. This could cause vapor maldistribution from one tray to propagate several trays upward.

Maldistribution varies the liquid to vapor (L/V) ratio from pass to pass. Lessons from packed tower maldistribution (20-22) extend to multi-pass trays (15, 20). Pinching, and therefore very poor separation, occurs in the low L/V regions, while the excessive L/V ratios in other regions only slightly improve separation. Overall, separation suffers and observed efficiencies diminish, sometimes grossly. Lockett and Billingham (22) show that the efficiency loss depends on the degree of L/V unevenness, as well as the proximity of the pinch.

Tools for reliably predicting multi-pass tray maldistribution are scarce. To our knowledge, no commonly available tray hydraulics rating tool will predict maldistribution in 2-pass trays. It is our hope that the findings and tools in our paper will promote understanding and good design to circumvent the turndown efficiency loss.

Conclusions

This study is the first to report the multiplicity of steady states in 2-pass valve trays at turndown, in the region where some valves open and others are closed. In this region the dry pressure drop becomes constant while the liquid head slowly declines with an increase in vapor load. There are generally three steady states, one with vapor split equally between the passes, while the other two produce stable but maldistributed hydraulic steady states. In the maldistributed steady states, the number of open valves differs from panel to panel, with a corresponding variation in vapor loads. This may cause the vapor to become highly maldistributed between the panels. This multiplicity can lead to unexpectedly reduced turndown in 2-pass valve trays. This multiplicity is found at loadings for which other predictive hydraulic metrics are satisfied.

According to our model, maldistribution due to multiplicity occurs without being induced by maldistribution of adjacent trays or feeds. It is possible whenever the vapor load is such that some valves are open while others are closed. It occurs for pairs of trays which are hydraulically linked by a shared center downcomer.

The distribution ratios (the ratio of the maximum pass L/V to the minimum pass L/V) produced by the maldistributed steady states can exceed 2.5 (a maximum of 1.2 is considered good). An even split of vapor between the two panels (with a distribution ratio of 1), is always one of the steady states. The distribution ratios for the two other maldistributed steady states vary strongly with the vapor rates, approaching unity near the open and closed balance points, and peaking where about half the valves are open.

The multiplicity and distribution ratios are insensitive to weir loads and to weeping. The multiplicity appears to be a vapor-driven phenomenon with little effect on or influence by the liquid.

The maldistribution can be largely alleviated and the distribution ratios reduced by using two valve weights, with 16-gauge and 14-gauge being the most common. For all but the extremely high L/V ratios, this practice would largely reduce the distribution ratios in the regions where valves open and close. However, this practice can also cause a wider range of loadings to experience some maldistribution.

We expect the steady state multiplicity reported here to extend to multi-pass trays. When the distribution ratios are high, we expect a possible significant efficiency loss upon turndown.

References

1. Kister, H. Z., "Distillation Operation", McGraw-Hill, NY, 1990.
2. Kister, H. Z., P. Mathias, D. E. Steinmeyer, W. R. Penney, and J. R. Fair, “Equipment for Distillation, Gas Absorption, Phase Dispersion, and Phase Separation”, Section 14, in R. H. Perry and D. Green “Chemical Engineers’ Handbook”, 8th Ed., 2008.
3. Summers, D. R., L. Spiegel, and E. Kolesnoikov, “Tray Stability at Low Vapor Loads”, Proceedings of “Distillation and Absorption 2010”, Eindhoven University of Technology, The Netherlands, p.611, 2010.
4. Kister, H. Z., R. W. Dionne, W. J. Stupin, and M. Olsson, ”Preventing Maldistribution in Multi-Pass Trays“, Chem. Eng. Prog., p.32, April 2010.
5. Yanagi, T., and M. Sakata, “Performance of a Commercial Scale 14% Hole Area Sieve Tray”, Ind. Eng. Chem. Proc. Des. Dev. 21, p.712, 1982
6. Glitsch, Inc. “Glitsch Ballast Trays A-1, V-1, V-0, Experimental test Data”, Bulletin Nos. 159/160, Dallas, Texas, 1983 (currently obtainable from Koch-Glitsch LP, Wichita, Kansas)
7. Kister, H. Z., “Effect of Design on Tray efficiency in Commercial Towers”, Chem. Eng. Progr., June, p. 39, 2008.
8. Klein, G. F., “Simplified Model Calculates Valve-Tray Pressure Drop”, Chem. Eng., p. 81, May 3, 1982
9. Glitsch, Inc., “Ballast tray Design Manual”, Bulletin 4900, 6th Ed., Dallas, Texas 1993
10. Brierley, R. J. P., P. J. M. Whyman, and J. B. Erskine, “Flow Induced Vibration of Distillation and Absorption Columns”, IChemE. Symp. Ser. 56, p. 2.4/45, 1979.
11. Bolles, W., and J. R. Fair, in J. M. McKetta (Ed.) “Encyclopedia of Chemical Processing and Design”, vol. 16, p. 86, 1982.
12. Colwell, C. J., Ind. Eng. Chem. Proc. Des. Dev. 20 (2), 298, 1981.
13. Buckley, P. S., R. K. Cox, and D. L. Rollins, "Inverse Response in a Distillation Column," Chem. Eng. Prog. 71(6), 1975, p. 83.
14. Summers, D. R., Private Communication, January 3, 2017
15. Bolles, W. “Multipass Flow Distribution and Mass Transfer Efficiency for Distillation Plates”, AIChEJ, 22 (1), 153, 1976
16. Summers, D. R., “Designing Four Pass Trays” Chem. Eng. Prog., p.26, April 2010
17. Bolles, W. L., "Estimating Valve Tray Performance", Chemical Engineering Progress, 72 (9), p. 43, 1976.
18. Colwell, C. J., and J. T. O’Bara, “Estimate Sieve Tray Efficiency in the Weeping Region”, Paper presented at the AIChE Spring Meeting, April 1989.
19. Lockett M. J., S. Banik, Ind. Eng. Chem. Proc. Des. Dev. 25, p.561, 1986.
20. Kister, H. Z., “Distillation Design”, McGraw-Hill, NY, 1992.
21. Pilling, M. “Ensure Proper Design and Operation of Multi-pass Trays”, Chem. Eng. Prog., p.22, June 2005.
22. Lockett, M. J., and Billingham, Trans. IChemE., 80, Part A, p. 373, May 2002; Trans. IChemE., 81, Part A, p. 134, January 2003.

AIChE Spring Meeting 2017
26th to 30th Mar 2017, San Antonio, Texas

SULZER

Troubleshooting EB/SM Splitters: How Can A Maldistribution Analysis Help?

Senthil Krishnamoorthy, Sulzer Chemtech USA Inc.
Lee Ming Yang, Sulzer Singapore Pte. Ltd.

Abstract

The ethylbenzene (EB) / styrene (SM) splitter (a.k.a. the EB recovery column) is the biggest column in the styrene production process. It can be considered a mega column: column diameters are often larger than 5 m, column heights are taller than 60 m, and 6 beds of structured packing are usually employed.

Generally, packed columns with large diameters are susceptible to large-scale maldistribution. For the EB/SM splitter, the low relative volatility (α-value) of the binary system, and the stringent product specifications, make the column extremely sensitive to maldistribution. Therefore, if an EB/SM splitter is exhibiting poor separation efficiency, a maldistribution evaluation is strongly recommended as part of the troubleshooting evaluation.

This paper will review the various sources of maldistribution susceptibility, which can stem from design, installation, and operational issues. Sometimes, the problem source(s) can only be exactly determined after opening up the column during plant shutdown. As the shutdown time frame is usually limited, there is only a narrow window period to both identify and resolve the source(s) of column underperformance.

Fortunately, it is possible to narrow down the possibilities prior to plant shutdown, via methodologies that encompass process, hydraulic and mechanical considerations. Some important steps include: simulation match back to operating data, hydraulic evaluation of the column internals, and maldistribution sensitivity analysis of the column. After opening up the column, further methodologies can help to pinpoint the problem source(s). Some important steps include: visual inspection of the column internals, and liquid distributor water tests.

The troubleshooting process will be illustrated via several troubleshooting case studies. They will examine the source(s) of column underperformance, and the subsequent corrective actions to return the column to satisfactory operations. These experiences should help engineers to perform troubleshooting in a structured manner, and allow plant owners to optimize their resources in a troubleshooting activity.

1. Introduction

The EB/SM splitter, otherwise called the EB recycle column, is a mega-column, the tallest and widest column in the distillation train. As the name implies, this column separates between ethylbenzene (light key) and styrene (heavy key), which together comprises more than 95% of the feed to the column, as shown in Figure 1.

Component	Composition [wt.%]
Benzene and Toluene	< 3
Ethylbenzene	35 to 40
Xylenes	< 0.05
Styrene	60 to 65
α-Methylstyrene	< 0.5
Other Heavies	< 3
Inhibitor	< 0.1

Figure 1: Typical composition ranges of the components in the EB/SM splitter feed stream.

Before the mid-70s, EB/SM splitters were equipped with trays, which presented two issues: (1) high liquid hold-up and residence time, and (2) high column pressure drop leading to high column bottom temperature. Both issues can cause serious polymerization problems. Due to the presence of the vinyl group (−CH2=CH3), the styrene monomer can undergo polymerization reactions.

From 1977 onwards, EB/SM splitters were progressively revamped with first generation structured packing, such as Sulzer Mellapak™. From the 2000s onwards, EB/SM splitters were again progressively revamped with second generation structured packing, such as Sulzer MellapakPlus™ (Fasseler, Kolmetz, Ng, Senthil, & Tau, 2005). Structured packing reduces both pressure drop and residence time in the column dramatically, mitigating polymerization and reducing Opex.

In addition to the above advantages, structured packing, especially the second generation versions, offer higher separation efficiency, or hydraulic capacity, or both, compared to trays. This results in significant Capex savings as well.

Over the span of 40 years of using structured packing in EB/SM splitters, there have been several columns that did not perform to expectations, usually in terms of separation efficiency. Sulzer was involved in troubleshooting many of these columns, and these industry experiences tell us that maldistribution is the primary culprit behind column underperformance. In these columns, the design efficiency was restored after rectifying or eliminating the source(s) of maldistribution. In troubleshooting, the real challenge often is: How do we identify and rectify the source(s) of maldistribution?

2. Understanding Maldistribution in an EB/SM Splitter

The light key, ethylbenzene, has a normal boiling point of 136°C, whilst the heavy key, styrene, has a normal boiling point of 145°C. With a boiling point difference of ~ 9°C, and close to ideal liquid interactions, this is considered a low relative volatility system, and $\alpha_{EB,SM} < 1.4$.

The low $\alpha_{EB,SM}$ value results in a high number of theoretical stages (NTS) required for separation. As the bottom product specification is more stringent, more NTS and packed beds are required.

Parameter / Specification	Value	Unit
Column top pressure	100 to 400	mbar
Industrial scale column diameter	3 to 10	m
NTS in the rectifying section	30 to 40	
Number of packed beds in the rectifying section	2	
NTS in the stripping section	60 to 80	
Number of packed beds in the stripping section	3 to 5	
Styrene impurity content in the top Ethylbenzene product	1 to 2	wt.%
Ethylbenzene impurity content in the bottom Styrene product	50 to 500	ppm (wt.)

Figure 2: Key column design parameters and product specifications.

2.1 A Thought Experiment: Two Packed Beds vs. Six Packed Beds

The importance of good, even distribution can be demonstrated via a thought experiment. Consider a binary system with 40 wt. % ethylbenzene and 60 wt. % styrene in the feed stream. The product specifications are 1.5 wt. % styrene in the top product and 200 ppm (wt.) ethylbenzene in the bottom product. This can be accomplished with 30 NTS in the rectifying section and 60 NTS in the stripping section.

Ideally, this system could be designed with 1 packed bed each in the rectifying and stripping sections. In reality, however, 2 beds are used in the rectifying section, and 4 beds are used in the stripping section.

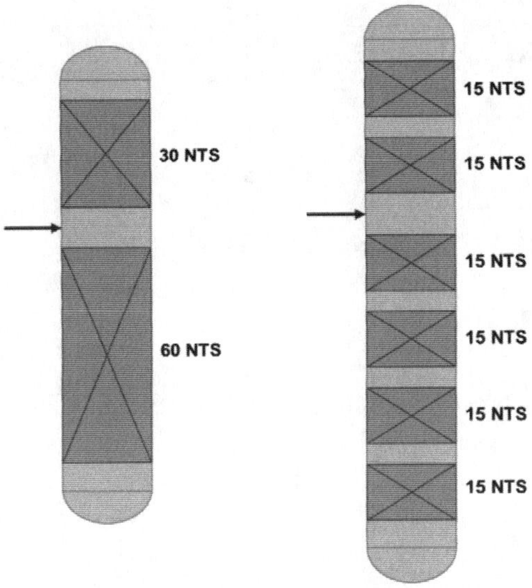

Figure 3: An industrial EB/SM splitter is designed with 6 packed beds, requiring significantly more column height and column internals (e.g. liquid distributors).

The use of 6 beds, instead of 2 beds, is entirely driven by the need to mitigate the negative effects on maldistribution on the "apparent NTS" and hence product purities. These effects are shown in Figures 4a and 4b, for the two-bed design and the six-bed design respectively.

The x-axis represents the degree of maldistribution, λ, specifically, λ_{liq} for liquid maldistribution. λ is a fraction – a value of 0.06 indicates 6% maldistribution. The y-axis represents the "apparent NTS", the actual NTS achieved due to the negative effects of maldistribution. At λ=0, the apparent NTS is equal to the design NTS. NTSD (green points) refers to the rectifying section, whereas NTSB (red points) refers to the stripping section.

In Figure 4a, the apparent NTS reduces drastically with increasing maldistribution, with the effect more pronounced in the stripping section (NTSB). There is always some liquid maldistribution even from a well-designed liquid distributor. Yet at 5% liquid maldistribution, the column will already be operating at half the design NTS, and the ethylbenzene content in the bottom product will increase to almost 3 wt.% (a factor of increase of 150 times!).

In contrast, in Figure 4b, the reduction in apparent NTS with maldistribution is less pronounced. At 5% maldistribution, the apparent NTS is 84, compared to the design NTS of 90. The ethylbenzene content in the bottom product will increase to ~500 ppm (wt.), but this can be resolved (relatively) easily with additional packing and/or increased boil-up. In the six-bed design, inter-bed liquid collection and redistribution allows equal concentration equalization in the liquid, preventing the propagation of liquid maldistribution that would have occurred in the two-bed design.

The conclusion of the thought experiment is: liquid maldistribution, and by extension, vapor maldistribution can have a pronounced effect on the separation efficiency and final product purities in an EB/SM splitter. In other words, the EB/SM splitter is extremely sensitive to maldistribution.

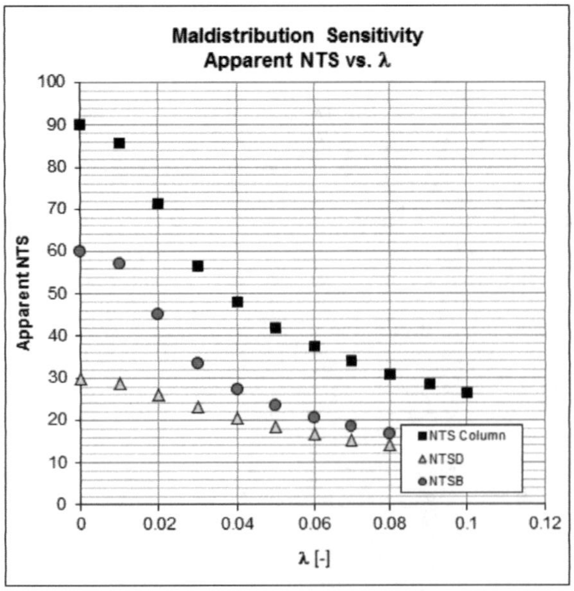

Figure 4a: Plot of apparent NTS against liquid maldistribution, λ_{liq}, for the two-bed column design.

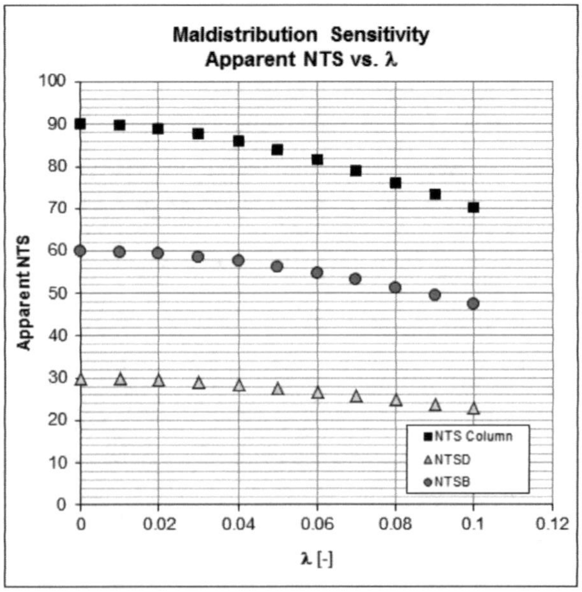

Figure 4b: Plot of apparent NTS against liquid maldistribution, λ_{liq}, for the six-bed column design.

2.2 Sources of Liquid and Vapor Maldistribution

Maldistribution, in both the liquid and vapor phases, is a hydraulic phenomenon, and can be induced due to various sources. Some of the commonly encountered sources, relevant to the EB/SM splitter, are listed below.

- Design Issues
  - Excessively tall packed beds.
  - Inadequate distribution quality and drip point density of liquid distributors.
  - Inadequate design of vapor distribution devices, e.g. feed inlet devices and chimney trays.
- Fabrication and Installation Issues
  - Incorrect fabrication of column internals, e.g. liquid distributors. (Structured packing is usually manufactured correctly.)
  - Improper installation of the structured packing, e.g. lack of proper sealing of the packing against the column wall.
  - Damage to the structured packing.
  - Uneven installation of the liquid distributors.
  - Deviation of the column and/or support structures from verticality.
- Operational Issues
  - Polymerization, leading to fouling in the structured packing.
  - Polymerization, leading to fouling in the liquid distributor.
  - Presence of foreign objects, e.g. rust, leading to plugging issues.

Maldistribution can occur from one of the above sources, or sometimes from a combination of these sources. Maldistribution can also occur at different locations in the column, and sometimes in multiple locations as well. Theoretically, it would be possible to inspect and investigate every part of the column on-site, but it would be laborious, time-consuming and cost-prohibitive for a mega-column. Instead, a systematic troubleshooting procedure, including a maldistribution analysis, can help to eliminate and hence narrow down possibilities. Ultimately, the troubleshooting process can be expedited by reducing amount of field work required. In Section 3, the overall troubleshooting methodology will be examined, illustrating how the location and source of maldistribution can be identified.

3. Troubleshooting Methodology for an EB/SM Splitter

The EB/SM splitter column is the heart of every styrene production plant – plant operators cannot afford for this critical column to fail. As seen from Sulzer's various troubleshooting experiences, plant operators will be under immense pressure to resolve any underperformance issue with the EB/SM splitter as soon as possible. If there is a plant shutdown to rectify the column, the shutdown time frame will be limited, adding further constraints to the troubleshooting process. However, there is no magic wand which can identify and pinpoint the source of column underperformance. Therefore, a well-organized, systematic approach, together with cooperation with the plant operators and responsive corrective actions, will be critical to the success of any troubleshooting activity.

3.1 Validation of Column Design

It is always worthwhile to perform a validation of the column design, as the first step of any troubleshooting activity. The column design is typically includes process design and equipment mechanical design.

Design validation further includes comparing the current column operation against the design conditions. Various parameters could have been adjusted or simply changed over time, e.g. feed composition, feed throughput, and column top pressure. The impacts of these changes on the column operation must be established. For instance, changing feed composition can result in increased reflux to achieve the same product specifications. It is thus important to ensure that the increased reflux flowrate falls within the operating range of the liquid distributors.

3.2 Verification of Manufacturing and Installation

The EB/SM splitter is a mega-column, which requires several thousand cubic feet of structured packing and very large liquid distributors, collectors and other column internals. Thus, it is important to verify manufacturing quality records of the packing and column internals for potential non-conformities. If there are photos taken during the fabrication and testing processes, they should be examined for any potential problems.

In particular, every non-identical liquid distributor should have been water tested at the workshop, to ensure good liquid distribution quality. There are three different test methods available and the liquid range test with level measurement is the most commonly adopted method. At the design volumetric flow rate, liquid level measurements are taken at all the arm troughs. The relative standard deviation, considering all the measurement points, is subsequently calculated. This relative standard deviation represents the extent of non-uniformity in liquid distribution, and should be below 5% for a well-designed distributor (as noted in Section 2.1).

Figures 10a and 10b: Photos showing a liquid distributor under water testing for distribution quality.

Similarly, on-site inspection and installation records can provide clues to potential non-conformities as well. It is always good practice to have a qualified field supervisor from the column internals vendor to be present during installation. In addition to overseeing and ensuring proper installation, an experienced field supervisor will be able identify and rectify most on-site problems. This can prove especially valuable in a revamp project, where complications can readily arise, e.g. when an EB/SM splitter is revamped from trays to structured packing. The entire installation timeline and all anomalies would also have been recorded in the field supervisor's inspection and installation records.

3.3 Maldistribution Analysis Based on Column Operating Data

A process simulation match back, accurately describing the current column operation is critical towards understanding and identifying the problems plaguing the column. A detailed methodology has been published by Daniel R. Summers (Summers, 2009). This paper will discuss only the key aspects salient to the EB/SM splitter column.

Plant operating data are usually available in abundance in an EB/SM splitter. Typically, instrumentation will be available to give the following data:

- Feed flowrate and temperature.
- Top and bottom product flowrates.
- Reflux flowrate and temperature.
- Reboiler heating steam flowrate and temperature.
- Temperature and pressure at the column top, column bottom, and every inter-bed spacing.
- Pressure drop across each individual packed bed.

Lab sampling is conducted regularly as well, usually at least once a day. Thus, data on the compositions of the feed, top ethylbenzene product, bottom styrene product, and sometimes the reflux and inter-bed liquids, will be available too. Some plants even install a sight glass at every inter-bed section. The photos and videos taken are critical in determining the hydraulic phenomena in the operating column.

When analyzed in conjunction, these data usually allow the process engineer to come up with a simulation that matches the plant operation closely. One possible method to perform the simulation match back is:

- Fix the column top pressure and the column pressure drop.
- Fix the top and bottom product specifications.
- Fix the reflux and reboiler heating steam temperatures.
- Vary the NTS in the rectifying section and the stripping section until a reasonable match to the reflux and reboiler heating steam flowrates are achieved.
- Ensure that the mass balance is closely matched by comparing the top and bottom product flowrates and compositions.
- Ensure that the temperature profile is closely matched by comparing the temperatures at the column top, column bottom, and inter-bed spacing.

The above method gives the process engineer a good feel for the current column operation. However, to isolate the location of the maldistribution, further match back to each individual packed bed will be required, to determine the NTS that each bed is achieving. This can be performed in several ways: (1) matching back to the pressure profile, (2) matching back to the temperature profile, and (3) matching back to both the pressure and temperature profiles. The first two methods are relatively less demanding but less accurate. The reverse is true for the third method.

In general, the temperature indicators will be more accurate than the pressure indicators and the pressure drop indicators (Summers, 2009). However, due to the close boiling points of ethylbenzene and styrene, despite a complete change in the composition profile, the temperature difference between the column top and column bottom is not huge. The temperature and concentration profiles of an EB/SM splitter operating at 400 mbar (a) are shown in Figure 10. The column top temperature (excluding the condenser) is ~104°C, while the column bottom temperature is ~121°C. The overall temperature difference of 17°C results in a temperature difference of <0.2°C per NTS! This makes it challenging to pinpoint the transition from one packed bed to the next.

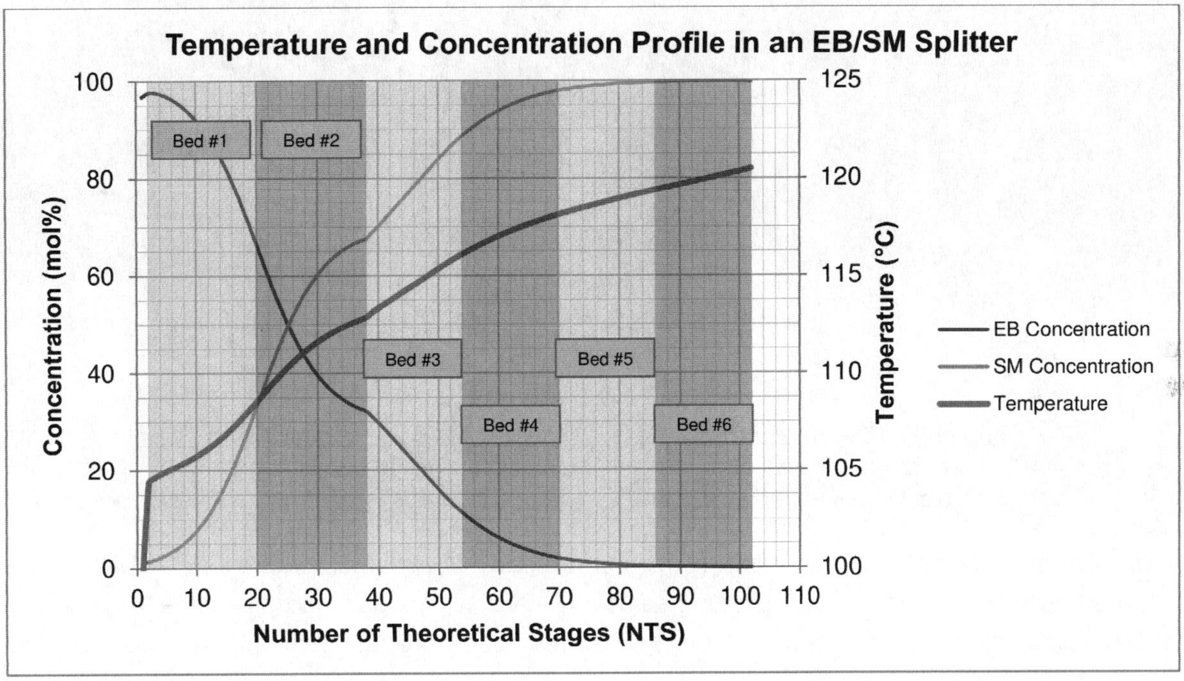

Figure 11: Temperature and concentration profiles of an EB/SM splitter operating at 400 mbar (a).

Pressure drop is, once again, an important topic. The temperature profile is a function of both the composition profile and the pressure drop profile. For the EB/SM Splitter, a significant change in the column pressure drop should result in a visible change in the temperature profile as well. A high pressure drop across a particular bed should raise alarms. It indicates several possibilities. There could be some fouling in the structured packing, obstructing vapor flow. The higher temperature, especially near the column bottom, facilitates polymerization, which propagates fouling. Fouling can cause redirecting of liquid and vapor flows, leading to maldistribution. A high pressure drop, however, is often not the result of maldistribution alone. On the contrary, maldistribution may result in a pressure drop *lower* than the design value.

A maldistribution sensitivity study can be performed, in a similar manner as in Section 2.1. The column is assumed to be equally divided into two parts, across the column cross-section, where one part receives more liquid and less vapor, and vice versa for the other part. However, the degree of liquid and vapor maldistribution, λ_{liq} and λ_{vap}, will have to be adjusted accordingly for each packed bed, to fit the plant operating data and the results of the above maldistribution analysis. The maldistribution sensitivity study will help to qualify the extent of maldistribution occurring in the packed beds.

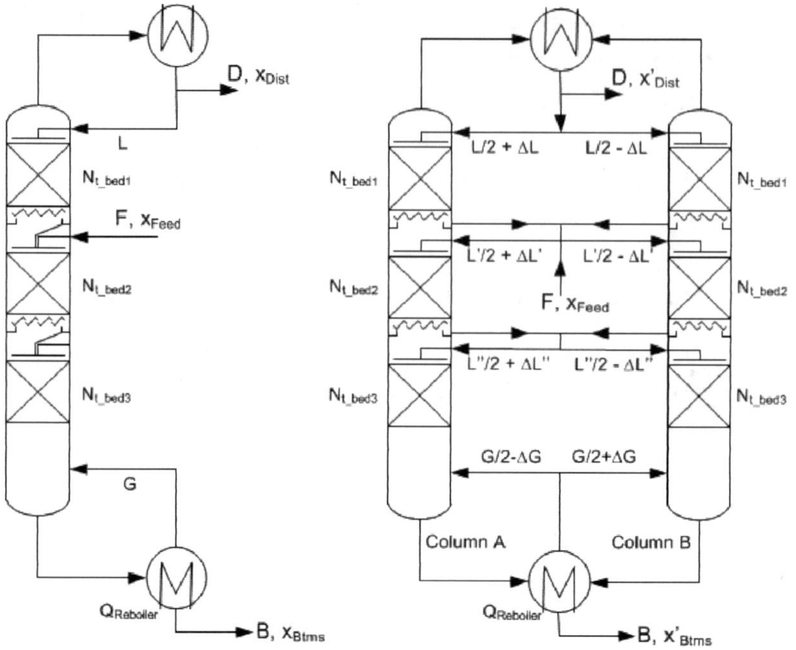

Figure 12: A maldistribution sensitivity study can be performed using the two column model, to assess the impact of maldistribution, λ_{liq} and λ_{vap}, on column separation efficiency.

It may also be advisable to perform a gamma scan, which can visualize the quality of radial liquid distribution across each packing section. However, issues with costs, column access and limitations on the use of specific scanning methods may pose restrictions to use this technique.

3.4 Visual Inspection and Liquid Distributor Water Test

Sections 3.1 to 3.3 detail the various activities that can be conducted while the column is in operation. In essence, they answer the questions:

- What could *possibly* go wrong??
- Where in the column could something *possibly* go wrong?
- How bad is the maldistribution to cause the observed reduction in efficiency?

These answers help us to isolate the potential problematic areas to examine in detail during the limited shutdown window.

Depending on the findings from Sections 3.1 to 3.3, a combination of the activities like visual inspection, water test and cleaning of packing / internals may be conducted.

4. Case Studies

Three case stories on troubleshooting EB/SM splitters are presented.

4.1 Wear and tear can break a column

A customer in the Far East is operating an EB/SM splitter with a top operating pressure of 100 mmHg. The column is equipped with 5 beds of Mellapak 250.Y packing. There are 2 beds in the rectifying section and three beds in the stripping section. The column has been in operation for more than 20 years.

The column was previously equipped with VRG type distributors. The VRG is a tubular type distributor with ground holes for liquid discharge. The customer experienced the problem of the distributors getting plugged frequently, and decided to upgrade to the VEP type distributor.

The VEP liquid distributor is a trough type distributor, comprising many arm troughs with splash baffles, fed by a main trough parting box which promotes good liquid mixing. The liquid is distributed over the packing via the arm troughs and splash baffles that are attached to the external walls of the troughs. In addition, the lateral, elevated position of the distribution orifices tolerates fouling, and the shape of the splash baffles minimizes liquid entrainment.

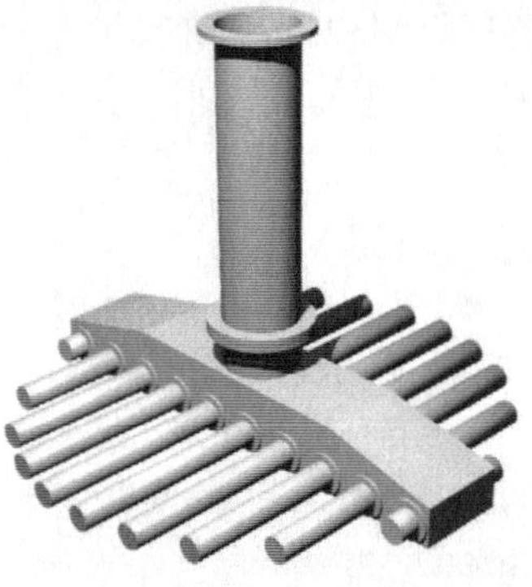

Figure 13a: Picture of a VRG liquid distributor.

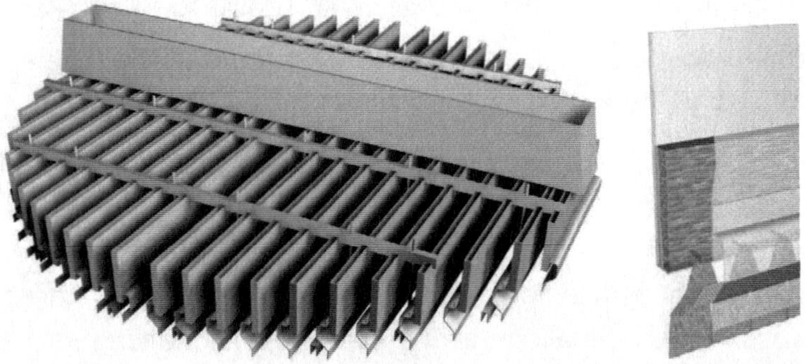

Figure 13b: Picture of a VEP liquid distributor, together with the illustration of the liquid parabola spreading effect due to the splash baffle.

The three distributors in the stripping section were replaced during a regular plant shut down. After start-up, the column experienced an underperformance issue with poorer separation efficiency. The styrene content in the distillate was increased to ~6.0 wt. % to keep the ethylbenzene content with in the acceptable limits of ~ 250ppm. This resulted in greater product losses and recycling, and greater energy consumption.

Section	Packing type	Design NTS	Observed NTS
Bed 1 (Top)	Mellapak 250.Y	100%	100 %
Bed 2	Mellapak 250.Y	100%	92 %
Bed 3 (Feed)	Mellapak 250.Y	100%	76%
Bed 4	Mellapak 250.Y	100%	70%
Bed 5 (Bottom)	Mellapak 250.Y	100%	69 %
Total		100%	79%

Figure 14: Comparison of the actual, observed NTS with the design NTS for individual packed beds in the first case study.

The customer was wary of the new distributor design and questioned the suitability of the VEP distributor for this application. However, Sulzer was confident in the VEP distributor, and undertook a troubleshooting initiative to diagnose the column.

Sulzer first investigated the process design and mechanical engineering aspects, and eliminated those possibilities. A simulation match back to operating data revealed that the NTS reductions were seen primarily in the stripping section (see Figure 14), which looked incriminating for the VEP distributors. However, a review of the field inspection and installation records revealed the real problem. The field supervisor observed flattened packing channels on the top packing layer of the packed beds in the stripping section. This occurred due to frequent dismantling of the old VRG distributors for cleaning. During the installation period, our supervisor had advised to replace the top packing layers of the beds; however, the customer dismissed the advice.

As a result, the blocked packing channels created maldistribution at the top of each bed, which propagated along the entire bed. This resulted in some 25% to 30% loss in separation efficiency in each bed (see Figure 14). Subsequently, Sulzer supplied one packing layer for each bed for immediate replacement in an emergency shut down. After the packing replacement, the column was back to normal operations with products meeting the design specifications.

Case 2: Never underestimate the power of nature.

Another customer in the Far East is operating an EB/SM splitter with a top operating pressure of 60mm Hg. The column is equipped with 6 beds of Mellapak 250.Y packing. The column has been in operation for more than a decade. Due to the fluctuation in the styrene market demand, the plant was shut down for an extended period of time.

When the market recovered, the customer wanted to change the packed bed below the feed location (Bed #3) prior to restarting column operation. The customer did not have specific revamp objectives. Thus, Sulzer proposed and conducted the revamp to MellapakPlus 252.Y, which will achieve the same efficiency but deliver higher capacity and/or a lower pressure drop.

After the revamp, the customer suffered a drastic deterioration in product purities. Based on the operating data, the column was simulated to check the packing efficiency in various sections. The temperature and pressure profile of the column helped to provide an estimation of observed efficiency in each section of the column.

Section	Packing type	Design NTS	Observed NTS
Bed 1 (Top)	Mellapak 250.Y	100%	81%
Bed 2	Mellapak 250.Y	100%	71%
Bed 3 (Feed)	MellapakPlus 252.Y (revamped from Mellapak M250.Y)	100%	80%
Bed 4	Mellapak 250.Y	100%	81%
Bed 5	Mellapak 250.Y	100%	40%
Bed 6 (Bottom)	Mellapak 250.Y	100%	56%
Total		100%	68%

Figure 15: Comparison of the actual, observed NTS with the design NTS for individual packed beds in the second case study.

From Figure 15, it was clear that that the column had lost efficiency in every bed. The loss of efficiency was especially severe in the bottom two beds. Upgrading one packed bed from Mellapak to MellapakPlus could not have been the reason for an overall deterioration in column performance. It is highly likely that there were undesirable mechanical conditions inside the column. Given that all beds exhibited substantial underperformance, Sulzer recommended an extensive column inspection with a longer shutdown window.

During the inspection, it was found that the most of the distributors were plugged with rust particles and other debris. In particular the main channel of bottom two distributors had accumulated large quantities of rust particles. Subsequently, a water test was conducted by circulating the water through the reflux pipe. The bottom two beds were found to be giving the worst quality of distribution. In general, the distribution quality was not in the desirable range for each of the top four distributors as well.

Figure 16a: Large amounts of rust particles and debris were discovered in the bottom two liquid distributors.

Figure 16b: A liquid distributor water test was conducted on-site in the column.

The distributors were dismantled for cleaning. In addition, a few layers of packing sections were removed to check for any traces of polymerization. The packing and distributors were reinstalled, and a second water test was performed to verify the quality of distribution. After start-up the column was found to achieve its required performance.

This plant was located very close to the seashore. The column shell was constructed using carbon steel material. During the long shutdown, the customer had kept the manhole open to atmosphere, causing significant rusting on the column inner walls. When the column is started up, most of this rust was washed by the continuous flow of liquid, and accumulated in various column internals, particularly the liquid distributors in the bottom two beds.

Case 3: Gaskets – The Trouble Makers.

A customer in the Americas is operating an EB/SM splitter with a top operating pressure of 60 mmHg. The Column is equipped with 5 beds of MellapakPlus packing, with 2 beds in the rectifying section and 3 beds in the stripping section.

The customer had revamped the plant from Mellapak to MellapakPlus. However, the column was not started up for a few years, due to reductions in styrene demand.

When the plant was eventually started up, the column achieved poor performance from the Day one. The customer provided the operating data to evaluate the performance of the column. A process simulation was performed to match the (top and bottom) product purities and the reflux flow rate by varying the number of theoretical stages. The pressure and temperature profile helped to estimate the efficiency of each packed section. The design and operating number of stages is tabulated as follows:

Section	Packing type	Design NTS	Observed NTS
Bed 1 (Top)	MellapakPlus	100%	87%
Bed 2	MellapakPlus	100%	67%
Bed 3 (Feed)	MellapakPlus	100%	61%
Bed 4	MellapakPlus	100%	47%
Bed 5 (Bottom)	MellapakPlus	100%	84%
Total		100%	69%

Figure 17: Comparison of the actual, observed NTS with the design NTS for individual packed beds in the third case study.

Based on our evaluation, the bottom section has lost efficiency to a large extent, when compared to the top section. Regardless, there is gradual reduction in packing efficiency in every packed section. The customer decided to shut the plant to perform column inspection and distributor water testing.

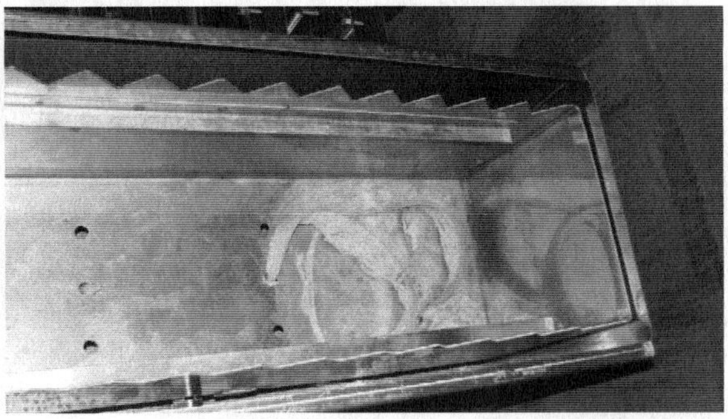

Figure 18a and 18b: Disintegrated gasket pieces falling off from a downpipe and accumulating a liquid distributor main channel.

During visual inspection, the distributor main channels were found to be plugged with broken pieces of gaskets in almost all distributors. Due to the blockage of orifice holes in the main channels, liquid flow to the certain arm channels were blocked, while the other arm channels received higher liquid flows. This is a classic case of maldistribution.

The distributors were cleaned and a water test was performed to ensure the quality of distribution before closing the manhole. When the plant was started up again; the EB/SM splitter achieved its required performance. It is recommended to avoid using gaskets for internal downpipe connections inside the EB/SM splitter column.

The three case studies illustrate the importance of every detail contributing to successful performance, much like how a kink in the cogs of a machine can cause it to fail. There is a lot of science behind the troubleshooting of an EB/SM splitter, but it is also an art. It takes a lot of imagination to visualize all possible causes and eliminate the improbable ones.

References

Fasseler, P. W., Kolmetz, K., Ng, W., Senthil, K., & Tau, Y. (2005). Design Guidelines for Distillation Columns in Ethylbenzene and Styrene Monomer Service. *AIChE Spring Meeting 2005.* Atlanta, Georgia.

Summers, D. R. (2009). How to Properly Evaluate and Document Tower Performance. *AIChE Spring Meeting 2009.* Tampa, Florida.

Paper 34a
(Proceedings 479206)

Vacuum Distillation:

Wash Zone Configuration

Andrew W. Sloley

Advisian (WorleyParsons Group)
Energy Center II
575 N Dairy Ashford St
Houston, TX 77079

Prepared for Presentation at the
American Institute of Chemical Engineers
2017 Spring Meeting
San Antonio, TX March 26-30, 2017

UNPUBLISHED

1. Summary

Refinery vacuum units include a deentrainment, or wash, zone for removal of heavy material from desired products. Wash zone reliability problems often shut vacuum units down, causing significant losses. Factors that strongly influence wash zone reliability include vapor and liquid distribution, upstream flow regimes, and wash zone efficiency and pressure drop. For fuels vacuum units, the wash zone should have the minimum amount of packing consistent with acceptable deentrainment. The minimum wash bed has lower capacity and higher reliability.

One major factor that must be evaluated on a case-by-case basis is phase separation in the transfer line. Even small amounts of phase separation will feed superheated vapor to the wash zone. Superheated vapor dramatically raises the amount of wash oil required for wetting the wash bed. Properly installed and operated, some vacuum towers have demonstrated trouble-free performance for 20+ years. Others, with improper understanding of the operation, suffer from one to two year run lengths, dramatically reducing refinery profits.

2. Introduction

This discussion has the intent of covering process considerations for wash zone operations. However, equipment constraints and process requirements interact. Process conditions must remain within equipment operating constraints for reliable and effective operation. Equipment constraints will be mentioned as required to explain the process limits and their importance. Previous work has focused on equipment constraints and includes more detail on them [1].

3. Operating Objectives

Refinery main fractionators include a wash section immediately above a two-phase feed. These towers have very high inlet velocities and high kinetic energy in the feed. The high inlet velocity and energy dissipation in the feed zone of the tower easily creates small liquid droplets. The droplets entrain into the rising vapor. Without a wash zone, the liquid droplets end up in the products above the feed.

The wash zone is a deentrainment device to remove the liquid droplets. The vacuum distillation tower works in a severe operating environment. Figure 1 illustrates a typical refinery vacuum tower and its wash zone. The wash zone is a packed bed above the tower feed.

The wash section has the following major objectives:

1. Deentrain liquid from the rising vapors
2. Maximize yields by maintaining low pressure drop and low wash liquid rate
3. Maximize reliability
4. Fractionate volatile contaminants (metal compounds)

These objectives compete with each other. Wash zones that meet some objectives tend to have problems with the other objectives.

4. Terminology

Figure 2 shows the relevant components of the wash bed, and the external and internal streams. The definitions of the streams for this discussion are:

Table 1 Wash Zone Streams

Stream	Definition
Wash (or wash oil)	Liquid placed on top of the wash bed to irrigate the bed to prevent coking.
Overflash	Distillate liquid leaving the bottom of the wash bed. This is material that was vaporized in the flash zone.
Entrainment	Liquid droplets moved up the tower by high vapor velocity. This is material that was not vaporized in the flash zone.
Slop wax [1]	The combination of entrainment and overflash that is collected on the tray below the wash bed. Slop wax = overflash + entrainment
Condensate	Liquid condensed on the bottom of the collector tray because the overflash is colder than the feed vapor. This stream is shown for completeness. In most units it has a very low flow rate.

---

1 The term slop wax comes from lubricants vacuum units. The bottoms lubricant cut had a maximum allowed viscosity or viscosity index. The bottoms of the tower went to processing for bright stock (cylinder oil) recovery with a minimum viscosity specification. The slop wax was a drag stream between the two products that removed material with too high a viscosity for the lubricant draw but too low a viscosity for the bright stock.

If the entrainment rate is high, most of the liquid gathered by the slop wax collector is entrainment, not a recoverable distillate.

5. Fractionating Versus Deentraining Wash Zones

Most refinery vacuum units are for fuels production. The vacuum tower products go to downstream processing for producing fuels. Many fewer vacuum units produce lubricant base-stocks. Lubricant production is only a small fraction of total vacuum unit capacity (<10%). This discussion focuses on fuels vacuum units. Nevertheless, many of the concepts apply to lubricants vacuum units.

Fuels vacuum units may have two purification objectives for the wash zone:

1. Remove mechanical entrainment of small liquid droplets from the vapor, or
2. Fractionate volatile materials to improve the lower product (HVGO) quality.

Mechanical entrainment puts vacuum residue in the distillate product. This can cause large problems with asphaltenes, metals, and carbon residue qualities of the HVGO.

The most common volatile contaminants of interest are vanadium and nickel porphyrins. High porphyrin levels are most common in heavy crudes (API gravity <= ~21.9°). Lighter crudes tend to have much lower porphyrin concentrations. Crudes well known for volatile metal compounds include Bachaquero, Leona, Isthmus, and Maya, as well as many others.

When the wash bed objective includes fractionation, fractionation requirements set the packed bed efficiency and liquid requirements. This efficiency and liquid rate tend to be much higher than the requirements for a wash bed intended to meet a deentrainment specification.

This discussion focuses on wash beds for deentraining services.

6. Conventional Configuration

Figure 3 shows a conventional configuration for a wash zone. Generic elements include:

1. Wash oil distributor to put liquid on the top of the wash zone packing.
2. Wash zone packing to act as a deentrainment device.
3. Collector under the wash zone to collect overflash and distribute vapor entering the wash zone.

Other elements in Figure 3 may vary between different conventional vacuum units but have little effect on the overall analysis of wash zone operation. In this case:

1. Overflash is recycled to the tower boot with an external pump. Other options include recycling the overflash upstream and allowing internal overflow of the overflash. Overflash disposition has a negligible effect on the current analysis.
2. The vacuum tower has no stripping trays. Stripping trays use steam to increase overall product yield. The presence of stripping trays has only a minor effect on the wash zone analysis here.

7. Conventional Operations

7.1. Ideal Wash Zone

Figure 4 compares an ideal flash zone to an achievable flash zone. Actual flash zone performance never achieves perfect deentrainment. Close-to-perfect deentrainment requires low vapor velocity to allow droplet settling and very few small liquid droplets (< 100 microns). Capital costs keep flash zone sizes small. Some entrainment always occurs.

A deentrainment or wash section above the feed removes the entrainment. To prevent coking, the deentrainment equipment requires net liquid flow at all points. The distillate liquid entering the wash zone is wash oil. The liquid leaving the wash zone is material that had to be flashed "over" the requirement to meet the yields observed. Hence the name overflash for distillate liquid leaving the wash zone.

With deentrainment as the primary objective, the ideal wash zone would have:

1. Minimum fractionation efficiency (zero if possible).
2. Minimum pressure drop (zero if possible).

No physical device can attain perfect performance. The comparison basis used here has been defined as:

1. One theoretical stage flash zone.
2. Minimum pressure drop (2 mmHg) across the slop wax collector tray.
3. Minimum pressure drop (1 mmHg) across the wash zone.

Various curves are presented to show how yield and unit capacity change with wash zone operation. Yield is quantified by cutpoint, the atmospheric equivalent temperature that would have the same distillate yield as the operating unit as on the ASTM D2892 [2] test. High cutpoints equal more distillate yield from the same feed. Capacity factors are relative unit feed rate. High capacity factors equal a higher feed rate to the vacuum unit.

7.2. Operating Limits

Minimum wash oil rates vary as a complex function of:

- Tower diameter
- Fractionation across the wash zone
- Mass flow rate
- Pressure drop across the wash zone
- Choice of mechanical equipment used

The objective of the wash oil is to create a targeted overflash rate. The liquid rate leaving the bottom of the bed strongly correlates with coking rates and reliability. Due to energy balance and fractionation effects, the wash oil rate will always be more than the overflash rate. The wash bed dries out from the top to the bottom [3]. The wash bed has the highest liquid rate at the top of the bed and the lowest liquid rate at the bottom of the bed. The dry-out ratio is the ratio of the liquid volume at the top of the bed to the liquid volume at the bottom of the bed:

$$dry-out\ ratio = \frac{liquid\ volumetric\ rate\ at\ the\ top\ of\ the\ wash\ bed}{liquid\ volumetric\ rate\ at\ the\ bottom\ of\ the\ wash\ bed}$$

The dry-out ratio is often quoted using standard liquid volume rates. However, here the actual flowing liquid volumes are used.

If the overflash rate is too high, yield drops as excess overflash increases the bottoms rate. If the overflash rate is too low, the wash bed can coke and the unit is forced to shut down. Both problems have a high economic cost.

This creates the fundamental operating challenge for vacuum tower wash zones:

1. The economic cost of using too high a wash rate is high.
2. The economic cost of using too low a wash rate is high.
3. Operating conditions are severe.

Overflash can be defined as a fraction of feed to the unit or a liquid load based on wash bed cross-sectional area. Older references often take the first approach, using a defined percent of unit feed. The most quoted range of overflash rates are based on Watkins [4, p. 110] and use 1 to 2 % overflash for conventional crudes and up to 4% overflash for high-metal content crudes. It is defined as:

$$overflash\ percent = 100\frac{overflash\ rate,\ std\ volume}{feed\ rate,\ std\ volume}$$

Many authors have used 2% as the starting point for unit optimization studies. This is consistent with the design curve included in Watkins.

The second approach uses overflash rates defined by actual flowing volumes and tower diameter. Various authors have suggested minimum overflash loads varying from 0.125 to 0.180 gpm/ft^2 of cross section area at the bottom of the wash bed (5.09 to 7.33 lpm/m^2) based on volume at conditions. The examples used here are based of 0.150 gpm/ft^2 (6.11 lpm/m^2) overflash load. The examples use a maximum vapor rate equivalent to a C-factor of 0.45 [1].

Figure 5 plots overflash percent versus overflash rate. The plot shows that a liquid load of 0.15 gpm/ft^2 (6.11 lpm/m^2) corresponds to 4.3% overflash for the example case used here. Thus, operation at the minimum recommended liquid load has 2.15 times the conventional recommended overflash percent. Clearly, the recommendations are highly inconsistent.

Operating pressure is the likely source of the difference. The worked example in Watkins uses 2% overflash and 42 mmHg flash zone pressure. Many modern vacuum towers run down to 20 mmHg (or lower) flash zone pressure. At low pressures, vacuum tower diameter increases. The larger vacuum tower needs a higher volume of overflash to get the minimum required rate at the bottom of the bed.

Extensive industrial experience has shown that targeting a minimum load (gpm/ft^2) is a fundamentally more sound approach to unit design and operation. Overflash percents should not be used.

Figure 6 and Figure 7 show operating changes versus overflash rate. As overflash rates increase, capacity decreases and yield drops.

7.3. Fractionation

The more fractionation in the wash bed,

1. The larger the composition gradient across the bed.
2. The larger the temperature gradient across the bed.
3. The larger the dry-out ratio from thermal effects.
4. The higher the revaporization rate and the larger the tower diameter for the same capacity or the lower the capacity for the same diameter.
5. The higher the wash oil rate for the same overflash rate.
6. The lower the HVGO draw temperature for the same yield (minor effect).

In general, an ideal wash bed would deentrain but have zero fractionation efficiency. Figure 8 illustrates changes in dry-out ratio, tower capacity, and relative wash oil rate with changes in the number of theoretical stages in the wash zone. Higher fractionation efficiency increases dry-out ratio, reduces unit capacity, and increases the needed wash oil rate. Figure 9 shows cutpoint (yield) versus wash zone stages. The more wash zone stages, the lower the cutpoint.

The largest change is moving from zero theoretical stages to one theoretical stage in the wash zone. While more stages do increase the wash oil rate for the same overflash, the increase is minor.

7.4. Pressure Drop

Pressure drop with no liquid contacting the rising vapor increases the amount of vapor superheat at the top of the bed. Higher vapor superheat increases the wash oil rate required. Some of the wash oil must vaporize to get the vapor down to its dew point.

Higher pressure drop has two boundary conditions:

1. Higher pressure drop with a fixed flash zone pressure causing the pressure at the top of the wash bed to be lower.
2. Higher pressure drop with a rising flash zone pressure and a fixed pressure at the top of the wash bed.

The first case, higher pressure drop with a fixed flash zone pressure, approximates a design case with a target yield defined by flash zone pressure and temperature. Figure 10 and Figure 11 show the dry-out ratios, relative capacity, and cutpoint for varying wash zone pressure drops for the first case.

The second case, higher pressure drop with a rising flash zone pressure, approximates an operating case limited by vacuum system capability. Figure 12 and Figure 13 show the dry-out ratios, relative capacity, and cutpoint for varying wash zone pressure drops for the second case.

The effect of wash zone pressure drop on dry-out ratio is small. The design case with a lower pressure at the top of the wash zone has a large impact on unit capacity (or size). The hydraulic limit in the wash zone is nearly always at the top of the bed. Therefore, reducing the pressure at the top of the bed creates a large capacity drop.

Figure 12 shows increasing capacity with pressure drop across the wash bed. The calculations include changes in transfer line operation with changing flash zone pressure. This capacity increase is due to changes in the transfer line. Section 7.8 includes more detail on transfer line effects.

7.5. Chimney Tray

Figure 1 and the other figures all show vacuum towers with a collector tray underneath the wash zone packing. Some vacuum towers do not have a collector tray. Without the collector tray, yields tend to be lower and deentrainment performance worse.

The collector tray:

1. Prevents the colder slop wax from quenching the flash zone vapors. This increases overall yield.
2. Reduces slop wax re-entrainment. This improves product quality.
3. Improves vapor distribution to the packing. This reduces local areas with high vapor rates and helps reduce localized coking rates. The result is more reliable operation.
4. Collects slop wax for possible metering and sampling. This gives the plant potential tools for improved optimization and increased yields.

The collector tray should be installed. However, this comes with more than an equipment cost. The collector tray creates pressure drop. In general, the higher the pressure drop, the better the vapor distribution above the tray. The collector tray pressure drop can be calculated using an orifice equation. These calculations are usually accurate for the HVGO and LVGO distributors above the wash section. However, for the slop wax collector, measured pressure drops are routinely twice the calculated value.

The higher measured value is caused by vapor flow maldistribution. Computational fluid dynamic analysis of tower entry and collector systems has shown complex flow patterns. These patterns often have severe local variations in velocity and direction [5]. If the chimney tray pressure drop is low and the feed entry poor, parts of the chimney tray may even have vapor flowing down instead of up through the chimneys. A better feed entry device and a higher collector tray pressure drop can dramatically improve vapor distribution to the bed above.

With a good feed entry device, the target pressure drop for the collector tray should be 1 mmHg. In service, the expected pressure drop is closer to 2 mmHg. The author has seen units with as much as 10 mmHg pressure drop across the collector tray.

Pressure drop across the collector tray has nearly identical consequences as extra pressure drop across the wash zone packing.

7.6. Loading (C-factor)

Vapor-to-liquid mass ratio affects the dry-out rate. Section 7.2 discusses the reduced dry-out ratios with higher liquid rates. This is because the higher liquid rate has a higher L/V ratio in the wash section. Less of the wash oil vaporizes to absorb the energy from the cooling vapor.

Lower vapor rates have the same results. Low vapor rates give low C-factors in the packing. Figure 14 and Figure 15 show the dry-out rate, capacity factor, and cutpoint for a constant 0.150 gpm/ft^2 (6.11 lpm/m^2) overflash rate at different C-factors. Lower C-factor reduces the dry-out ratio but also reduces the yield.

Figure 16 shows how the column area, wash oil rate, and overflash rate change as the C-factor changes for a constant feed rate. Higher C-factors are equivalent to higher mass flow rates. The change in mass flow rate is the underlying cause of the change in dry-out ratio. While the higher C-factor produces a higher dry-out ratio, it also requires a smaller area. The change in area is more important than the change in dry-out ratio.

This is also the reason why the higher operating pressure mentioned in Section 7.2 gives a lower overflash percent. The change in wash oil rate is the product of two factors:

1. Higher operating pressure increases the density and increases the max rate per unit area, increasing the dry-out ratio.
2. Higher operating pressure decreases the column cross section area, decreasing the irrigation area.

$$Wash\,Oil\,Volumetric\,Rate = Overflash\,Load \;\times\; Dry-out\,Ratio \;\times\; Cross\,Section\,Area$$

Dry-out ratio and cross section area change in opposite directions when pressure changes. The change in cross section area is more important than the change in dry-out ratio. Hence, as pressure increases, the total volume flow of wash oil drops.

7.7. Entrainment

Entrainment can be the source of liquid on the wash bed. In some units the overflash rate is approximately zero (< 0.02 gpm/ft^2, 0.81 lpm/m^2). In spite of this, multiple-year runs without coking the bed were achieved.

With no overflash, the slop wax is 100% entrainment. In these units, the slop wax rate was measured and was found to be very high (over 10% of the total residue rate from the bottoms). The crude was clean, light, and thermally stable.

Under these conditions, entrainment irrigated the bottom of the wash bed. The small amount of overflash used irrigated the top part of the bed. Entrainment reached far enough into the wash bed that it overlapped the region where overflash was present. The entire bed was irrigated and did not coke up. Figure 17 illustrates this operation. Intentionally operating in this mode is not recommended. The probability of coking the bed is high. Only small changes in either wash rate or entrainment rate can lead to coking or severely off-specification HVGO product.

7.8. Transfer Line Phase Separation

In the early 1990s the author and others reviewed several severe coking incidents where everything indicated that the wash bed should have been sufficiently irrigated. However, plant data was showing very little overflash. In these cases, the transfer line allowed for phase separation between the vapor and the liquid.

With phase separation, the depressurizing vapor becomes superheated. The superheat is not available to vaporize more liquid in the transfer line. The limited surface area between the vapor and liquid limits the amount of heat transfer. Figure 18 illustrates this operation.

The basic technique of accounting for phase separation in the transfer line has been referred to as the non-equilibrium model for transfer lines [6]. The author uses a derivative of the basic approach and prefers to call this the phase-slip model because phase-slip avoids confusion about the topic. The term equilibrium model is already in wide use in the discussion of equilibrium-stage distillation models versus rate-based distillation models.

As well as reducing yield, the superheated vapor entering the vacuum tower increases the dry-out ratio. The superheated vapor vaporizes some of the overflash as the vapor-liquid mixture reaches equilibrium. Figure 19 and Figure 20 shows the effect of phase slip.

Phase slip index (φ) has been defined as:

$$\varphi = \textit{phase slip index} = 100\,\frac{\textit{pressure drop total} - \textit{pressure drop before phase separation}}{\textit{pressure drop total}}$$

Transfer line phase separation is one of the most difficult factors to evaluate when designing a unit. With data, the results of phase separation are clear. Getting the data, however, can be very challenging. The base configuration used for all the comparisons had a transfer line with a φ=29. Values of 10 to 30 are typical for many operating units.

Based on data, unit operations can cover the entire range. Some units have achieved φ of 0. Others have demonstrated yields and required wash oil rates consistent with φ of over 90. Phase separation is the single largest factor in uncertainty in understanding wash zone operation on many units.

Transfer line phase pressure drop and phase separation are highly dependent on transfer line geometry, overall pressure profile, and operating conditions. Nevertheless, factors that make transfer line separation more likely include:

1. Low heater outlet pressure.
2. Low vacuum tower pressure.
3. Large diameter transfer lines.
4. Reduced rate operation.

7.9. Residence Time

Time and temperature combine to create coke. Keeping the oil hot for a longer period of time increases the risk of coking the wash bed. Residence time is not directly included in ideal calculations, but should never be overlooked. Higher irrigation rates (higher wash oil rates) reduce the residence time in the wash bed.

7.10. Interactions

The author has covered major process issues in the wash zone that affect performance. Each has been covered independently. In actual units, the factors combine. Usually, higher efficiency beds:

1. Have higher pressure drop (use more, or more efficient, packing).
2. Have more liquid hold-up because of the higher pressure drop resulting in longer residence time.
3. Use smaller packing sizes and aggressive surface treatments (to increase efficiency) so they are more likely to trap coke in the packing.
4. Have more stages, so they are more subject to local dry out problems due to limits in vapor and liquid distribution.

Combined with transfer line phase separation and problems with feed characterization [7] and molecular weight prediction [8], dry-out ratios can be high. Field data have shown some units have dry-out ratios as high as 15:1.

8. Hot-Wash Configuration versus Cold-Wash Configuration

Hot wash units use the direct liquid drawn from the collector above the wash section as the wash oil. This is normally an HVGO stream. The hot wash is a bubble point liquid on the collector tray and remains a bubble point liquid when it is returned to the top of the wash packing.

Cold wash units use a liquid that comes from downstream in the HVGO cooling circuit as the wash oil. The cold was stream is a subcooled liquid when returned to the top of the wash packing.

Figure 21 and Figure 22 illustrate the hot wash and cold wash configurations.

As long as the liquid rate leaving the bottom of the wash bed is the same, the amount of liquid on the top of the bed is not particularly relevant to meeting process objectives. However, a number of practical concerns favor hot wash.

The higher liquid rate to the top of the bed in hot wash allows for larger openings in the liquid distributor. This makes the distributor less likely to plug or foul. Additionally, subcooled liquids sprayed into a rising vapor cause some of the vapor to condense. The condensing vapor can pull the liquid cones in and collapse part of the cone. Spray cone collapse can lead to portions of the top of the bed having less liquid than expected. Liquid rates in the wash packing are so low that even small deviations in liquid distribution can cause accelerated coking in the wash bed. Hot wash should be used instead of cold wash.

9. Wash Oil Distributors

Two types of liquid distributors are used in wash oil services:

1. Spray header distributors.

2. Trough distributors.

Mechanical and service details of these have been previously covered [1]. In addition to mechanical considerations, process performance of the wash oil distributors affects wash zone performance.

Data currently available do not clearly support one type over the other. Both can be used. Both have benefits and costs. Overall, the spray distributor is usually cheaper to purchase, and easier and cheaper to install. Unless a specific reason is present, the spray distributor is the recommended choice.

Table 2 summarizes the distributor characteristics. Major factors include:

1. Entrainment of distributor liquid before it reaches the packing.
2. Distribution quality and fractionation performance of the distributor.
3. Heat-transfer performance of the distributor.
4. Size of distribution holes.

Table 2 Distributor Characteristics

	Spray Distributor	Trough Distributor
Entrainment of wash liquid	Small droplets created at spray nozzles. Liquid splashing when entering packing.	Vapor channeling due to high vapor velocity when it passes between troughs can entrain liquid from the packing.
Distribution Quality	Low quality distribution reduces fractionation effectiveness (desired) but can lead to low liquid in some areas that might accelerate coking.	Smooth distribution of liquid increases fractionation effectiveness. This is undesired unless fractionation for metals control or viscosity control is required.
Heat-Transfer Performance	Sprays are excellent heat transfer devices. This is undesired.	No heat transfer above the packing. But the good liquid distribution increases heat transfer in the packing.
Size of Holes	Holes in spray nozzles slightly larger.	Very small holes for most services. This is a potential problem even with deflector plates used to increase hole sizes.
Other	Lower cost and relatively simple installation.	Higher cost. Installation can be labor intensive in large-diameter columns. More likely to be damaged by pressure surges.

10. Spray Chambers

In a column without internals, there is nothing for coke to attach to. Spray chambers use one to three rows of spray nozzles. The wash liquid contacts the rising vapor. The small rising droplets combine with larger droplets coming from the spray nozzle. Liquid drops agglomerate by direct impact and by Brownian motion. Delayed coker main fractionators routinely use spray chambers. The delayed coking service is an extremely severe coking service.

Figure 23 illustrates a vacuum tower with a spray chamber. When trays were still used in vacuum towers, having a spray chamber could significantly reduce flash zone pressure. Additionally, trays have a poor record of rapid coking and relatively high entrainment. Trayed towers often used wire-mesh pads [9]–[15]. The mesh pads then created their own problems with coking, high-pressure drop, and collapse. With annual turnarounds common in the 1950s, plants could tolerate the mesh pads coking. With current multiple-year runs, mesh pads have no place around the wash section of vacuum towers.

In spite of the problems with trays and mesh pads, they had better deentrainment performance than spray chambers. To get adequate deentrainment, spray chambers need to operate with an excellent inlet feed distributor and low C-factors. An upper limit on C-factor of 0.20 should be used with spray chambers. A more prudent limit would be a C-factor of 0.18 or below. A spray chamber is really a separator drum. For comparison, most separators without internals are designed for equivalent C-factors of 0.15 to 0.175.

11. Recycle Wash Systems

Some units have recycle wash systems to increase the amount of wash oil while reducing the overflash rate. These systems take some of the slop wax and add it back to the wash oil. Figure 24 shows a one-stage recycle wash system. Figure 25 shows a two-stage recycle wash system. The vacuum tower has two separate wash beds. The upper wash bed uses HVGO as the wash oil, the same as a conventional system. The lower bed uses 100% slop wax as the wash oil. The wash oil from the upper bed falls down into the lower bed. This creates a purge for the system.

The refiner must keep in mind that the recycle contains entrainment. Potential problems with this include:

1. Increased residence time at a high temperature for the recycled entrainment – possibly accelerating coking.
2. Fouling and plugging of the packing or distributors with solids (scale, clay, rust, iron sulfide) contained in the entrainment.
3. Impurities in the HVGO from entrainment of the wash oil return.

Recycle systems have few benefits. The situations where they work well include units running relatively clean and thermally stable crudes. With very few exceptions, the recycle wash systems analyzed by the author could have been replaced with better or lower cost options.

12. Case Study 1

A refiner converted a trayed vacuum tower into a packed vacuum tower. Figure 26 illustrates the history of modifications made. The trayed tower had three trays in the wash zone. The original configuration is shown in detail '0' in Figure 26. The tray spacing was:

Table 3 Vacuum Tower Tray Spacing

Tray	Elevation	Spacing	Spacing
3	31'-0"		
		5'-6"	1.676 m
2	25'-6"		
		5'-6"	1.676 m
1	20'-0"		
		7'-6"	2.286 m
Feed	11'-6"		

The large tray spacing between the second and first tray was intended to reduce the amount of entrainment. Deentrainment improvements with extra tray spacing dramatically decrease once tray spacing exceeds 4 ft (1.219 m). The extra spacing probably had minimal effect on tower performance.

12.1. Revamp 1: Replace Trays With Packing

The refiner was attracted by the benefits of lower pressure drop and higher capacity of structured packing. Trays 1-3 were removed and replaced with a combined packing-grid bed. Total bed depth was approximately 11' (3.35 m). The bottom 5' (1.52 m) was grid and the upper 6' (1.83 m) was structured packing. A spray header was installed for liquid distribution. Based on expected packing efficiency the bed was estimated to have four ideal stages of separation. The estimate of four theoretical stages assumed that the spray header would perform as a high-quality liquid distributor.

The installation was mechanically convenient. The lower elevation was the existing Tray 1 support. No slop wax collector was installed. Figure 26 shows these modifications in detail '1'.

The wash oil rate was set based upon a 2:1 dry-out ratio. This was a common basis used for older vacuum towers with trays. No data were available from the unit as overflash flowed internally into the bottoms.

Performance did not meet the expected goals. Entrainment became severe at a C-factor of 0.38 compared to the predicted capacity limit of 0.5. Additionally, after approximately 16 months of operation, bed pressure drop started to rise. Entrainment rapidly increased as the pressure drop increased. The unit was shut down after 24 months.

12.2. Revamp 2: Add a Slop Wax Collector and Slop Wax Recycle

The possible causes of the rapid coking and lower than expected capacity were a matter of contentious debate. No data were available on actual overflash rates. The decision was made to install a slop wax collector, increase the wash oil rate, and recycle the slop wax to the heater feed. Figure 26 detail '2' shows these modifications.

The reasoning included:

1. A higher wash oil rate would reduce the coking rate.

2. Recycling the slop wax would allow for recovery of distillate in the slop wax. Therefore, the higher overflash rate would not reduce yields.
3. The collector and pump would allow for sampling the slop wax. This would give the plant better data for optimization.
4. The chimney tray would reduce re-entrainment of liquid by separating the slop wax from the vapor in the flash zone.

That the chimney tray might improve wash zone performance by improving vapor distribution was not considered. Later analysis would show that the improved vapor distribution was important.

The modifications did improve the operation. Capacity rose to a C-factor of 0.42 at the same level of entrainment seen after the 1st set of modifications. While improved, this was still not the promised value of 0.5.

Samples revealed nearly zero overflash in the slop wax. Both an asphaltene balance and a sodium balance showed that the slop wax collected was over 95% entrainment. This was true with double the wash oil rate used in the previous run.

If the slop wax is nearly 100% residue, then it contains no distillate to recover by recycling it to the furnace feed. Operating events supported this analysis. The slop wax pump had an operating upset and the pump had to be removed from service. Unit yields did not change when the pump was out of service.

Based on matching a measured overflash in the slop wax, the wash bed was estimated to have a 15:1 dry-out ratio. This seemed very high. Referring to Figure 8, at three theoretical stages the dry-out ratio is less than 7:1 and the incremental change per stage is already very low. It was difficult to understand how 15:1 could occur. The most popular suggestion was that small errors in laboratory technique were causing the dry-out ratio to be over-estimated.

At the start, gas oil quality was very good. The gas oil had a 2 color and no significant asphaltene content. Quality stayed at this level until the pressure drop started to increase.

No significant improvement in run length was seen. After approximately 16 months, pressure drop started to rise again.

12.3. Revamp 3: Reduce Wash Bed Height

At this point the system was clearly not understood and the path forward was not clear. Several other units were suffering from similar wash zone problems at this time. Different strategies were being used at different plants to understand the problem.

Since product quality had been so good at the start of the run, the idea was that less packing was needed. The perceived benefit was that less packing would reduce residence time in the bed and reduce the coking rate.

Figure 26 detail '3' shows the modification made. Wash bed height was reduced from 11' (3.35 m) to 5' (1.8 m). The bed included roughly 40" (1 m) of grid and 20" (0.5 m) of structured packing.

This change resulted in a lower initial pressure drop (less packing), and HVGO quality still met downstream processing requirements. The quality was better than that seen after the first revamp, but not as good as after the second revamp. This was as expected.

Run length did increase. The unit ran for approximately 20 months before pressure drop started to rise. The unit reached 36 months before needing to be shut down for packing replacement. The extra run-length came from a combination of three factors:

1. A lower starting pressure drop (less packing depth).
2. A longer duration before pressure drop started to rise (20 months versus 16 months).

3. A slower rate of increase (due to lower coking rate from lower residence time).

Sampling of the slop wax showed a 7:1 dry-out ratio. This seemed high for the one theoretical stage expected from the reduced packing height.

Modeling with calculations forced to match the dry-out ratio showed that the maximum operation had a 0.45 C-factor, not a C-factor of 0.42. The higher dry-out ratio creates a higher calculated C-factor at the top of the bed.

Experience from other units supported the 0.45 C-factor limit for acceptable entrainment. The original 0.5 C-factor estimate for this crude was over-optimistic.

During this run, data from other units was showing that phase separation in the transfer line was becoming important. Due to the cost of modifying the transfer line, no modifications were proposed.

12.4. Revamp 4: Relocate Wash Oil Distributor

The packing still had insufficient wash oil. Calculations with an assumed phase slip of 50 (φ) matched a 7:1 dry-out ratio. Modifications were planned to increase the wash oil rate to the 7:1 dry out ratio. Spray header modifications were implemented in the 4th revamp.

During this work, it was noticed that the inside of the tower showed evidence of significant liquid spray on the vessel wall. The previous revamp (#3) had reduced the packing height, but not relocated the spray header. A quick piping modification allowed the spray header to be dropped to the correct elevation above the packing.

Plant data after these modifications showed the expected dry-out ratio to match the prediction. Gas oil quality was excellent and the full desired run length of 5 years was achieved.

12.5. Revamp 5: Update Feed Entry

To reach the maximum capacity, a new feed entry device was proposed for the fifth shutdown. The idea was that if less entrainment was carried into the wash bed, vapor rates could be pushed to a higher C-factor. Some entrainment would occur, but product quality would remain acceptable.

After installation of the new feed entry device, rates could be pushed as high as a C-factor of 0.47. However, minor operating changes would cause the tower to flood rapidly. The decision was made to limit feed rates to the 0.45 C-factor.

The new feed entry device improved the HVGO quality slightly and also reduced slop wax rates a small amount.

12.6. Overview of Case 1

Modifications unfolded over a number of years. At the start, the potential for wash zone coking was not well understood. The unplanned factors that contributed to wash zone coking were phase separation in the transfer line and high efficiency in the packed bed. The phase separation required higher wash oil rates than originally planned. The wash bed was reduced in height to reduce the number of theoretical stages. After modifications were made, run length met the original targets. Further developments that improved operations included adding an improved feed entry device.

13. Case Study 2

A refiner upgraded a vacuum tower wash zone from trays and wire-mesh mist eliminators to a packed bed using a 3rd generation random packing. The packing was 6' (1.83 m) of rings. The planned operation was to run with 0.15 gpm/ft^2 (6.11 lpm/m^2) of overflash. The slop wax was collected and recycled to the furnace feed. The overflash pumps always had reliability problems and rarely remained in service for a complete 4-year run of the vacuum tower.

The packing coked during every run. Plant data showed high pressure drops in the packed bed starting approximately 6-8 months after the run started. The bed was always coked when examined at the end of a run. The bed was replaced in-kind in subsequent runs. In spite of the coking, the wash oil rate was not increased.

Even though the bed coked, the rings had superior performance (lower pressure drop and higher capacity) than the wash trays and mist eliminator pads that were removed.

The current work was intended to upgrade the vacuum tower from the 3rd generation random packing to a mix of structured packing and grid. The wash bed was upgraded during modifications that added a diesel product to the vacuum tower. The new wash bed used ~4 ft (1.22 m) of a mix of grid and structured packing. A new collector tray, slop wax draw nozzle, and slop wax pumps were installed. The slop wax pumps were converted to use variable frequency drives to control flow rate instead of using a control valve. This improved pump reliability by reducing average suction specific speed of the pump [16]. Wash oil rates were re-set based on an approximate estimate of a φ of 50 and a targeted overflash rate of 0.15 gpm/ft^2 (6.11 lpm/m^2).

After modifications, laboratory testing of the overflash showed rates slightly above 0.15 gpm/ft^2 (6.11 lpm/m^2). Over a period of weeks the rate was optimized. Assuming a φ of 30 matched the plant data.

Operation has been consistent and no evidence of coking has been seen from the operating data.

14. Conclusions

Wash zones remain areas where refiners can make, or lose, large profits. Reliable operation allows for high capacity, yield, and quality. With sufficient wash oil, packed beds are very reliable in wash zone service. Factors influencing the right wash oil rate remain unknown to many operators. This review covered a partial list of the factors including:

- Efficiency
- Pressure drop
- Operating rate
- Transfer line phase separation
- Proper liquid and vapor distribution

Properly installed and operated some vacuum towers have demonstrated trouble-free performance for 20+ years. Others, with improper understanding of the operation, suffer from one to two year run lengths, dramatically reducing refinery profits.

15. References

[1] A. W. Sloley, "Refinery Vacuum Towers: State of the Art," presented at the AIChE National Meeting, San Francisco, CA, 2013.

[2] ASTM, "D2892-05, Standard Test Method for Distillation of Crude Petroleum (15-Theoretical Plate Column)," West Conshohocken, PA, 2005.

[3] A. W. Sloley, "Atmospheric Distillation Process: Fundamental Concepts," presented at the AIChE Spring Meeting, New Orleans, LA, 2014.

[4] R. N. Watkins, *Petroleum Refinery Distillation*, 2nd ed. Houston, TX: Gulf Publishing Company, 1969.

[5] K. Kolmetz, W. Kiong, K. Ng, Y. T. Senthil, A. W. Sloley, and T. M. Zygula, "Optimize process operation with new vacuum distillation methods," *Hyd Proc*, vol. 84, no. 5, pp. 77–85, May 2005.

[6] S. W. Golden, V. Shah, and J. P. Kovatch, "Improved Flowsheet Topology for Vacuum Distillation Simulations," presented at the Canadian Chemical Engineering Conference, Calgary, ALB, Canada, 1994.

[7] D. C. Villalanti, J. C. Raia, and J. B. Maynard, "High-Temperature Simulated Distillation Applications in Petroleum Characterization," in *Encyclopedia of Analytical Chemistry*, Chichester, UK: John Wiley & Sons, 2000, pp. 6726–6741.

[8] K. H. Altgelt and M. M. Boduszynski, *Composition and Analysis of Heavy Petroleum Fractions.* New Orleans, LA: Marcel Dekker, Inc., 1994.

[9] A. H. Orr, "Use of Wire-Mesh Sections in Asphalt Entrainment Control in Vacuum Pipe Stills, Part 1... Case History - Cosden Petroleum Corp.," *OGJ*, pp. 84–85, Nov. 1953.

[10] D. H. Stormon, "Use of Wire-Mesh Sections in Asphalt Entrainment Control in Vacuum Pipe Stills, Part 1... Case History - General Petroleum Corp., Torrance, Calif.," *OGJ*, pp. 87, 121–122, Nov. 1953.

[11] H. S. Fourer, "Use of Wire-Mesh Sections in Asphalt Entrainment Control in Vacuum Pipe Stills, Part 1... Case History - Sincalir Refining Co., Marcus Hook, Pa.," *OGJ*, pp. 86–87, Nov. 1953.

[12] J. S. Corlew, "Use of Wire-Mesh Sections in Asphalt Entrainment Control in Vacuum Pipe Stills, Part 1... Case History - Sinclair Refining Co., Sinclair, Wyo.," *OGJ*, pp. 82–84, Nov. 1953.

[13] J. W. Russell, "Use of Wire-Mesh Sections in Asphalt Entrainment Control in Vacuum Pipe Stills, Part 2... Case History - Ohio Oil Co., Robinson, Ill.," *OGJ*, pp. 95–96, Nov. 1953.

[14] C. O. Anderson, "Use of Wire-Mesh Sections in Asphalt Entrainment Control in Vacuum Pipe Stills, Part 2... Case History -Rock Island Refining Co. Indianapolis," *OGJ*, p. 99, Nov. 1953.

[15] J. H. Fisher and L. R. Bunger, "Use of Wire-Mesh Sections in Asphalt Entrainment Control in Vacuum Pipe Stills, Part 2... Case History - Sinclair Refining Co., E. Chicago," *OGJ*, pp. 96–99, Nov. 1953.

[16] A. W. Sloley, "Consider VFDs for Centrifugal Pumps," *Chem Proc*, vol. 76, no. 4, pp. 44–45, Apr. 2014.

16. Tables and Figures

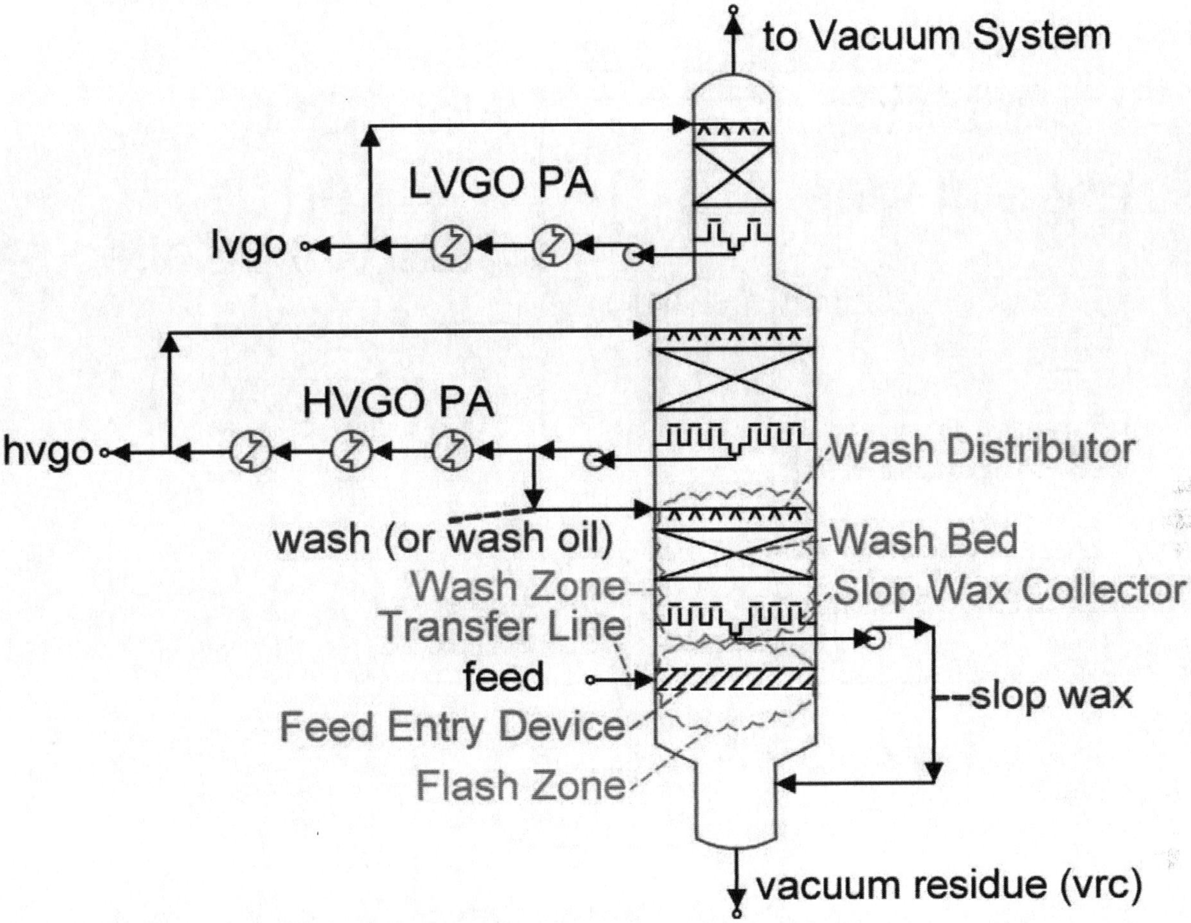

Figure 1 Conventional Vacuum Tower

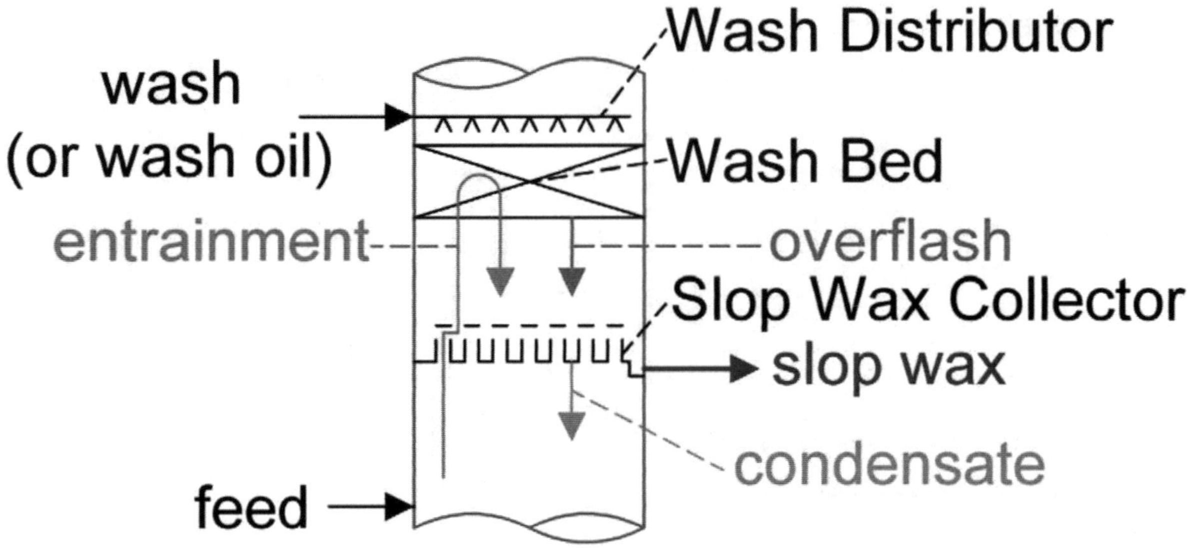

Figure 2 Wash Zone Configuration and Terminology

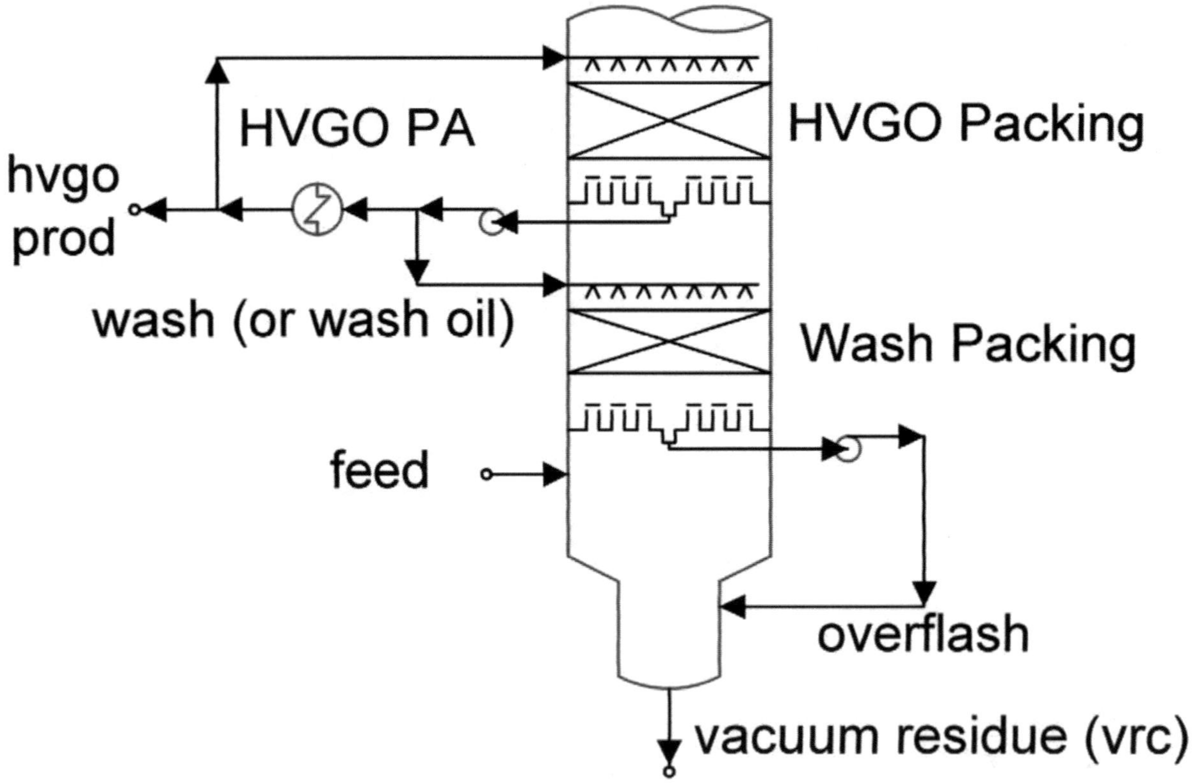

Figure 3 Conventional Wash Configuration

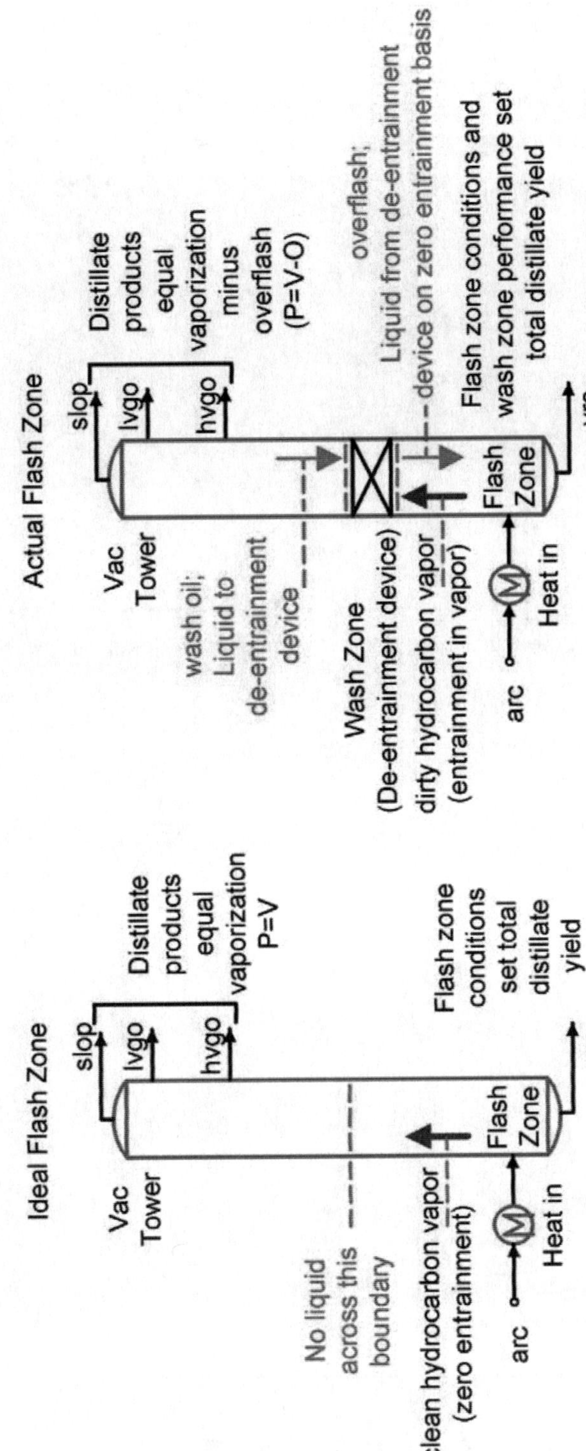

Figure 4 Ideal versus Actual Flash Zones

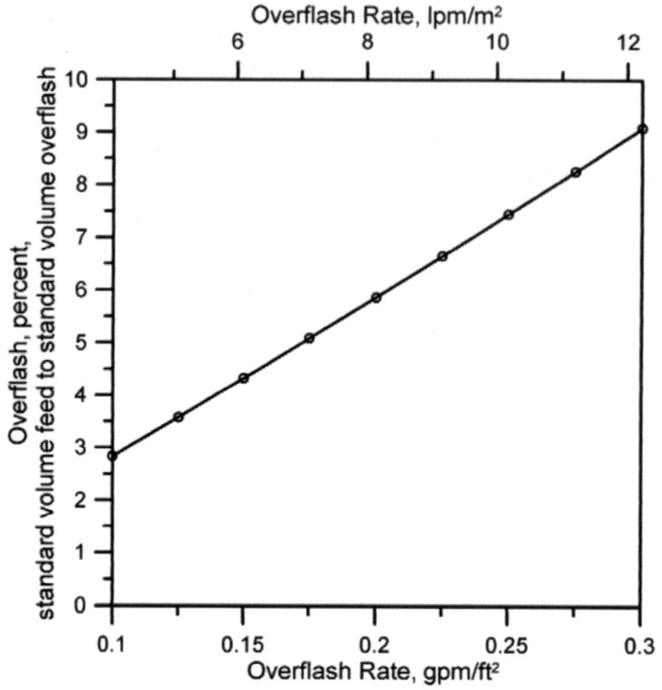

Figure 5 Overflash Percent versus Overflash Rate

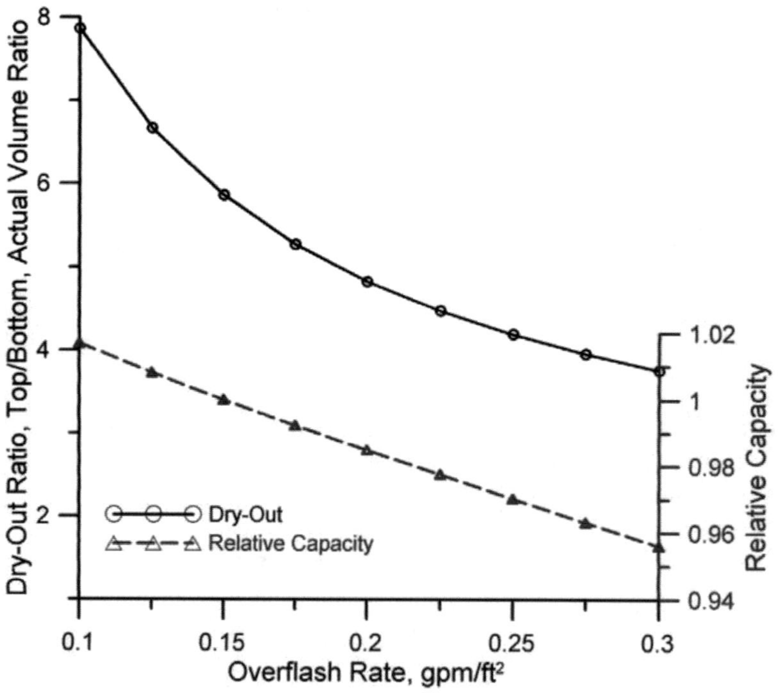

Figure 6 Dry-Out and Capacity versus Overflash Rate

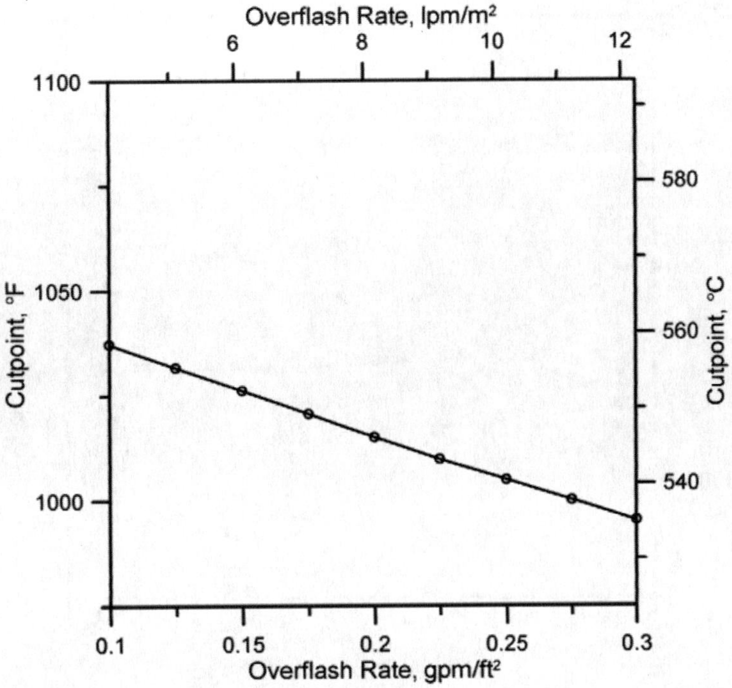

Figure 7 Cutpoint versus Overflash Rate

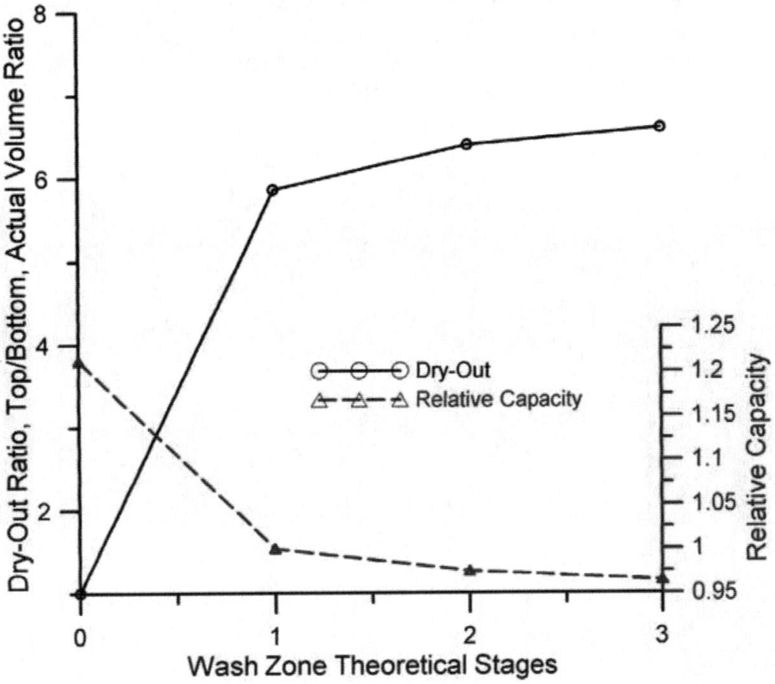

Figure 8 Dry-out and Capacity versus Wash Zone Stages

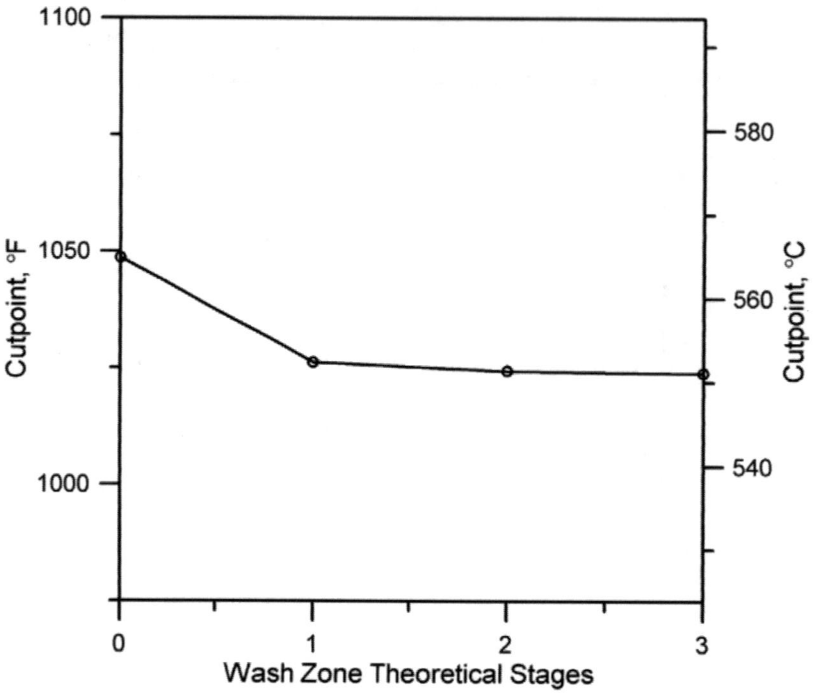

Figure 9 Cutpoint versus Wash Zone Stages

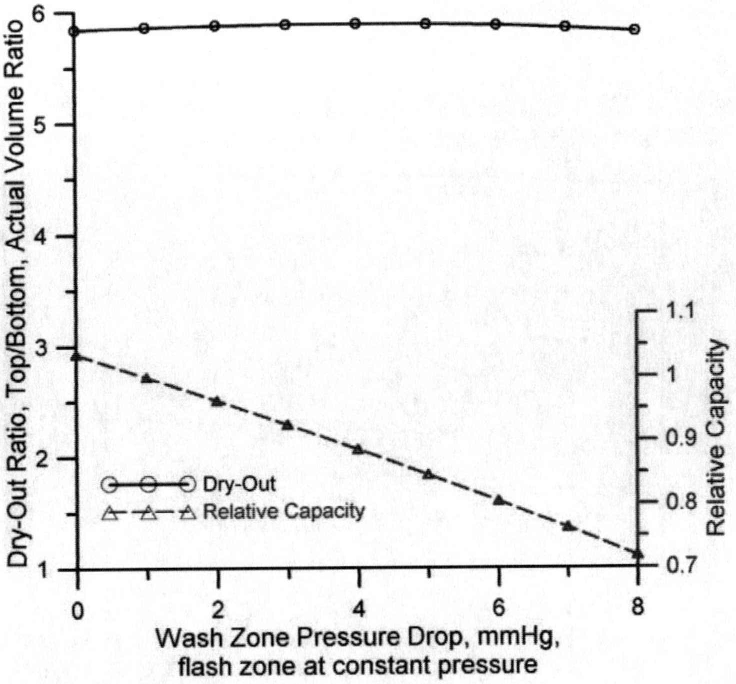

Figure 10 Dry-Out and Capacity versus Wash Zone Pressure Drop

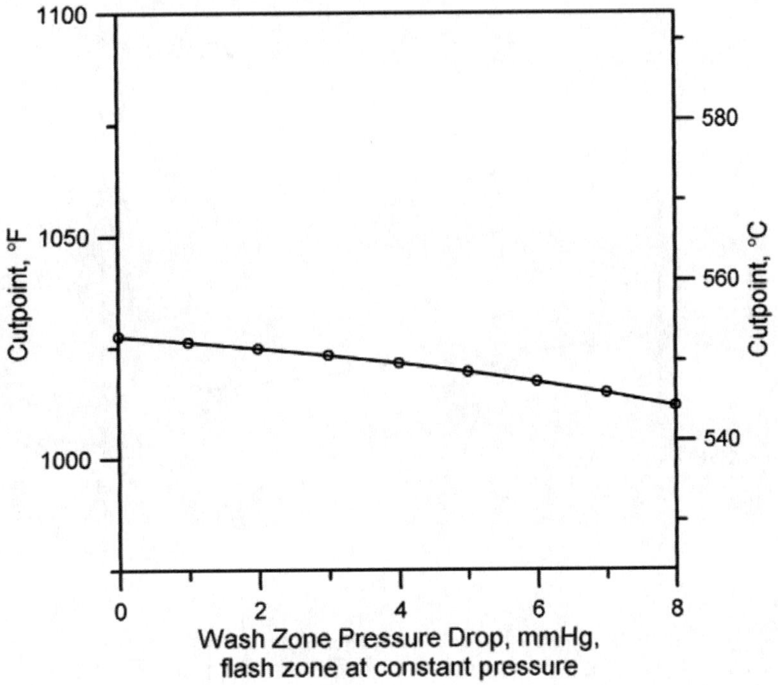

Figure 11 Cutpoint versus Wash Zone Pressure Drop

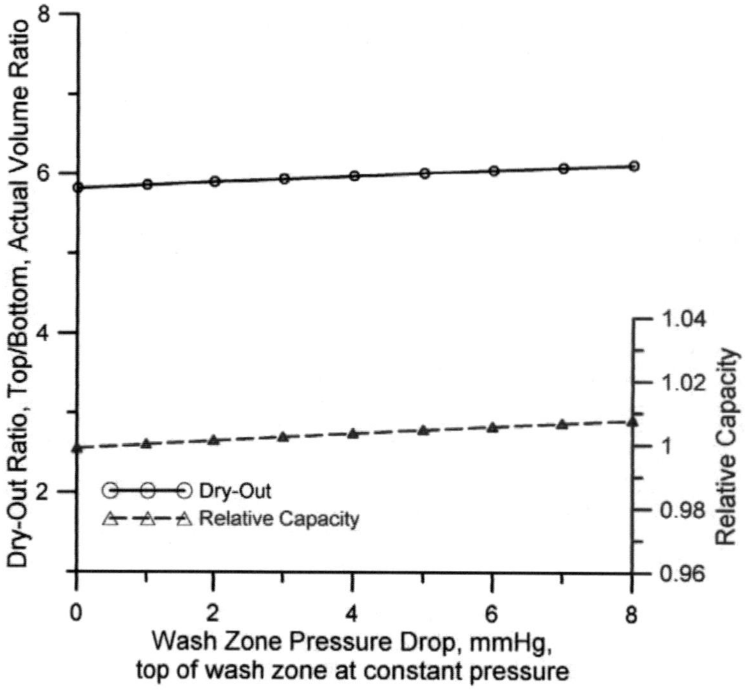

Figure 12 Dry-Out and Capacity Versus Wash Zone Pressure Drop

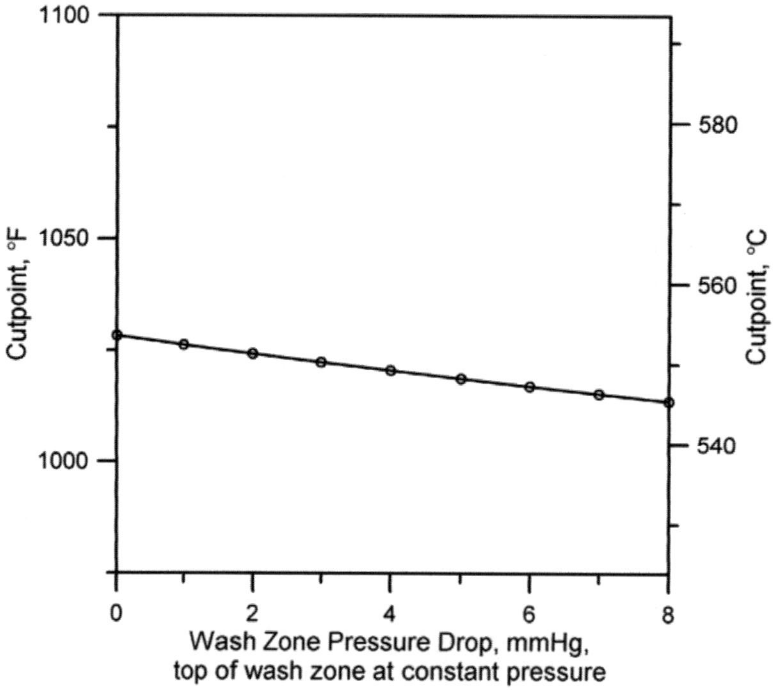

Figure 13 Cutpoint Versus Wash Zone Pressure Drop

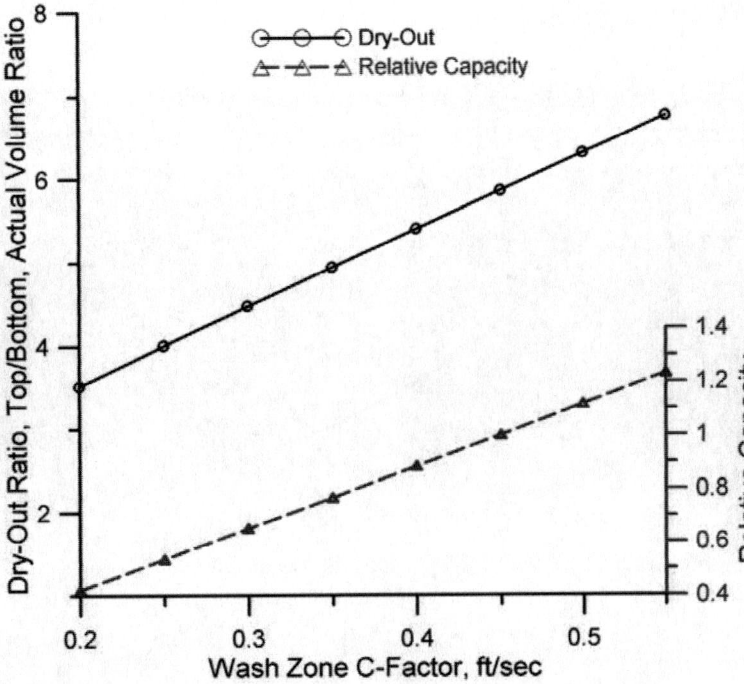

Figure 14 Dry-out Ratio and Capacity Versus C-factor (Constant Feed Rate)

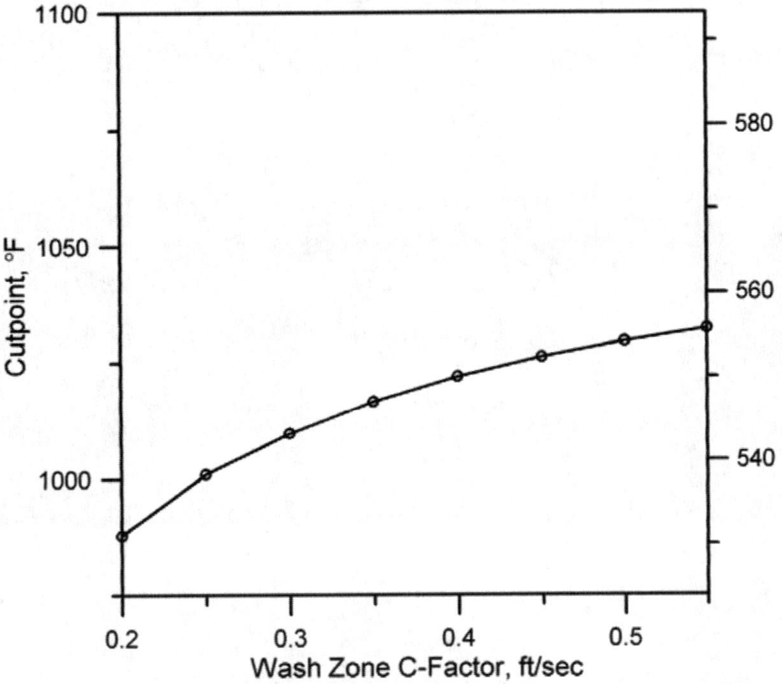

Figure 15 Cutpoint Versus C-factor (Constant Feed Rate)

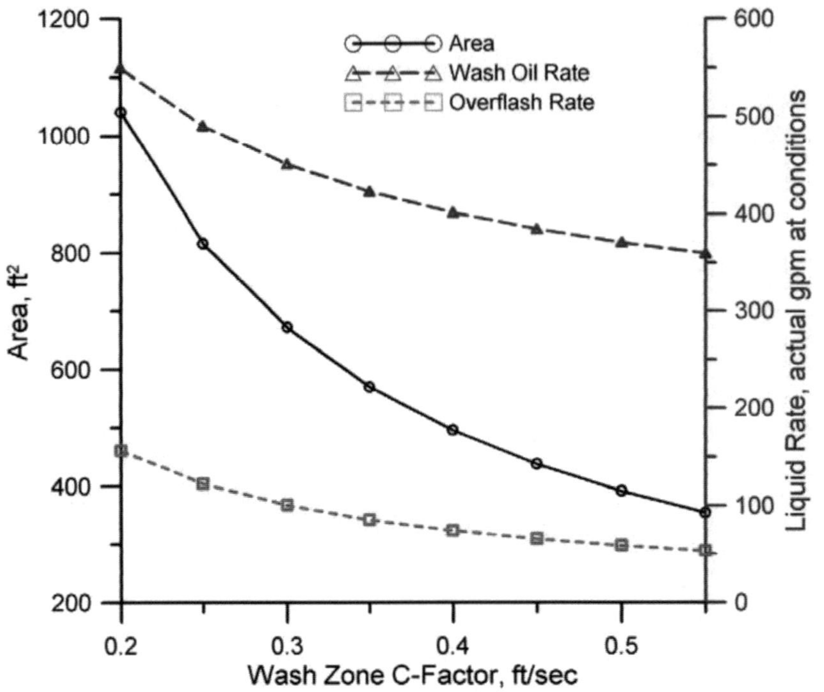

Figure 16 Area and Liquid Rates Versus C-factor

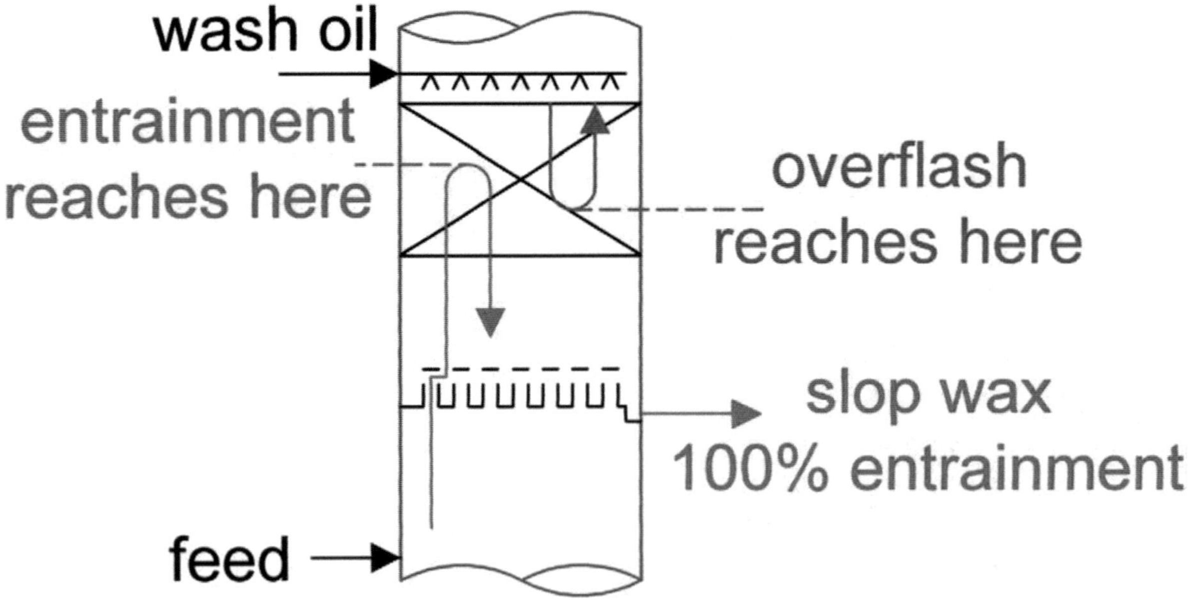

Figure 17 Zero Overflash Operation: Entrainment and Overflash Overlap

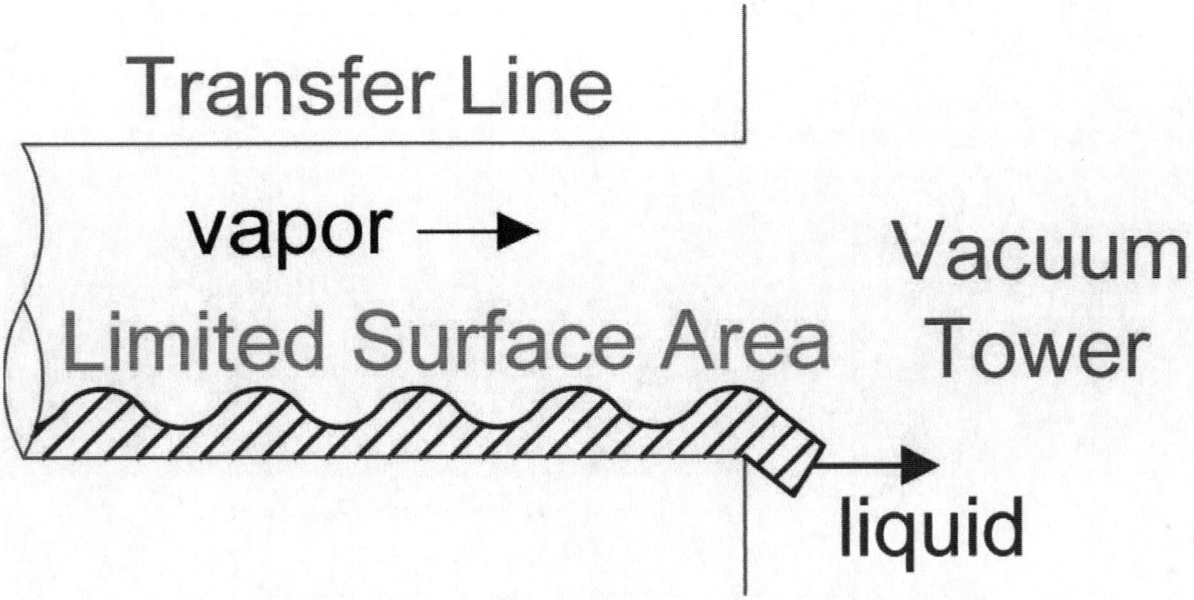

Figure 18 Transfer Line Phase Separation

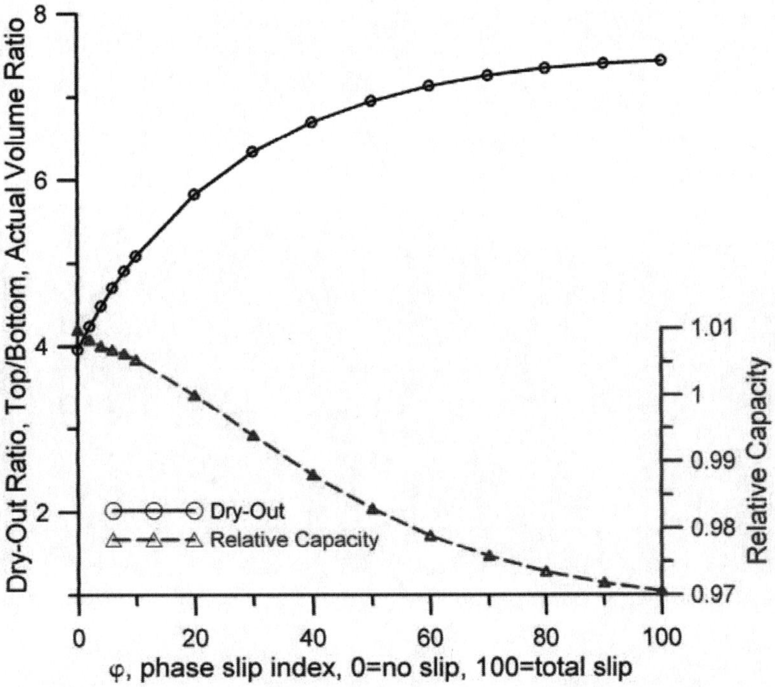

Figure 19 Dry-out Ratio and Capacity Versus Transfer Line Phase Slip

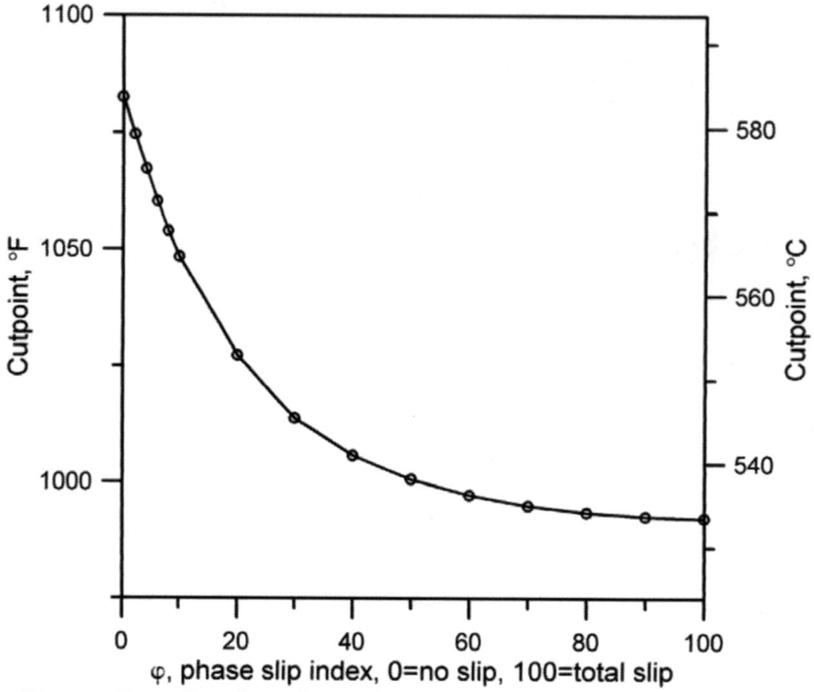

Figure 20 Cutpoint Versus Transfer Line Phase Slip

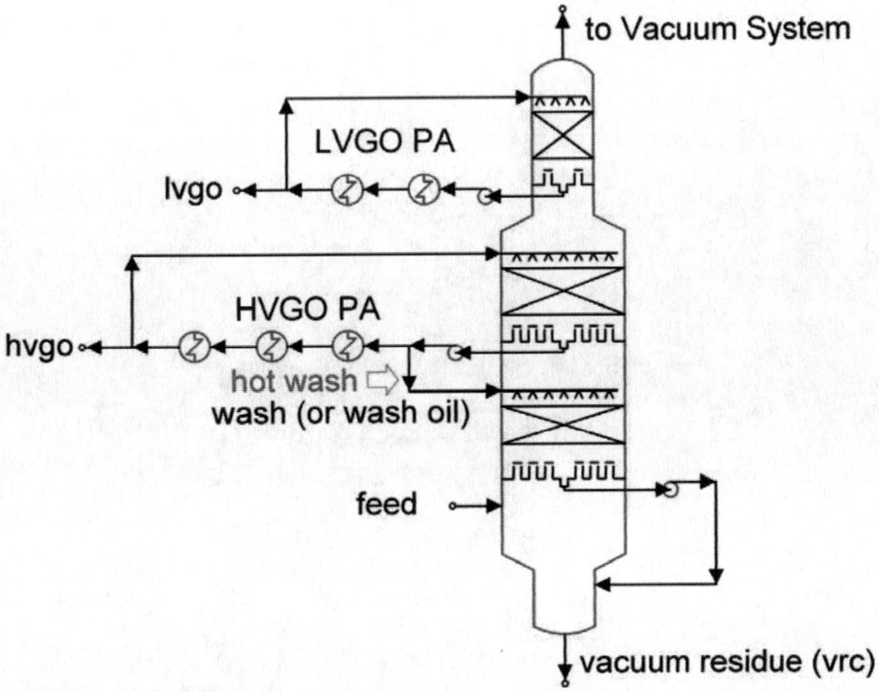

Figure 21 Hot Wash Configuration

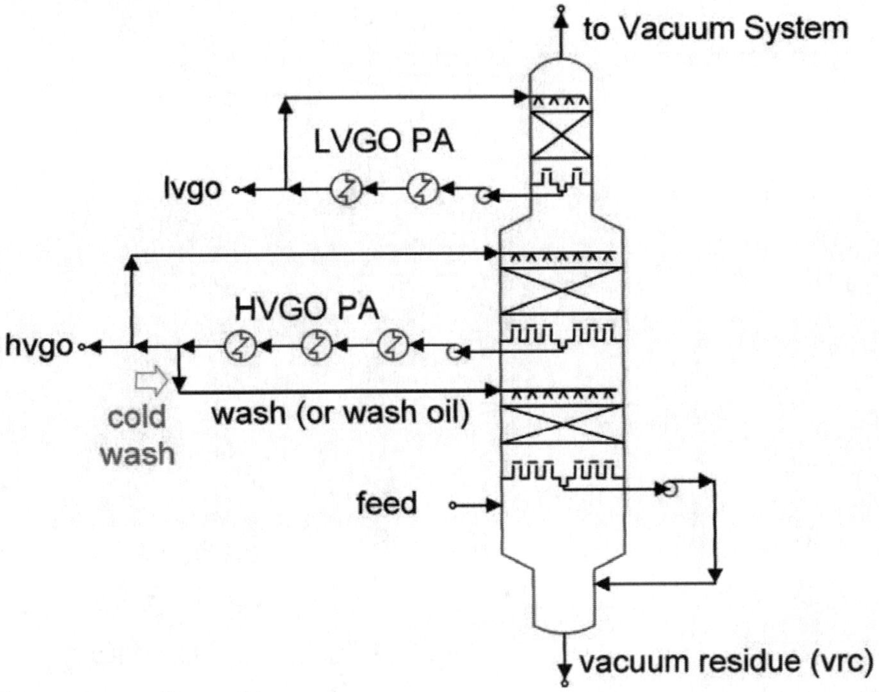

Figure 22 Cold Wash Configuration

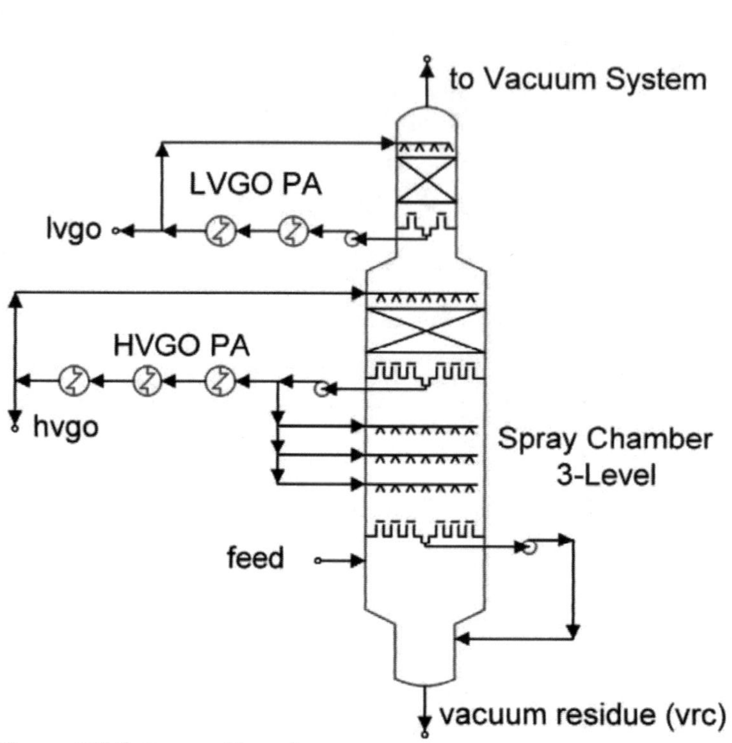

Figure 23 Vacuum Tower With Spray Chamber

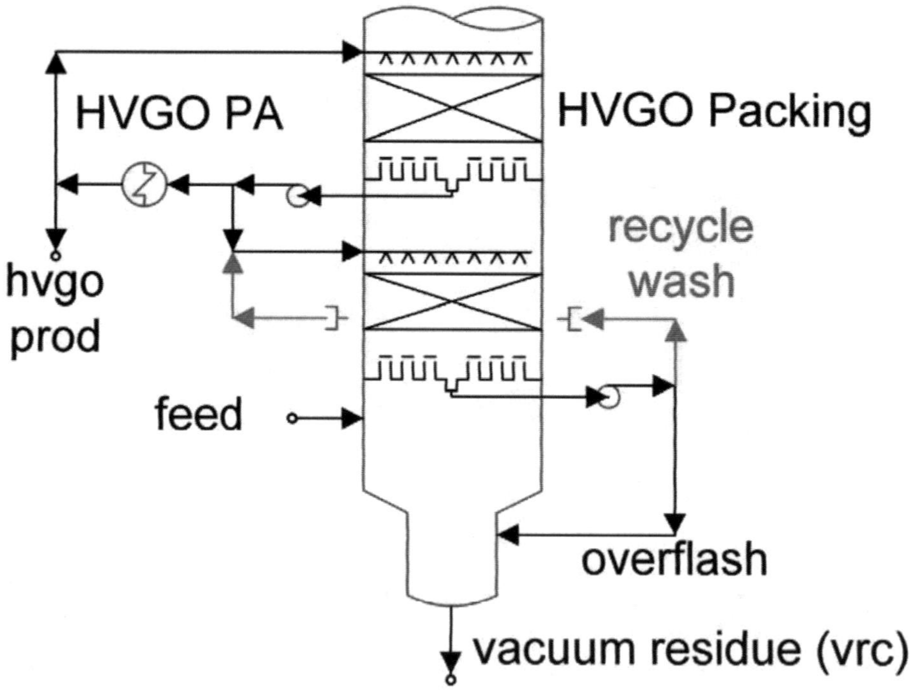

Figure 24 Recycle Wash System: One Stage

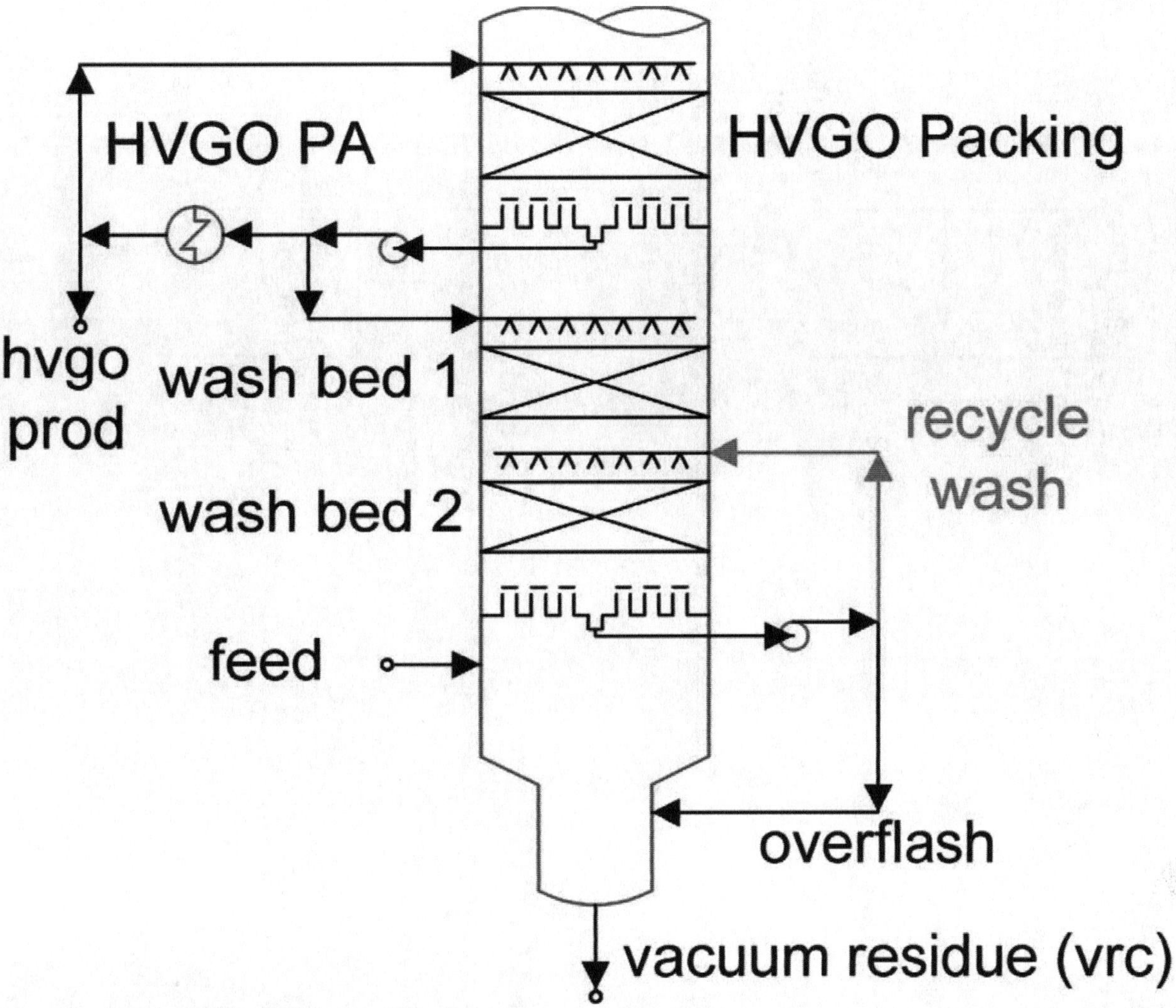

Figure 25 Recycle Wash System: Two Stage

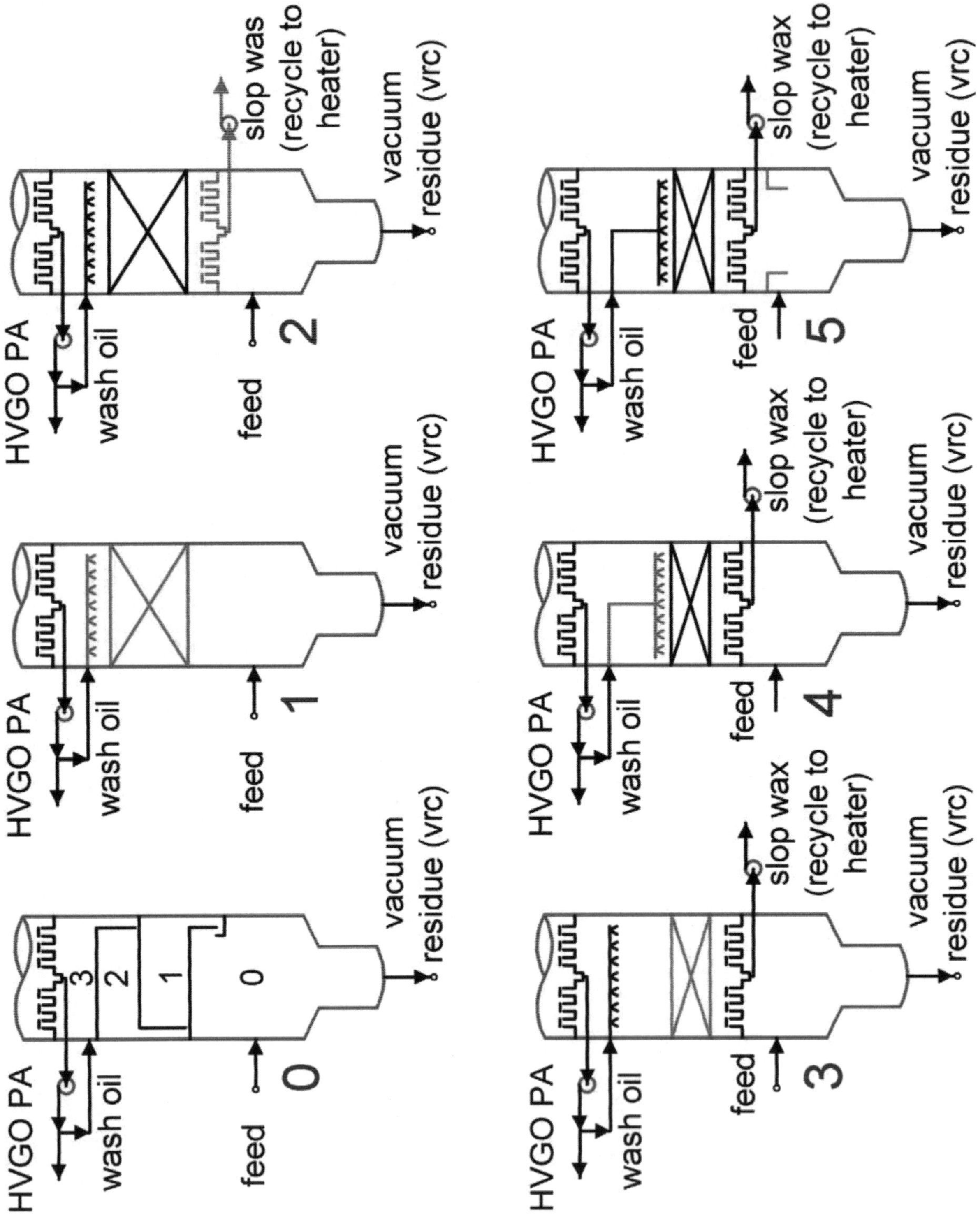

Figure 26 Case History 1: Upgrading a Trayed Vacuum Tower

OPTIMIZING PUMPAROUND BALANCE BETWEEN ENERGY AND FRACTIONATION EFFICIENCY

Soun Ho Lee
GTC Technology US, LLC
Euless, Texas, United States

Presented at AIChE 2017 Spring Meeting
San Antonio, TX
March 25-30, 2017

"Kister Distillation Symposium"
(#34) - Optimum Design and Operation in Refinery Distillation

March 27, 2017

INTRODUCTION

In a petroleum refinery, multi-product fractionators, including crude atmospheric columns, crude vacuum columns, FCC main fractionators and coker main fractionators are core pieces of equipment. Most multi-product fractionators are equipped with multiple pumparound circuits for an inter-condenser system. Compared to the pumpback method, the pumparound method achieves more uniform vapor/liquid traffic through the column. Multiple pumparound circuit locations and balances influence not only the unit heat balance but also column fractionation efficiency. Inappropriate pumparound system design can cause capacity limitation, low energy or downgraded fractionation efficiency.

This paper will discuss critical issues for optimum pumparound design between energy and fractionation efficiency requirements. The case study will demonstrate how pumparound circuit design is modified for optimum energy and fractionation efficiencies.

PRINCIPLE OF PUMPAROUND BALANCE

To minimize liquid traffic variations in the refinery multi-product fractionator, inter-condenser design philosophy is implemented. Inter-condensers can be configured as either pumpback or pumparound circuits. The pumparound reflux method achieves more uniform liquid balancing through the column compared to the pumpback method. This uniformity of liquid enables the column to be sized at a smaller diameter. The higher pumparound draw temperatures increase the opportunity of heat recovery for lower energy consumption. In addition, the overhead condenser size is reduced. The main trade-off is that the pumparound circuit design requires more trays and/or packing for heat transfer performance and can result in higher column elevation. In summary, the advantages of the pumparound reflux arrangements far outweigh any disadvantages, such that it has replaced the pumpback reflux method in most modern refinery multi-product fractionator designs.

Figure 1 illustrates a fractionation section positioned in-between two pumparound sections. If column heat balance is shifted toward the lower positioned pumparound, more valuable heat can be recovered through the lower pumparound circuit. However, it reduces the internal reflux rate for the fractionation section; fractionation efficiency between adjacent products can be downgraded.[1] On the other hand, shifting heat balance toward the upper pumparound can improve fractionation through a higher internal reflux rate. However, column energy efficiency would be downgraded. Moreover, higher overhead condenser duty can result.

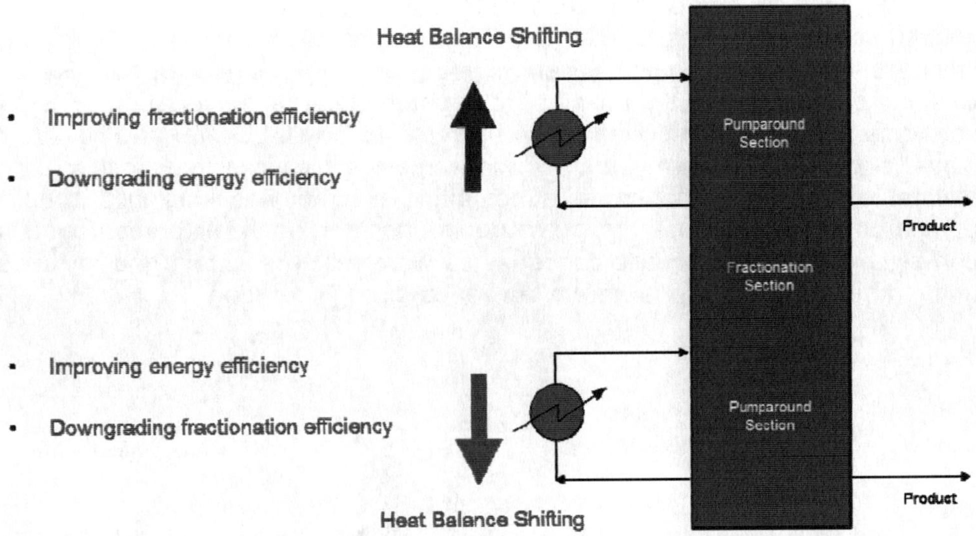

Figure 1: Pumparound Balance: Energy Efficiency vs. Fractionation Efficiency

Column internal vapor and liquid traffic heavily relies on pumparound circuit locations. The number and location of these circuits are determined by feed composition, product yield patterns, fractionation requirements, overhead condenser size and other factors. The refinery multi-product fractionator is designed to provide the best performance for specific ranges of feed compositions and product yields. Therefore, a large change from design conditions may induce performance downgrading. Fractionation performance between adjacent products requires specific design internal reflux at a given number of fractionating trays or packed bed depth. Overhead condenser duty is also determined by design heat balances. In most cases, actual feed conditions processed in the column often deviate from original design ranges. To maintain desired unit performances at changed feed conditions, most refiners adjust and rearrange pumparound balances. These operation parameter changes shift column traffic through the crude atmospheric column. The pumparound rate change impacts neighboring fractionation section internal reflux rates so fractionation performance is affected. Unbalanced column traffic often results in unit capacity limitations. Adjusting and re-arranging pumparound balances at actual process conditions are necessary to achieve desired performance.

The actual revamp case study described below demonstrates how pumparound balance is optimized at actual process conditions.

CASE STUDY: BACKGROUND

Figure 2 illustrates the crude distillation unit under discussion. This crude atmospheric column separates crudes to intermediate products: (1) Unstabilized Naphtha (2) Kerosene (3) Light Diesel (LD) (4) Medium Diesel (MD) (5) Heavy Diesel (HD) and (6) Reduced Crude (R/C). Unstabilized naphtha is then fed to the naphtha stabilizer to separate LPG and naphtha. The kerosene stream is transported to the hydrotreating unit. LD, MD and HD are combined to the diesel pool after hydrotreating. Reduced crude is transported to either conversion units or blended to fuel oil. The crude atmospheric column was originally constructed with four pumparound circuits: one top pumparound and three diesel pumparound circuits. Various conventional crude slate and condensates were mixed and charged to this crude distillation unit. The feed compositions were often varied during operation.

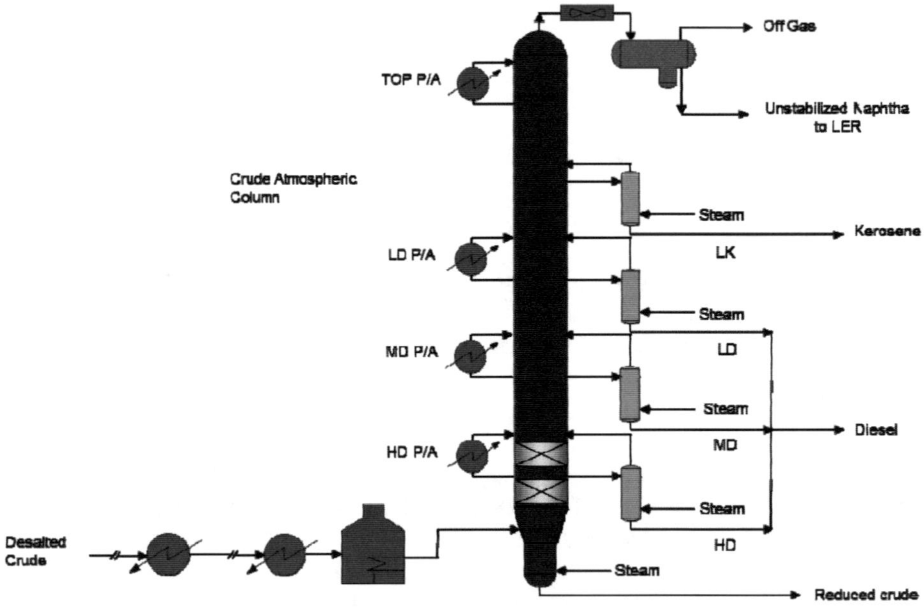

Figure 2: Original Crude Distillation Unit Configuration

The unit limitation facing the refiner was that the crude atmospheric column had difficulty processing crude slates containing a high percentage of kerosene boiling range materials. In the winter season, kerosene product is usually more valuable than diesel and naphtha. Relieving any limitations on kerosene production was necessary to improve unit economics.

CASE STUDY: PRE-REVAMP TEST RUN

Prior to any process evaluation, a dedicated crude unit test run was conducted to gather pertinent operating data. The importance of a test run cannot be overemphasized. Daily operating data do not always provide all the needed information for reliable evaluations. A crude slate containing high kerosene boiling range material was selected for the test run, as this represented one of the typical operating crude slates in the unit. For equipment limitation checking, the test run charge rate was determined as the maximum crude charge rate in which the crude unit operated without loss of fractionation efficiency. In gathering pertinent data, all associated instruments were calibrated prior to the test run. The measured flow rates were verified via flow meter orifice calculations and storage tankage levels to establish whether mass balance closure error was within suitable for reliable modeling.[2] Traditional laboratory methods do not provide appropriate characterization for heavy oil boiling range material. The ASTM D7169 method was used for crude feed, heavy diesel, and reduced crude stream laboratory test to obtain better heavy boiling range material characterization.

Since distillation equipment pressure drop is related to column traffic, a survey of measured pressure drop across the column was required. This is a useful tool in column troubleshooting and evaluation. In particular, the survey allows the troubleshooter to pinpoint equipment locations that require further evaluation. The measured pressure drop profile through the column showed that the top pumparound pressure drop was over twice as high than the LD and MD pumparound section pressure drop on the same basis. In addition, the measured HD packed bed pressure drop was much higher than expected. This pressure drop survey indicated that column top and bottom sections might be more loaded or fouled compared to the middle sections.

CASE STUDY: PROCESS EVALUATION

This crude unit was originally designed and constructed to process high diesel and low kerosene content crude slates. As a result, three pumparound circuits were located in the diesel section while no pumparound circuit was included in the kerosene section. This design philosophy in this crude unit showed the best performance with diesel rich and kerosene poor crude slates.

For the first step of root cause identification, the feed composition for the test run was compared with the original design crude composition. The composition spectrums are displayed in Figure 3. These compositions were obtained through blending programs. The blending program predicts recoverable material yield with a clear cut basis between products and does not take into consideration actual distillation column performance. Nevertheless, this crude structure comparison helps comprehend the deviation between the original design and the pre-revamp test run. These spectrums show that the test run crude slate contains much more kerosene boiling range materials compared to the original design crudes. Petroleum gas and naphtha contents were also increased in the test run crude slate while diesel boiling range material percentages are almost same between the two crude slates.

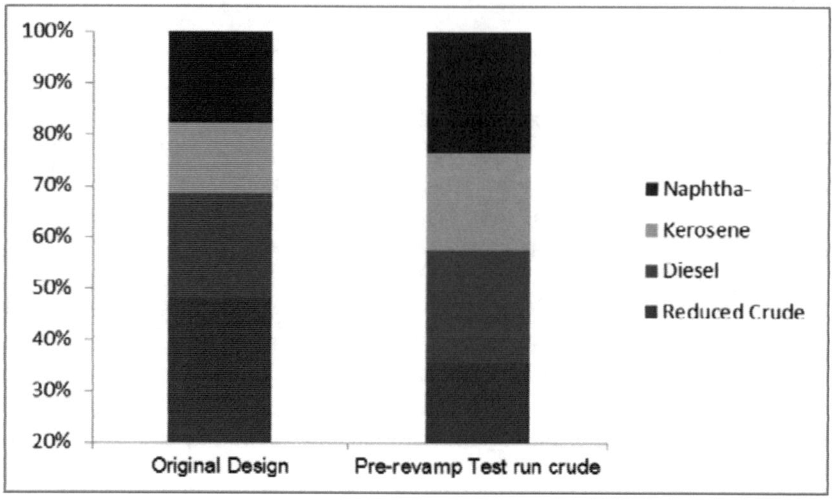

Figure 3: Feed Composition Comparison

To scrutinize unit limitation and identify the root cause, the column traffic profiles were reviewed at the pre-revamp test run conditions. A simulation model was constructed and validated with the test run conditions in order to quantify column traffic conditions.

Reliable simulation modeling is another cornerstone of successful process evaluation. Throughout the modern hydrocarbon industry, simulation is widely used to design and/or analyze distillation column performance and has become a basic tool for process engineering. A lot of design firms rely heavily on simulation modeling to establish heat and material balances. Commercially available steady-state process simulators are regularly upgraded with regards to the thermodynamic package, component and property data base, properties calculation, numerical method algorithm, software interface and other capabilities. Nevertheless, selecting reputable simulation software does not guarantee the reliability of simulation modeling. Accurate simulation modeling still requires extensive knowledge and understanding of process and equipment which need to be modeled. Inherent gap between actual conditions and theoretical simulation modeling should not be overlooked. It has been observed that unreliable simulation modeling which neglects these issues can lead to design flaws and performance deteriorations.

In an earlier paper[3], the discussion included that conventional flow-sheeting topology does not adequately address refinery fractionators such as crude vacuum distillation column. Actual flash zones operate at non-equilibrium state, and fluid mixing through transfer lines is highly non-ideal. Conventional simulation topology commonly used in the industry does not predict crude column performance properly. This especially applies to the flash zone, transfer line, wash and stripping sections. In this crude atmospheric column simulation modeling, it was observed that conventional topology did not simulate energy loss through transfer line properly. Therefore, process solutions based on conventional simulation potentially under- or over-predict operating parameters. Such results may induce performance short-fall or failure into a retrofit solution.

To obtain reliable simulation modeling results, this crude atmospheric column was modeled using modified simulation topology. This updated modeling procedure dissected the column into multiple blocks in order to model the crude atmospheric column, the furnaces and the transfer line properly. The interval of pseudo components was adjusted to match the obtained laboratory distillation temperature span. Tray efficiency for each fractionation section was determined through various sensitivity analyses.

Simulated pumparound capacity profiles at the pre-revamp test run conditions are displayed in Figure 4. This plot shows that capacities at four pumparound circuits were severely imbalanced, like a "dumbbell". The top and HD pumparound sections were run at high capacities while the LD and MD pumparound sections were run at lower capacities. This capacity profile is also in accordance with pre-revamp test run pressure drop measurements.

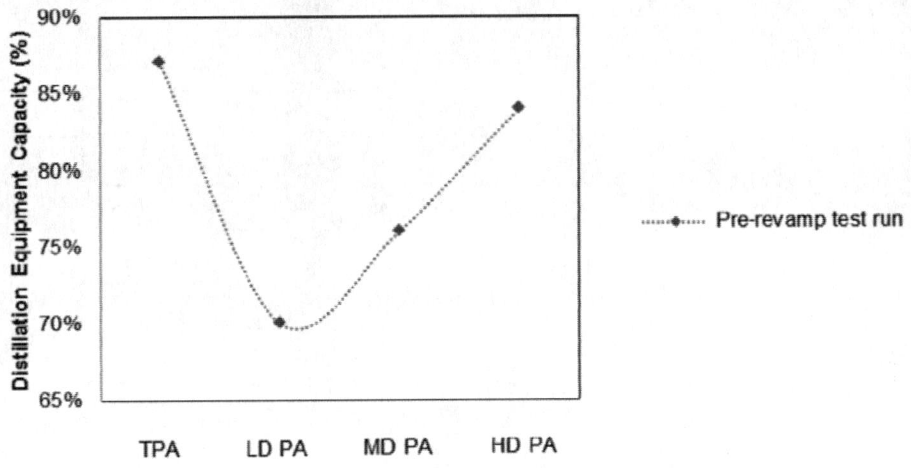

Figure 4: Pre-revamp Pumparound Capacity Profile

One of the options to improve pumparound balancing is to shift Top and HD pumparound loads to LD and MD pumparound circuits. However, simple pumparound rate redistribution will impact fractionation in the existing column arrangement. Vapor and liquid traffic is decreased above the section where pumparound rates are increased. In this particular column, increasing LD and/or MD pumparound rates would result in the reduction of the internal reflux of the naphtha-kerosene fractionation section and downgrade the separation between these products. One possible result is a reduced kerosene flash point and difficulty in meeting the specification. Simulation modeling for the test run case also verified that fractionation sections in the crude atmospheric column were overly sensitive to internal reflux rates. To compensate for this sensitivity, the crude column had been operated with high top pumparound rates and high internal reflux for naphtha and kerosene fractionation. It was necessary to optimize crude column design including pumparound rearrangement at current crude slate conditions, for maximum crude unit capacity and performance.

CASE STUDY: REVAMP DESIGN

The process evaluation revealed that there was a mismatch between the pumparound locations and actual process conditions. To increase kerosene yield with a minimum mechanical modification scenario, the existing LD pumparound service was converted to a new heavy kerosene pumparound service. In this conversion, the number of kerosene draws was increased from one (1) to two (2) while diesel draws are reduced from three (3) to two (2). A modified schematic of the crude distillation unit is described in Figure 5. This revamp strategy did not require any column nozzle and external piping configuration changes.

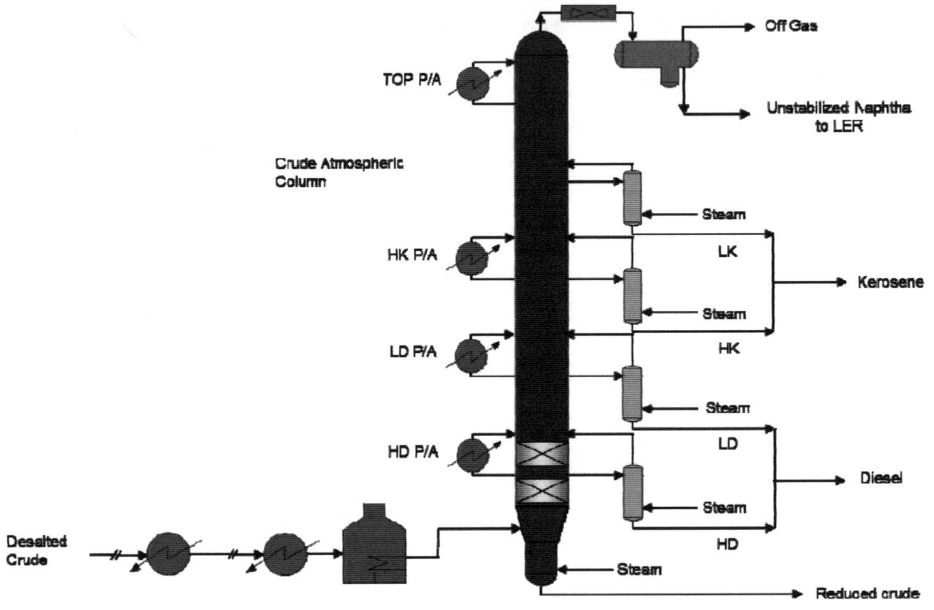

Figure 5: Revamp Crude Distillation Unit Configuration

These product and pumparound rearrangements resulted in a lower logarithmic mean temperature difference (LMTD) of the new heavy kerosene and new LD pumparound circuits compared to previous services. A careful evaluation of the preheat train was required to make sure that the new configuration did not result in lower pre-heat temperature to the furnace and / or possibly reduced recovery or higher energy consumption. Some modifications to the pre-heat train and exchangers were completed to optimize the heat recovery at the new conditions. An overall heat balance check including furnaces showed the desired feed temperature would be achieved at current retrofit design condition.

To maximize tray efficiency and capacity, all trays in the crude atmospheric column and the side strippers were replaced by GT-OPTIM™ high performance trays.

CASE STUDY: POST-REVAMP PERFORMANCE

This crude unit was successfully revamped and post-revamp test runs were conducted with two sets of crude slates. The post-revamp test run product yield patterns are compared to the pre-revamp test run value in Figure 6. Post re-vamp test run 1 and 2 indicate operating data at kerosene-rich and diesel-rich crude slates respectively.

Overall unit capacities have been substantially expanded. However, these product yield patterns indicate that post-revamp crude slates were much heavier than pre-revamp crude slates. Therefore, overall capacity gain does not represent actual column capacity gain. Only distillate yields need to be compared to gauge actual column capacity gain.

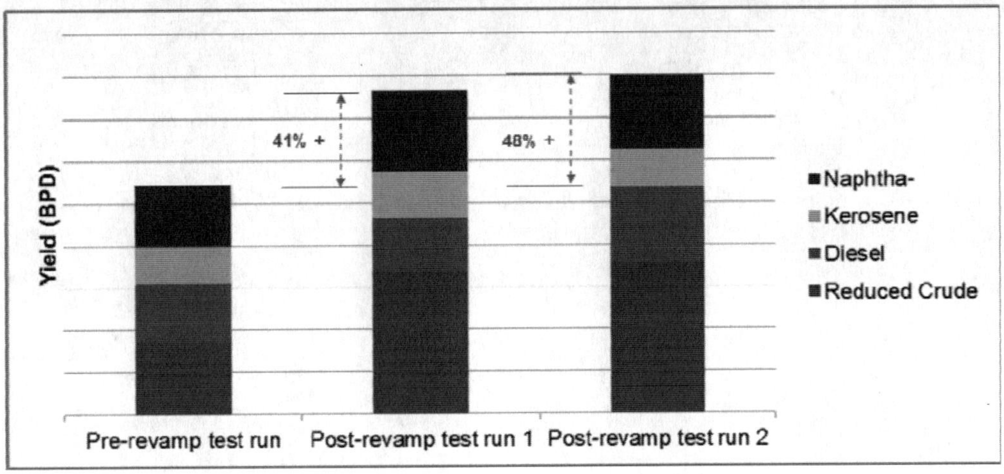

Figure 6: Pre- & Post-revamp Total Product Yield Pattern

Figure 7 highlights distillate yield patterns among the cases. Total distillate yields are increased and kerosene-rich crude slates are processed without any limitation. Although the number of diesel pumparound circuits was reduced, the crude unit produces much higher diesel yields.

Post-revamp pumparound capacity profiles are added and compared with pre-revamp test run conditions. In Figure 8, the blue line indicates simulated pumparound capacity at the post-revamp test run condition 1. Capacities at post-revamp test run 2 are plotted in yellow. Recognizable improvements in pumparound balancing are demonstrated in this graph.

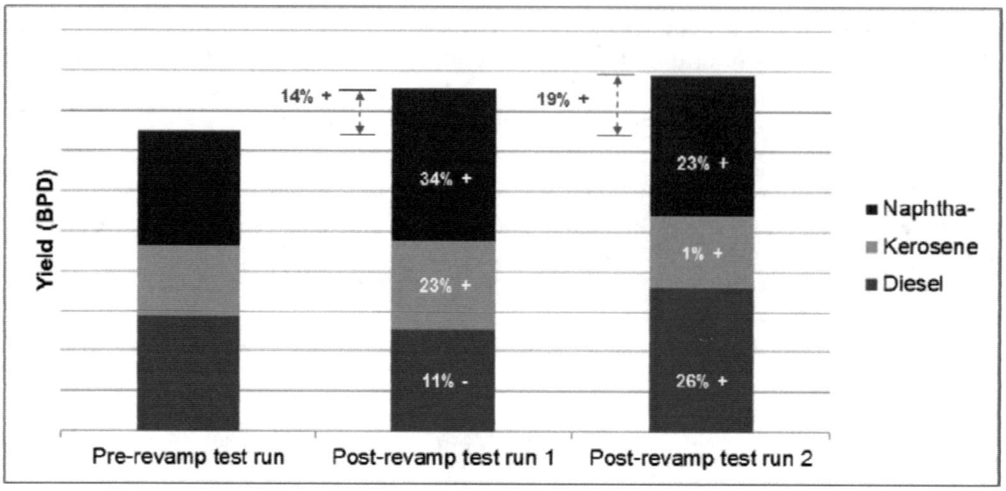

Figure 7: Pre- & Post-revamp Distillate Yield Pattern

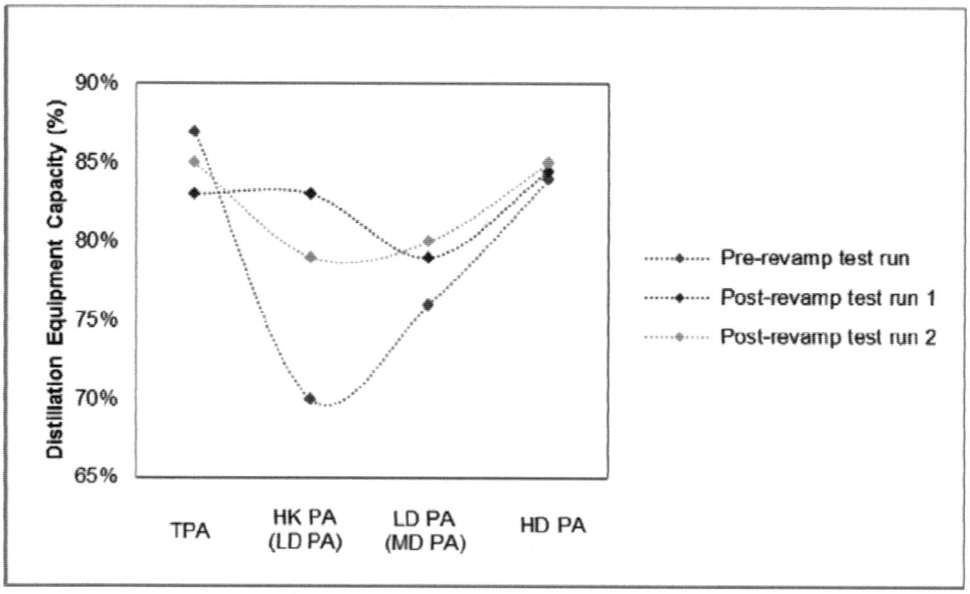

Figure 8: Pre- & Post-revamp Pumparound Capacity Profile

Column fractionation performances are also reviewed. Figures 9-11 show fractionation performance with a simulated fractionation section internal reflux L/V (Liquid / Vapor) ratio. Figure 9 indicates a naphtha-kerosene distillation gap and kerosene flash point values are enhanced in spite of the reduced fractionation section internal reflux L/V ratio. Improved kerosene-light diesel fractionation performances are also shown in Figure 10. The post-revamp test run crude slates are actually heavier feed stocks containing more atmospheric residue boiling range material. Nevertheless, Figure 11 shows that the reduced crude 5% distillation temperature values are increased, indicating improved diesel recovery. Simulated internal reflux L/V ratio at the wash section top was increased but the unit steam consumptions for the bottom stripping section were noticeably improved.

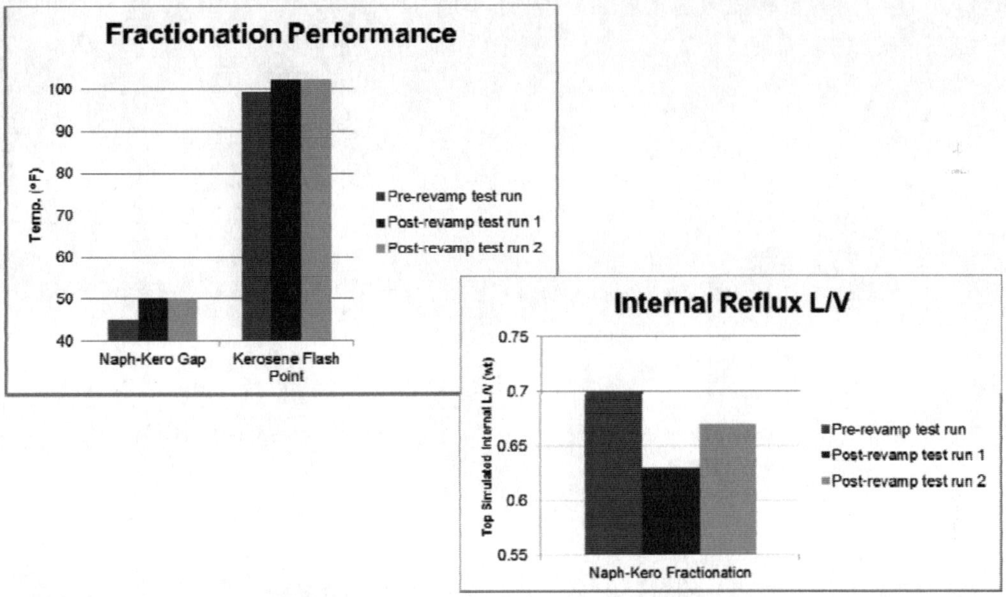

Figure 9: Pre- & Post-revamp Fractionation Performance: Naphtha-Kerosene

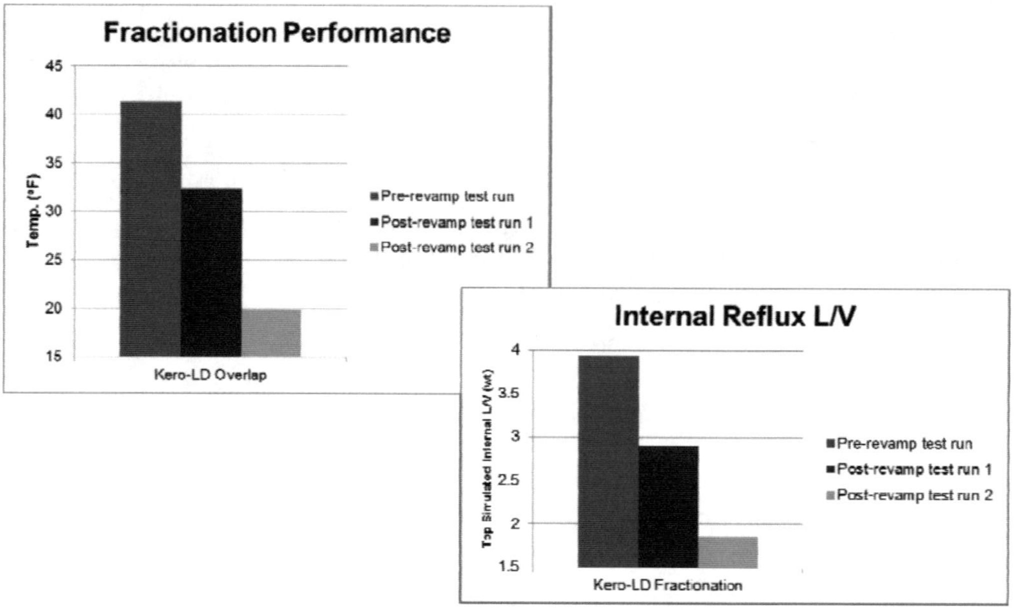

Figure 10: Pre- & Post-revamp Fractionation Performance: Kerosene-LD

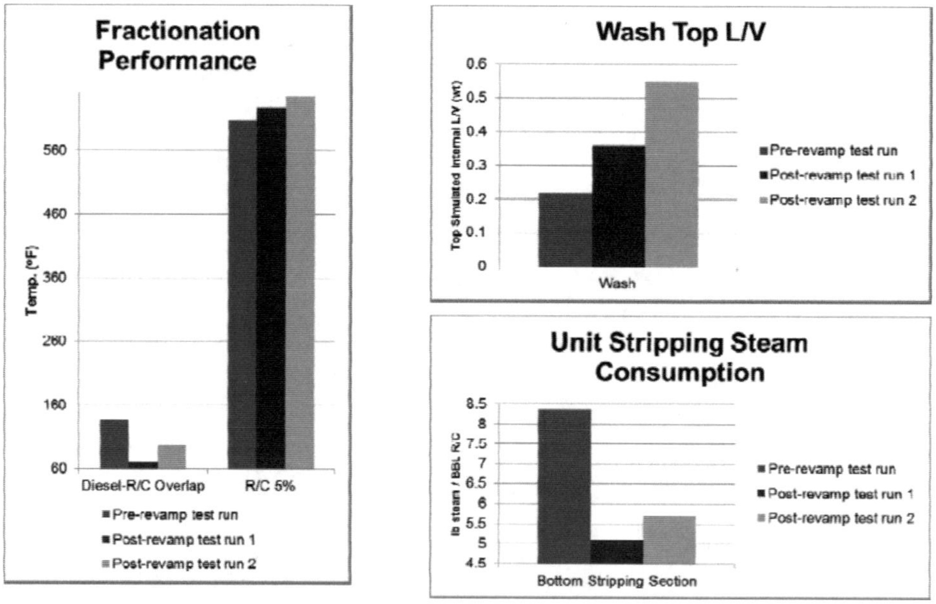

Figure 11: Pre- & Post-revamp Fractionation Performance: Diesel-R/C

Table 1 tabulates simulated efficiencies of the fractionation sections. Fractionation efficiencies are improved. The aforementioned GT-OPTIM™ high performance tray implementation contributed to the fractionation efficiency improvement.

Table 1: Fractionation Efficiency Comparison

Fractionation section	*Pre-revamp Test Run*	*Post-revamp Test Run*
Naphtha-Kerosene Fractionation	*~62%* note1,2	*~70%* note1,2
Kerosene-LD Fractionation	*~40%* note1,2	*~60%* note1,2
Bottom Stripping	*~20%* note1,2	*~40%* note1,2

Note

1. *Simulated value*
2. *Overall efficiency (number of theoretical stages / number of fractionating trays)*

CONCLUSION

Significant process condition deviations from the original design conditions can cause non-optimum pumparound balance and downgrade either energy efficiency or fractionation performance. Unfortunately energy efficiency and fractionation performance have an inverse relationship in pumparound balance. Therefore, evaluating and optimizing pumparound balance at actual process conditions are essential to achieve desired refinery multi-product fractionator performance.

REFERENCES

1. Libermann N P, "Troubleshooting Process Operations," Pennwell Publishing Company, 3rd Edition.

2. Kister, H.Z, "Distillation operation," McGraw-Hill Company, 1990.

3. Golden S, *et al*, "Improved Flow Sheet Topology for Petroleum Refinery Crude Vacuum Distillation Simulation," 44th Annual CSChE Conference, 1994.

ACKNOWLEDGEMENTS

GT-OPTIM™ is a trademark of GTC Technology US, LLC

Marathon Petroleum Company LP

Shell ConSep™ Trays Improve Diesel Recovery and Naphtha Quality in a DHT Stripper

Gregory A. Cantley – Marathon Petroleum Company LP, Russell, KY
Waldo De Villiers – Shell Global Solutions (US) Inc, Houston, TX
Eric Vos – Shell Global Solutions International BV, Amsterdam, Netherlands
Grzegorz Kamyszek – Sulzer Chemtech USA, Tulsa, OK

2017 AIChE Spring National Meeting, San Antonio, TX, Paper 2c (Unpublished)

DHT History

- Originally built in mid-70s
- Revamped to Low Sulfur Diesel operation in early 90s
- Revamped to accept Coker Naphtha in early 00s
- Rates on LP Stripper were fairly constant until the Coker Naphtha revamp when overhead rate increased from 1% to about 7.5% of charge rate
- Charge Rate has crept up about 30% from the original design to the Pre-Revamp rates

2

DHT Stripper Pre-Revamp

- Prior to revamp the dP in the top 6 trays showed that the rectifying section was flooded at 0.20 psi per tray.
- Lab analysis showed the D3710 98% endpoint was around 440°F.
- The overhead naphtha was routed back to the crude unit to be reprocessed.

3

DHT Stripper Pre-Revamp

- After a pre-revamp catalyst change, the unit charge was increased 5%.
- The dP in the top trays increased to 0.22 psi per tray.
- The D3710 98% endpoint was approaching 460°F.
- A plan was needed for the next catalyst change!

4

DHT Stripper

- A 7'-0" diameter rectification section with 6 floating round valves trays.
- The open area on the tray was 9.6%
- The downcomers are sloped.
- 24" Tray Spacing

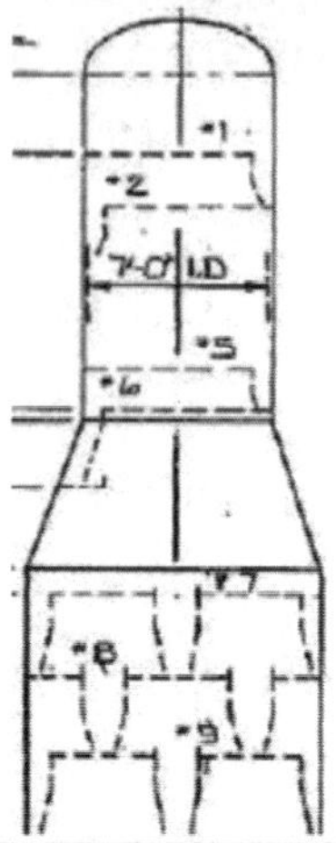

DHT Stripper

- The top section was severely flooded.
- Also, the refinery wanted to continue rate increases on the unit.
- With a system limit of 94% at Base+5% charge rate, high capacity trays would not stay un-flooded for long.
- **MPC had not installed any Beyond System Limit trays ever!**

	Original Design Ratings	Pre-revamp Tray Rating
Equation 13 (%)	73	127
Downcomer flood (%)	75	87
DC backup (in liq)	8.6	15.5
Dry tray dP (in liq)	3.1	10.4
DC backup (%TS+W)	32	57
System Limit (%)	50	94

Shell ConSep™ tray overview

- **Con**tact and **Sep**aration
- Combines a conventional contacting deck with a separation deck utilizing centrifugal force to separate entrained liquid from the vapor
- Allows operation above the jet flood point, but permits normal downcomer flow as well
- Tray arrangement can be a conventional single pass, multiple pass, or high capacity Shell HiFi tray
- Separation is performed using high capacity swirl tube separators
- Can be retrofitted to existing towers with tray spacing as low as 18 inches
- Operation reverts to that of a normal tray at turndown conditions

7

Tested at FRI

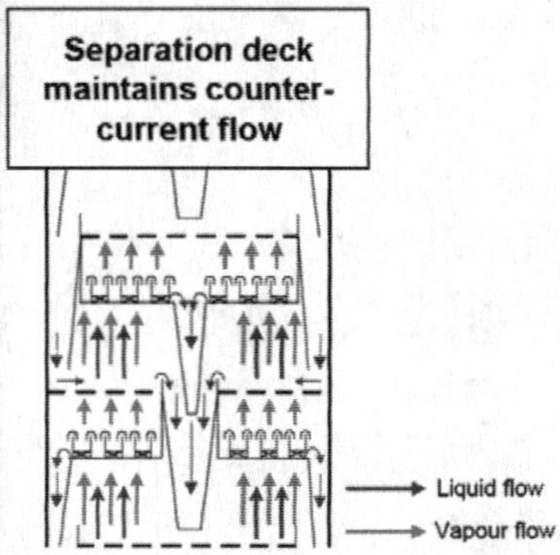

8

View showing HiFi tray downcomers & swirltubes

9

Tray Ratings

	Original Design Ratings	Pre-Revamp Tray Ratings	Shell ConSep™ Design Rates*
Equation 13 (%)	73	127	146
Downcomer flood (%)	75	87	100
DC backup (in liq)	8.6	15.5	20
Dry tray dP (in liq)	3.1	10.4	13.7
DC backup (%TS+W)	32	57	74
System Limit (%)	50	94	108

*Based on the original conventional tray design.

10

Shell ConSep™ Tray Ratings

	ConSep Design Ratings
Fraction Maximum useful capacity (%)	90
Downcomer Froth Backup (%)	59
Downcomer Clear Liquid (inch)	10.9
Tray Pressure drop (psi)	0.198

11

Revamped DHT Stripper Layout

- Five new Shell ConSep™ trays above the feed.
- The top tray is a false downcomer with a swirl tube deck to prevent liquid entrainment to the overhead.
- A seal pan collection system was installed to direct liquid into the feed tray false downcomers.
- Trays were supplied by Sulzer in 2013.
- Trays were installed on existing and expansion rings without welding to the column shell.

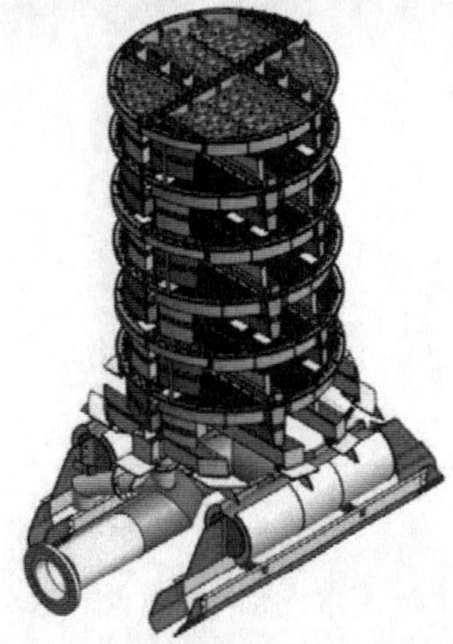

12

Shell ConSep™ trays installed in DHT Stripper

13

Shell ConSep™ Trays installed in DHT Stripper

14

15

ConSep Trays installed in DHT Stripper

16

Original Tray Ratings

	Original Design Ratings	Pre-revamp Tray Ratings*	Shell ConSep™ Design Rates*	Post-revamp Tray Ratings*
Equation 13 (%)	73	127	146	192
Downcomer flood (%)	75	87	100	150
DC backup (in liq)	8.6	15.5	20	33.8
Dry tray dP (in liq)	3.1	10.4	13.7	22.5
DC backup (%TS+W)	32	57	74	125
System Limit (%)	50	94	108	138

*Based on the original conventional tray design

17

Shell ConSep™ Tray Ratings

	ConSep™ Design Ratings	Post-revamp ConSep™ Ratings
Fraction Maximum useful capacity (%)	90	116
Downcomer Froth Backup (%)	59	90
Downcomer Clear Liquid (inch)	10.9	18.7
Tray Pressure drop (psi)	0.198	0.33

18

DHT Stripper – hydraulic analysis

- Rating calculations indicate that the ConSep trays have exceeded the typical capacity limits expected, including the maximum allowed swirl tube vapor capacity, downcomer backup and downcomer inlet velocity.
  - At moderate to high pressures, the maximum capacity of Shell ConSep™ trays is normally set by a downcomer liquid backup constraint. At lower pressure this limit is clearly conservative.
  - For this application, the liquid load to the swirl deck is still relatively low and therefore the swirl tubes have achieved higher vapor load factors than for what is typically designed.
  - As the liquid captured by the swirl deck is sufficiently degassed before being returned to the downcomers below, high downcomer velocities can be tolerated.

19

Results

- The overhead Naphtha stream is consistently below 400°F ASTM D3710 98% endpoint.
- The Diesel bottoms stream is consistently greater than 140°F flash point.
- Optimization of the unit after the Shell ConSep™ trays were installed resulted in ~40 vol% reduction of Naphtha recycle to the crude unit with ~120% increase in reflux.
- This change in the Naphtha distillation subsequently allowed the naphtha to be routed to the reformer, instead of back to the crude unit.
- The charge rate of the unit was increased ~5% right after the revamp. An additional ~10% unit charge rate increase test run was completed this year.

20

Questions

Rigorous Troubleshooting Identified Crude Column Constraints

Multiple major column problem areas correctly identified and resolved during a turnaround

Jill Brown Burns, Valero Energy, San Antonio, TX

Kristen Becht, Valero Energy, Ardmore, OK

Brandt Mueller, Valero Energy, Ardmore, OK

Presented at the AIChE Spring Meeting, Distillation Conference, San Antonio, Texas, March 27, 2017

Rigorous Troubleshooting Identified Crude Column Constraints

Jill Brown Burns, Kristen Becht, and Brandt Mueller; Valero Energy Corporation

Troubleshooting operating problems is a daily part of the job description for a refinery process engineer, especially when charge rate to the refinery is inhibited. The successful identification of column operating problems becomes critical when the resolution includes modification to distillation tower internals that can only be executed during a turnaround outage. "Downtime is expensive. The costs of misdiagnosing a problem is equally enormous."[1] Likewise is failing to diagnose the problem all together.

Development of turnaround work scope challenges every refinery process engineer, especially as turnarounds seem to grow further and further apart throughout the life cycle of a refinery process unit. While this practice reduces major maintenance costs, extending turnaround cycles consequently minimizes the opportunities to address unit operating problems that require downtime (without the enormous cost of an unplanned outage). Thus the engineer needs to effectively identify resolutions to operating problems, and develop the RIGHT scope to address them during the outage. The consequences of misjudging the fix can be significant, as the whole team is left scrambling in the midst of major maintenance work to retrace troubleshooting steps and to order equipment with a hefty emergency expediting price tag.

While many great papers have been presented that highlight interesting and fascinating discoveries found inside a column during a turnaround, in this paper the authors will describe a case study where all the critical discoveries were predicted and anticipated through good engineering practices. Troubleshooting fundamentals like energy balance and operating data analysis were employed along with utilizing more rigorous tools such as isotope scanning to conclusively identify three separate and unrelated areas of limitation in the crude column. We will also discuss an unconventional hot tap bypass that allowed the unit to remain in operation until the planned turnaround, as well as outline the design changes that were implemented during the outage to prevent and address future limitations. The authors will finally present turnaround findings – "validations" rather than "discoveries" - that prove the conclusions drawn from operational and troubleshooting data analysis were correct, the appropriate equipment was ready and onsite, and the planned work scope was spot-on for addressing the tower problems.

Background

Valero Ardmore Refinery operates a 90 MBD Crude and Vacuum distillation unit, charging primarily mid-continent crude through the Cushing terminal. The crude tower is a typical atmospheric operation (630°F, 40 psig flash zone) with 42 trays, a packed gas oil wash section, and a shrouded stripping section. Overhead light naphtha is routed to a gas plant, and atmospheric tower bottoms are routed partially to a gas oil hydrotreater (CFHT) with the balance to the vacuum tower. Five side draw products from the crude tower are further processed within the refinery. The tower has four pumparounds (Heavy Naphtha, Kerosene, Light Diesel, and

Atmospheric Gas Oil) utilized to control product cutpoints according to downstream specifications.

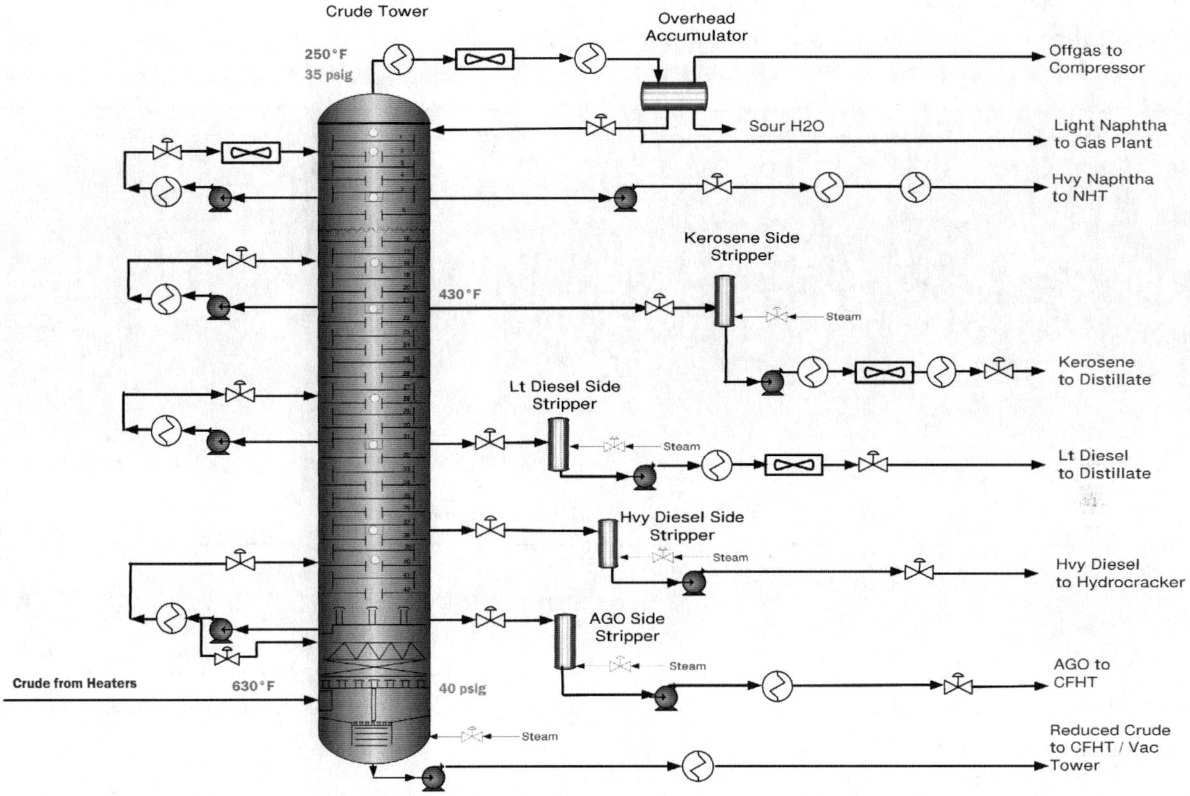

Figure 1. Schematic of Ardmore Crude Tower

This crude tower, like many atmospheric crude units, has a history of corrosion in the top of the tower. The top of the column was overlaid through Tray 4 with high alloy during a series of outages between 1998 and 2011. Overhead condensers experienced short run lengths, attributed to amine salt under-deposit corrosion. The overhead line was replaced, and the trays were renewed with high alloy during the 2011 crude unit turnaround. A chemistry change was made in 2012 to address the root cause of the corrosion, and to eliminate the high salt-point amine injection for neutralization of hydrolyzed chlorides in the overhead. The refinery expected the high alloy and revised chemistry to improve the reliability in the top of the tower, but scheduled an isotope scan as a "health check". Isotope scans (both active area chords and center downcomers) in October 2012 and July 2013 showed normal tray loadings throughout the tower. A high froth height was observed on Tray 3 during the July 2013 scans, potentially signaling the early stages of column fouling/corrosion. However, operation remained normal.

In August 2013, several key heat exchangers in the crude unit preheat were cleaned. After this point, instability in the level of the kerosene side stripper was noticed. The main column remained stable until November 2013, when the column overall pressure drop began to show intermittent periods of liquid accumulation (flooding).

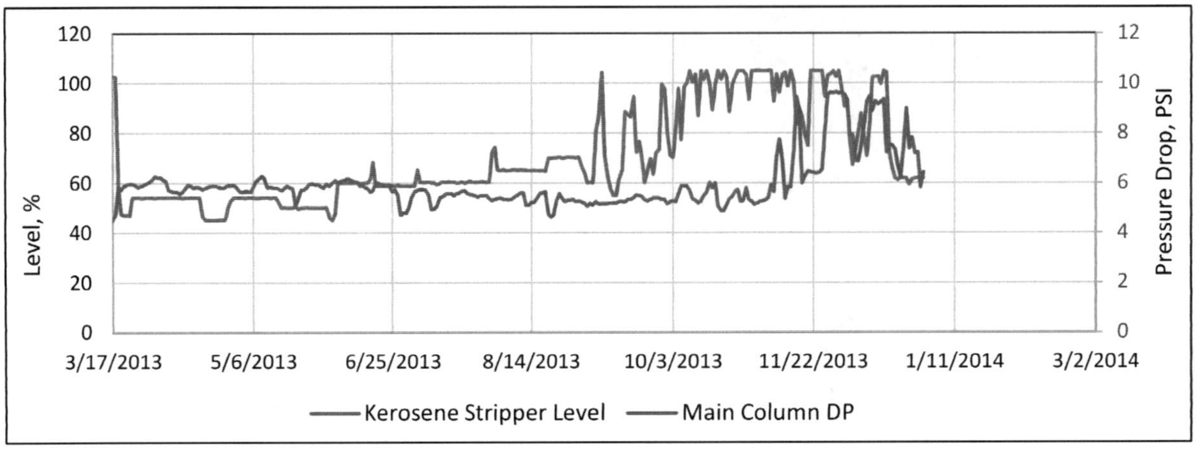

Figure 2. Kerosene Stripper Instability

Given the history of corrosion in the tower's top section, it was hypothesized that the top naphtha section was flooding, holding up liquid, and subsequently limiting liquid available to the kerosene draw.

When the instability occurred, the pressure drop in the tower increased by 3 psid, which has an equivalent head of ~10 feet of naphtha-ranged hydrocarbon. With the pressure drop indication on the overall tower, the location of the accumulation must be determined by reviewing other data. The 10 feet of liquid head could be localized to the top head of the tower or distributed amongst 20 trays. Further investigation was required.

Troubleshooting the Crude Tower

Engineers have a lot of tools available to diagnose column flooding. Isotope scanning is a powerful tool, and was an integral part of troubleshooting this tower. However, the authors started with the fundamentals - reviewing data (field and historian) and evaluating the tower energy balance.

Using Engineering Fundamentals to Troubleshoot Tower Flooding

Ardmore had altered the heat removal in several of the pumparound circuits by cleaning crude preheat exchangers. The authors recognized that redistributing the heat load in the tower impacted internal reflux. If a trayed section operates near a hydraulic limit, the change in internal loading could push that section into flood, with resultant high pressure drop and instability observed at a product draw. Pumparounds, tower temperatures, and product overlaps were analyzed to understand the impact of exchanger cleaning on the tower's internal traffic.

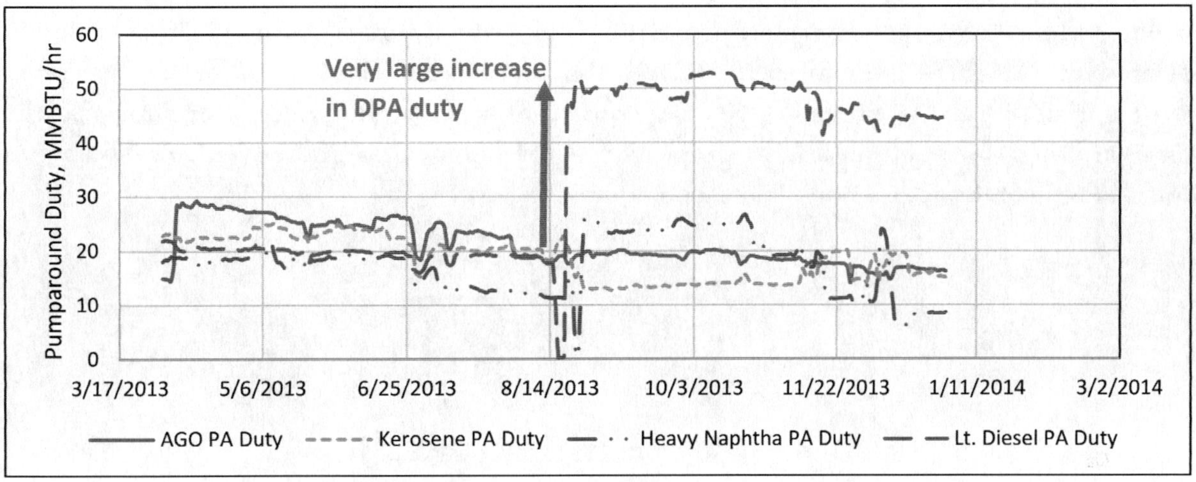

Figure 3. Pumparound Heat Balance Shift

Diesel pumparound duty had increased by over 200% (19 MMBTU/hr to 50 MMBTU/hr), significantly shifting heat removal lower in the tower. The kerosene-light diesel overlap increased with the higher diesel pumparound duty, indicating reduced internal reflux traffic

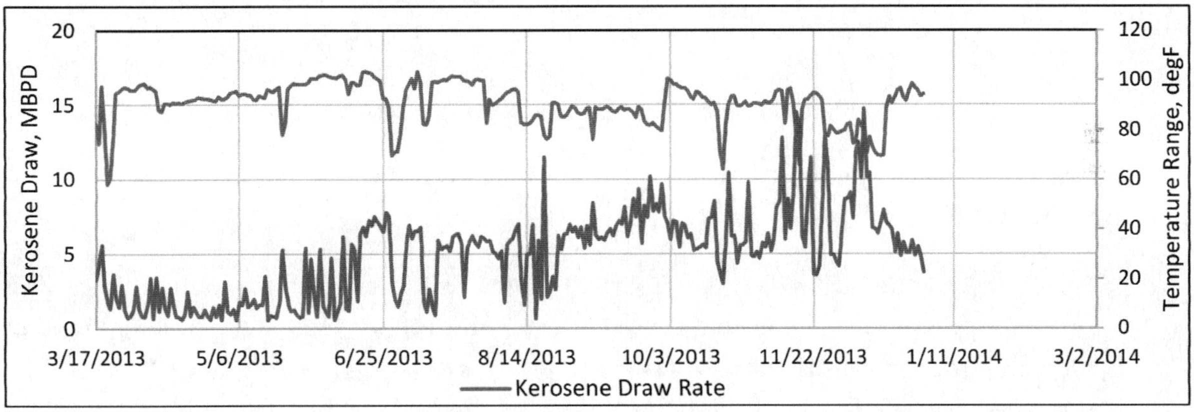

Figure 4. Kerosene Draw Data

below the kerosene draw. The temperature indication in the vapor space of Tray 21 (directly below the kerosene draw sump) was observed to oscillate. This unstable behavior preceded the exchanger cleaning by two months. After the instrument was checked in the field by a technician, it was hypothesized that this temperature instability may be symptomatic of flooding, where hold-up robs cooler condensed internal reflux from the trays below the kerosene draw.

The kerosene draw is configured with a center downcomer draw sump. Pumparound and product are drawn from the sump and split in external piping where the pumparound stream is routed to a pump, and the product drains to the side stripper on stripper level control. The liquid condensed by the pumparound plus the internal reflux from the trays above can either be drawn as product or overflow the sump onto Tray 21. The skirt panels of the Tray 20 center downcomer were not submerged to provide a positive liquid seal at high draw rates, indicating that the original design intended for a large proportion of the liquid to overflow the sump and act as internal reflux for fractionation of kerosene from light diesel.

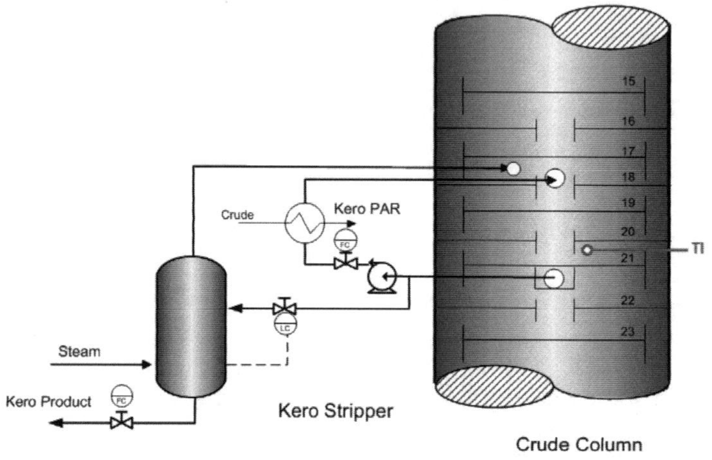

Figure 5. Kerosene Section Configuration

Several step tests were conducted to test the sensitivity of the hydraulics in the kerosene pumparound section as follows:

1. The pumparound exchangers were partially bypassed to allow for a higher circulating kerosene pumparound rate for a fixed duty. The higher circulation rate induced a pressure drop increase in the tower.
2. Kerosene product draw rate was increased. A higher product flow control setpoint caused loss of level in the side stripper, indicating that liquid was not available to the draw.

Because of the sensitivity to tower heat balance, a simulation was developed to understand the liquid internal reflux contribution to the loading in the kerosene section. An increase in heavy naphtha pumparound duty would have directionally increased that internal reflux traffic, while an increase in diesel pumparound duty would have lowered the vapor traffic through that section and directly reduced the potential load that could be internal reflux. The model tuned to a steady state operating case indicated that the internal reflux from the naphtha section contributed roughly 50% of the total liquid load to the kerosene pumparound – a significant percentage.

The internal loads generated by the simulation were utilized to rate check the existing 2-pass, round valve trays with conventional, straight downcomers to determine whether any of the conditions both before and after the shift in pumparound duties exceeded hydraulic capacity.

	Before Cleaning *Normal Ops*	After Cleaning *Unstable Ops*
Glitsch Eq. 13[2]	65.3%	65.6%
Downcomer Loading, gpm/ft2	164	143

Figure 6. Tray Hydraulics in Kerosene Pumparound

The liquid loading of Trays 19-20 was typical for a pumparound system, with high circulating liquid rates. The downcomer top areas of these two-pass trays were not balanced, so the side downcomers had preferentially higher liquid flux rates and operated near the top of what is commonly considered best practice for this service. Yet the shift to pull more heat out of the diesel pumparound directionally lowered the liquid rate through these trays. Based on a theoretical "clean" tray, the kerosene pumparound should have ample capacity for handling higher vapor loads consistent with more naphtha yield. However, the operating data indicated that these trays appears particularly sensitive to vapor load changes.

Some additional field checks were conducted to understand the unstable temperature and level behavior in the kerosene section, but yielded little clarity to the problem. Pressure taps were not available to narrow the column pressure drop to a specific area, but cutouts were made in the column insulation to collect skin temperature data that would directionally indicate flooded trays.

Using Isotope Scans to Troubleshoot Tower Flooding

After the foundation was laid with the basic engineering methods, isotope scanners were engaged to provide simple gamma scan chords of the two active areas and center downcomers while the column was loaded/flooded. These basic scans showed several major shifts in liquid accumulation from the previous scans conducted only 4 months prior.

The tower was entirely full of dense liquid from Tray 20 (kerosene draw) to the top. Some clearing (return toward clear vapor) was indicated on Trays 2-3. The tray immediately below (Tray 21) was very lightly loaded with imbalanced froth height between the two passes. The authors were also surprised to note almost 2-ft of accumulation both above and below the AGO wash packed bed.

The center downcomers from Tray 18 to the top of the tower were full of dense liquid. However, the draw sump from Tray 20 was full, yet less dense, and not consistent with the heavy absorption seen on the active area.

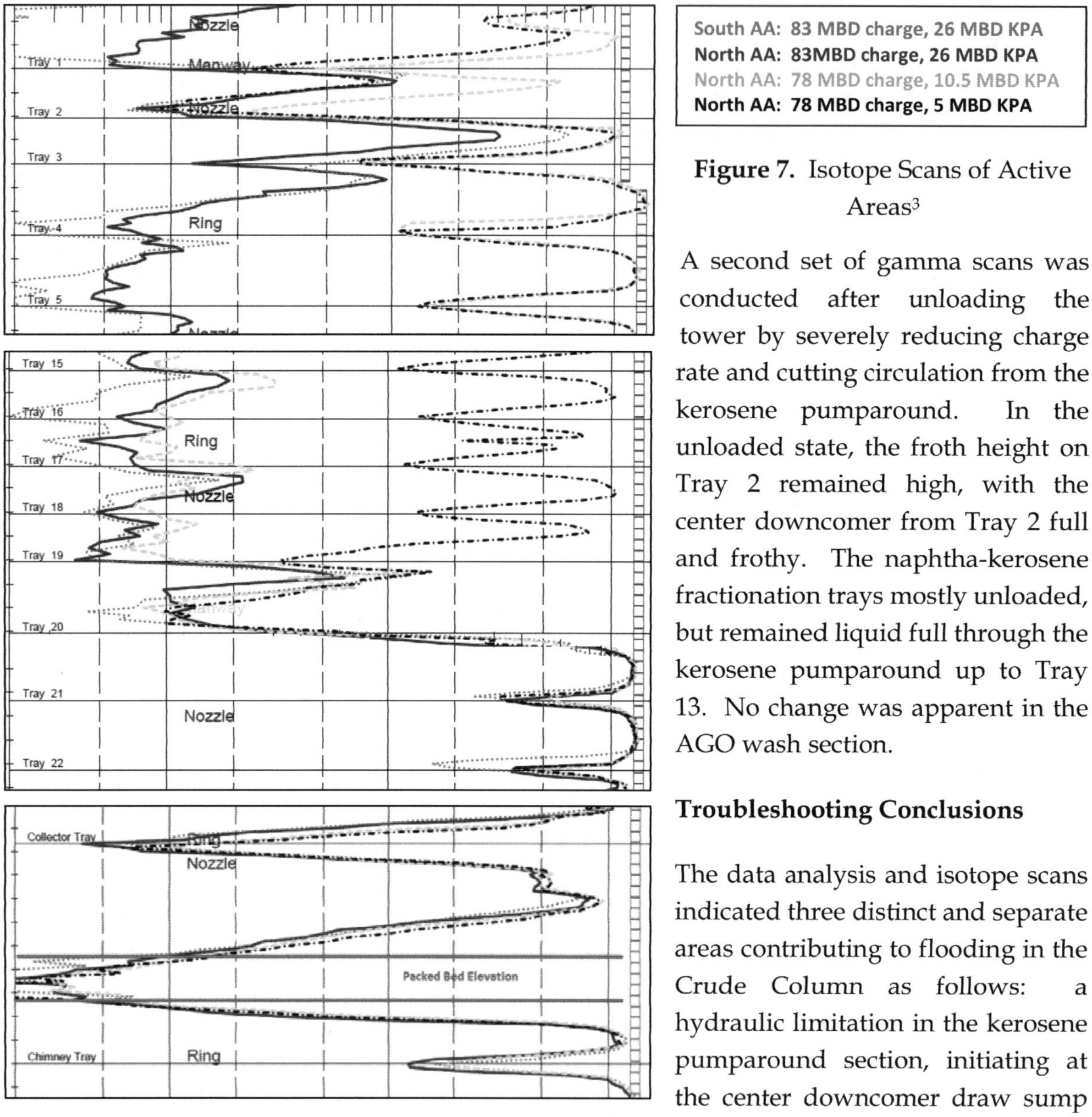

Figure 7. Isotope Scans of Active Areas[3]

A second set of gamma scans was conducted after unloading the tower by severely reducing charge rate and cutting circulation from the kerosene pumparound. In the unloaded state, the froth height on Tray 2 remained high, with the center downcomer from Tray 2 full and frothy. The naphtha-kerosene fractionation trays mostly unloaded, but remained liquid full through the kerosene pumparound up to Tray 13. No change was apparent in the AGO wash section.

Troubleshooting Conclusions

The data analysis and isotope scans indicated three distinct and separate areas contributing to flooding in the Crude Column as follows: a hydraulic limitation in the kerosene pumparound section, initiating at the center downcomer draw sump on Tray 20; a hydraulic limitation in the top reflux trays in the column; and accumulation at the wash bed.

Even with severely-reduced loading, the second isotope scan indicated continued accumulation in the kerosene pumparound, thus this limitation was considered the primary capacity constraint. Based on the light absorption in the downcomer, it appeared that vapor was potentially bypassing up the downcomer. This type of phenomena is commonly related to one of three issues as follows:

1. Open area restriction on the tray above inducing vapor to preferentially push up the downcomer
2. Overdrawing of kerosene product rate to unseal the downcomer
3. Damage to the draw sump that impacted the clear liquid seal in the downcomer.

The authors have seen an increasing trend of crude column capacity restrictions initiating in the kerosene distillation range throughout the Valero refinery system. This issue has been tied to phosphorous in the crude feed. During the August 2013 exchanger cleaning, several exchanger deposit samples were analyzed and indicated almost 5% of P_2O_5. Historical crude receipts were also analyzed, showing a period immediately preceding the column problems with up to 3-5 ppm phosphorous. The hydraulic restrictions in the column also resembled similar cases where the open area was fouled by phosphorous. While the lack of positive seal in the draw pan generated concerned, the kerosene draw rates had not exceeded normal range of operation. No operational upsets experienced would have been likely culprits for damage. The authors thus concluded that the kerosene pumparound section may have been exposed to fouling related to phosphorous from the crude.

To unload the kerosene pumparound trays, heat duty was shifted away from the kerosene pumparound to the other pumparounds and overhead reflux. Unfortunately, the naphtha pumparound and reflux were also constrained, according to operating data and isotope scans. Because this section of the tower experienced historical corrosion and fouling issues associated with free water and salt deposition, this seemed the most likely culprit.

Unlike the naphtha and kerosene sections, there was no internal experience with AGO wash bed fouling within the Valero circuit. A field walk revealed that according to pressure drop measurements, the AGO wash oil strainers had failed and exposed the wash spray nozzles and packed bed to fouling and maldistribution of liquid. The atmospheric flash zone is not typically prone to the level of coking that a vacuum tower is, but entrainment from the feed can contribute to coking with sufficient residence time. The isotope scan showed a high level on the overflash collector tray and a simple two phase flow calculation indicated high momentum in the feed nozzle that exceeds most conventional best practices for operation without a vapor horn or other centrifugal inlet device. It was thus concluded that entrainment was occurring and could not appropriately drain from the wash section. The overflash liquid hold-up and residence time in the wash zone may have contributed to coking of the bed.

As the cycle progressed, the AGO limitations became more of an operating constraint when the bed accumulation periodically flooded up to the AGO collector tray and resulted in high endpoint "black" AGO and heavy diesel.

Actions Employed to Regain Stability, Extend Run Length

The Ardmore Crude unit operation was severely constrained by the three hydraulic restrictions identified during the troubleshooting process. The presumed fouling in the kerosene pumparound limited both the vapor and liquid rates through those trays. Pumparound duty was maximized lower in the column, and naphtha pumparound / reflux was constrained by the hydraulic restriction in the reflux trays. Naphtha endpoint was uncontrollable, impacting the downstream units. This ultimately resulted in reduced charge rate to the unit.

Chemical treatment was considered to mitigate the effect of phosphorous flooding in the column, but rejected based on the hazard of moving the foulant to downstream fixed catalyst beds. A mechanical solution was preferred.

Constraining Unit Operation

The three different flooding mechanisms in the naphtha, kerosene, and AGO sections made it difficult for operations to properly respond to upsets in the column. A move to decrease flooding in one section exacerbated flooding in another, which made normal adjustments to the column heat balance very limited. Increased reflux or heavy naphtha pumparound rates would initiate liquid carryover into the overhead system. Increased internal reflux in the kerosene section would flood kerosene up to the heavy naphtha section and rob internal reflux from the light diesel section. Increased AGO pumparound rates would flood the AGO packed bed and cause resid entrainment into the AGO and sometimes up to the heavy diesel draw.

The overhead pressure was increased and heater coil outlet temperature lowered to reduce vapor traffic in the column above the flash zone and ultimately lower the impact of the vapor-induced flooding in the kerosene section. The lower coil outlet temperature also reduced the momentum (ρV^2) of the tower inlet, which limited resid entrainment into the AGO. Pumparound exchangers were cleaned throughout the tower to aid in heat removal (trading circulating flow for more ΔT).

The kerosene pumparound flow was minimized to lower liquid loading on the trays. This area had the highest level of flooding and the largest impact to stable column operation. At times that the kerosene section would flood, kerosene would build up to the heavy naphtha draw and the light diesel pumparound had a high potential of cavitation. Rate-of-change alarms were set on the light diesel pumparound to indicate loss of flow and hold-up in the tower.

Hot tap bypass

With the aid of a simulation model, the authors evaluated different projects to mitigate the flooding in the kerosene section by rerouting liquid or vapor in the tower, and ultimately concluded that the most effective solution was to reroute the internal reflux liquid from several trays above the kerosene draw directly to the kerosene stripper through a hot-tapped bypass.

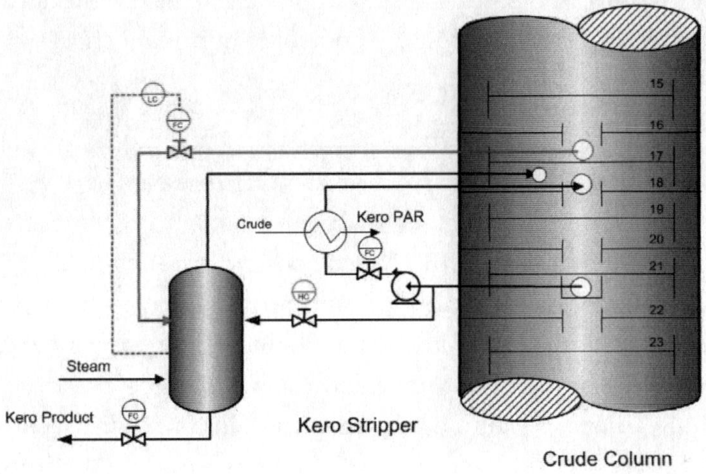

Figure 8. Kerosene Section with Hot Tap Bypass

A nozzle was first hot-tapped into the center downcomer from Tray 16 on the Crude Column. Piping, along with a butterfly valve, was installed from here to a second tap located above the liquid level on the top tray of the Kerosene Side Stripper. As the column restriction became severe with liquid accumulation occurring the majority of the time, the level on the side stripper was set up to control the withdrawal of liquid from Tray 16, with the normal kerosene product draw off of Tray 21 fixed in manual. Kerosene product flow control was set at a maximum up to a kerosene flash limit to remove as much liquid from the column as possible to prevent flooding from propagating into the naphtha section.

Development of Turnaround Scope and Design

Although Operations became proficient at managing tower stability within the new hydraulic constraints, ultimately the tower continued to degrade and the crude unit team began looking toward options to address the problems during a unit outage. When replacing tower internals, it often bears a similar cost to upgrade the features for either improved capacity or in the case of the Ardmore crude Column, improved reliability.

The authors carefully evaluated each section of constraint in the tower to arrive at a final turnaround scope that would ultimately result in full resolution of the column problems.

Naphtha Section

The reflux trays were heavy-gauge carbon steel, which bears a high risk of corrosion in an environment with acidic chloride attack. The presence of iron scale in pumparound exchangers and the flooding indicated by isotope scans suggested that the trays were fouled or corroded. The authors elected to upgrade the tray metallurgy, add anti-fouling features into the tray design, add a high efficiency coalescing element internal to the overhead reflux drum to remove free water returning to the column, and engineer a modification to the naphtha pumparound return to combine with cold true reflux on temperature control to eliminate shock condensation.

Kerosene Section

The kerosene pumparound trays were carbon steel with moveable valves, which have little forgiveness to the type of fouling consistent with phosphorous. The authors designed anti-fouling trays with fixed valves having a high net rise and multichordal side downcomers to maximize available active area for the kerosene pumparound trays. In anticipation of fouling, the plan included replacement of all tray active panels with new panels having fixed valves on the four trays above the kerosene pumparound and the six trays below the pumparound draw. These trays operated at or near the temperature regime where other towers have experienced phosphorous fouling. Similarly, the kerosene side stripper was retrayed with similar anti-fouling features.

AGO Section

While bypassing of the wash strainers may have resulted in plugging of the spray header and subsequent dryout of the wash bed, several inherent design issues were also identified with this section and were thus added to the turnaround scope. The column has a 192" diameter, yet the wash spray header had only seven nozzles. In the authors' experience, this often does not provide enough coverage nor adequate overlap to fully wet a wash bed. The wash spray header was redesigned for nineteen nozzles, the wash bed replaced with structured grid packing, and the overflash collector tray downpipe enlarged to accommodate additional liquid rate that appeared to be accumulating on the tray.

Turnaround Execution Plan

Because of the extensive work to the key columns in the unit, the authors' developed a turnaround plan to ensure proper execution of the scope as follows:

- Protocol to address issues, problems, or changes should they arise, and how to communicate those issues across shifts within the turnaround team (process engineers, turnaround maintenance, and the tower contractor).

- Process inspection sheets, so that all engineers across all shifts were checking, measuring, and inspecting the correct components using "cheat sheets" that contain pertinent tower internal information.
- Initial entry by process engineering to facilitate early identification of scope creep that included collecting samples and properly documenting the as-found state of the column.
- Final inspection and punchout procedures to assure that all work was properly executed and prevent performance issues during the run cycle due to poor installation.

Turnaround Discovery

Upon initial entry, the state of all key areas of the column were confirmed to reflect the authors' predictions.

Naphtha Section

As predicted, the top 4 trays were found to be severely corroded, with the remnants of the top tray severely restricting Trays 2 and below.

The scope was executed as planned to replace the top four trays with upgraded metallurgy, and to integrate the heavy naphtha return piping with the reflux return to Tray 1. The top temperature controls the heavy naphtha pumparound rate, and the combined return temperature controls the bypass around the heavy naphtha pumparound cooler.

Kerosene Section

As predicted, both the kerosene pumparound and side stripper had severe iron phosphate deposits, which obstructs movement of the tray valves and accumulates on the tray decks.

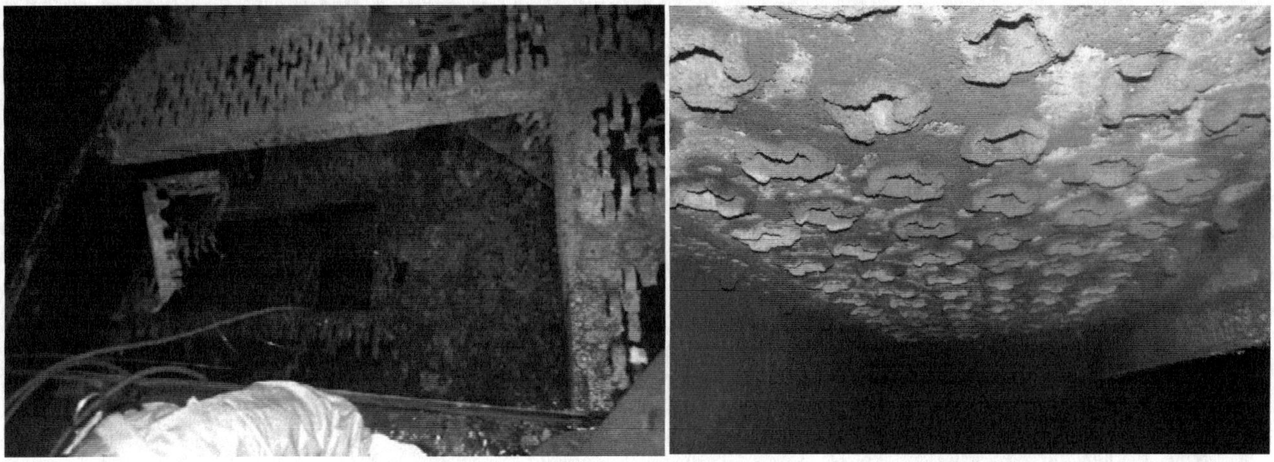

Figure 9. As found conditions of Reflux and Kerosene Pumparound Trays

The tray replacement scope was executed as planned, with additional modification to submerge the center downcomer skirt panels at the kerosene draw tray to provide a positive seal to prevent vapor bypassing with high kerosene draw rates.

AGO Section

The AGO wash bed experienced significant coking as predicted. The wash distributor and nozzles were also found to be partially plugged. The overflash downbox was plugged with coke and completely obstructed.

Figure 10. As-found Conditions of Overflash Collector and Downbox

Deconstruction of the wash bed took longer than anticipated due to the heavy coke build-up. Replacement standard Type-3 grid packing was installed, along with holdown grid and through-rods for uplift resistance. To improve spray cone overlap and reliability to prevent future coking, the existing 7-nozzle spray header was replaced with a newly designed wash distributor containing 19 spray nozzles. The overflash overflow downbox was replaced to increase drainage capacity while maintaining an exit velocity sufficient to prevent plugging.

Conclusions

Through hard work and solid engineering, the Ardmore refinery crude unit team was able to identify and resolve three separate and distinct limitations to operating capacity of the towers. The design improvements built into the new tower internals will increase reliability for the next operating cycle, and the operating best practices established during the troubleshooting process will improve overall product quality.

Note: The views expressed in this paper are that of the authors and do not necessarily reflect the views of Valero Energy Corp.

REFERENCES

1. Pinto, R. Duarte, M. Perez, H.Z. Kister, "Combine Temperature Surveys, Field Tests, and Gamma Scans for Effective Troubleshooting," Hydrocarbon Processing, April 2003, p. 69.
2. Koch-Glitsch. "Bulletin 4900. Ballast Tray Design Manual." 2013 Anniversary Reprint.
3. Isotope scans by Tracerco.

Correct Problem Diagnosis Debottlenecks a Gas Plant with No Hardware Changes

By

Henry Z. Kister, Fluor, Aliso Viejo, CA, USA

Presented at the Distillation Topical Conference, AIChE Spring Meeting, San Antonio, Texas, March, 2017

Summary

This paper describes a case study of a gas plant (designed by others) that was bottlenecked by the reboil system of a deethanizer stripper. The reboiler system that consisted of two parallel once-through thermosiphon circuits experienced two different steady states. When operated at low rates, an undesirable steady state was established with maldistribution between the two circuits, in which one was circulating and the other remained liquid-full. This steady state persisted when the tower feed rates were raised until the circulating reboiler reached its heat transfer limitation, still with the other circuit liquid-full.

Pressure drop calculations, close design reviews, examination of operating charts, and gamma scans, led to the correct diagnosis that enabled debottlenecking the reboiler circuits and the entire gas plant with no hardware modifications. The desired steady state was established by an operation that temporarily generated higher liquid circulation and pressure drop through the reboiler circuits.

This case teaches a number of very important key lessons:

- With a parallel reboiler system, maldistribution between parallel circuits does occur and can cause severe bottlenecks. To the best of our knowledge, this is the first time that such maldistribution is reported in the literature.
- Pressure balances need to be considered at the design not only for the maximum design rates, but also at the minimum expected operation rates.
- It is important to provide flexibility in once-through reboiler systems that will allow operating as a circulating thermosiphon systems, at least at low rates.
- Finally, good troubleshooting can provide low cost (in this case, no cost) debottleneck of an entire gas plant.

Background

This newly-built gas plant was bottlenecked by the deethanizer stripper. The tower bottleneck appeared in the reboiler circuit. In early April, 2014, Fluor, who had no involvement in the tower or reboiler design, but has extensive experience in distillation design and troubleshooting, was requested to assist with investigating the root cause of the bottleneck and in finding ways of overcoming it. A task force of operating company process and operation engineers, and Fluor's distillation specialists, was formed. The joint team identified the root cause. Based on the root cause, a solution was successfully implemented. After this the deethanizer stripper and the gas plant were in full operation achieving the design rates. This paper describes our troubleshooting investigation and the lessons learnt.

Process Description

The deethanizer stripper is a 15'-6" tower equipped with 4-pass conventional valve trays with moving round valves. The tower base contains a preferential baffle that divides the bottom sump into a separate reboiler draw compartment and a separate bottoms draw compartment. Liquid leaving the bottom valve tray is diverted to the reboiler draw compartment (Figure 1). From there it flows into a battery of four horizontal thermosiphon reboilers. The bottom tray liquid is diverted to the reboiler draw compartment by notched bottom seal pans. As shown in Figure 2, the 8" deep notches in the seal pans are located right above the reboiler draw compartment. There are no notches above the bottom draw compartment. The purpose is to feed all the bottom tray liquid to the reboilers.

The vapor-liquid mixture returning from the thermosiphon reboilers enters above the bottoms draw compartment via two opposite 24" nozzles. Each nozzle is equipped with a pipe that extends to near the centreline of the tower and has a 4'-6" cutout in the bottom quadrant that is directed towards the bottom draw sump (Figures 1 and 2). The purpose is to have all the reboiler return liquid enter the bottom draw compartment, and none enter the reboiler draw compartment. The vapor returning from the reboilers ascends to the bottom tray, while the liquid returning from the reboilers descends to the bottoms draw compartment. Any excess bottom tray liquid not boiled by the reboilers overflows the baffle into the bottoms-draw compartment.

This design generates a once-through[1] thermosiphon reboiler system. Such a system intends to capture all of the liquid leaving the bottom tray and feed it directly to the reboiler. The reboiler return liquid is entirely drawn as bottoms product. None of the reboiler return liquid is recycled back to the reboiler, hence the name once-through.

Once-through systems can achieve one full theoretical stage of separation, if the trapout tray and draw pan do not leak[1]. With the cold bottom tray liquid being fed to the reboiler, it can also maximize the reboiler temperature difference and capacity. Reboil vapor is limited to about 40% of the bottoms product rate, due to the normal limitation of 30% maximum vaporization (by weight) in thermosiphon reboilers[1]. This makes once-through thermosyphon systems appropriate only for low-boilup systems such as strippers. Many deethanizer-strippers use once-through thermosiphon reboiler systems.

Figure 1. Deethanizer Stripper Reboiler Circuits

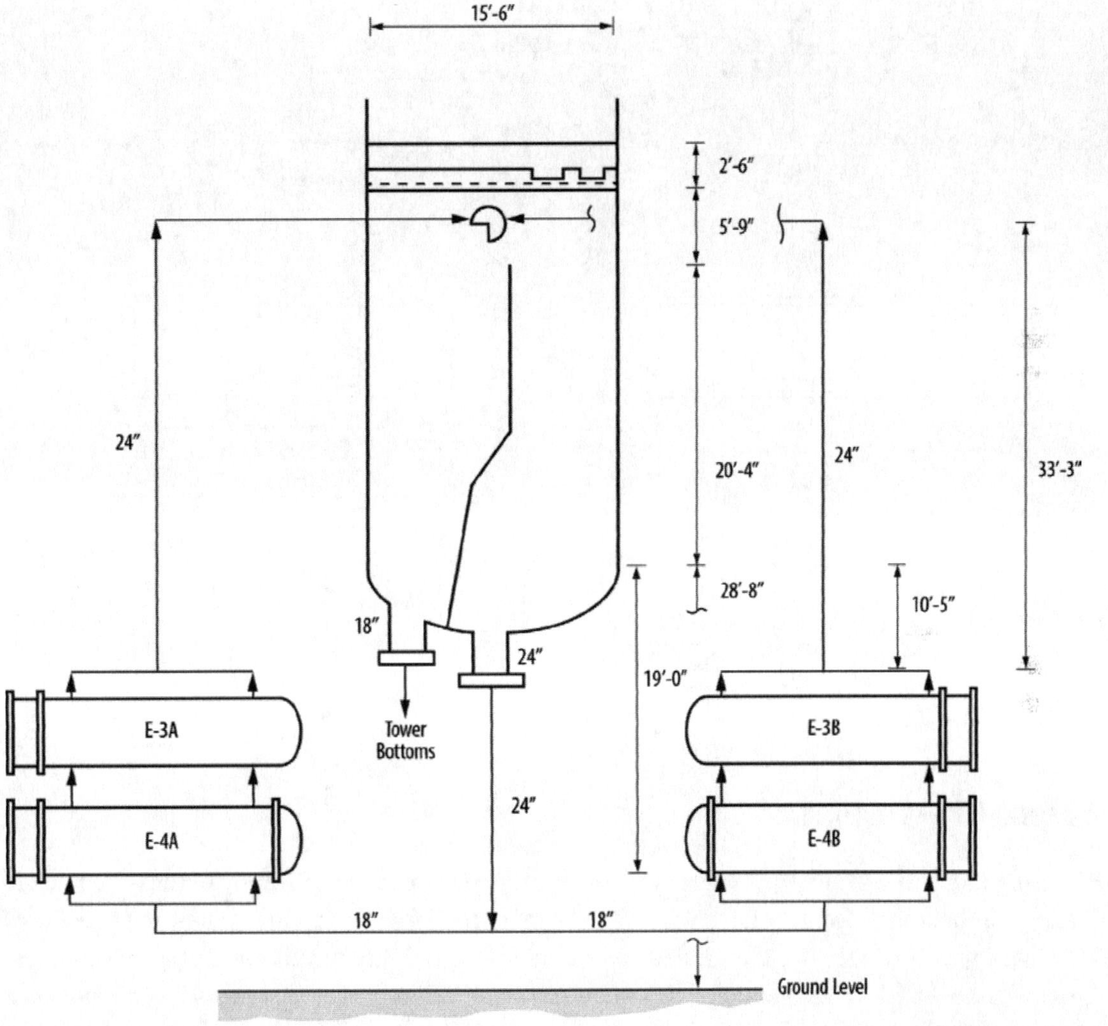

This deethanizer stripper has four horizontal reboilers, arranged in two parallel circuits, "A" and "B" (Figure 1). In each circuit, the lower reboiler (E-4A or E-4B) is heated by hot naphtha liquid, while the upper reboiler (E-3A or E-3B) is heated by high-pressure steam. In each circuit, the naphtha and steam reboilers are designed to provide duties of 32 MM Btu/h and 40 MM Btu/h, respectively. On the boiling side, liquid to the reboilers is provided from the reboiler draw sump via a 24" pipe which splits into two 18" pipes, each going to one circuit. The reboiler effluent from each circuit returns to the tower via a 24" line.

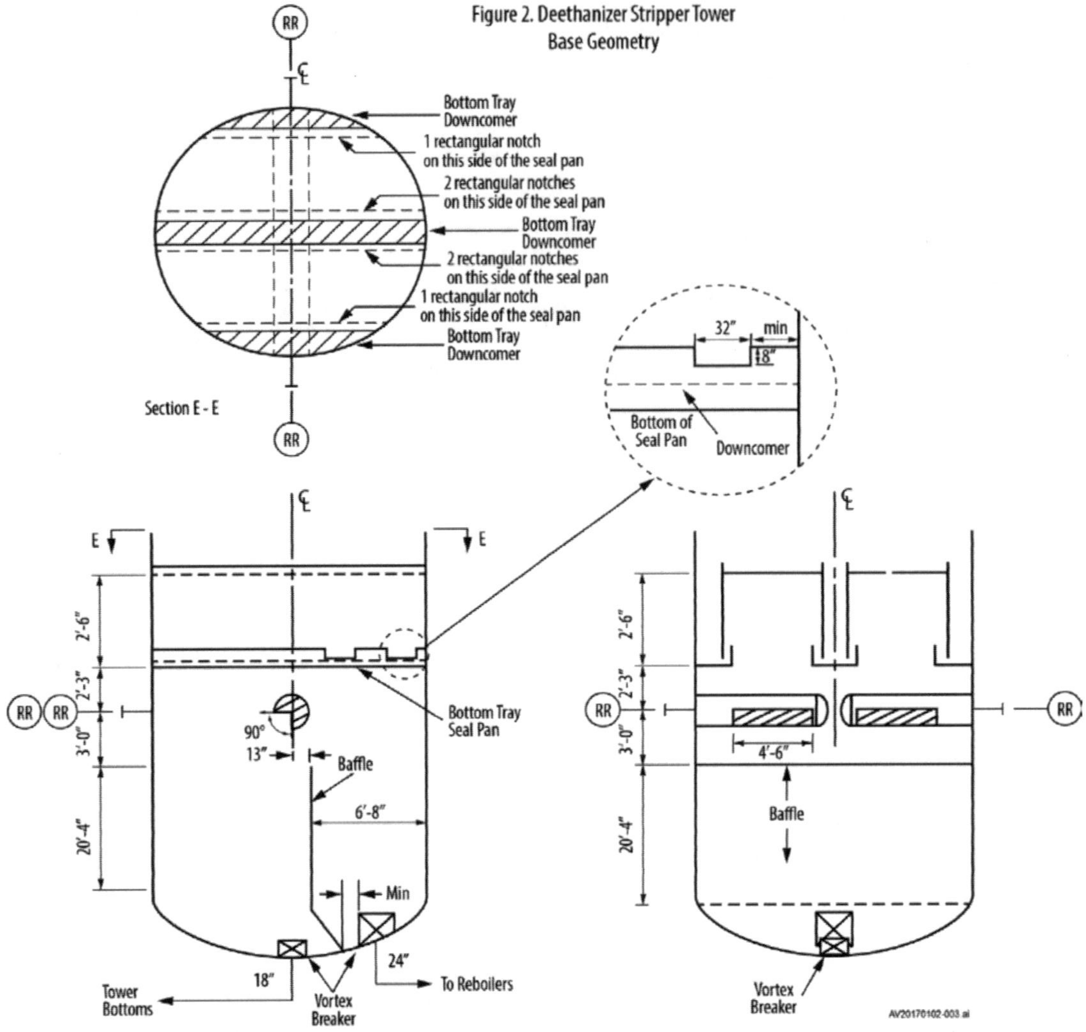

Figure 2. Deethanizer Stripper Tower Base Geometry

Initial Performance

During the start-up, the plant was at low rates and to improve tower stability it was decided to take the steam out of reboiler E-3A. This was successful at creating a steady steam flow through the E-3B reboiler circuit and provided tower stability but the plant has been unable to re-establish flow through the E-3A/E-4A set of reboilers ever since. For the next six months or so, the E-3B/E-4B circuit did all the reboiling, while the deethanizer bottoms liquid circulation has been difficult to establish through the E-4A/E-3A circuit.

Figure 3 shows typical operating data from 6 months after start up and compares them to the design values. The zero steam flow rate through E-3A and the absence of significant temperature difference across the heating side of E-4A indicate no boiling and no significant circulation through the A circuit. To maximize duty, steam to E-3B was raised to the maximum, well above the design steam rate for this exchanger, but still not high enough to make up the lost duty of E-3A. This restricted the feed flow rate through the tower and bottlenecked the entire gas plant. There was a need to get the stripper bottoms liquid to start flowing through the E4A/E-3A circuit.

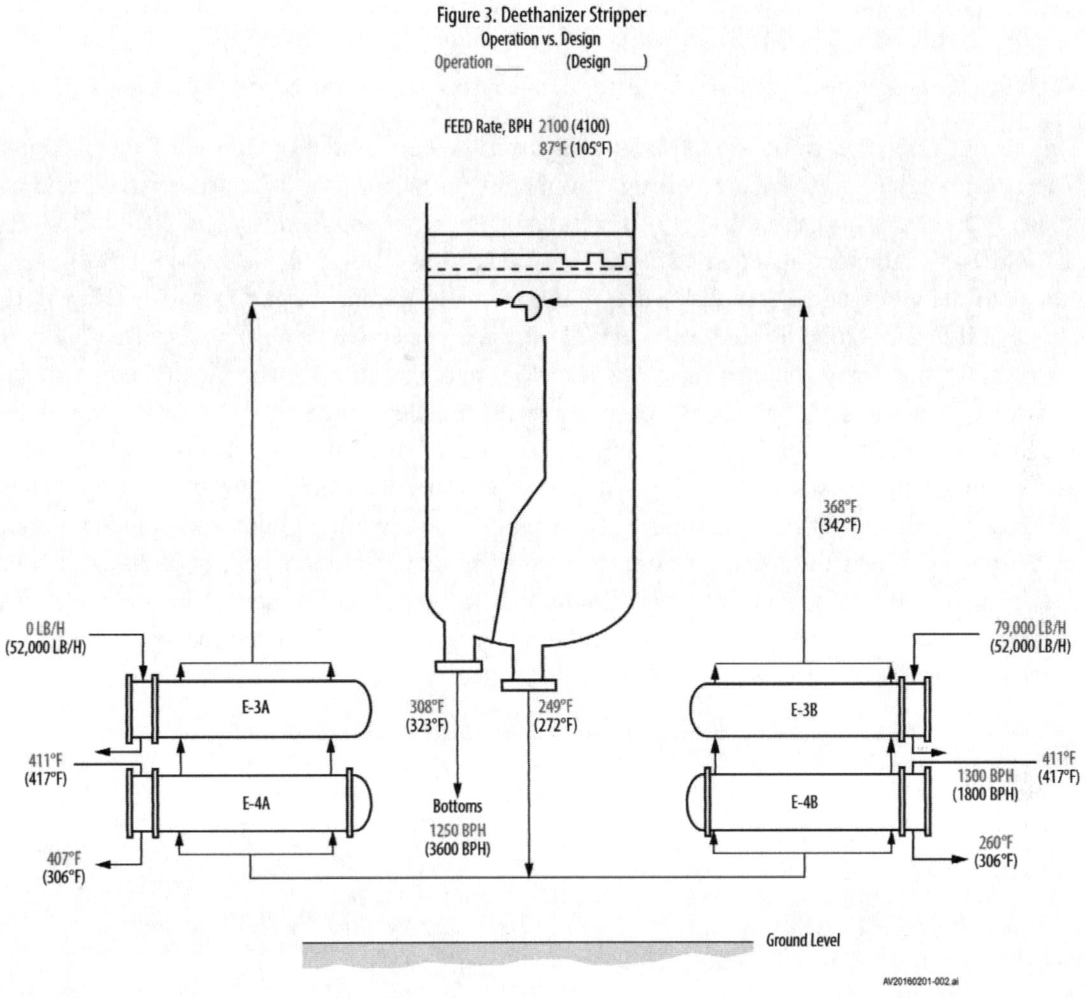

Troubleshooting

Initial troubleshooting endeavors focused on initiating flow through the A circuit by way of inducing a gas-lift. One possibility was that the lights were stripped out of the A-circuit reboilers, leaving behind heavy hydrocarbon liquid. Heating this heavy liquid may not have produced enough vapor to initiate reboiler action. Injecting a light gas like nitrogen could possibly gas-lift the heavy liquid and initiate thermosiphon circulation. This technique has been successful in the past for initiating circulation in liquid-full thermosiphon reboilers[2-4].

A source of nitrogen was connected to a bleeder valve immediately downstream of E-3A. Nitrogen was injected for 4-6 hours in an attempt to provide lift and jump-start the thermosiphon effect. This action was unsuccessful in initiating circulation through the A circuit.

This came as a surprise, as in other reboilers[2-4] a similar action had been successful. There were questions as to whether enough nitrogen was applied, whether it was applied at the correct point, and whether the liquid in the reboilers was just too heavy and needed draining and replacing by fresh liquid before the gas lift attempt. Addressing these problems would have required some painful modifications or operations with no assurance of success. The validity of the theory that heavy liquids in the A-circuit are the root cause was questioned. With these in mind it was decided to perform more troubleshooting before trying out a fix.

One useful reference point was process data from the very initial operation of the stripper. The operating charts from shortly after initial start up (Figure 4), at a stripper feed rate of about 1700 bph, show both the A and B circuits operating simultaneously and stably. Both E-3A and E-3B reboilers operated stably with a steam flow rate of about 32,000 lb/h to each. For both the E-4A and B reboilers, the temperatures of naphtha entering and leaving were about 400°F and 320°F, respectively. At 22:40 the stripper feed rate plunged, steadying at about 1200 bph forty minutes later. To improve stability, the plant manually reduced the steam flow rate to E-3A to about 8000 lb/h, with the steam rate to E-3B remaining at its previous value. The naphtha temperatures into and out of E-4A became the same, about 360°F, suggesting it stopped heating. At the same time the temperature outlet from E-4B decreased to about 260°F, suggesting that this reboiler was working harder than before. Both of these suggest that the A-circuit stopped circulating. Early on the next day, the feed rate to the stripper rose again, to 1500 bph, but this time the A circuit remained inactive. It was common practice to attempt to restart circulation by opening and closing the steam valve on the stalled set but circulation remained stalled.

Figure 4. Process Data for Initial Operations of Stripper Reboiler Circuits

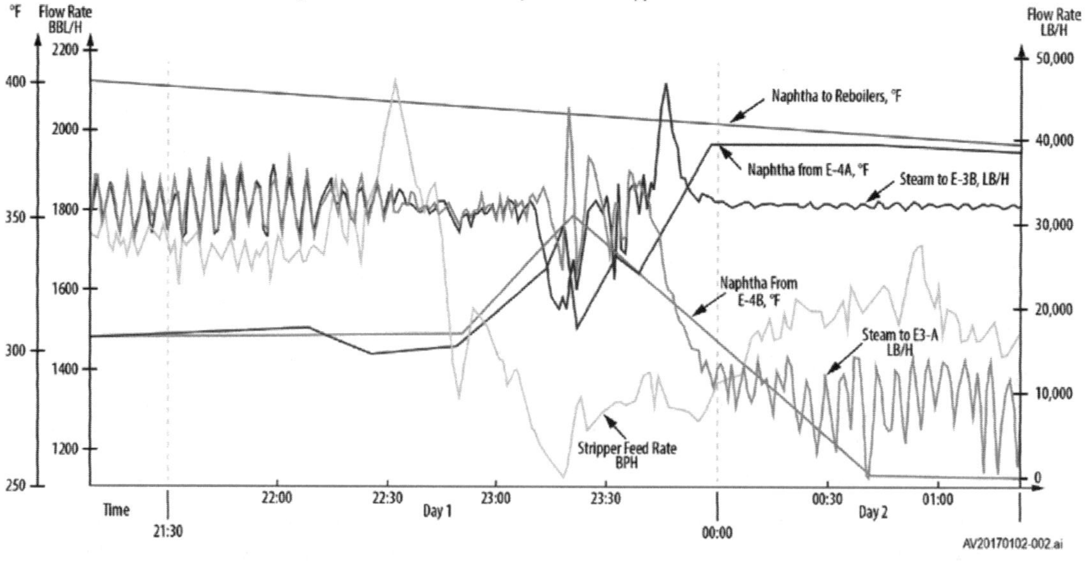

This sequence of events suggests the possibility of the existence of two hydraulic steady states, with an easy switchover from one to the other. This is likely to take place when the reboilers pressure drop on the boiling side is low.

Calculations performed during the design by the engineering contractor that designed the circuits showed a total pressure drop of 1 psi for the exchangers in each circuit at the design conditions. The contractor's calculated pressure drop were 0.4 psi and 1.3 psi for the reboiler liquid inlet line and the reboiler return lines, respectively. The total pressure drop is equivalent to 11 feet of liquid head.

Figure 1 shows the elevations of the reboilers circuits. The diagram shows that not much liquid head is required to overcome the pressure drop in the reboilers circuit. The liquid level in the reboiler draw compartment will therefore be near the bottom of the compartment, even at the design rates. At the low rates, the pressure drop will be lower still, and the liquid level will drop further, possibly dropping out of the reboiler draw compartment into the reboiler liquid inlet pipe. Further, at low boilup rates the bottom tray may weep, sending liquid directly into the bottom draw compartment, and causing the liquid level to be lower still.

At a low liquid head, fluids tend to form their favorite flow patterns. Two steady states can form: one, with operation of the two reboilers simultaneously, as intended. The other has all the boiling taking place in one circuit, and the other circuit filling with liquid to the same height as the liquid in the reboiler liquid inlet pipe. Once this steady state forms, the steam that is still going in (the 8000 lb/h going to E-3A after midnight on the second day in Figure 4) will strip the lights out of the near-stagnant liquid. Before long, heavy liquid will form in this circuit, and this liquid will be difficult to boil. This will stabilize this steady state and it will persist.

The proposed theory states that the liquid level in the reboiler compartment would be very low, possibly absent, and will not approach the overflow baffle. This can be easily tested using gamma scans. The tower base was gamma scanned both on the bottom compartment side and on the reboiler compartments side. In addition, neutron backscatter scans were performed to confirm or deny the gamma scan results.

The gamma and neutron scans showed the following:

- Nothing abnormal showed up on the inner active area scans for the bottom 6 trays. Other trays were not scanned.
- Both the gamma and neutron scans showed that the liquid level in the bottom draw compartment was 10-12 feet above the tangent line. This was the same as the level indicated by the level instrument.
- Both gamma and neutron scans had difficulty finding a liquid level on the reboiler draw compartment. There definitely was no liquid level above the bottom tangent line. The gamma scans showed some absorption at the tangent line, about the same absorption as the froth on the trays, but this could have been due to the welds at the tangent line. The neutron scans showed no liquid at all in the reboiler draw compartment.

The absence of liquid level in the reboiler draw compartment gave strong support to the two steady-states theory.

Solution

With the theory verified, the challenge was now shifting from the steady state with the stagnant "A" reboilers back to that of the two reboiler circuits operating simultaneously.

In a circulating thermosiphon reboiler (as distinct from once-through thermosiphon, like the stripper reboiler), increasing the liquid head in the tower generates more liquid circulation through the reboiler circuit and reduces the fractional vaporization[4]. Considering the pressure balance and assuming a constant heat duty, the higher liquid head supplied to the reboiler is matched by the additional friction head due to the enhanced liquid circulation rate plus the additional static head due to higher amount of liquid in the reboiler shell and outlet line. Extending this reasoning to the stripper reboiler, circulating liquid and increasing the liquid head in the reboiler compartment will raise the pressure drop through the active reboiler circuit. This higher pressure drop will induce some of the liquid to follow the easier path (less pressure drop) through the inactive reboiler. Once this liquid gets heated, it will start boiling, and the thermosiphon will be established in the previously-inactive circuit.

One way of generating circulation and increasing the level in the reboiler draw compartment is by raising the level in the bottom draw compartment until it overflows the baffle, and keep the bottom draw compartment full during the circulation. The risk is that the liquid level may ascend to the trays and cause flooding and possible damage there. It is important to avoid excessive buildup of liquid up the tower.

In principle, the level transmitter can be watched during the operation to avoid excessive liquid level. This was difficult in this deethanizer stripper. Since the reboiler return pipe is only two feet above the baffle (Figure 2) and the vapor from the reboiler return blows downwards causing a frothy dispersion, the frothiness may reduce the specific gravity and lead to optimistically low level measurement. Also, the high tap of the level transmitter is just less than two feet above the baffle, which may make it difficult to see when the level rises towards the trays. Fortunately, there is a differential pressure transmitter across the stripper that comes from the upper tap of the level transmitter and therefore can monitor any accumulation of liquid above this.

The plan was to raise the liquid level in the bottom draw compartment to the level of the reboiler return, an increase in height of about 20 feet. When the level was first raised, the level was controlled using the differential pressure transmitter. Subsequently a pair of electronic pressure gauges was added to the reboiler side of the column to monitor the level directly, an additional gauge was added to the product outlet line to monitor the level on the bottom draw side of the baffle. In order to avoid damage due to entrainment of liquid through the trays the dP across the bottom trays was closely monitored.

To raise level on the reboiler side of the baffle, the bottom draw side liquid was overflowed to the reboiler side by reducing the bottom product flowrate. The level rose until it reached the top of the baffle when it 'flat lined' as it overflowed.

After one hour overflowing the baffle, level had increased on the reboiler side by 13 feet. At that time, the following two process indications provided evidence that the "A" side reboilers were starting to circulate liquid:

(1) A rapid decrease in steam consumed by 3B reboiler suggested a decrease in liquid flow to the "B" side reboilers, and

(2) A rapid decrease in 4A hot side outlet temperature suggested an increase in duty in 4A reboiler.

Steam was quickly opened to 3A to ensure we maintained "A" side reboiler circulation. Almost immediately, even split and duties were confirmed between "A" and "B" side reboilers. Bottoms outlet flow was increased to stop overflowing the baffle, and bottoms level was returned to normal.

The thirteen foot level increase was probably not enough to get the level of weathered liquid (the liquid that was sitting in the A reboiler circuit for a lengthy period) up to the reboiler return nozzle on the tower. But the fresh liquid that had entered the reboiler as level was being raised provided enough vaporization to gas-lift the weathered naphtha into the tower and start circulation.

The column remains prone to losing reboiler duty. This is believed to be because a portion of the bottom tray liquid bypasses the reboiler side of the baffle. The bottom tray is an active tray. It remains unclear whether the tray weeps enough liquid to cause the problem or whether some of the liquid directed to the reboiler draw side of the baffle falls into the product side of the bottom. To fix this in the future the bottom tray will be replaced with a total trapout tray to ensure all bottoms liquid must pass through the reboilers.

Until the plant gets to an opportunity to install the chimney tray, the reboiler continues to stall periodically and raising the liquid level has become a standard procedure that has been executed several times since by operations to regain lost reboiler circulation

Lessons Learnt

The investigation highlights the importance of good troubleshooting and problem diagnosis for overcoming bottlenecks. In our case, pressure drop calculations, close design reviews and examination of operating charts, and gamma scans, led to the correct diagnosis that enabled debottlenecking an entire gas plant with no hardware modifications.

Maldistribution between parallel reboiler circuits has received little attention in the literature, with no published case studies picked up by Kister's survey[2]. To the best of our knowledge, this is the first time this type experience is reported. This maldistribution is especially severe at low rates, when the pressure drops in the reboiler circuits are low. Due to the low pressure drop, the fluids find their favourite paths and steady states, and can easily swing from one steady state to another, promoting instability. With parallel reboiler circuits, careful analysis is important of the pressure balance both at the normal and at a reduced rates operation, and should be performed at the design. This case study demonstrates that increasing the pressure drops in these circuits can help route the fluids to the preferred patterns and counter this maldistribution.

Baffled systems have their benefits[1], but can also generate or amplify hydraulic problems, and need to be designed carefully. In the case of the deethanizer stripper, the baffle was far taller than it needed to be. Had it been even a couple of feet shorter, it would have been easy to raise the liquid level in the bottom compartment above it and achieve better operational flexibility.

It is important to provide flexibility when designing once-through thermosiphon circuits. As pointed out[1,2,4], they can be quite troublesome, and simple features that can convert them temporarily to circulating thermosiphon circuits can go a long way towards easing startups and avoiding process bottlenecks.

This case teaches a number of very important key lessons:

- With a parallel reboiler system, maldistribution between parallel circuits does occur and can cause severe bottlenecks.
- The pressure balance both at high and at low rates needs to be considered at the design.
- It is important to provide flexibility in once-through reboiler systems that will allow operating as a circulating thermosiphon system, at least at low rates.
- Finally, good troubleshooting can provide low cost (in this case, no cost) debottleneck of an entire gas plant.

References

1. Fractionation Research Inc. Design Practices Committee “Reboiler Circuits for Tray Columns”, Chemical Engineering, p.26, January 2011.
2. Kister, H. Z. , “Distillation Troubleshooting”, Wiley-Interscience, NJ, 2006.
3. Kister, H. Z., T. C. Hower Jr., P. R. de M. Freitas, and J. Nery, "Problems and Solutions in Demethanizers with Interreboilers," Proc 8th Ethylene Producers Conf, New Orleans, Louisiana, 1996.
4. Kister, H. Z., "Distillation Operation", McGraw-Hill, NY, 1990.

www.clarksolutions.com.br

Clark Solutions for AIChE's Kister Distillation Symposium 2017

On the Role of Proper Mist Elimination System for Distillation Tower Performance

Bruno B. Ferraro[1], Vitor A. Sturm[1], Michael D. Montani[1], João P. Torgal[1,2], Alexandre C. Bastida[1], Nelson P. Clark[1]

[1]*Clark Solutions, São Paulo, SP, Brazil*
[2]*University of São Paulo, São Paulo, SP, Brazil*

Abstract

Mist eliminators usually represent a low percentage of column internals overall cost, nonetheless these devices can increase tower capacity and separation efficiencies. A proper understanding of mist eliminators fundamentals, design and installation promotes better distillation operating conditions. This paper presents basic operating mechanisms of employed mist eliminators, basic design equations and installation techniques to ensure proper operation of these devices that also play a role in a fine separation process.

Keywords: Mist Elimination, Entrainment, Mist Formation

1 INTRODUCTION

Engineering will always find room for improvement as ideal solutions will never be met. Improvements can be cost-effectively achieved by means of design enhancements, technologies combination or innovations.

Mass transfer applied to distillation is recognized as a thoroughly developed field in regard to technologies and understanding – and its barriers are well known and studied. This article purpose is to advise on the understandings and limitations of one distillation process barrier – entrainment.

The studies for improving capacity or efficiency require the evaluation of basic mechanisms. In distillation, designers and operators research for internals that will produce finer gas bubbles, higher gas-liquid contact at the packing interface and thinner film interfaces, among others.

As a process consequence, entrainment is natural wherever gas and liquid phases are in close contact. As driving forces are further increased inside distillation columns – higher velocities, turbulence and contact, thinner films – to enhance production, an increased entrainment will be onset, and usually finer entrained particles will be generated.

Mist, fine liquid particles laden by a gas stream, is mainly generated by two major mechanisms: mechanical entrainment as a result of combined actions of dynamic shear stress due to forced contact between the liquid and gas phases and solid surfaces with liquid drag by the gas stream, generating droplets of micrometric diameters; and condensation due to thermodynamic changes in the system or chemical reactions in the gas stream, generating droplets of submicron diameters.

Mist eliminators in a properly designed and operated distillation plant will support operation and uptime. A proper selection and design of mist eliminators for an adequate removal of liquid carryover will increase purity and recovery lying in the fact that liquid entrained in process gas causes equilibrium disturbance on downstream stages, hindering proper mass transfer.

2 MECHANICAL ENTRAINMENT

Entrainment is the draw of liquid droplets of varying diameters alongside the gas flow. When two fluids flow together, their interface can become unstable depending on flow velocity and fluid properties such as densities, viscosities and surface tension. This instability may form waves and generate droplets in a wide range of sizes.

For absorption and stripping towers entrainment is the partial transition from countercurrent flow into concurrent flow for a liquid load fraction, partially affecting process efficiency, product composition and recovery.

Inter-tray entrainment is an often discussed subject mainly because it has a direct impact over tray efficiency. Between stages in a distillation there is a higher less-volatile components concentration at the bottom tray. The less-volatile and nonvolatile components can eventually be carried over as entrainment to the tray above and can consequently hinder mass-transfer and affect tray separation efficiency.

Several studies have been conducted over the past decades to determine which operational and design parameters are responsible for entrainment inception, quantity and effects. Colburn (1936) studied the effects of entrainment on tray efficiency. Fair (1961) correlation for crossflow sieve, valve and bubble-cap trays predicts entrainment flooding, including at froth regime for lower tray spacings. Kister and Haas (1987) correlations are good fits for collected data at spray regime for both sieve and valve trays. These studies and correlations have been developed to address not only process conditions but also design parameters such as orifice diameter, tray spacing and clear liquid height. These empirical models rely on large database and do not physically describe entrainment behavior.

After film rupture at the gas-liquid interface, a droplet is propelled at a given angle, rotation and velocity subject to drag, weight and buoyancy. The droplet vertical position will respect the following equation derived from applied position-dependent forces:

$$\boldsymbol{\rho_L V_L \frac{d^2\vec{x}}{dt^2} = \frac{\rho_G}{2}\left(\frac{d\vec{x}}{dt} - \vec{v_G}\right)^2 C_d A_{L,p} + \rho_G V_L \vec{g} - \rho_L V_L \vec{g}} \ (1)$$

To numerically solve Equation (1) one must know the initial droplet vertical velocity ($\frac{d\vec{x}_0}{dt}$) at the gas-liquid interface ($\vec{x}_0 = \vec{0}$) and properly correlate drag coefficient ($\boldsymbol{C_d}$). Gas and liquid densities ($\boldsymbol{\rho_G}$ and $\boldsymbol{\rho_L}$), gas bulk velocity ($\boldsymbol{v_G}$), droplet volume ($\boldsymbol{V_L}$) and projected area ($\boldsymbol{A_{L,p}}$) are known. The solution of this equation can show whether a droplet of given diameter ($\boldsymbol{d}$) is ascending or descending depending on its travelled distance ($\vec{x}$). In other words, this equation determines whether or not a droplet reaches the downstream equipment (or tray above). Froth or liquid jets are not described by Equation (1), which addresses only spray regime.

Drag coefficient studies were reviewed by Aggarwal and Peng (1995).

Holbrook and Baker (1934) studied the effect of slot velocity for different tray spacing in a bubble-cap column. They observed that at lower tray spacing there is a carry-over increase for higher slot velocities whereas the inverse was observed at higher tray spacing. Their results can be partially explained by Equation (1). In both trays spacing at higher slot velocity droplets are assumed to be generated at an averaged higher diameter and quantity because of higher available kinetic energy. Independent to tray spacing, initial entrainment will be the same for the same slot velocity, but accordingly to Equation (1) droplets will tend to decelerate proportionally to diameter ($\boldsymbol{d}$) as they travel upward ($\vec{x}$) until they either achieve an ascending terminal velocity ($\frac{d^2\vec{x}}{dt^2} = \vec{0}$) and are carried over or they fall back ($\frac{d\vec{x}}{dt} \cdot \vec{v_G} < 0$), reducing entrainment proportionally to droplet distribution as gas rises. A finer distribution would reach higher than a coarser distribution, but a coarser distribution could still be carried over and reach the tray above if spacing is small enough as it was observed by Holbrook and Baker (1934). Therefore, Equation (1) partially explains entrainment stability and disengagement for any defined set of already suspended droplet distribution.

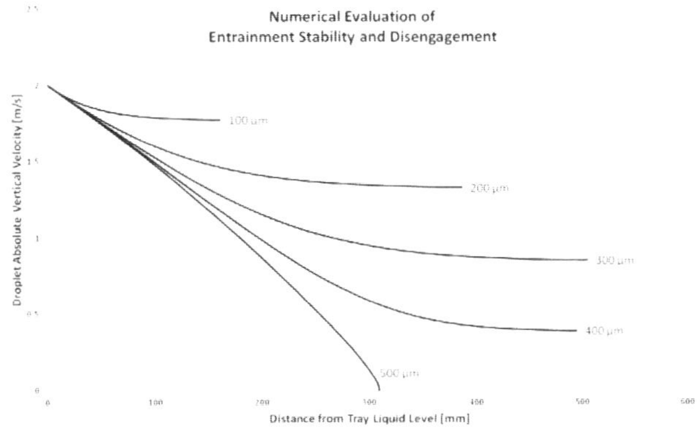

Figure 1. Numerical evaluation of entrainment stability and disengagement (air-water system at 2 m/s).

As vapor flows, gravitational, buoyancy and drag forces acting on droplets tend to equilibrium. The equilibrium point is represented by Equation (1) at $\frac{d^2\vec{x}}{dt^2} = \vec{0}$. A special case is when both $\frac{d^2\vec{x}}{dt^2}$ and $\frac{d\vec{x}}{dt}$ tend to zero.

The Souders-Brown equation (Souders and Brown, 1934) represents this position-dependent force balance special case. Souders-Brown was originally formulated to account for entrainment and capacity on design of fractionating columns. The equation was derived from the force equilibrium between drag ($\boldsymbol{F_d}$), weight ($\boldsymbol{F_w}$) and buoyancy ($\boldsymbol{F_b}$) applied on an averaged droplet diameter and drag coefficient to determine maximum allowed gas bulk velocity in a process:

$$\boldsymbol{F_d} = \boldsymbol{F_w} - \boldsymbol{F_b} \tag{2}$$

Or,

$$\frac{\rho_G}{2}(v_G)^2 C_d A_{L,p} = \rho_L V_L g - \rho_G V_L g \tag{3}$$

Souders-Brown Equation is then stated as:

$$v_G = \sqrt{\frac{4gd}{3C_d}} \cdot \sqrt{\frac{\rho_L - \rho_G}{\rho_G}} \tag{4}$$

And further simplified as:

$$v_G = K\sqrt{\frac{\rho_L - \rho_G}{\rho_G}} \tag{4}$$

The K-factor is often employed to set velocities for sizing a number of processes and internals, among then, trays and mist eliminators.

As a simplification for the de-entrainment problem, the K-factor is widely applied as a pressure dependent constant, relying in an averaged droplet diameter and drag coefficient which does not properly correlate for a wide range of processes and conditions. Notwithstanding its limitations, K-factor is a practical design parameter.

As continuation of Souders and Brown original work, the K-factors, remodeled as Flooding Capacity factors (C_{SB}), are frequently correlated for different operating conditions and design parameters. Equation (5) describes Kister and Haas (1987) correlation:

$$\boldsymbol{C_{SB} = 0.0277(d_o^2\sigma/\rho_L)^{0.125}(\rho_G/\rho_L)^{0.1}(TS/h_{ct})^{0.5}} \qquad (5)$$

While Souders-Brown correlation gives a generalized approach, Kister and Haas (1987) empirical correlation also attempt to implicitly evaluate droplet generation with orifice diameter (d_o), surface tension (σ), liquid density (ρ_L) and vapor to liquid density ratio (ρ_G/ρ_L) as well as implicitly evaluate droplet distribution with tray spacing ($\boldsymbol{TS}$) and clear liquid height at the froth-to-spray regime ($\boldsymbol{h_{ct}}$).

Equation (1) only physically describes droplet distribution variation between trays for a given set of droplets.

None of these models properly describe droplet generation initial distribution neither coalescence nor atomization.

3 CONDENSATION

In a distillation column, pressure and temperature gradients applied over a gas stream may cause condensation with respect to composition. Although condensed liquid is consequently a form of liquid entrainment with a different composition compared to the gas stream composition, the gas-condensed liquid stream overall composition that reaches the next stage tray will be the same as the gas stream composition that left the previous stage tray unless there is disengagement. The condensed liquid entrained in the gas stream is composed mainly of less volatile components and would already have been condensed and separated on the next stage tray.

This entrainment is initially made of sub-micron droplets, differently from large droplets observed from liquid carryover. These droplets tend to increase in size depending on pressure and temperature gradient, gas and liquid composition and presence of condensation surface such as liquid flow, existing droplets or vessel wall. High levels of condensation cause internal reflux.

4 MIST ELIMINATION DEVICES

Process engineers currently have a wide range of mist elimination equipment to select depending on generated mist particle size distribution.

The simplest form of current mist eliminators employed in industry are sets of crimped parallel plates which cause gas direction change along channels, capturing particles by inertial impaction at every deflection. Chevron or vane mist eliminators parallel plates are separated by distances that typically vary from 10 to 50 mm with 3 to 6 passes.

Chevron mist eliminators bring lower efficiency than other mist eliminators, but also lower pressure drop and higher liquid draining capacity. Chevrons are usually the best option for high fowling operation as blades support higher drainage and allow simple washing when plugged.

Figure 2. MaxiChevron® mist eliminator

Clark Solutions MaxiChevron® Double-Pocket™ is similar to a conventional chevron. However, a set of pair of blades with draining slots replaces conventional blades, promoting draining. Therefore, this equipment is able to operate at even higher vapor capacities than those achieved by conventional chevron in the same service at the expense of higher pressure drop and lower fowling capacity.

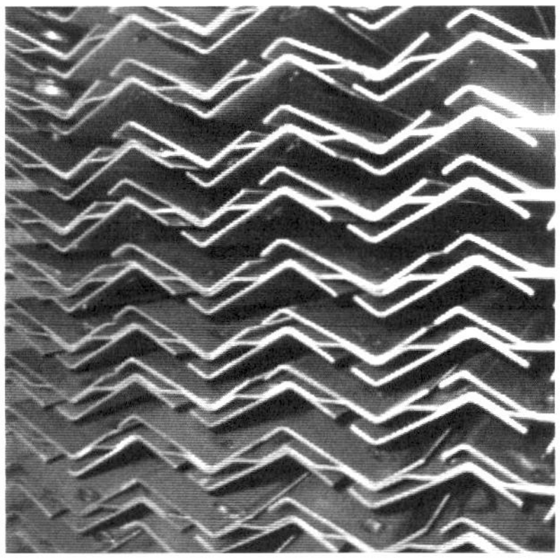

Figure 3. Clark Solutions MaxiChevron® Double-Pocket™

For higher efficiency, fine metal wire provides much smaller and difficult to avoid obstacles than large metal sheet plates. Wire mesh pads are sets of knitted wire mesh, arranged in particular geometries to capture small particles.

Figure 4. MaxiMesh® mist eliminator

Originally developed in the early 1950's with few different mesh densities, wire mesh pads have developed substantially in the last 70 years. Today, they can be offered in a wide variety of geometries, wire sizes and knitting with or without polymers which will allow capture of finer particles in the micrometric range.

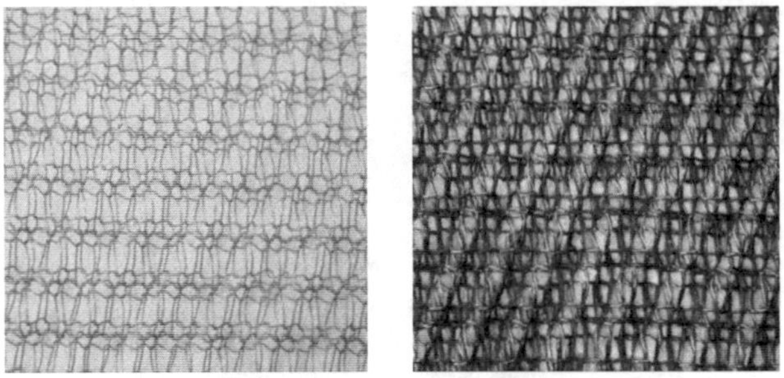

Figure 5. Knitted and Co-knitted mesh

Plain metal wire meshes are designed to achieve capture of particles with few micrometers. More efficient designs will generally require a co-knit mesh, but will also increase pressure drop. Special applications may require special alloys or plastics for the wire mesh structure.

There are other types of mist eliminators such as candle filters and swirl flow cyclones. However, chevron and wire mesh mist eliminators represent most applications in distillation. This paper discusses wire mesh mist eliminators because of its broadest use.

5 DE-ENTRAINMENT THEORY

De-entrainment is the disengagement of suspended droplets within gas flow. Equation (1) describes de-entrainment as an inflexion of the droplet vertical velocity vector in respect to gas flow ($\frac{d\vec{x}}{dt} \cdot \overrightarrow{v_G} < 0$). But de-entrainment can be further increased by mist elimination collection mechanisms.

The drawback of mechanically achieving de-entrainment is pressure drop. Studies have been conducted to describe and correlate collection mechanism efficiency, overall efficiency and pressure drop from dry to flood operations.

Different devices exist in order to separate the entrained liquid from the gas phase, making use of three major mechanisms to separate and coalesce droplets.

Inertial impaction occurs when a gas stream carrying liquid particles passes through an obstacle. The gas will avoid the obstacle, but particles with sufficient inertia will be projected from the gas flow and impact on the obstacle. Inertial Impaction will be as pronounced as the gas velocity raises, as liquid to gas density ratio increases and as target gets smaller (*i.e.* as steep is the gas trajectory change).

Direct interception occurs when particle diameter is small enough to follow gas streamlines, but big enough to collide on obstacles' edges. This effect will increase as the target gets smaller and the mist eliminator void fraction is reduced.

Brownian diffusion occurs with extremely small particles which acquire a Brownian movement caused by energy transfer on collision with randomly moving gas molecules. This movement will drive these particles to eventually collide with the mist eliminator targets. Brownian diffusion capture will be favorably affected by the reduction in target diameter and bed void fraction. Lower gas velocities will increase particle residence time within the mist eliminator bed, resulting in a higher probability collection, increasing the overall efficiency of the mist eliminator designed for this latter capture mechanism

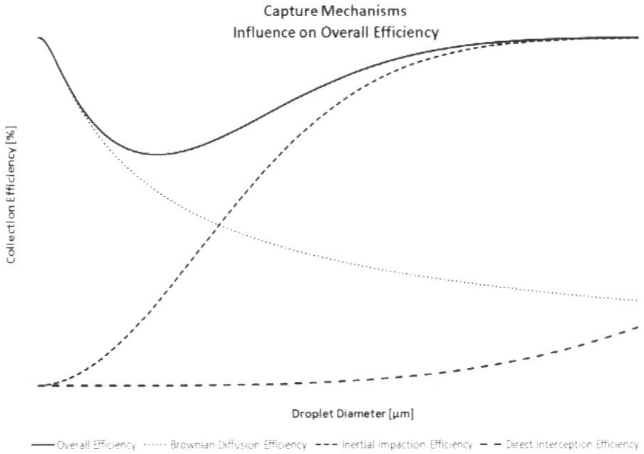

Figure 6. Collection efficiency for different capture mechanisms

(Co-)Knitted wire mesh pads mist eliminator devices collects droplets primarily via inertial impaction, thus, the droplet size range collected is specific. Since entrainment raised from mechanically dragged droplets in distillation towers contains size distributions fit for this equipment efficiency, wire mesh pads are the first de-entrainment devices to be considered – unless fowling is a primary concern.

The efficiency of a filtering media can be expressed as the fraction of the particle distribution that is collected. One way to account for efficiency is to sum contributions of individual wires or fibers, assuming that they are equal in size and provide the same individual collection efficiency for the uniformly packed bed. This can be expressed as:

$$\boldsymbol{E = 1 - exp\left[-f\sum \eta\right]} \qquad (6)$$

The above equation is derived originally from the Kleinschmidt equation (Kleinschmidt, 1939) designated to calculate particle collection efficiency of Pease-Anthony cyclonic scrubbers, and was later used for particle collection on fiber filters, where the exponential term is also referred as Penetration (Pan and Zhong, 2006). The equation was formulated to account for single filament efficiency and generalized for the equipment extension.

This equation (6) can be used for wire mesh mist eliminators, with f representing the effective contact area expressed as $\frac{-A_e LF}{\pi}$. Mesh pad thickness (L), specific surface area (A_e) and fraction of wires in an adequate collection position (F) are usually known.

The efficiency sum in equation (6) can be stated as the contribution of inertial impaction (η_{II}), direct interception (η_{DI}) and Brownian diffusion (η_{BD}) minus the interfering efficiency of interacting mechanisms (η_{IM}):

$$\boldsymbol{\sum \eta = \eta_{II} + \eta_{DI} + \eta_{BD} + \eta_{IM}} \qquad (7)$$

For wire mesh pads, the most substantial contribution for equation (7) is from inertial impaction.

This approach, although a good method, may sometimes be misleading for some mesh-type mist eliminator devices since specific geometric parameters (*e.g.* layering of the knitted meshes and crimps) are not taken in consideration, oversimplifying the mesh pads into a bed of wires.

Therefore efficiency models must be evaluated before applying on wire mesh mist eliminators, since some equation premises are not always met. Mead-Hunter et al. (2014) reviewed several efficiency models from literature. One widely used single target efficiency model was developed by Langmuir and Blodgett (1946).

More specific equations were developed for wire mesh mist eliminators overall equipment efficiency, such as the semi-empirical model of Carpenter and Othmer (1955) and the mechanistic models of Brunazzi and Paglianti (1998 and 2000). These models rely on single target efficiency models such as Langmuir and Blodgett (1946) to predict overall equipment efficiency along all wire mesh layers. Brownian diffusion and direct interception can be neglected for the evaluation of wire mesh mist eliminators efficiency as shown by Holmes and Chen (1984).

Both Carpenter and Othmer (1955) semi-empirical model and Brunazzi and Paglianti (1998) assume no re-entrainment and no liquid hold-up. While Carpenter and Othmer (1955) consider droplets remixing along wire mesh layers, Brunazzi and Paglianti (1998) consider no mixing in their multi-cell efficiency model.

Operating outside creeping flow droplets may not experience all cells as assumed by Brunazzi and Paglianti (1998). Droplets may be disturbed and eventually follow interstitial paths between cells, which can explain Brunazzi and Paglianti (1998) overestimated efficiency.

Most correlations for single fiber efficiency through different mechanisms assumes specific flow fields that exist when diameter-based Reynolds numbers are very low, such as Stokes or Kuwabara Flow Fields (Kuwabara, 1959). For typical vapor velocities the flow field around mesh wires may differ significantly from creeping flow. For these reasons, many wire mesh applications may be misrepresented.

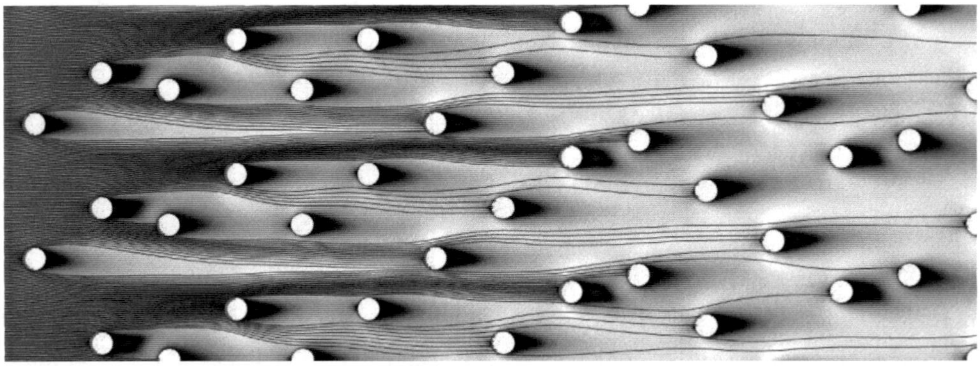

Figure 7. Computational Fluid Dynamics for wire arrays (approximation velocity = 2 m/s, gas density = 1.2 kg/m³, particle density = 1.000 kg/m³, wire diameters = 0.3 mm, particle diameter = 10 µm, void fraction =0.96) – layers are not always aligned, but wire spacing inside layers are the same

When designing wire mesh mist eliminators, efficiency and pressure drop balance is sought. Various factors influence equipment pressure drop (*e.g.* operational conditions, mesh thickness, mesh geometry, wire diameters and void fraction). Generally, higher efficiencies bring higher pressure drop. Co-knitting allows for a higher increase in efficiency to pressure drop ratio. Multi-zones allow increase in capacity as lower zones can be designed with lower density mesh and upper zones with co-knitted higher density mesh.

Figure 8. Clark Solutions MaxiMesh® with dual zones (plain metal bottom drainage zone and fiberglass co-knitted high efficiency zone)

Efficiency and pressure drop are often correlated under specific conditions, processes and equipment styles. Many of these correlations are proprietary models.

El-Dessouky et al. (1999) presents a set of correlations for capacity, efficiency and pressure drop built from a series of experiments in typical mesh pads specifically used in multi-stage flash desalination units. Al-Dughaither (2010) studies the behavior of mesh pads on low velocity bubble columns.

6 PRACTICAL DESIGN

Although technology develops and more products and tools become available, many designs still follow previous standards. Souders-Brown equation and empirically determined K-factors are a dominant design technique.

On recurrent cases, and when pressures are near atmospheric, simple standard equipment such as plain metal wire mesh pads and single pocket chevrons can be calculated using Souders-Brown equation with known K-factor values.

As an example, a Maximesh® Style 431 wire mesh pad for an air and water system with an actual gas flow of 10,000 m^3/h, operating at 1 bar and 25°C, can have its ideal area calculated following Souders-Brown equation as:

$$K_{431,1\,bar} = 0.106[m/s]$$

$$V \cong 0.106[m/s]\sqrt{\frac{997[kg/m^3] - 1.2[kg/m^3]}{1.2[kg/m^3]}} \cong 3.0[m/s]$$

$$A \cong \frac{10{,}000[m^3/s]}{3.0[m/s]} \cong 0.9[m^2]$$

Clearly, K-factors should be corrected for high pressures and more precise correlations could retrieve a more compact equipment area. Pressure changes, substantial gas or liquid density variations, excessive liquid loads and other factors can lead to a different choice of mist eliminator design and area.

This ideally calculated area may be modified through more detailed calculations to achieve desired droplet diameter range removal efficiency. Nonetheless, this is a valuable widely used method for estimating mist eliminator sizes.

7 INSTALLATION

Proper equipment installation is a major requirement for mist eliminator system performance.

Wire mesh and chevron mist eliminators are designed to fit normal sized manholes of 600 mm (24"). Also, sections should not weight more than 20-25 kg to allow for a single man to hold and manipulate the individual sections. As a general rule, sections width will be around 350-500 mm.

Depending on tower diameter the number of sections will change. They can be full length, half length or even cut in thirds, fourths or sixties, depending on sections weight and tower diameter, similarly to structured packing and trays.

Mist eliminator arrangement and flow direction is chosen depending on calculate area, space availability and geometry as well as operating conditions.

Inside distillation, absorption or stripping towers, mist eliminators installation are mainly horizontal as gas vertically flows.

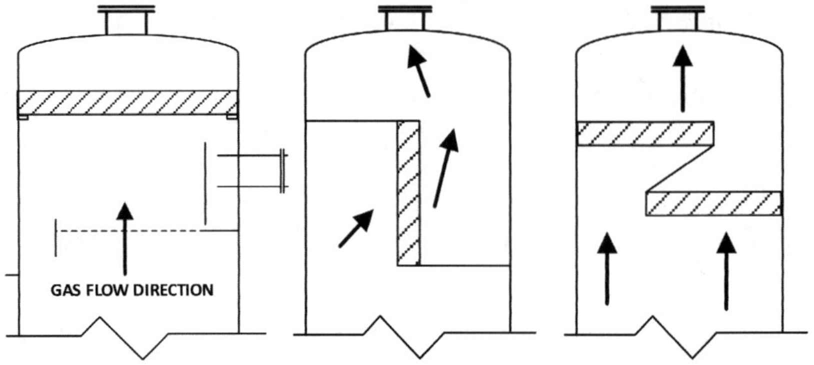

Figure 9. Mist eliminators possible arrangements

However, other arrangements are possible, such as vertical installation with horizontal flow and Z-shapes, depending on design requirements. Computational Fluid Dynamics can assist on design evaluation and improvement of mist eliminators geometry and installation, assessing mist eliminators effective area in respect to flow velocity fields.

For applications where operating pressures are close to atmospheric, some rules of thumb can be applied regarding mist eliminator location.

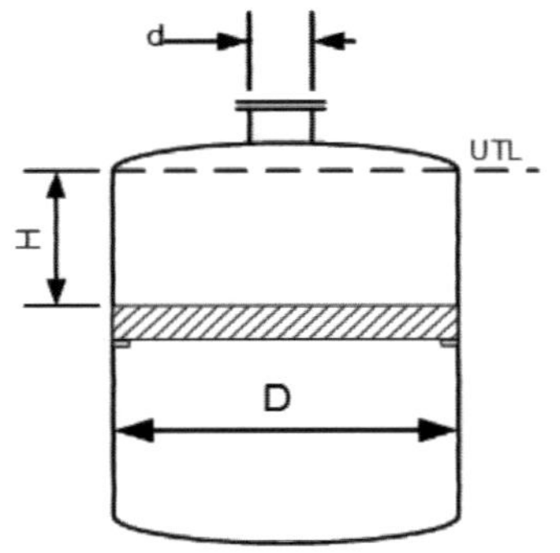

$$H \geq \frac{D-d}{2}$$

$$H \geq \frac{D+d}{2}$$

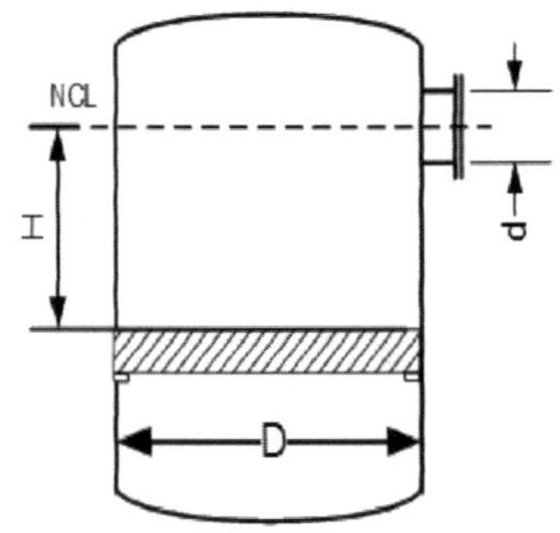

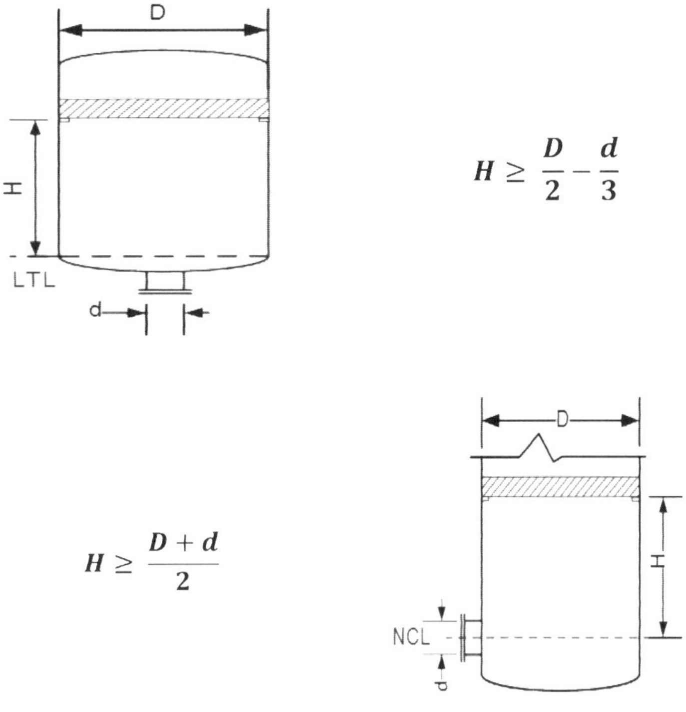

$$H \geq \frac{D}{2} - \frac{d}{3}$$

$$H \geq \frac{D + d}{2}$$

Figure 10. Mist eliminators positioning practical standards

Simulations can be made to evaluate and achieve more compact vessels. Proper installation is sought to enhance gas distribution across mist eliminator surface as uneven flow distribution will result in uneven velocity distribution and compromise performance.

There is a wide range of fastening options to install mesh pads. Tie wires fasten mist eliminator sections to a support ring welded or attached to the vessel wall. The ring contains holes that are usually spaced by 200-250 mm, depending on mist eliminators sections width.

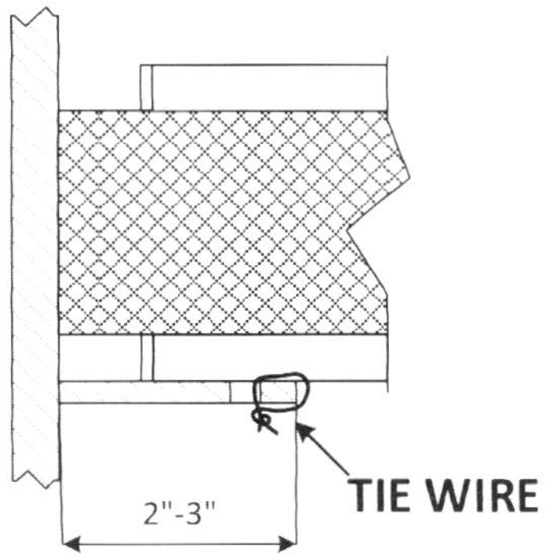

Figure 11. Tie wire fastening

Tie wires can be applied provided a non-corrosive and low pressure service. High pressure systems may submit tie wires to stresses that eventually result one or more sections displacement.

On corrosive or more demanding applications, J-Bolts are the preferred choice. J-Bolts offer a more resistant and stronger fastening between the mist eliminator and the support ring.

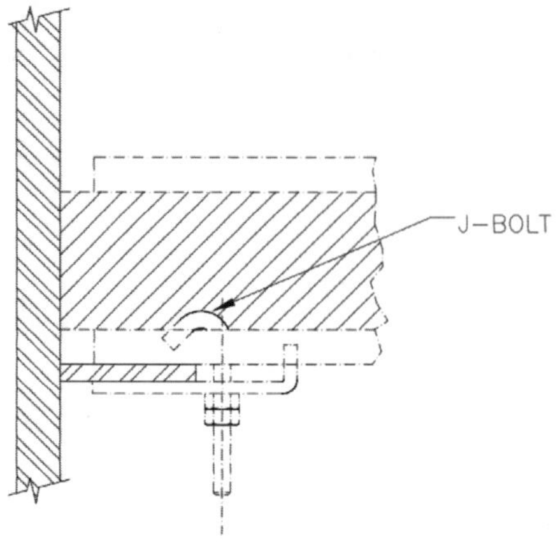

Figure 12. J-bolt fastening

If bottom access is difficult, a third option is the L-Bolt. Similarly to a key, once the section is in place, turning the bolt to the proper position will firmly fasten the section.

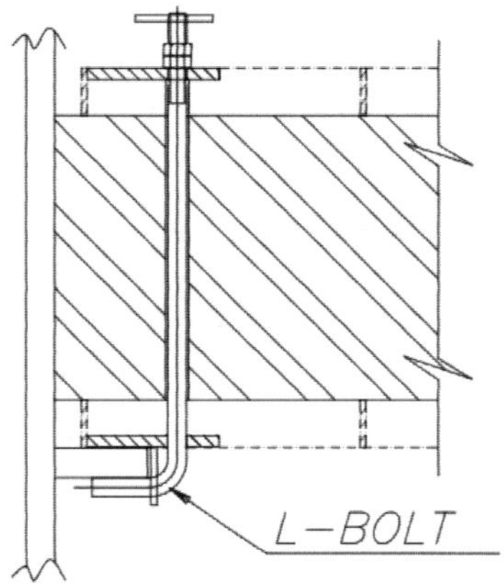

Figure 13. L-bolt fastening

There are other ways to install mist eliminators (*e.g.* between rings with removable sections or using upper support bars). Selection is usually made based on operating conditions and maintenance comfort.

8 APPLICATIONS

Typically the rectifying section is above the feed tray, where top vapor is corrected to achieve desired composition. On this section both foaming and entrainment are less significant due to lower vapor density and velocity. As such, tray spacing can be reduced at the rectifying section with lower compromise of system capacity.

The stripping section, on the other hand, experiences higher vapor density and velocity and as a consequence an increased foaming and entrainment. Solids, when present, are also mainly observed on the stripping section. Therefore, higher tray spacing may be used to guarantee system capacity and avoid mixing between trays. An alternative solution is to provide entrainment removal between trays. Some applications, such as a whiskey still, utilizes an entrainment removal section only between the stripping and rectifying sections, consisting solely of a plate.

But for some operations the feed tray is above the rectifying section, such as the wine feed on sugarcane alcohol distillation. Therefore both liquid and solid entrainment must be removed from the phlegm lateral removal below feed. In some cases a wave-plate mist eliminator is positioned between the distillation column and the rectification column in order to prevent entrainment to reach the rectification column feed.

There are plenty of other well proven typical mist eliminators applications such as wire mesh pads installed at the gas exit of overhead condensers to protect downstream compressors and increase condenser's throughput; chevron or fast draining wire mesh pads such as Maximesh® Style 353 installed on vacuum tower's steam stripper section; and very high efficiency, co-knitted wire mesh pads installed on glycol, amine and solvent drying and absorption towers to reduce carryover losses.

As a practical example one can consider a Jet-A side stream stripper producing 15,000 BPD of Jet-A. Assuming a 0.5% carryover from the stripper the amount of Jet-A which was carried to the Naphtha section would be 75 BPD of Jet-A, roughly $ 125,000 in losses per year considering a $ 5.00 price difference per barrel between Jet-A and Naphtha. A wire mesh pad of 5 square meters for this service will cost around $ 5,000. Payback will therefore be less than 30 days.

This same approach can be applied for other applications. While the economics may not be as favorable, payback is typically lower than 3 to 4 months.

9 CONCLUSION

The growth in mist elimination science reflects the perceived interest in attainable product recovery and purity with these low investment equipment. Proper selection and design of mist eliminators are a cost-effective approach to increase operational revenue and may support capacity increase.

Although entrainment is known as a distillation barrier for nearly a century, mist elimination is still a developing science. New studies and modern computational tools offer new perspectives in the understanding of mist eliminators as new products and developments are established.

10 REFERENCES

Aggarwal, S. K. and Peng, F. 1995. A Review of Droplet Dynamics and Vaporization Modeling for Engineering Calculations. Journal of Engineering for Gas Turbines and Power 117: 453-461.

Al-Dughaither, A. S.; Ibrahim, A. A. and Al-Masry W. A. 2010. Investigating droplet separation efficiency in wire-mesh mist eliminators in bubble column. Journal of Saudi Chemical Society 14(4): 331-339.

Al-Sheikh, J. N. 1959. A Study of Entrainment in Distillation Column for Sieve Plate. Master Thesis, Rice University, Houston. 59 p.

Bujake, J. E. 2007. Beverage Spirits, Distilled. Kirk-Othmer Encyclopedia of Chemical Technology.

Brunazzi, E. and Paglianti, A. 1998. Design of wire mesh mist eliminators. American Institute of Chemical Engineers Journal 44(3): 505.

Brunazzi, E. and Paglianti, A. 2000. Design of complex wire-mesh mist eliminators. American Institute of Chemical Engineers Journal 46(6): 1131-1137.

Carpenter, C. L.R. and Othmer, D. F. 1955. Entrainment removal by a wire-mesh separator. American Institute of Chemical Engineers Journal 1(4): 549-557.

Colburn, A. P. 1936. Effect of Entrainment on Plate Efficiency in Distillation. Industrial and Engineering Chemistry. 28(5): 526-530.

Cristini, V. and Renardy, Y. 2006. Scalings for Droplet Sizes in Shear-Driven Breakup: Non-Microfluidic Ways to Monodisperse Emulsions. FDMP 2(2): 77-93.

El-Dessouky, H. T.; Alatiqi, I. M.; Ettouney. H. M. and Al-Deffeeri, N. S. 2000. Performance of wire mesh mist eliminator. Chemical Engineering and Processing: Process Intensification 39(2): 129-139.

Fair, J. R. 1961. How to predict sieve tray entrainment and flooding. Petro/Chem. Eng. 33(10): 45-52.

Holbrook, G. E. and Baker, E. M. 1934. Entrainment in a Bubble-Cap Distillation Column. Industrial and Engineering Chemistry 26(10): 1063-1067.

Holmes, T. L. and Chen, G. K. 1984. Design and selection of spray/mist elimination equipment. Chemical Engineering 91(21): 82-89.

Ishii, M. and Grolmes, M. A. 1975. Inception Criteria for Droplet Entrainment Two-Phase Concurrent Film Flow. AIChE Journal 21(2): 308-318.

Kister, H. Z. and Haas, J. R. 1987. Sieve tray entrainment prediction in the spray regime. Institution of Chemical Engineers Symposium Series No. 104.

Kister, H. Z. and Haas, J. R. 1990. Predict Entrainment Flooding on Sieve and Valve Trays. Chemical Engineering Progress 86(9): 63-69.

Kleinschmidt, R. V. 1939. Factors in spray scrubber design. Chem. Met. Eng. 46: 487.

Kuwabara, S. 1959. The forces experienced by randomly distributed parallel circular cylinders or spheres in a viscous flow at small Reynolds numbers. Journal of the Physical Society of Japan 14(4): 527-532.

Langmuir, I. and Blodgett, K. 1946. A mathematical investigation of water droplets trajectories. Collected works of Irving Langmuir. Pergamon Press 10: 348-393.

Pan, N. and Zhong, W. 2006. Fluid transport phenomena in fibrous materials. Textile Progress 38(2): 1-93.

Pyott, W. T.; Jackson, C. A. and Huntington, R. L. 1935. Factors Affecting Entrainment in Bubble-Cap Columns. Industrial and Engineering Chemistry 27(7): 821-825.

Souders Jr., M. and Brown, G. G. 1934. Design of Fractionating Columns: I. Entrainment and Capacity. Industrial & Engineering Chemistry 26(1): 98-103.

Ujile, A. A. and Iminabo, J. 2014. Evaluating Sieve Tray Flooding in a Distillation Column Using Kister and Haas; and Fair's Correlations. Chemical and Progress Engineering Research 25: 16-23.

Troubleshooting Tower Carryover

By

Ed Grave, ExxonMobil Upstream Research Company, Spring, TX, USA
Norman Yeh, ExxonMobil Upstream Research Company, Spring, TX, USA
Juan C. Juarez, ExxonMobil Upstream Research Company, Spring, TX, USA

Presented at the Kister Distillation Symposium, AIChE Spring Meeting, San Antonio, Texas, March, 2017

Introduction

For most traditional TEG contactors, high capacity structured packing is the ideal mass transfer device due to the low liquid rate in this service. However, structured packing should only be used in a fairly clean system as it is susceptible to plugging. For dirty systems, bubble cap trays have been traditionally applied. A capacity gain of 200-300% over bubble cap trayed towers is possible without any loss of mass transfer efficiency (1). Other high capacity trays such as, Shell Swirl Tube Trays™, that incorporate integral cyclones can achieve a capacity boost of 400% over bubble cap trays albeit with higher glycol carryover (1). There are a few high capacity tray technologies available that can be used to achieve a capacity boost over bubble cap trays albeit with higher glycol carryover. These new high capacity mass transfer devices exceed the capacity of most existing de-entrainment equipment, such as wire mesh and vane packs. The capacity of demisting devices is further reduced with increased entrained liquid loading as the high capacity devices push to higher superficial velocities. Thus, TEG carryover can ultimately dictate the capacity of the TEG contactor if the de-entrainment internals are not specified correctly or not modified during a revamp to handle the additional TEG carryover.

High concentrations of H2S and CO2 and even small amounts of contaminants such as heavy hydrocarbon liquids, iron sulfides, salts and other organics in the feed gas can also present operational difficulties. These contaminants can be absorbed in the TEG, accumulate, foul, foam and react to form other byproducts (2). Therefore, the cleanliness of the system must be considered when selecting internals.

This article summarizes the lessons learned in debottlenecking a “dirty” TEG contactor at our LaBarge Black Canyon Facility in Wyoming US. This work was performed in progressive steps between 2001 and 2004.

Facilities Overview

The TEG Facility at the Black Canyon Facility was originally designed to process 480 MMSCFD in two 240 MMSCFD Trains (Process 1 and Process 2). It is a part of the overall LaBarge Operation that consists of a gathering system, dehydration, gas purification, and sales. The TEG System at Black Canyon Facility has three main functions:

- Prevent corrosion in the 28 inch carbon steel Feed Gas Trunkline feeding Shute Creek Gas Facility 38 miles away
- Prevent hydrate formation upstream of the Shute Creek Facility
- Prevent well field contaminants from entering the Shute Creek Facility H_2S Removal Process

The facility is unique as it dries gas containing 65% CO_2 and 5% H_2S at over 1300 psig with a very low TEG circulation rate (about 0.6 gal TEG / lb H2O removed). The gas density in the contactor is over 10 lb/ft^3 (2).

The LaBarge wells produce no liquid hydrocarbons to carry contaminants out of the gas stream. The contaminants are heavy sulfur and organic compounds and elemental sulfur which build-up over time. Dibenzothiophene (DBT) is a major contaminant in the feed gas to the process facility. Fouling is quite severe over the entire TEG system and accelerates with increasing gas processing rates.

The accumulation of impurities also corresponds to increased TEG losses that contaminate the solvent systems at the Shute Creek Facility. The result is increased operating costs at both locations.

Annual turnarounds were required to clean the system and recharge the TEG. The shutdowns were usually staggered between the two process trains.

Improvements to the inlet filtration and separation facilities to minimize fouling in the TEG system have been studied, but no economical options have been identified to date.

Original TEG Contactor and Outlet Scrubber Design

The towers were originally equipped with 12 single pass bubble cap trays. The trays were designed with a downcomer inlet sump in order to provide static liquid downcomer sealing and extend operating range. This design element was commonly done in low liquid rate applications. However, it should never be applied in "dirty" services as these pots can be filled with debris and potentially block liquid flow exiting the downcomers.

The feed gas was introduced into the tower via a flush nozzle. Hence, there was no vapor distributor. The bottom of the nozzle was only 4 inches above the HLL, which likely contributed to liquid re-entrainment from the bottom liquid surface to the bottom tray. The small clearance not only reduced capacity, it perpetuated foaming and fouling on the tray. Fouling was very severe as can be seen in Figures 1 and 2 after only one year of operation.

Figure 1 - Bottom bubble cap tray after one year in service

Figure 2 - Bottom tray outlet picket fence weir

At the top, the tower was equipped with a vertically oriented (horizontal vapor flow) vane pack mist eliminator with two windows. It was a simple vane pack with 5/8" (16mm) blade spacing with no pockets. The vanes drained via a downpipe behind the inlet weir of the first tray. The mist eliminator was undersized with a k-factor of 0.8 ft/second at 240MMSCFD. The gas outlet scrubber was also equipped with a similar vane pack. At the time, this scrubber was found to be sized also too small and some of the gas was being bypassed due to excessive pressure drop.

As illustrated in Figure 3, some fouling can be seen in the vane pack, but the fouling was less severe than in the bottom of the tower.

Although each tower was designed to process 240 MMSCFD, it could actually process around 280 MMSCFD when the tower was clean with TEG losses at around 0.6 gal/MMSCF. Capacity would drop by 50% within 9-12 months with excessive pressure drop due to extreme fouling and foaming on the trays. The fouling was restricted mostly to the bottom of the vessel and the bottom three (3) trays and decreased moving up the column. Foaming was clearly visible in the level bridles, causing low level control trips and upsets throughout the system. Gamma scans also verified foaming in the bottom section of the tower. Fouling also restricted the TEG circulation rate.

Figure 3 - Gas outlet vane pack mist eliminator at the top of the tower

Debottlenecking to 750 MMSCFD (375 MMSCFD each process train)

Future operations required the total throughput at the Wyoming facilities to increase to 375 MMSCFD per process train. Also, annually planned shut downs were to be eliminated. Reduction in TEG carryover was also desired as the TEG would have to be recovered downstream at the Shute Creek Facility before entering the gas treating units. The recovered TEG is processed and trucked back to the Black Canyon Facility.

The following modifications were performed in Process Train 1, August 2001

- Replaced all bubble caps with Provalve®. Provalve® are fixed valves (no movable caps). The orifice opening is similar to a moving valve, but the fixed cap pushes the vapor out more horizontally, reducing the accumulation of fouling material on the tray deck. Structured packing was not considered due to the heavy fouling service.
- Removed all downcomer inlet sumps.
- Removed bottom seal pan and extended the downcomer from the bottom tray to the bottom liquid level to avoid trapping fouling material in the seal pan.
- Installed Koch-Glitsch Inlet Vane Diffuser at the gas inlet nozzle. See Figure 5.
- Note – no modification work was performed on the vane pack mist eliminator at the top of the tower.

Figure 3- New Provalve®

Figure 2- Koch-Glitsch Inlet Vane Diffuser

The tower operated for an entire year at 375 MMSCFD with lower pressure drop and equivalent water removal and circulation rate. Liquid level at the bottom of the tower was fairly steady. The tower was pushed to 400 MMSCFD for about 6 hours and plant stability was maintained with no visible hydraulic limitations. However, the TEG carryover was high at 5.7 gal/MMSCF at 375 MMSCFD. Due to the very low circulation rate and high gas velocities, the trays were operating in the spray regime and therefore the existing vane pack was being overwhelmed with liquid and small droplets due to some foaming. The k-factor through the vane mist eliminator was at 1.17 ft/second.

The tower was opened in the summer of 2002 and as can be seen in the picture in Figure 6, the amount of fouling on the Provalve® was considerably less than with bubble caps, even though it was operating at 34% higher gas rates. Fouling material was present throughout the trays, but most of the vapor passages through the valves were not blocked.

The bottom of the tower (sump) was now "dirtier" with more solids. The inlet vane diffuser also collected some solid material with most of the solids observed at the end of the distributor with about one quarter to one third of the vanes blocked.

Cleaning the Provalve® was much easier because this task could be done in situ; instead of removing all of the bubble caps because of heavy fouling on the inside of the caps.

During this turnaround in the summer of 2002, the same modification was done in the TEG contactor in Process 2. There was one exception; the inlet vane diffuser had fewer vanes, in other words, larger vane spacing. This change was made to reduce the amount of fouling material collected in the vane and facilitate cleaning.

Figure 4 - Bottom tray after one year of operation with Provalve®

The Problem – Liquid Carryover

The TEG carryover into the Shute Creek facility was higher than expected and added considerable operating cost to its recovery. The high entrainment rate was attributed to two main sources:

- The "dirty" system contaminated the TEG causing higher degree of foaming and small droplets were now being entrained in greater amounts due to the high superficial vapor velocity at 375MMSCFD.
- The mist eliminators vane packs were too small for high gas rates and were not able to remove the entrained liquid. The downstream knock-out drum was not collecting any liquids and was being entirely bypassed.

During the turnaround in 2002, a Calgon Carbon bed filtering system was installed to treat the entire TEG circulation rate in Process 1. This allowed a direct comparison between a "dirty" system prior to the turnaround and a comparison with Process 2, which had no carbon filtration. The carbon filtration had a significant impact on the entire operation of the TEG system. Better stability, less to no foaming, and higher circulation rate. The entrainment rate dropped from 5.7 gal/MMSCF to 1.8 gal/MMSCF. For further reading regarding the implementation and testing of the carbon filtration system see reference #2.

Table 1 - Process 1 Performance Test TEG Entrainment Summary with Carbon Filtration

	Gas Rate	TEG Circ.	TEG Losses	
	MMSCFD	gpm	Gal/MMSCF	% of circ. TEG
Dirty Tests	319.4	10	1.0	2%
	369.4	12	5.7	12%
Clean Tests	319.7	12	0.8	1.5%
	375.1	12	1.8	3.9%
	374.9	18	2.5	3.6%

In the meantime, Process 2 was already seeing increased carryover and overall system instability only 2 months after the shutdown. Foaming could be seen in the level bridles. To be fair, Process 2 has always performed worse than Process 1 as the gas is fed first to Process 1 off the manifold. This gas is cleaner as most of the solids will continue down the line to Process 2.

Although adding the carbon filtration on the TEG was a huge improvement, the carryover was still unacceptable especially when the TEG circulation rate was increased to 18 gpm. The carryover increased further as the flow rates were pushed beyond 375 MMSCFD to 400-425 MMSCFD.

It was decided to replace the existing top mist eliminator vane pack with demisting cyclones at the next opportunity in Process 1 and Process 2 in May and October of 2003.

Troubleshooting - First Fix Attempt - Demisting Cyclones Installation

ExxonMobil had successful experience with demisting cyclones at high pressure in a hydrocarbon service. However, this application would be a first time application of demisting cyclones in a TEG service. The high CO_2 concentration in the feed gas also made it a special application because the conditions were outside the range of available vendor data. Nonetheless, TEG losses were expected to be less than 1.0 gal/MMSCF after installation.

After discussions with the vendor CDS (now FMC Technologies), the number of cyclones and mechanical implementation of the cyclones were agreed as shown in Figure 7.

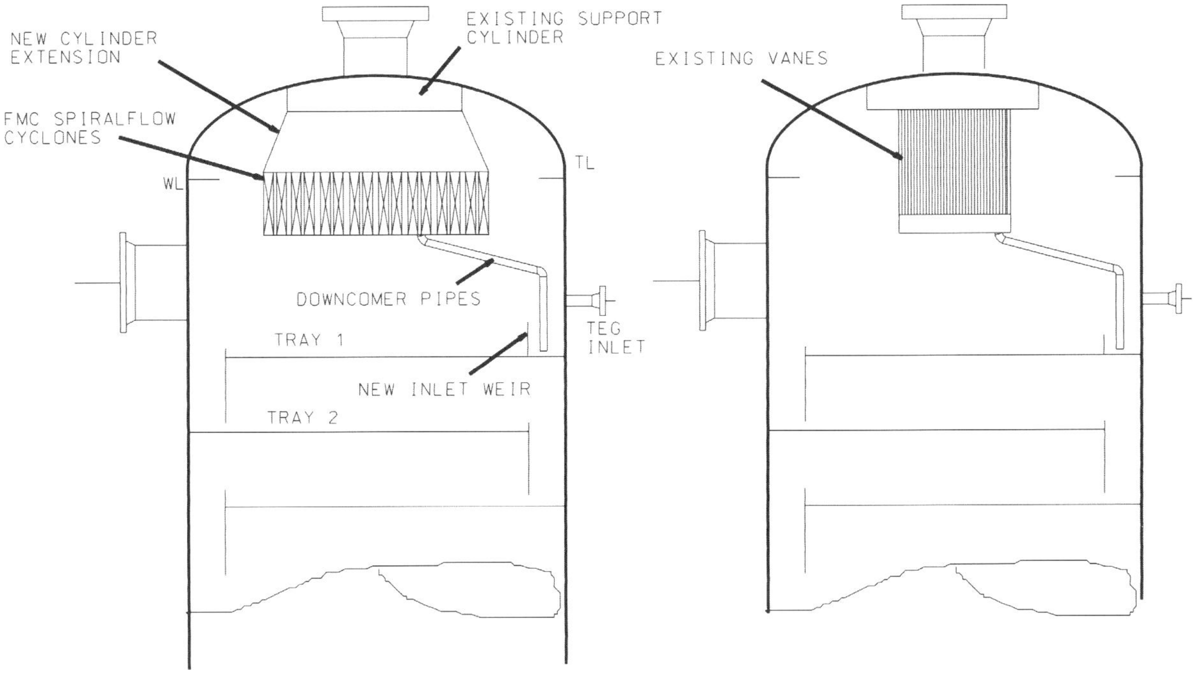

Figure 5 - TEG Contactor Mist Eliminator Modifications

The demisting cyclone deck was able to be installed without any welding to the vessel shell by modifying the existing vane pack internal support structure. The inlet weir on the top deck was also modified to provide sufficient head to seal the cyclone drainage pipes.

Upon start-up, the performance of the demisting cyclones was similar to the vane packs at low gas throughputs of 240-280 MMSCFD. However, as flow rates increased, performance deteriorated for both contactors to an unacceptable level.

As can be seen in Figure 8, Process 1 TEG losses climbed rapidly from 0.6 to 10.6 gal/MMSCF between 290 MMSCFD and 328 MMSCFD, only a 13% increase in flow. The carryover stayed relatively flat with increased gas flow rate. In the meantime, the TEG carryover in Process 2 was less than in Process 1 and followed a more familiar parabolic function typically associated with re-entrainment. Historically, Process 1 always performed slightly better than Process 2, as the latter receives more solids. So, two different phenomena were possibly being faced.

In addition, the observed pressure drop for both units was about 15-20% higher than calculated, with Process 2 operating with a slightly lower pressure drop.

Troubleshooting - Development of hypothesis

A number of theories were developed:

1. The new trays with increased flow rate were operating in the “blowing regime”. Essentially all liquid is being entrained, leaving the tray decks dry. Provisions were made to minimize this event from occurring by blocking the outlet weir with pickets and providing sufficient hole area. This operating regime is also more likely to occur at lower circulation rate. For all the performance tests, the circulation rate was increased to 18 gpm. In terms of mass transfer performance, the tower was performing as expected with acceptable gas dew point. Therefore, this scenario was not very likely.

2. The trays were operating in a spray regime such that a significant amount of liquid was being entrained and overwhelming the cyclones. High entrained liquid rate could explain the high pressure drop.
3. Foaming on the trays will have similar consequences as point two above. These towers have a history of foaming, but after the carbon bed filtration was added, the TEG had a smaller tendency to foam. In addition, cyclones should be able to break foam much better than the vane packs.
4. High viscosity liquid and high vapor density were causing film instability in the cyclone resulting in surface re-entrainment.
5. Insufficient drainage head. Due to the unexpected high liquid load to the cyclones the required drainage head was too high and the cyclones were unable to drain properly.
6. Syphoning. The high pressure drop was causing some liquid to be siphoned from the tray below.
7. Downpipes for drainage were not sealed properly and vapor was bypassing. This can occur with leakage on the tray deck or around the inlet weir.

Looking at Figure 8 below, there was something unusual about Process 1. The quick jump in entrainment rate within a relatively small increase in gas rate is an indication of either liquid syphoning or gas bypass through the downpipes. With gas bypass, we would also expect a decrease in pressure drop, which was not the case during this series of tests.

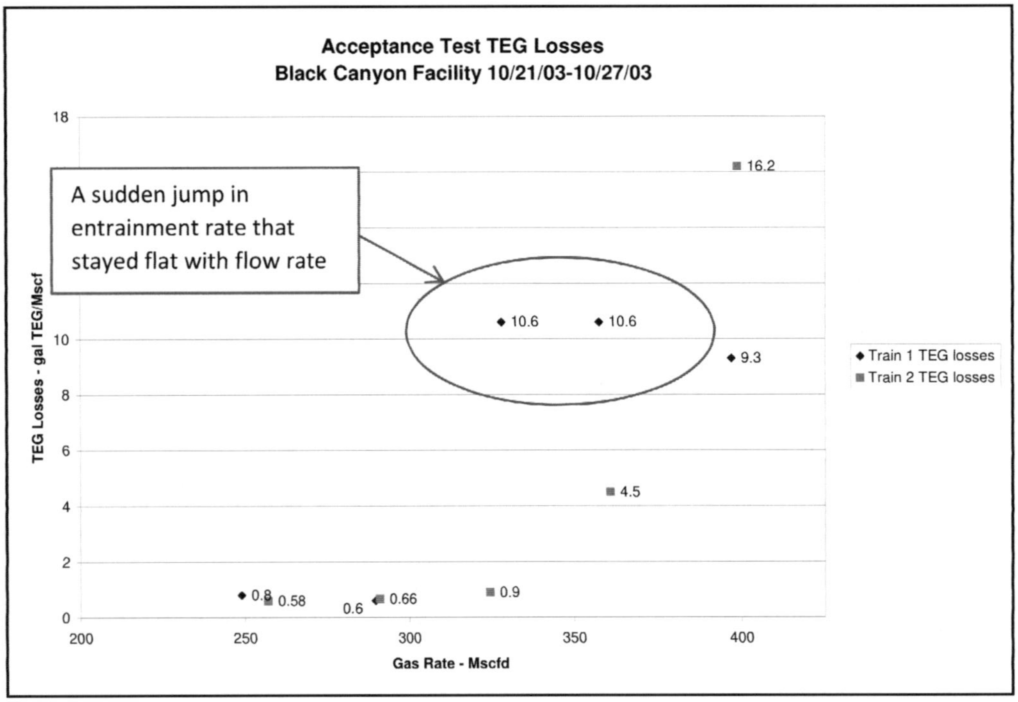

Figure 6- TEG Contactor Demisting Cyclones Performance Data

Troubleshooting - Second Fix Attempt

Based on the high pressure drop and the quick spike in entrainment rate in train 1, the lead hypothesis was that the available head was not sufficient for drainage. In November of 2003, the Process 1 contactor was inspected and no mechanical damage was found. No fouling was found on the tray, in the cyclones, or in the cyclone drain header. It was noticed, however, that the inlet weir was not properly sealed against the vessel wall. Dark discoloration on the vessel wall also suggested that some liquid was bypassing the inlet weir. It was unclear if the amount of leakage was significant enough to lose seal around the downpipes.

During this downtime, the sealing of the downpipes were moved from behind the inlet weir of tray 1 to the downcomer of tray 4 as shown in Figure 9. To make sure that sufficient liquid height existed to seal the downpipes, a seal cup was added to assure sufficient static seal. Special modifications were required to allow the downpipes to pass through the downcomer floor of tray 2. Seal plates were seal welded around the downpipes and downcomer floor to prevent leakage. Downcomer capacity was also checked to make sure that the additional space being taken up by the seal cup and downpipes would not control capacity of the tower. The outlet weir on tray 3 was modified to make certain that liquid would fall into the seal cup. Prior to start-up, the seal cup was filled with TEG.

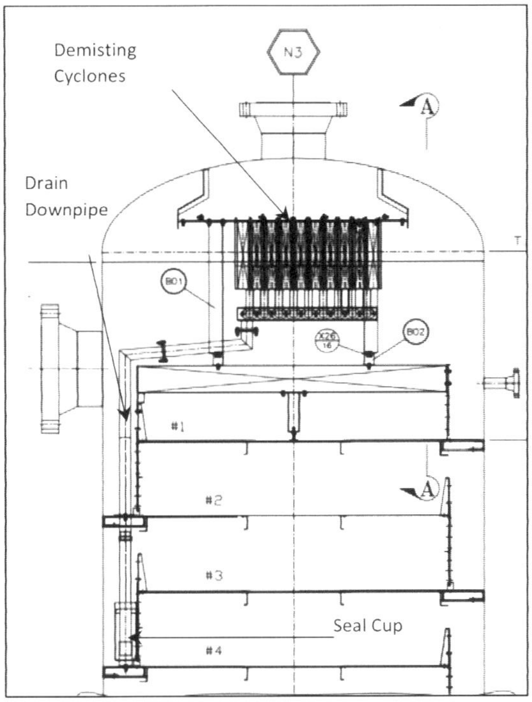

Figure 7 - Downpipes Seal Cup Modification

As can be seen in Figure 10 below, the measured TEG losses were significantly reduced after the modifications were made to the Process 1 TEG Contactor; however, these losses were still not acceptable. The process guarantees and the expected cyclone performance was less than 1.0 gal/MMSCF.

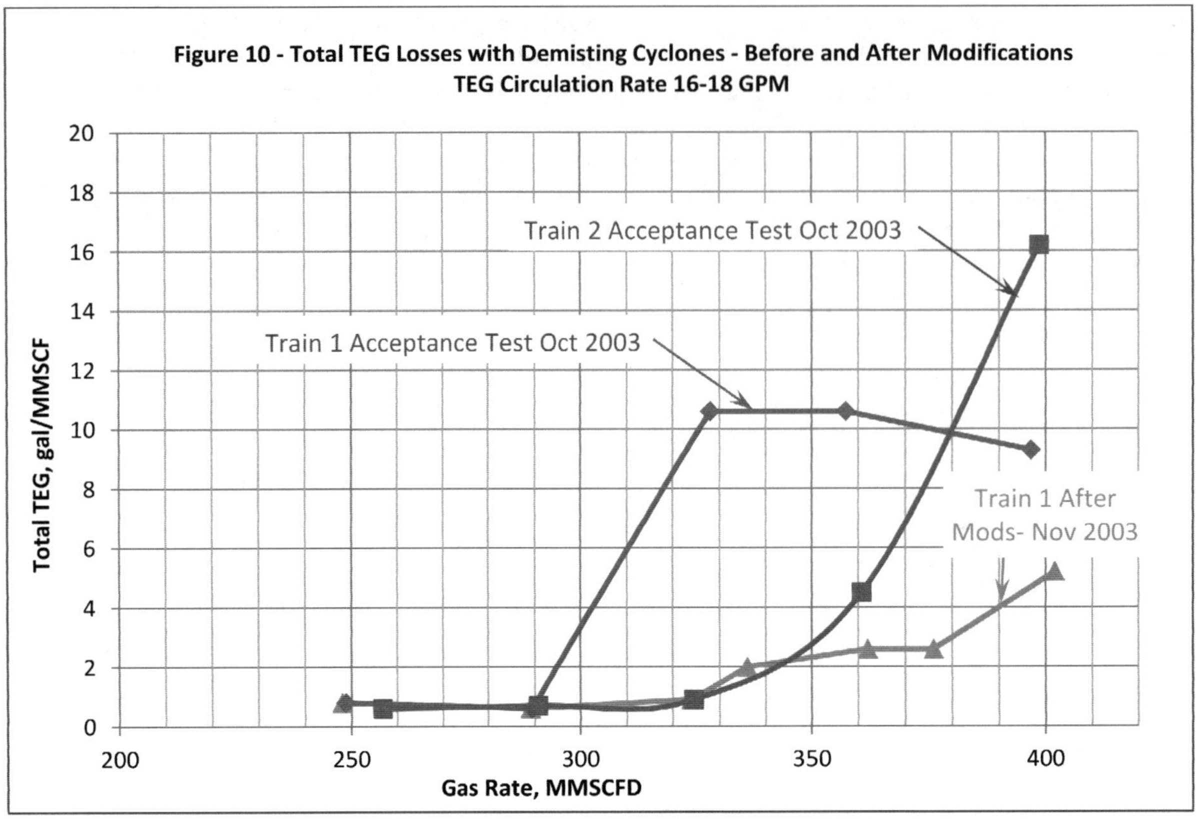

Troubleshooting – Back to Hypothesis

After a re-evaluation of the new data, there were two leading causes for the high entrainment:

1. Insufficient number of cyclones for this particular system (high viscosity and high vapor density) causing film instability and re-entrainment.
2. Liquid flux rate hitting the cyclones from the trays is too large at the higher gas rates. The high pressure drop is an indicator that the cyclones are seeing a larger liquid load than the original design assumptions.

The next scheduled turnaround was to be in September 2004, when a final fix would be attempted in Process 1 first.

Troubleshooting - Third Fix Attempt

Not knowing which of the two causes was controlling, it was decided to fix everything. More cyclones were added to reduce the momentum and at the same time decrease the liquid load per cyclone (Figure 11).

Figure 11 -Second Fix Demising Cyclones deck

This also reduced overall pressure drop and drainage head requirement. In addition, a new vane pack system was installed, covering the entire bubbling area, about 15 inches above the tray deck, to knock down bulk liquid. (Figure 12) This reduced the liquid load to the cyclones. The vanes have no pockets due to the known fouling associated with this application.

CDS (FMC Technologies) developed a clever design to integrate the vane pack into the tower with limited space. Since the vane pack occupies the entire bubbling area, it blocked the top vessel manway. The vendor developed a support system in which the vanes fitted into C-channels such that the vanes could easily be pulled out through the manway to allow access to the top tray. The vane support structure was fixed to the internal attachments supporting the demisting cyclones. Thus, no welding was required.

Figure 12 -Second Fix New Vane Pack Deck

In October of 2004, a performance test was done on the modifications. The TEG loses decreased from about 6 gal/MMSCF to 0.21 gal/MMSCF at 375 MMSCFD. To date the TEG losses are less than 0.6 gal/MMSCF at rates greater than 375 MMSCFD.

Conclusions

After completion of this work, the following conclusions can be drawn:

- A clean TEG system will minimize foaming and increase capacity as well as decrease maintenance, operating expense and downtime. As demonstrated, the full stream carbon filtration system reduced TEG liquid carryover by 65-70%.
- Although vendor data was available on demisting cyclones with glycol at elevated pressure and good experience was also available with cyclones in hydrocarbon service at high pressure, this application was outside the data range. The entrained liquid load is also an important parameter that often is ignored or is unknown.
- The combination of the high capacity trays and mist elimination equipment, along with the new solvent filtration system, allowed the gas capacity of the Black Canyon Facility to be almost doubled.

Acknowledgements:

The authors would like to thank our employer, ExxonMobil, for the opportunity and encouragement in the execution of this work. Also, we would like to recognize all of those who contributed, directly or indirectly, to this work, including: the operators and facility engineers at Black Canyon Facility and 3rd-parties such as Koch-Glitsch and FMC Technologies. They were extremely cooperative and followed through until the problem was fixed.

References

1. Polderman, H. et.al. "Experience with Debottlenecking of Gas Dehydration Plants, Laurence Reid Conference, February 26, 2006
2. MacFarland, S. et.al. *"Sour Gas TEG Dehydration: Water Content prediction, Wet Gas Sampling Technique, Solvent Contamination and Contactor Performance"*, Laurence Reid Gas Conditioning Conference, February 24, 2003.
3. Yang, M. and Kollinger, S. "The Evolution of Glycol Contactors", Sulzer Technical Article, February, 2016

A Better(?) Way to Calculate the Slope of the Equilibrium Line

Ross Taylor
Clarkson University
Potsdam, NY, USA

and

Markus Duss
Sulzer Chemtech Ltd.
Winterthur, Switzerland

**Presented at the Distillation Topical Conference,
AIChE Spring Meeting, San Antonio, Texas, March, 2017**

A Better(?) Way to Calculate the Slope of the Equilibrium Line

Ross Taylor, Clarkson University, NY, USA
Markus Duss, Sulzer Chemtech Ltd, Winterthur, Switzerland

Summary:

The calculation of tray efficiencies or the HETP of packed columns requires, among other things, the slope of the equilibrium line. There is, however, no simple definition of the slope for multicomponent systems because the vapor liquid equilibrium is represented by a surface (in hyperspace if there are more than 3 compounds). Thus, for multicomponent systems any pseudo-binary method for estimating tray efficiencies and HETP must rely, in part, on an approximate method of estimating the slope.

The objective of this paper is to present a new(?) method of approximating the slope suitable for modeling multicomponent systems based on numerical differentiation of the phase equilibrium surface with respect to the mole fraction of the *design component*. That component is defined as that compound we seek to eliminate from the mixture in a particular section of the column. Unlike the traditional concept of *key components* there is only one design compound per column section. The approach presented here can be easily applied to absorber and stripper columns, where the LK/HK approach is not applicable.

Introduction

The calculation of tray efficiencies or the HETP of packed columns requires, among other things, the slope of the equilibrium line. For binary systems the slope is unambiguously defined (although not always rigorously calculated). However, there is no easy way to use a definition of the slope for multicomponent systems because the vapor liquid equilibrium is represented by a surface (of dimensionality one less than the number of compounds).

Thus, for multicomponent systems any pseudo-binary method for estimating tray efficiencies and HETP must rely, in part, on an approximate method of estimating the slope. Information regarding thermodynamic interactions is lost using pseudo-binaries and the slope of the equilibrium line may, therefore, have very limited meaning for a multicomponent system.

Capek, Taylor, and Duss (2015) looked at several different methods for approximating the slope of the equilibrium line for multicomponent systems. It was shown that these different methods could lead to dramatically different estimates of the equilibrium slope.

Our purpose here is to take further that earlier work and propose a different and, we believe, definitive way to approximate the slope.

We note that for binary systems this method is exactly equivalent to the rigorous definition of the slope. We will further demonstrate that this method provides a sensible approximation in cases when other published methods have the potential to lead one astray.

Background

In the standard model for point, tray, and section efficiencies for distillation in tray columns due to Lewis (1936) the section efficiency is related to the tray efficiency by:

$$\eta_{Section} = \frac{\ln\left[1 + \eta_{Tray} \cdot (\lambda - 1)\right]}{\ln(\lambda)} \quad (1)$$

with the Murphree vapor tray efficiency obtained from:

$$\eta_{Tray} = \frac{e^{\lambda \cdot \eta_{Point}} - 1}{\lambda} \quad (2)$$

The point efficiency is found from

$$\eta_{Point} = 1 - e^{-N_{OG}} \quad (3)$$

where the overall number of transfer units for the gas, N_{OG}, can be expressed as follows [4]

$$\frac{1}{N_{OG}} = \frac{1}{N_G} + \frac{\lambda}{N_L} \quad \text{with} \quad \lambda = \frac{m}{L/G} \quad (4)$$

where $\lambda = mG/L$ is the stripping factor and m is the slope of the equilibrium line.

Inspection of Equations (1-4) shows that the slope of the equilibrium line always occurs together with the flow ratio in the stripping factor. Thus, the important parameters are the numbers of transfer units for gas and liquid phases and the stripping ratio. The way in which the efficiency depends on these parameters is shown in Fig. 3 in Duss and Taylor (2017) reproduced here for convenience as Fig. 1.

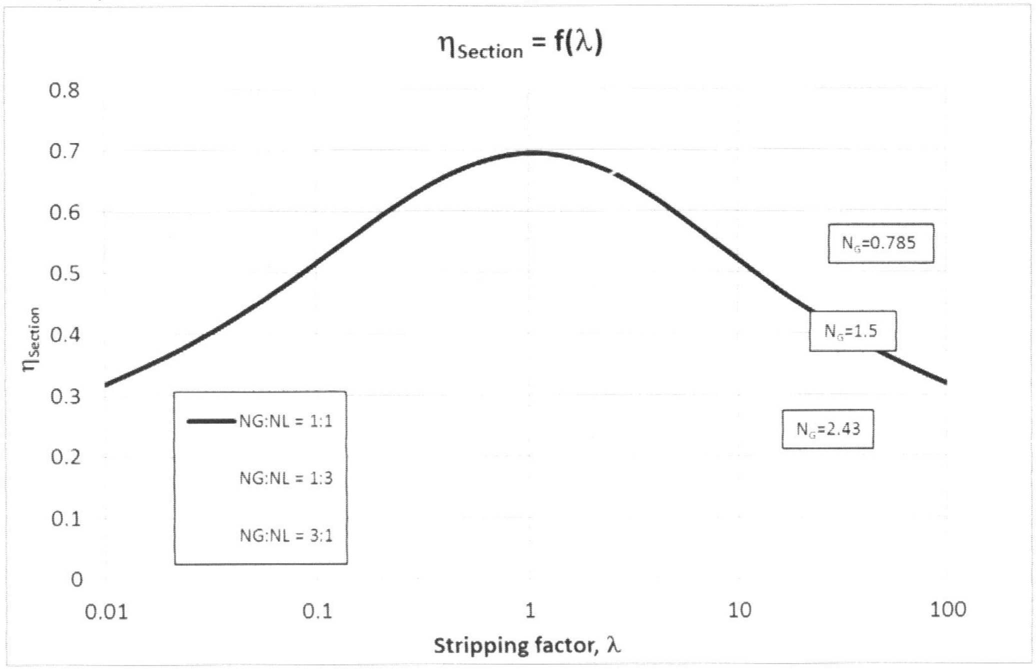

Figure 1: Section efficiencies for varying stripping factors obtained using equation 1. The following ratios of N_G and N_L were assumed: black line 1:1; green solid line 1:3; green dotted line 3:1. Values of N_G are adjusted to achieve the same maximum section efficiency (from Duss and Taylor, 2017).

The performance of packed columns often is be expressed using the HETP which is related to the height of packing by

$$HETP = H / N_{EQM} \quad (5)$$

where N_{EQM} is the number of equilibrium stages needed to accomplish the same separation possible in a real packed column of height *H*. The equilibrium stage model of distillation and absorption is described in a number of textbooks (see, e.g., Seader and Henley, 2006). If the equilibrium line may be assumed straight, the HETP may be related to the height of a transfer unit by

$$HETP = H_{OG} \frac{\ln(\lambda)}{\lambda - 1} \quad \text{with} \quad H_{OG} = H_G + \lambda H_L \quad (6)$$

the heights of a transfer unit (HTU) for the gas and liquid phases are H_G and H_L.

(7)

Figure 2 shows how $HETP / H_{OG}$ varies with the stripping factor.

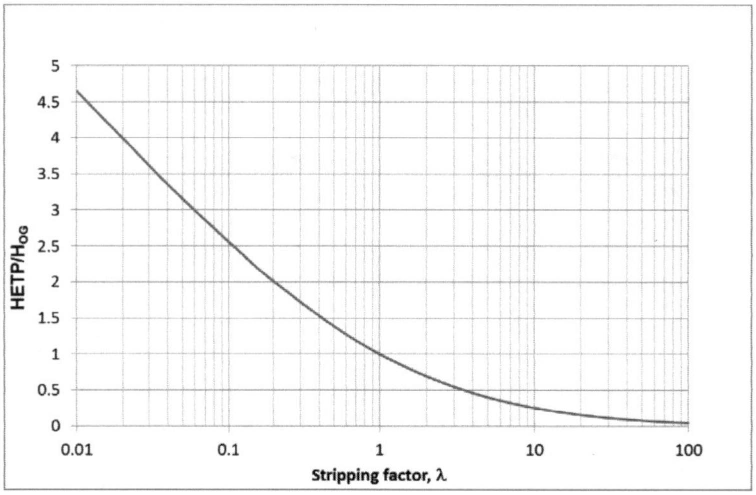

Figure 2: Dependency of HETP / HOG for packed columns from Equation (4).

The rigorous calculation of the efficiencies and HETPs in multicomponent systems is more complicated; modelling mass transfer in such systems requires the use of matrix arithmetic. Thus, for such systems, the slope is not a scalar as appears in Eqns. (4) and (6), but a square matrix of dimension c-1 where c is the number of components in the mixture. Convenient summaries of the equations for multicomponent systems and illustrations of their use can be found in books by Lockett (1986) and by Taylor and Krishna (1993).

Rigorous matrix models of the tray efficiency are rarely used in distillation simulation and design (for a comprehensive application of such models see, for example, Klemola and Ilme, 1996, and Ilme, 1999). It is much more likely that the equations presented above for a binary system will be applied to a representative pair of compounds (the keys) appearing in the multicomponent mixture and the resulting efficiency assumed to be the same for all species on a given tray.

We are not here concerned with the validity of many of the assumptions underlying the use of such a "binary" approach to modelling multicomponent systems. Most of the terms appearing in Equations (1-5) can be obtained without too many ambiguities. That cannot, however, be said for the slope of the equilibrium line. For a multicomponent system, there is no equilibrium *line*, there is, instead, a surface (in hyperspace for mixtures of more than 3 compounds). While there may be a rigorous – tensor based – definition of the slope of such a surface, this definition is of no practical use in the simpler pseudo-binary approach to the calculation of the efficiency that is most widely employed.

Imagine yourself at the top of the ski slope in Fig. 3. In what direction do you point the tips of your skis? The answer depends on how good a skier you are. If – like the first author of this paper – you were born in the UK and never learned to ski, then you would be tempted to take the route shown in the left side panel of Fig. 3. On the other hand, if you were born in Switzerland, home to some of the greatest downhill skiers in the world – like Markus, then you take the route shown on the right.

Figure 3: Left – Left – Route taken by Ross; Right – Route taken by Markus

In view of all of the foregoing we conclude: **slope matters!**

Estimating the Slope in Binary Systems

Binary Method 1 – Constant Relative Volatility Model

Vapor – liquid equilibrium in a binary system can be represented by:

$$y_A^* = \frac{\alpha_{AB} x_A}{(1+(\alpha_{AB}-1)x_A)} \tag{8}$$

where α_{AB} is the relative volatility between compounds *A* and *B*.

The slope of the equilibrium line then is obtained by differentiation. For constant relative volatility we obtain:

$$m = \frac{\partial y_A^*}{\partial x_A} = \frac{\alpha_{AB}}{(1+(\alpha_{AB}-1)x_A)^2} \tag{9}$$

Note that Equation (11) is valid only when the relative volatility is constant, whereas Eqn. (10) is true even when the relative volatility is variable (in general, it depends on temperature, pressure and composition).

The limits of this slope are given by

$$\left.\frac{\partial y_A^*}{\partial x_A}\right|_{x_A \to 0} = \alpha_{AB} \qquad \left.\frac{\partial y_A^*}{\partial x_A}\right|_{x \to 1} = \frac{1}{\alpha_{AB}} \tag{10}$$

and illustrated in Fig. 4.

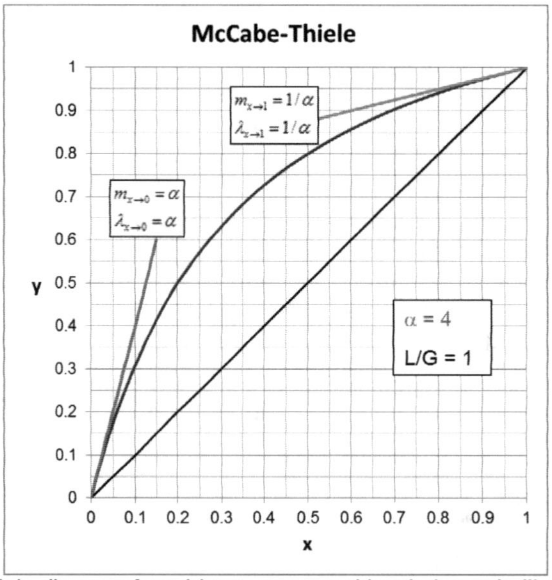

Figure 4: McCabe-Thiele diagram for a binary system with relative volatility equal to 4 and *L/G* = 1. The red line indicates the slope of the equilibrium line for x→0 and the green line for x→1.

If the relative volatility is a function of composition then the slope is given by:

$$m = \frac{\alpha_{AB} + x_A \frac{\partial \alpha_{AB}}{\partial x_A} - x_A^2 \left(\frac{\partial \alpha_{AB}}{\partial x_A} \right)^2}{(1 + (\alpha_{AB} - 1) x_A)^2} \tag{11}$$

We do not know how often this more rigorous calculation is performed, we suspect not often.

Binary Method 2 – Numerical Differentiation

A numerical approximation to the slope is easily obtained from

$$m = \frac{y_A^*(x_A + \Delta x_A) - y_A^*(x_A)}{\Delta x_A} \tag{12}$$

where $y_A^*(x_A)$ is the mole fraction in the vapour in equilibrium with the liquid of composition x_A. The two vapor phase mole fractions needed in this equation can be calculated from simple binary flash calculations: for the second flash, just perturb the mole fraction of species *A*, perturb the mole fraction of *B* by the same amount in the opposite direction and repeat the flash calculation to get a second value of the equilibrium vapor mole fraction. Then use Eqn. (12) to obtain the slope.

This method is completely rigorous no matter how complicated the dependence of the relative volatility on temperature, pressure and composition.

Estimating the Slope in Multicomponent Systems

Vapor-liquid equilibrium in the ternary system methanol – isopropanol – water is shown in Figure 5. While noting that this is not the surface whose slope we want to characterize (that would by the vapour mole fraction as a function of the liquid mole fraction) we can still pose this question: in what direction do we measure the slope? Before we provide our answer to this question we consider some of the methods that have been described in the literature.

Multicomponent Method 1 – Constant Relative Volatility Model

Equation (8) is easily adapted for multicomponent systems. Simply replace the components *A* and *B* with the light and heavy key components respectively:

$$y_{LK}^* = \frac{\alpha_{LK-HK} x_{LK}}{(1 + (\alpha_{LK-HK} - 1) x_{LK})} \tag{13}$$

The fact that this is easy to do does not, however, mean it is the most appropriate thing to do. The resulting slope is very sensitive to the particular pair of compounds identified as keys. While often it is easy to identify the keys, there are times when it is less straightforward. In addition, any information concerning molecular interactions other than of the key components is lost. Therefore, the so calculated slope of the equilibrium line is of very limited relevance.

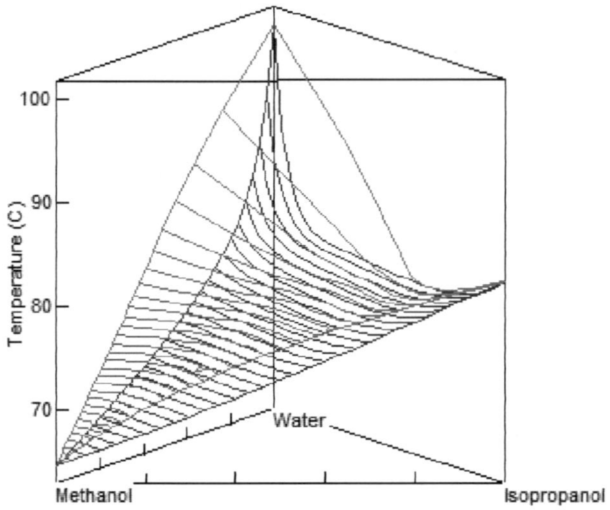

Figure 5: VLE diagram for methanol – isopropanol – water.

Multicomponent Method 2 – Numerical Differentiation

Equation (10) can be adapted in a similar way for multicomponent systems.

$$m = \frac{y^*_{LK}(x_{LK} + \Delta x_{LK}) - y^*_{LK}(x_{LK})}{\Delta x_{LK}} \tag{14}$$

However, in such cases, while it is clear how to perturb the mole fraction of species identified by *LK*, it is not at all obvious how the mole fractions of the other compounds should be adjusted so as to ensure that the mole fractions still sum to unity. It is necessary to do that before computing the equilibrium vapour mole fraction.

One possible resolution is to subtract the perturbation of *the light key* from the heavy key fraction. A variant is to conduct a binary flash containing *only* the light and heavy keys. There is, in that case, no ambiguity about the slope, but the influence of other compounds on the phase behaviour of the mixture as a whole vanishes entirely. Another possibility is to distribute the perturbation of the light key over the mole fractions of all of the other species. But, if that is done, how much of the perturbation do we assign to each of them? Still one more possibility is to distribute the perturbation over all species except *LK* and *HK*. We can continue to dream up methods of ensuring the mole fractions always sum to one; the point is that it is now obvious that there is no straightforward and unambiguous resolution to this problem. Moreover, however one chooses to proceed can have an impact on the calculated slopes and, hence, on the calculated efficiencies and any ultimate column design.

Multicomponent Method 3 – Method Described by Lockett (1986)

An alternative approach described by Lockett (1986) (although it must be noted that he makes no claim to have invented this approach) is equivalent to drawing a straight line between corners of the mass balance triangle as shown in on a pseudo-McCable-Thiele or Hengstebeck diagram such as that shown in Fig. 6. (Note for the pedantic: The Hengstebeck diagram uses lumped mole fractions and may not always be exactly the same as the Modified McCabe-Thiele diagram, as used by, for example, Seader and Henley, which divide the mole fraction of the light key by the sum of the mole fractions of the light and heavy keys.)

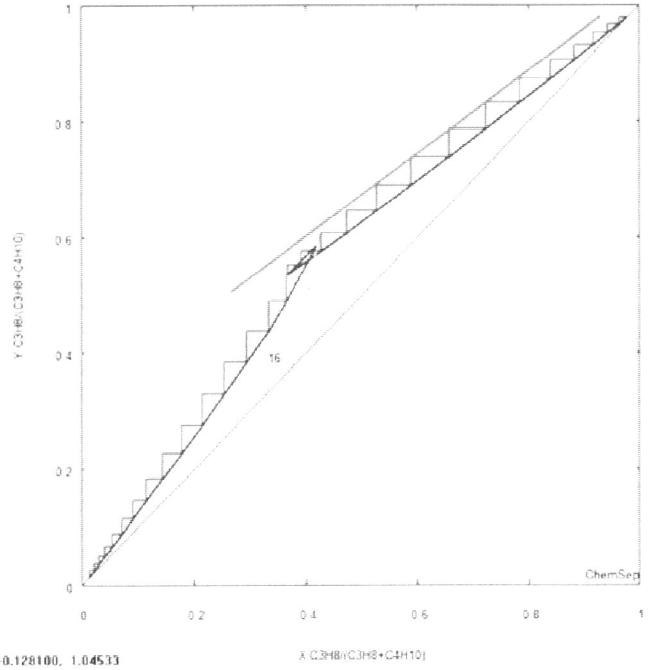

Figure 6: McCabe-Thiele diagram to illustrate slope estimation as per Lockett (1986).

One complication with methods that use the light and heavy key compounds is that the identity of the keys can change over the height of the column and this can lead to inappropriate values of the slope possibly leading, in turn, to negative efficiencies, a state of affairs that one should strive to avoid.

Negative slopes can also arise in the method described by Lockett (1986) because the pseudo-equilibrium line can fold back on itself. This situation is illustrated in Fig. 7. This situation can sometimes be avoided by switching the identity of the key components used for the creation of these diagrams, but that might require using compounds that are not true keys. This can have a significant impact on the computed estimate of the slope.

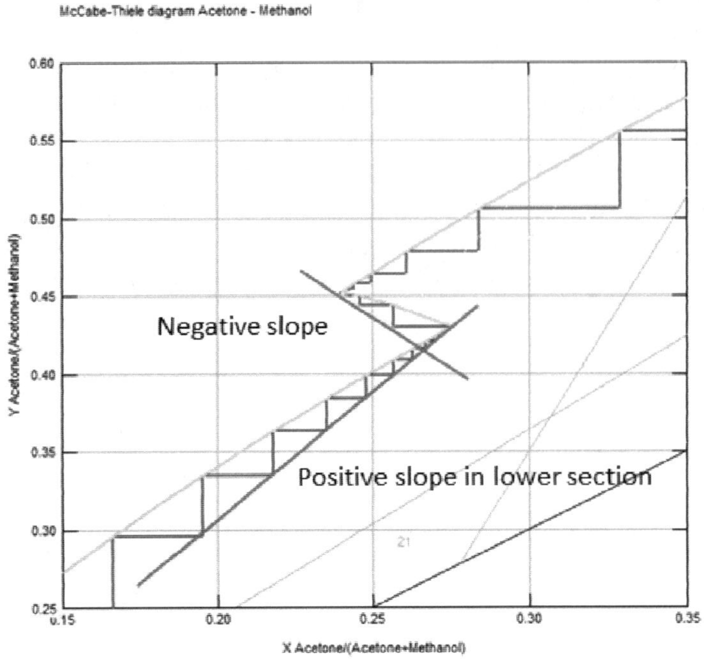

Figure 7: Fragment of McCabe-Thiele diagram to illustrate negative slopes that are possible with method in Lockett (1986).

It must be noted that the Lockett method as described here (and by Lockett himself) should not be applied to absorption and stripping operations (Lockett makes no claim that they should.) The same is true of the relative volatility model.

Multicomponent Method 4 – Proposed Method

Instead, we propose here a method that does not require us to identify light and heavy keys and always, in our experience, produces estimates of the slope that appear reasonable and meaningful.

Definition – The Design Component

We now introduce a new concept, that of the design component. The design component is defined to be that component which we seek to eliminate from the column section for which it is defined. (In that sense, it is analogous to a key component, but there is only one design component per section, not two). The concept of the "design component" can be interpreted as an extension of the HK/LK methodology to multicomponent systems with multiple product streams. As with the identification of key components, the identification of the design component will require the engineer to know what the column is designed to accomplish.

The slope is obtained by adapting Equation (12) as follows:

$$m = \frac{y^*(x+\Delta x_D) - y^*(x)}{\Delta x_D} \tag{15}$$

The procedure is as follows:

1. Carry out a bubble point calculation using, as a basis, the composition of the liquid on the tray in question. This will provide the equilibrium vapor composition $y^*(x)$
2. Add a small amount Δx_D to the mole fraction of the design component $(x+\Delta x_D)$
3. Normalize the liquid fractions
4. Carry out a second bubble point calculation to get $y^*(x+\Delta x_D)$
5. Compute the slope from Equation (15)

Some points to emphasize:

1. The bubble point calculation suggested in Step 1 of the above procedure is likely to be entirely unnecessary. With an equilibrium stage column simulation the equilibrium stage vapor composition is the same as that obtained from a bubble point calculation on the liquid composition. On the other hand, if the slope calculation is being done as part of a rate-based model simulation then the bubble point calculation is necessary.
2. The second flash calculation in step 4 of the above procedure is to be carried out at the same temperature used for the flash calculation in step 1.
3. In a (single) column section, the design component does not switch from one to another component.
4. This method is easily applied to absorption and stripping where two key components can be difficult or impossible to identify.

It is worth noting here that a rigorous rate based approach to column modelling does not require the calculation of the slope the equilibrium line, m. However, the first author of this paper is frequently asked to report a single representative HETP value. To extract such a representative HETP from the results out of a rate based simulation requires knowledge of the slope, m. Alert readers will have already noticed that, in general, there will be not one single HETP. Instead, a representative answer can only be provided for the design component(s).

Examples

In what follows we illustrate the identification of the design component and the resulting slope calculation by considering several examples taken from the literature (including four examples from the *Distillation* chapter by Seader (1997) in the 7th edition *of Perry's Chemical Engineers' Handbook*.

Case 1: **An Industrial i-Butane/n-Butane Fractionator**

Klemola and Ilme (1996) Ilme (1997) report data from an industrial *i*-butane/*n*-butane fractionator that is used here as the basis for this example.

The column has 74 valve trays; the feed being introduced onto tray 37. The feed is a mixture containing at least 8 identifiable compounds, but only two are present in significant amounts – isobutene and n-butane – and, therefore, this is essentially a binary separation. For binary (and near binary) mixtures, the identification of the key and design components is as straightforward as it can ever be. The design component in the rectification section is n-butane; in the stripping section it is isobutane.

The composition profiles for this column are shown in Fig. 8 (note that the mole fraction axis is logarithmic). The slopes of the equilibrium line computed from three different models and shown in Fig. 9 is entirely consistent with what one would expect given the relative volatility between the two C4s is nearly constant (as is clear from Fig. 10). The key components for the constant relative volatility and Lockett methods are isobutane and n-butane.

Interchanging the identities of the design components has no noticeable effect in this example. For true binaries the identification of the design component is irrelevant since the slope will be the same no matter what compound is selected. In this near-binary mixture the impact of the design component identity has no noticeable impact (as long as it is one of either n-butane or isobutane).

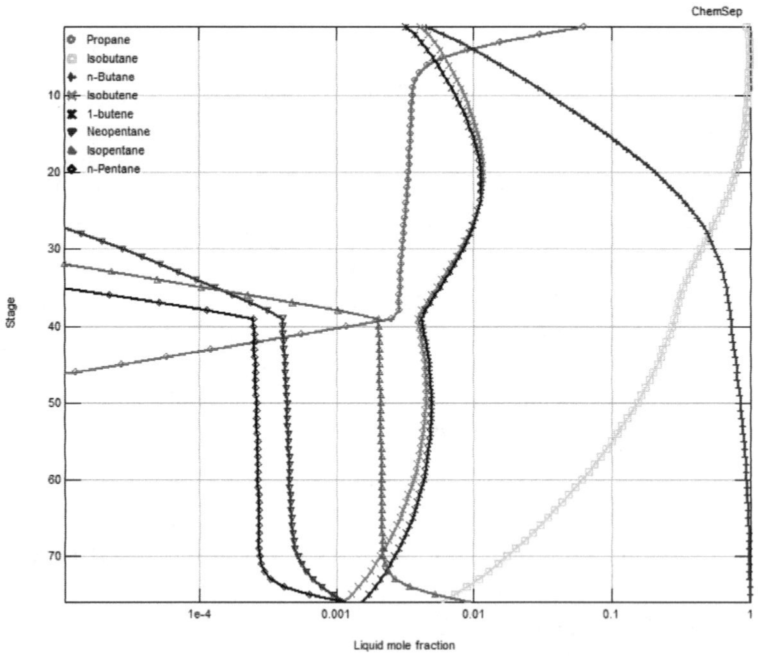

Figure 8: Composition and slope profiles in C4 splitter.

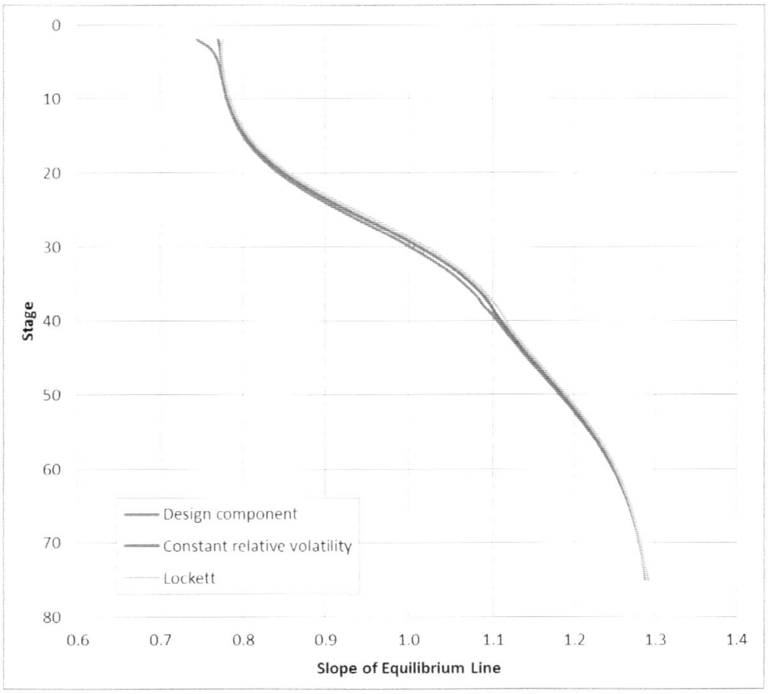

Figure 9: Slope profile in C4 splitter.

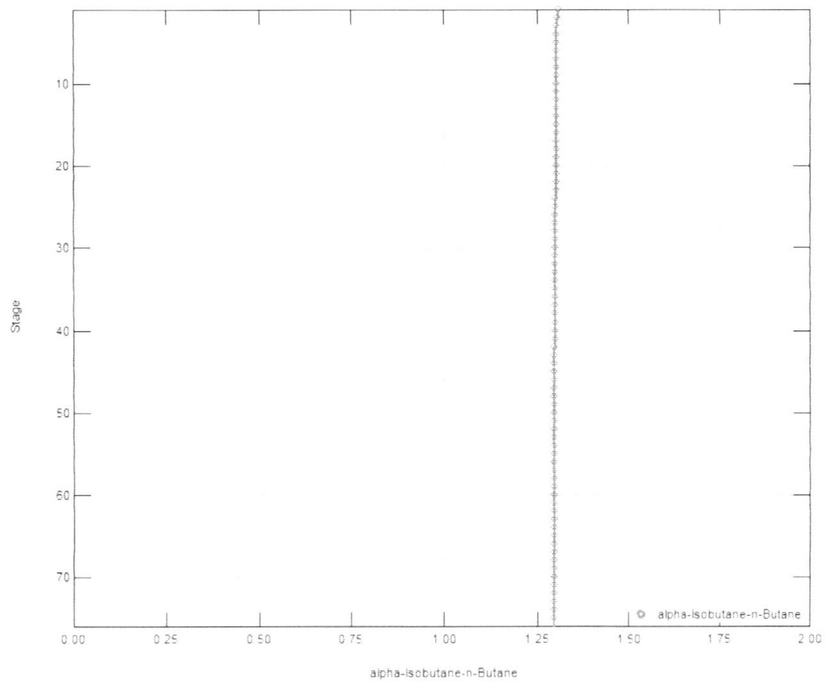

Figure 10: Relative volatility between isobutane and n-butane profile in C4 splitter.

Case 2: Multicomponent distillation column

Specified feed and computed product flows are shown in Fig. 11 for a hydrocarbon splitter from Seader (1997). This column is intended to split the mixture so that n-butane and lighter goes overhead and the pentanes become the bottom stream. The split is not especially sharp and can be improved with some simple changes in the specifications and additional stages in the top and bottom sections of the column. The design component in the rectifying section is the more volatile of the pentanes: isopentane. The design component in the stripping section is n-butane. The slopes for this case are shown in Fig. 12. They key components for the relative volatility and Lockett models are n-butane and isopentane. The models do not show a significant difference because this system does not deviate too strongly from one with constant relative volatility.

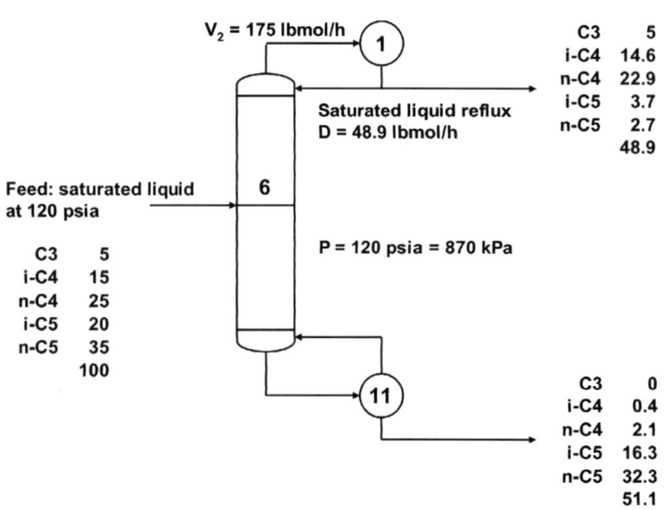

Figure 11: Specifications and calculated product stream flows for butane- pentane splitter. Flows are in lbmol/h.

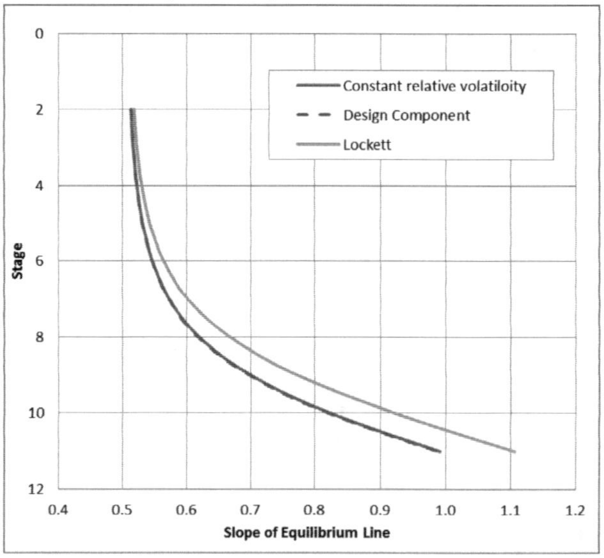

Figure 12: Slope profile for multicomponent distillation column.

Case 3: Nonideal Distillation

This example is taken from Doherty and Malone (2001) who refer to it as the De Rosier problem. The task is to design a distillation column to separate a feed of 20 mol/s methanol, 10 mol/s isopropanol, and 20 mol/s water. The bottom product is to contain no more than 0.5 mol% methanol and the distillate is to contain at least 99 mol% methanol but no more than 50 ppm water.

Following Doherty and Malone, the NRTL model was used for the activity coefficients. Doherty and Malone estimate the minimum reflux as 5; we used a value of 8 in this example and specified the bottoms product rate at 30 mol/s. Efficiencies of alcohol-water and alcohol-alcohol systems obtained experimentally in sieve tray columns vary from 60% to 100% (see Chapter 14 in *Perry's Chemical Engnrs. Handbook*, 7th Edition). After specifying an average efficiency of 80% we found that 99 equilibrium stages with the feed to stage 20 could provide a needed to get the distillate product below 50 ppm methanol.

Composition profiles are shown in Fig. 13. Based on this it seems reasonable to select the design components as isopropanol in the top and methanol below the feed. (These are the key components for the relative volatility and Lockett methods.) However, it is the specification on water in the distillate that is the key to this problem and so the design component in the top section of the column is chosen to be water.

Slope profiles based on these choices also are shown in Fig. 13. There is a greater difference between the models in this example than in the preceding cases, most especially for the Lockett model at the bottom of the column. The relative volatility and Lockett models will lead to lower efficiencies over the bulk of the column (this assumes the numbers of transfer units for the gas and liquid phases remain the same.) Thus, one might anticipate the final column designs to differ bearing in mind the requirement to reduce the water content in the overhead to less than 50 ppm. When we carry out simulations of the 99 tray column using the Chan and Fair model (for details of which see, for example, Taylor and Krishna, 1993) then the proposed method of estimating the slope leads to an estimate of the water content of 34 ppm. Use of the Lockett method, on the other hand suggests that the overhead content of water is 675 ppm. To get down to 50 ppm water in the overhead using the Lockett approach for the slope requires 136 stages. (These results should not be taken as evidence of the superiority of the proposed method of estimating the slope; what it does mean is, as we said in the introduction: **slope matters**.)

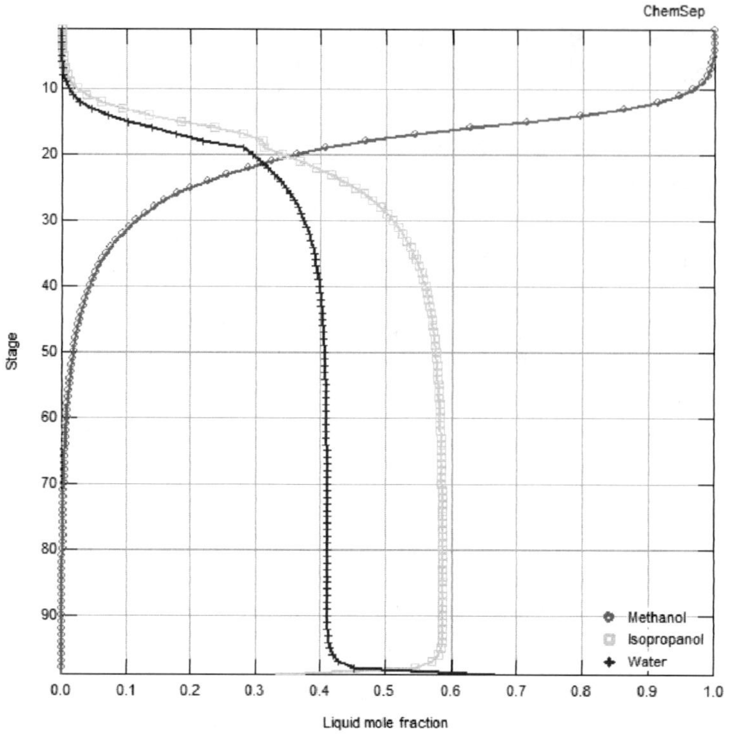

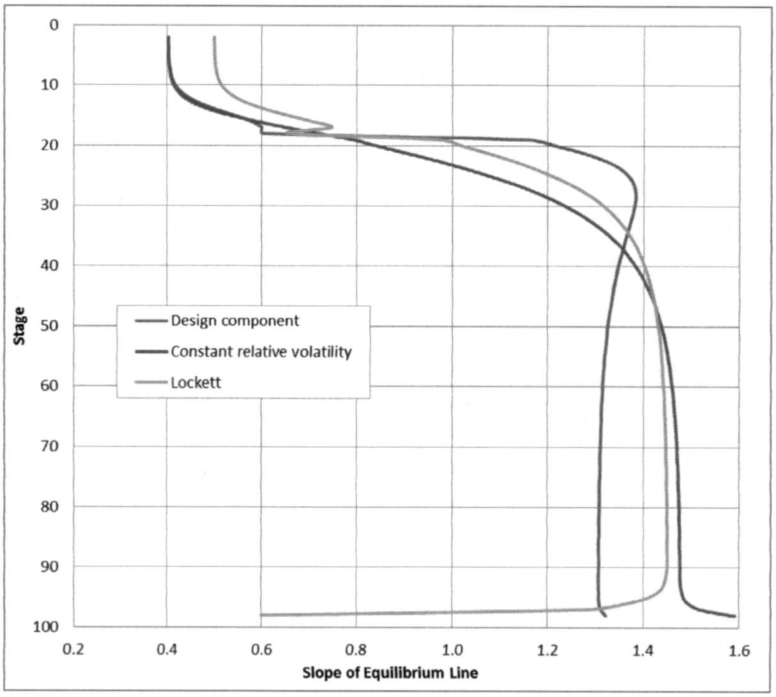

Figure 13: Composition (top) and slope profiles (bottom) in nonideal distillation process.

Case 4: Sidestream column

This more complicated example shown in Fig. 14 (from Seader (1997) features a partial condenser (with vapor product) and a vapor sidestream withdrawn from the 13th stage. The specifications were selected to obtain three products: a vapor distillate rich in C2 and C3, a vapor sidestream rich in *n*-C4, and a bottoms rich in *n*-C5 and *n*-C6. Further improvement in the purity of the sidestream as well as of the other two products could be obtained by increasing the reflux flow rate (or reflux ratio), and the number of stages in each section of the column.

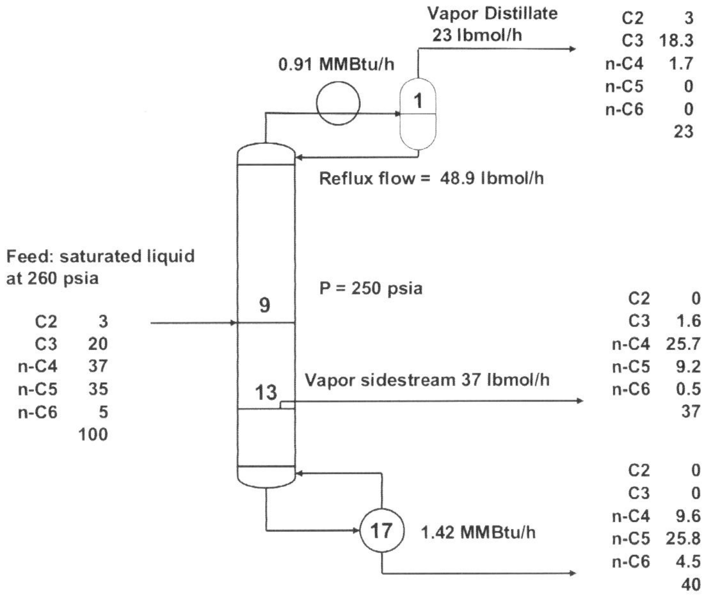

Figure 14: Specifications and calculated product stream flows and heat duties for light hydrocarbon still. Flows are in lbmol/h.

The design component in the rectifying section is n-butane.

To assign the design component in the mid-section (between the feed and the side-draw) we return to the feed stage and follow the liquid stream down. Ask what compound needs to be eliminated in order to recover *n*-C4 in the sidedraw: that would be propane. Finally, for the stripping section below the sidedraw, the design component is n-butane.

The slope of the equilibrium line computed from the three methods we are considering is shown in Fig. 15. The key components for the relative volatility and Lockett methods are propane and n-butane for the two sections of the column above the sidestream and n-butane and n-pentane for the section below the sidestream.

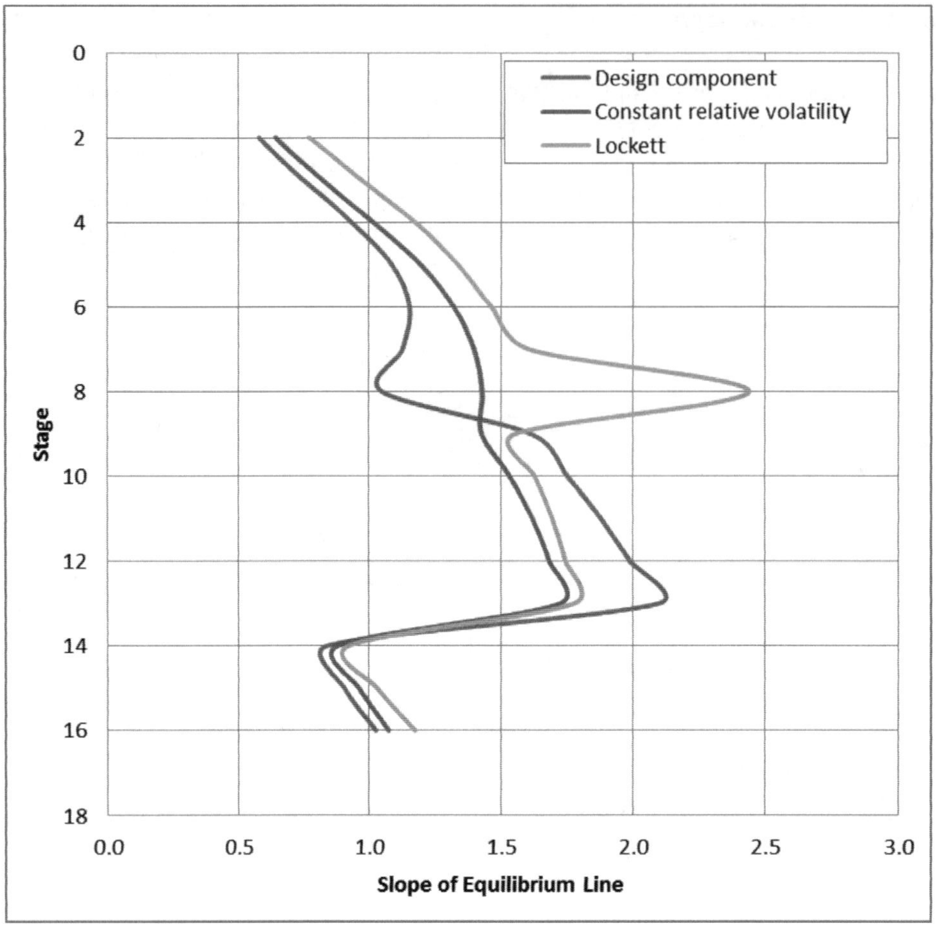

Figure 15: Slope profile in sidestream distillation column.

Notice that for the proposed method the slope is less than one in the top of the column before increasing to approximately 1.2 and falling again over stages 6 to 9 (the feed is on stage 9 and the sidedraw on stage 13). Thus, the rectifying section is functioning as a C3-C4 splitter and C5 goes along for the ride with C4. In the mid-section between the feed and the sidedraw the fraction of C4 falls, but not by much. There, the slope of the equilibrium line depends on the relative volatilities between C3, C4, and C5. Below the sidedraw there is essentially no C3 left, and this becomes largely a binary separation between C4 and C5 and the slope depends on the relative volatility between only those two compounds. This latter point explains why all three models show similar slopes for the bottom of the column. The Lockett model displays an unusually sharp peak on stage 8. This is due to the maximum in the mole fraction profile for n-butane and the fact that the feed stage is not optimized for the C3-C4 separation, more for the C3-C5 separation.

Case 5: Very Simple Absorber

This very simple absorber is based on Example 6.1 in Seader and Henley (2006). When molasses is fermented to produce a liquor containing ethyl alcohol and a CO2 rich vapor containing a small quantity of ethyl alcohol. The alcohol can be recovered in an absorber using water as illustrated schematically in Fig. 16.

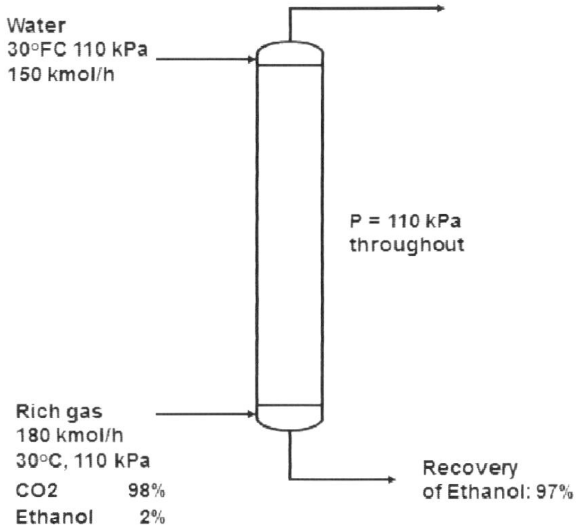

Figure 16: Specifications and calculated product stream flows and heat duties for absorber. Flows are in kmol/h.

The purpose of this absorber is to eliminate ethyl alcohol from the entering gas. Thus, the design component is ethanol. The equilibrium slopes are shown in Fig. 17. We have shown the slopes obtained from the relative volatility and Lockett methods even though neither method should be applied to absorption and stripping operations. For this illustration the key components in the relative volatility and Lockett methods are ethanol and water. Note that these two methods suggest slopes that are significantly different to that obtained using the proposed method. This means that the influence of the liquid phase resistance will be substantially higher when using these other methods. In turn this means higher HETP and taller columns to accomplish a required separation.

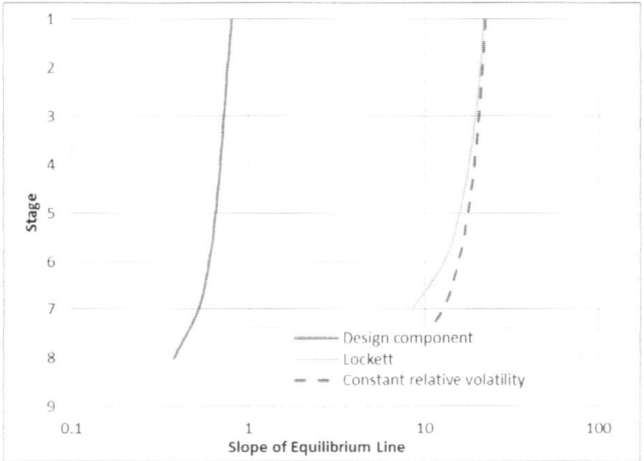

Figure 17: Slope profile in simple absorber.

Case 6: Less Simple Absorber

The absorber shown in Fig. 18 is Exercise 10.37 in Seader and Henley (2006). The oil is taken, as per their instructions, to be n-decane. The computed (rounded) product rates were obtained using the Peng-Robinson equation of state for the thermodynamic properties.

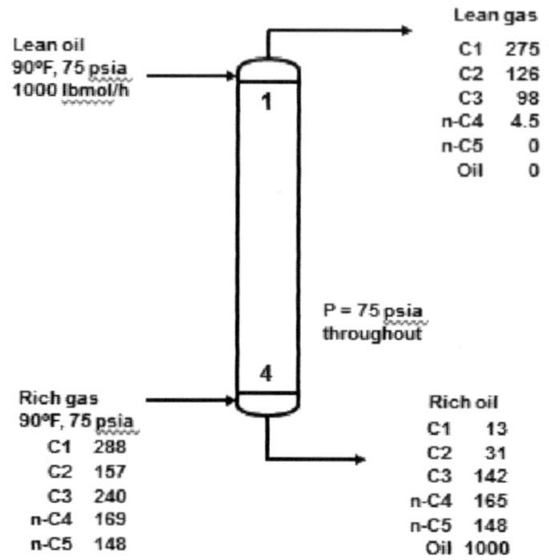

Figure 18: Specifications and calculated product stream flows and heat duties for absorber. Flows are in lbmol/h.

The absorber is designed to recover nearly all of the n-butane, while allowing significant amounts of propane to appear in both top and bottom product streams. Based only on this information it seems likely that the design component is n-butane; the slope for this assignment is shown in Fig. 19. We will return to discuss the relative vagueness of the identification of n-butane as the design component in the discussion later. Since the relative volatility and Lockett methods do not apply, we do not show them here.

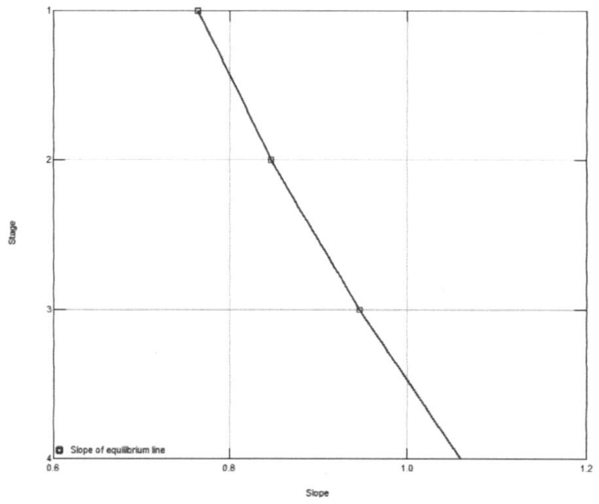

Figure 19: Slope for less simple absorber based on n-butane.

Case 7: Complicated Absorber

Consider the absorber specifications shown in Fig. 20 from Seader (1997). Note that a second absorber oil feed is used in addition to the main absorber oil and that heat is withdrawn from the seventh theoretical stage.

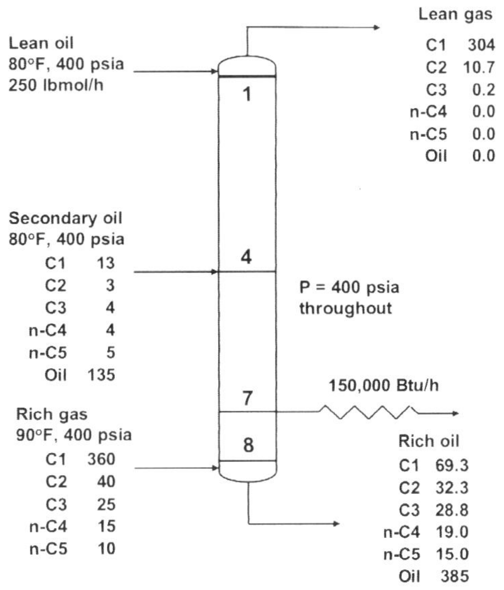

Figure 20: Specifications and calculated product stream flows
and heat duties for absorber. Flows are in lbmol/h.

The bottom of the column is where the bulk of the absorption takes place. We see that this column has been designed to allow almost all of the methane to pass out in the overhead while recovering as much as possible of the ethane and heavier compounds. Thus, the design component here is ethane throughout the column. The equilibrium slope is shown in Fig. 21. Since the relative volatility and Lockett methods do not apply, we do not show them here.

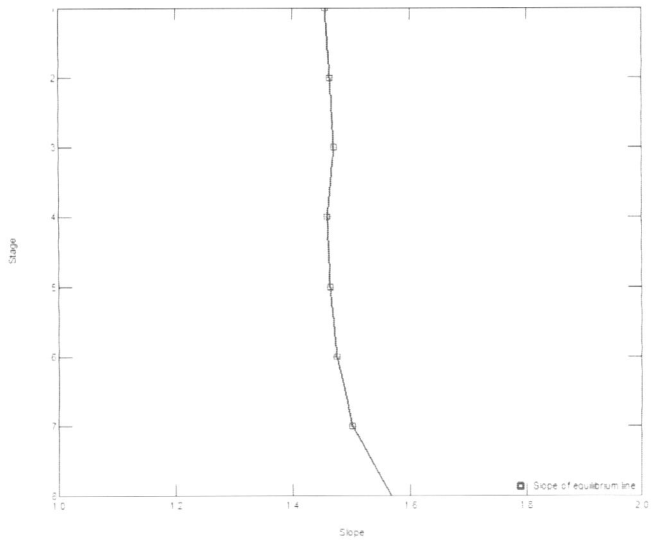

Figure 21: Slope for complicated absorber based on ethane.

Case 8: Reboiled Stripper

Specifications and computed flow rates for the reboiled-stripper from Seader (1997) are shown in Fig. 22.

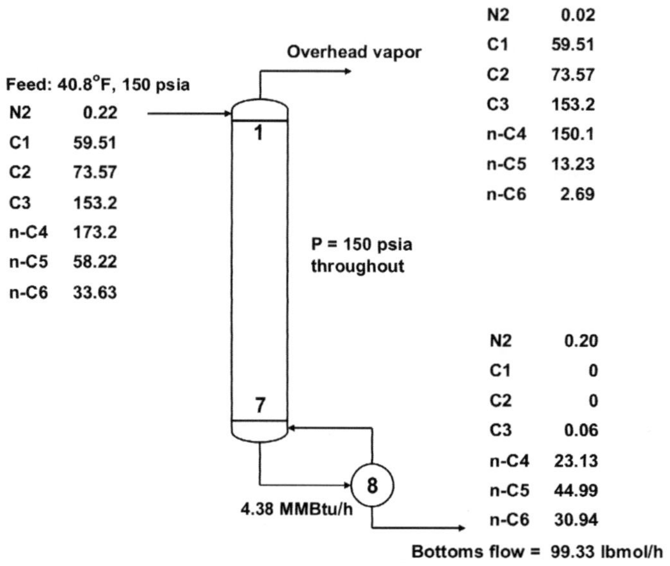

Figure 22: Specifications and calculated product stream flows and reboiler heat duty for a reboiled stripper. Flows are in lbmol/h.

The specifications are made so as to ensure that most of the *n*C5 and *n*C6 is recovered in the bottoms.
In this case it seems likely that the design component will be n-butane. The slope of the equilibrium line based on n-butane is shown in Fig. 23. As before, the relative volatility and Lockett methods do not apply, and so we do not show them here.

However, in general, the design engineer has to know the required specifications for n-butane and propane.

If the specification is for some 25mol-% n-butane in the bottom, then this is the design component for all stages. However, if the column has to achieve some 0.1 mol-% of propane, than this becomes the design component.

The HETP would be higher for propane than for n-butane (most likely), because the stripping factor is significantly higher for propane, and the packing height or number of trays must be adjusted accordingly.

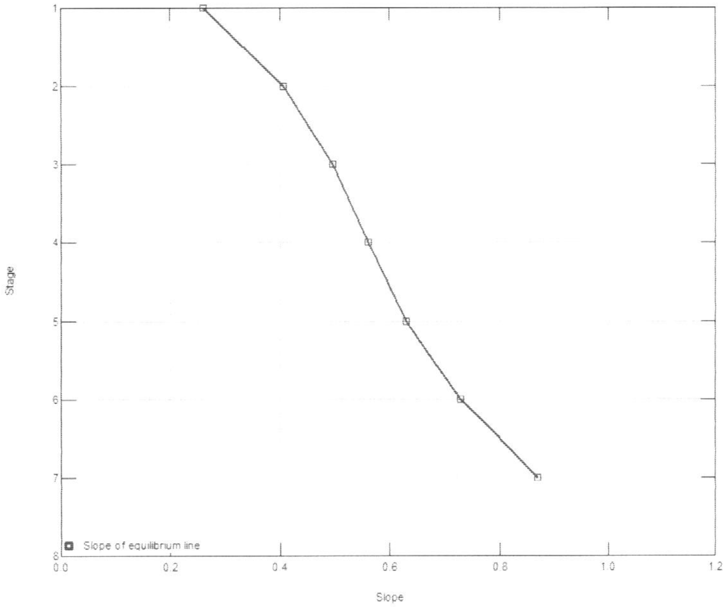

Figure 23: Slope for reboiled stripper based on n-butane.

If the specification is for the sum of the two components, then the situation is a bit more complicated. Actually, then the design has to be checked for both components and the result which requires more trays (or a higher packing height) would be the design component.

This is good illustration that demonstrates that design is not always straightforward. Let us assume that sum of nC4 and lighter is specified and the outcome was (just fantasy numbers here to illustrate the point):

nC4: 5 m packing (or e.g. 9 trays) for 25 mol-%

C3: 8 m packing (or 16 trays) for 0.1 mol-%

Now, the design for propane (8 m packing) would yield a n-butane concentration of, say, 15 mol-% nC4 (this is just an estimate). Thus, a design based on propane would achieve an overly good outcome that was too conservative (the sum of the concentrations is 15.1 mol%).

On the other hand, if you design for n-butane (5 m), then propane might be e.g. double the concentration as per simulation. However, this might not be a problem because an increase from 0.1 to 0.2 mol-% could be acceptable (the sum is still about 25 mol-%).

Conclusions / Recommendations

In this paper we have proposed a new method for estimating the slope of the equilibrium line for multicomponent systems. Strictly speaking, for such cases, phase equilibrium in multicomponent systems can be represented, not by a line, but by a surface (likely in hyperspace). However, there are many times when a chemical engineer needs an approximate value of the slope of such a surface. Examples include:

1. Estimating the overall mass transfer coefficient.
2. Estimating the HETP in packed columns.
3. Calculating the tray efficiency of tray columns.

We have proposed a new way to estimate the slope based on numerical differentiation of the phase equilibrium surface with respect to the mole fraction of the *design component*. That component, defined here for the first time is that which we seek to eliminate from the mixture in a particular section of the column. Unlike the traditional concept of *key components* there is only one design compound per column section. In addition, the new concept can be easily applied to absorber and stripper columns, where the LK/HK approach is not applicable.

We have illustrated the identification of the design component with the help of several examples taken from the literature on absorption and distillation. Further, we have also shown that the slope based on the design compound does not misbehave; this in contrast to some other methods that have been proposed in the literature. The method has been used for the design of packed columns by Sulzer for many years and has proven to be successful.

We have also illustrated the designer's dilemma that can arise when using equilibrium stages to model multicomponent systems: some "experience" or "intuition" required (in order to end up with an appropriate design).

Acknowledgements

This paper is based in part on work carried out during the sabbatical of Markus Duss at Clarkson University in 2016.

All simulations carried out specifically for this paper were performed using *ChemSep* (www.chemsep.com).

Skiing icons from www.flaticon.com.

Symbols

a_I	(m^2/m^3)	Specific vapor-liquid interfacial area
G	($kmol/(m^2 s)$)	For packing: Gas flow rate per cross sectional column area
		For trays: Gas flow rate per cross sectional bubbling area
H_G	(m)	Height of transfer unit in gas phase
H_L	(m)	Height of transfer unit in liquid phase
H_{OG}	(m)	Overall height of transfer unit
$HETP$	(m)	Height equivalent to a theoretical plate
k_G	($kmol/(m^2 s)$)	Gas side mass transfer coefficient
k_L	($kmol/(m^2 s)$)	Liquid side mass transfer coefficient
k_{OG}	($kmol/(m^2 s)$)	Overall gas side mass transfer coefficient
L	($kmol/(m^2 s)$)	For Packing: Liquid flow rate per cross sectional column area
		For trays: Liquid flow rate per cross sectional bubbling area
m	(-)	Slope of the equilibrium line
N_G	(-)	Number of transfer units in gas phase
N_L	(-)	Number of transfer units in liquid phase
N_{OG}	(-)	Number of transfer units overall gas
x_I	(-)	Mole fraction in the liquid

Greek

α	(-)	Relative volatility
η_{Point}	(-)	Murphree vapor point efficiency
η_{Tray}	(-)	Murphree vapor tray efficiency
$\eta_{Section}$	(-)	Section efficiency or overall efficiency
λ	(-)	Stripping factor = $m/(L/G)$

Literature

Capek, E., R. Taylor, & M. Duss, *Slippery Slopes, How Should We Approximate the Slope of the Equilibrium Line in Multicomponent Systems*, AIChE Spring Meeting, Austin, TX, 2015

Doherty, M. & M.F. Malone, *Conceptual Design of Distillation Systems*, McGraw-Hill, New York, 2001

Duss. M. & R. Taylor, Hidden ties – An explanation for O'Connell's success, AIChE Spring Meeting, San Antonio, Texas, March 2017

Ilme, J. *Estimating Plate Efficiencies in Simulation of Industrial Scale Distillation Columns*, Ph.D. Thesis, University of Lapeenranta, Finland (1997)

Klemola, K.T. & J. Ilme, *Distillation Efficiencies of an Industrial-Scale i-Butane/n-Butane Fractionator, Ind. Eng. Chem. Res.* , **35**, 4579 (1996)

Lewis, W.K., Rectification of Binary mixtures, *Ind. Eng. Chem.*, ***28*** (4), pp 399–402, 1936

Lockett, M.J., *Distillation tray fundamentals*, Cambridge University, New York, NY, 1986

Seader, J.D. & E.J. Henley, *Separation Process Principles*, 2nd Ed., Wiley, New York, 2006

Seader, J.D., *Distillation*, Chapter 13 in *Perry's Chemical Engineers' Handbook,* 7th Ed., R.H. Perry & D. Green, Eds., McGraw-Hill, New York, 1997

Taylor, R. & R. Krishna, *Multicomponent Mass Transfer*, Wiley, New York, 1993

Novel Optimization of a Dividing Wall Distillation Column

Bailee J. Roach, R. Bruce Eldridge - James R. Fair Process Science and Technology Center & University of Texas at Austin, Austin, Texas
Domenic Barsotti - Red Cedar Technology, East Lansing, MI

Biography

Bailee J. Roach, is currently finishing her PhD in the Process Science and Technology Center at The University of Texas at Austin. She received her undergraduate degree from Virginia Tech. Upon graduation, she will join Dow Chemical Company in Lake Jackson, TX.

R. Bruce Eldridge,

Domenic Barsotti,

Abstract

Originally patented in 1949, the dividing wall distillation column has gained renewed attention due to the potential for reduced energy consumption, an estimated thirty to fifty percent over traditional multiple column arrangements, and lower capital investment. Unfortunately, this new interest has not been accompanied by the development of rigorous design approaches that utilize the latest in computational methods. In the current study, optimization and process simulation software was seamlessly integrated to yield minimum energy consuming dividing wall column (DWC) designs over a range of operating conditions. The modeling effort was supported by data collected from the Separations Research Program (SRP) experimental column. The results yield a powerful methodology for future DWC retrofits or new column construction.

This research focuses on obtaining experimental pilot data and validating a process simulation. The simulation includes modeling the heat loss to the atmosphere and heat transfer across the dividing wall. With the accurate simulation on a 6" pilot scale column, the impact of heat transfer across the wall is able to be determined at a wide range. Additionally, the column is able to be scaled to determine the impact of heat loss and transfer on a commercially sized column.

Introduction

Traditionally, the production of three high purity products from a three component mixture requires the use of two distillation columns. A Dividing Wall Column (DWC) offers an alternative to this approach. (1) The DWC contains a vertical partition, dividing the column into two sides. The feed side (prefractionate) separates the lowest and highest boiling products and the product side (mainfractionate) separates the intermediate component. This configuration reduces capital costs by utilizing only one column, reboiler, and condenser. By partitioning the feed and side product, a reduction in thermodynamic losses is created, which can result in energy savings. Previous studies have found that a DWC can also produce an estimated 30% energy savings over a traditional multi-column scheme. (2) Despite these advantages, validated predictive models, which will facilitate widespread adoption of the technology, are lacking.

Pilot Column

A newly constructed six inch diameter column located at the James Fair Pilot Plant operated by the Separations Research Program (SRP). This column allows for collection of data to be obtained over a wide range of operating conditions. The pilot plant, shown in the left of Figure 1, was used to study the separation of a higher order alcohol system at vacuum conditions. The alcohol case separates n-hexanol (light key, or LK), n-octanol (middle key or MK), and n-decanol (heavy key or HK), which emulates the system studied with the Fieg column (3-5). The split between the n-hexanol and n-octanol is a more difficult split than the n-octanol and n-decanol.

The pilot DWC consists of six flanged sections: upper dividing wall, lower dividing wall, stripping, rectifying, and two connecting sections to the condenser and reboiler. Each section is fabricated with 6" schedule 40 pipe. The dividing wall is created with a welded wall, which divides the column's cross section area exactly in half. The wall is fabricated with a ¼" thick 304 stainless steel plate. The plate was internally welded to the 6" pipe to ensure the wall is well sealed. The column is packed with structured packing Sulzer Mellapak 500.Y.

The experimental flow diagram is shown in the right of Figure 1. The experimental system is composed of a feed system, dividing wall column, three product areas – overhead, sidedraw, and bottoms, and a liquid split at the top of the dividing wall. The column is fed with a through a series of tanks, which ensures an experimentally constant feed composition. A DWC includes additional degrees of freedom versus a standard distillation column, including the top of the wall liquid split and sidedraw product reflux. The top of the wall liquid split between the prefractionate and mainfractionate is controlled using two control valves. The liquid falling from rectifying section to the upper dividing wall collected on a total trap, which is fed to a tank, as shown in Figure 1. From the tank, the liquid is pumped to two control valves, which split the liquid at the desired amount. The liquid is reheated with a heat exchanger before returning to either the prefractionate (stream 630A) or mainfractionate (steam 630B). The sidedraw reflux (640B) is performed in a similar manner.

The column is allowed to reach steady state before samples are taken. Sample points are available at five locations within the system: feed entering the column, overhead product, top of wall liquid reflux, sidedraw product, and the bottom product. Various control methodologies were used to ensure steady state data was achieved. Steady state is defined as three samples of consecutive values over a minimum two hour window.

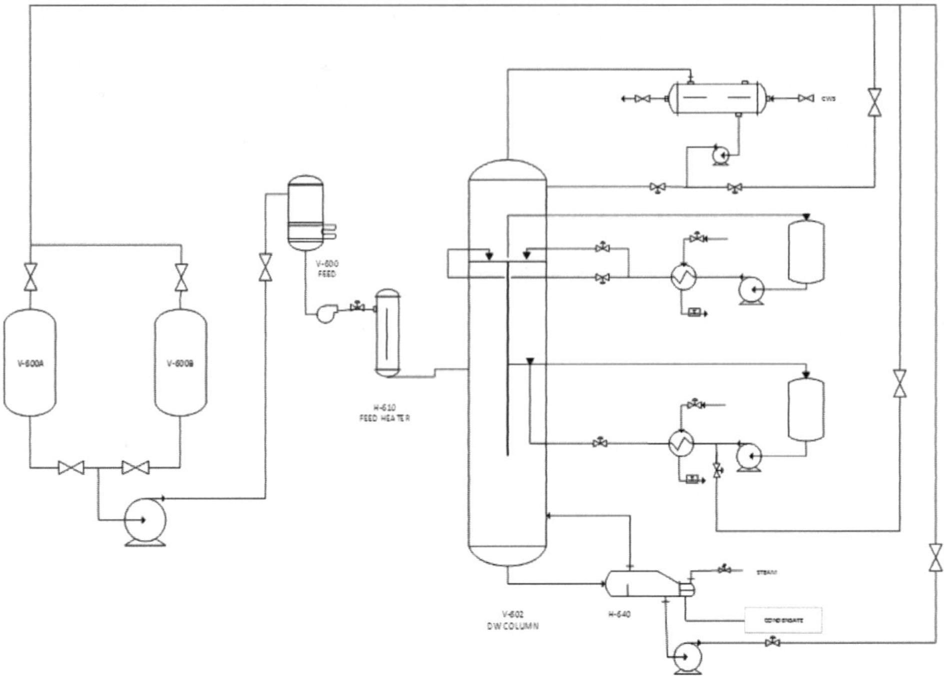

Figure 1. Left, 6" DWC Pilot Column located at the Pickle Research Center in Austin, TX; Right, Flow Diagram of the Pilot Column

Modeling and Optimization

Software

A DWC simulation was created using the software AspenPlus. AspenPlus is a widely used commercially available process simulator software. A general DWC model was created assuming negligible heat loss to the atmosphere and heat transfer across the dividing wall and was used to design and size the pilot DWC equipment and instrumentation. With results from the pilot experiments, the simulation was updated to reflect actual column operation.

In order to fit the AspenPlus model to the experimental conditions, the software HEEDS was used to describe heat loss and wall heat transfer in the pilot column. HEEDS is a design software with the ability to link Microsoft Excel and AspenPlus. HEEDS modifies the desired variables within the Aspen input file and results are extracted through the Aspen output file. The optimization method used in this research is HEEDS' simultaneous hybrid exploration method that is robust, progressive, and adaptive (SHERPA) method. The SHERPA method simultaneously uses multiple search methods in a single search by taking advantage of the best attributes of one method while minimizing the disadvantages of a different method by suppressing its participation in the search.

The general DWC flow diagram from AspenPlus is shown in Figure 2. The pilot DWC is represented with five columns. Current modeling software does not have the option to model a DWC with a single column shell. The initial approach involved using the four column method and then breaking the mainfractionate column into two separate columns to separate the upper and lower mainfractionate. This allowed the simulation to recreate the pilot column's total trap tray configuration. The simulations in this research were performed using an equilibrium stage approach based on the NRTL VLE model.

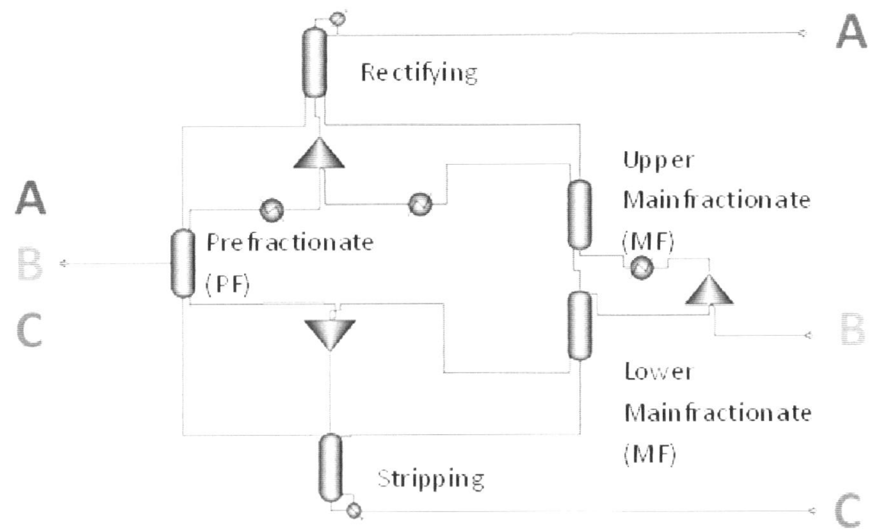

Figure 2. AspenPlus Simulation Diagram for Pilot DWC

Determining Experimental Heat Loss to the Atmosphere

Method

The Fieg pilot column (3-5) reported that for their 2.5" pilot column, a heat loss of approximately 20% of the reboiler duty was observed. In order to quantify the pilot column's heat loss to the atmosphere, total reflux tests were performed. During the pilot column operation in total reflux, no feed and products were added or taken off during operation. AspenPlus does not have capability to model total reflux without inputting a feed and product value. Total reflux was modelled by inputting a small feed flowrate in comparison to the overhead reflux flow rate, with the three product flow rates equally divided from this flowrate.

AspenPlus allows for the input of heat loss per column stage. The heat loss calculations were performed in Excel using the following equations. Heat loss to the atmosphere per stage is calculated using the heat loss equation:

$$Q_{loss,i} = U_{atm}A_{atm,i}(T_i - T_o)$$

The objective of the Heeds simulation is to minimize the difference in the experimental internal flows to the simulation by varying the heat loss coefficient, Uatm. This assumes the heat transfer area is equal to the outside column surface area, including the insulation area. The area is calculated using the outer diameter of the column and estimating the height of the stage with the HETP. T_o is defined as the outside temperature. The column's heat transfer across the dividing wall is negligible due to the column operating in total reflux, the temperature difference across the prefractionate and mainfractionate sections are negligible. The simulated U_{atm} value is inputted into the Excel sheet, which calculates the heat loss per stage, and is inputted into the Aspen input file. The experimental internal flowrates used for matching are the liquid flowing into the prefractionate and mainfractionate and the sidedraw reflux. The pilot column is equipped with flow transmitters on each of these streams.

The simulation optimizing is an iterative procedure. Because of the heat loss issues, the experimental reboiler duty could not simply be inputted as the heat input into the DWC simulation. The reboiler duty was used to match the overhead reflux flow.

Results

Figure 3 and Table 1 include the comparison of the pilot data with the simulated results. Figure 3A plots the comparison temperature profile of the pilot data and simulation with no heat loss. The profiles are not well aligned, especially in the upper regions of the column. simulated results show a different temperature on either side of the dividing wall, while the pilot results show the temperatures are aligned. Table 1 includes a summary of the simulated data. The internal flow rates for the pilot column do not match well with the simulation. The pilot column's sidedraw reflux rate is 45.99 lb/hr, while the simulation's sidedraw rate is 32 lb/hr. The product compositions are not significantly impacted by the heat loss.

Adding the average heat loss to the environment to the simulation results in a better match of the temperature profiles. Figure 3B plots the temperature profile, which is more aligned in the dividing wall section. The pilot's internal flow rates are well predicted with the simulation. An interesting observation is the impact of heat loss on the product compositions. The products are better predicted with the heat loss in the simulation, but are not significantly impacted.

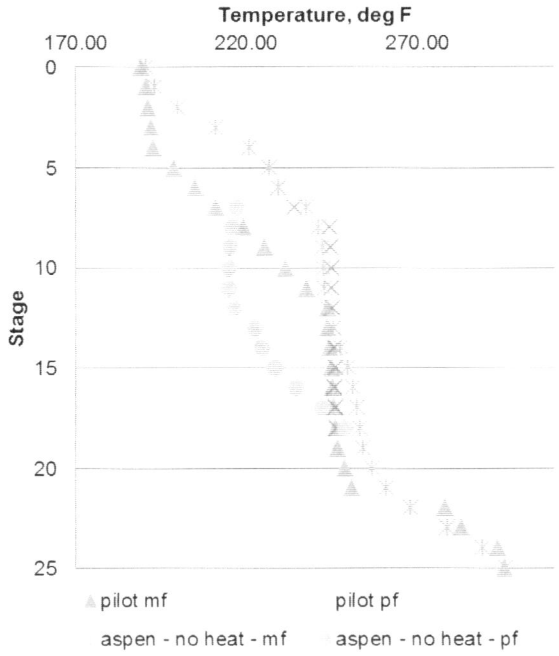

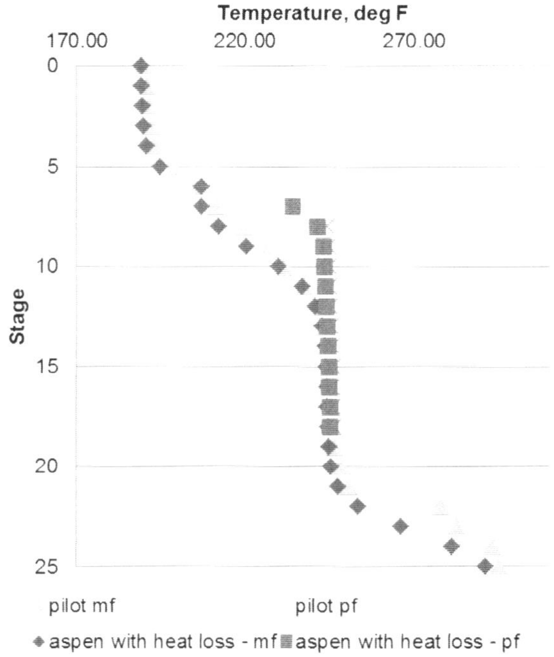

Figure 3. A) Left, Comparison temperature profile of Total Reflux Pilot Data and simulation results with no heat transfer across the wall; **B)** Right, Comparison temperature profile of Total Reflux Pilot Data and simulation results with heat transfer across the wall

Table 1. Comparison of Alcohol Total Reflux Experimental Pilot Data and Simulated Results

Date		9/3/2015		
Case		Total Reflux, Alcohol		
Liquid Split, βL	fraction	0.3		
Data		pilot	sim	sim
Uatm	Btu/ hr ft2 degf	-	-	2.8
overhead reflux	mass flow rate, lb/hr	23.04	23.04	23.02
630A		11.70	9.00	12.64
630B		27.32	24.00	29.50
640B		45.99	32.00	45.91
ovhd LK	mass fraction	1.000	1.00	1.000
630 LK		0.449	0.457	0.432
sidedraw MK		0.982	0.976	0.985
bottoms HK		0.971	0.956	0.972
Qreboiler	kBTU/hr	32.37	8.80	23.70

Determining Experimental Heat Transfer across the Dividing Wall

Method

Heat Transfer across the dividing wall baffle may impact column efficiency. Regardless of wall placement, on either side of the wall different separations are occurring. This can cause a temperature differential to occur across the dividing wall, which in turn can impact column efficiency. During the pilot experiments, heat differences across the dividing wall as high as 50°F were recorded. Current DWC models mostly assume there is no heat transfer across the dividing wall, making the column thermally equivalent to a Petlyuk configuration. The Fieg pilot column (3-5) reported for their experiments that the heat transfer across the wall ranged from 35-50% of the net reboiler duty. The reported value for Uwall was 700 $\frac{\mathrm{W}}{\mathrm{m^2K}}$ (123 $\frac{\mathrm{BTU}}{\mathrm{hr\ ft^2 {}^\circ F}}$).

In order to determine the heat transfer across the wall for the pilot simulation, additional heat transfer calculations were added to the model. The heat transfer across the dividing wall per stage is calculated using:

$$Q_{wall,i} = U_{wall}A_{wall,i}(T_{pf,i} - T_{mf,i})$$
$$Q_{wall,i} = U_{wall}A_{wall,i}(T_{mf,i} - T_{pf,i})$$

HEEDS allowed for the heat transfer calculation in Excel, with the results inputted into the Aspen simulation. HEEDS varied the value of Uwall and the reboiler duty to find the internal flowrates. The area of the wall was calculated using an average HETP and the internal diameter of the dividing wall column. This assumes the heat transfer area is equal to wall area. Experimentally, the wall is contacted by a percentage of vapor and liquid, but this exact value is not known.

Similar to the process to find the average Uatm, two cases were used to find Uwall. Each simulation used the average heat loss to the atmosphere values found in the total reflux cases. Case #1 used the reboiler duty as the independent variable to set the overhead reflux rate. The Uwall was changed to match the change in flowrates of the 640B and 630B streams. The second case involved the reboiler duty as the independent variable to set the 630B flowrate and the Uwall was changed to match the change in flowrates of the 640B and 630B streams.

Results

The results show a better fit with adding heat transfer across the dividing wall. Figure 4A shows the temperature profile comparison without heat transfer and Figure 4B includes heat transfer across the wall. The temperatures are better aligned by including heat transfer. Table 2 summarizes the results in a table. The product purities are almost consistent for each case. However, the internal flowrates are impacted by heat transfer, which can impact column design.

Table 3 includes a summary of Uwall for four cases, with an average Uwall of 281 $\frac{W}{m^2K}$ $(49.5 \frac{BTU}{hrft^2{}^\circ F})$. This value is between the values of heat transfer between a gas and stainless steel wall and a liquid and stainless steel wall. This value is lower than the Fieg column's reported value. The larger pilot diameter of 6" versus the Fieg column of 2.5" shows that a larger diameter will have less of an impact by the heat transfer.

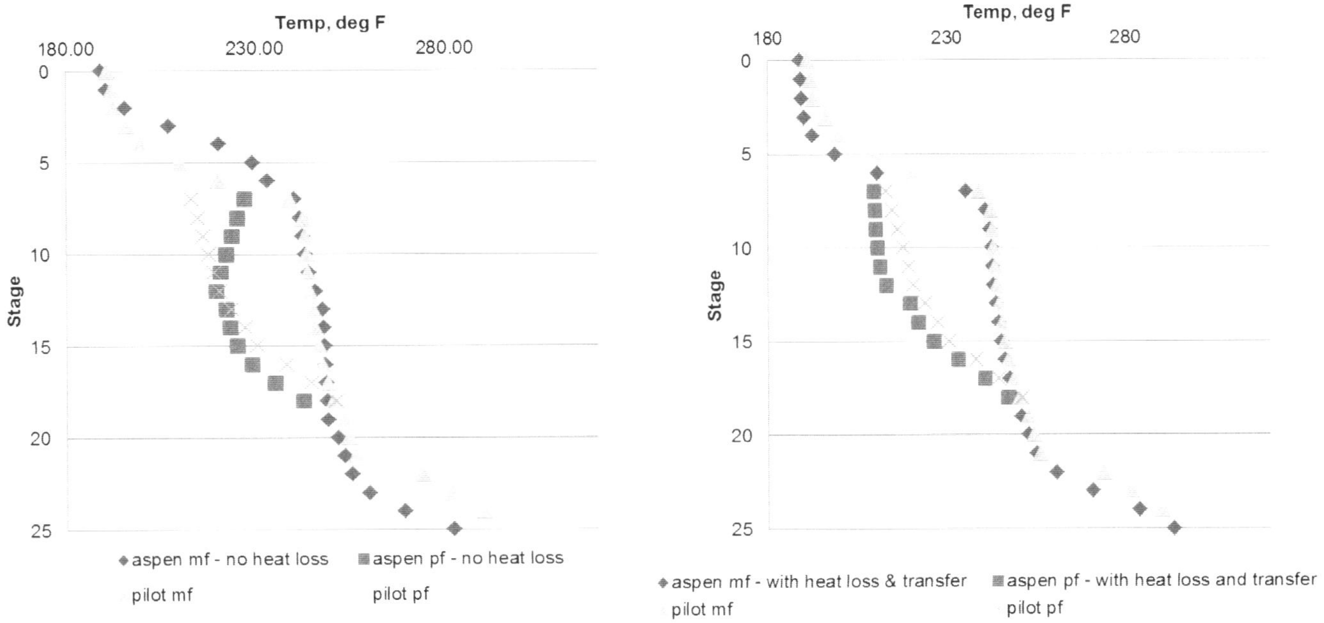

Figure 4. A) Left, Comparison temperature profile of Finite Reflux Pilot Data and simulation results with no heat transfer across the wall; **B)** Right, Comparison temperature profile of Finite Reflux Pilot Data and simulation results with heat transfer across the wall

Table 2. Comparison of Alcohol Finite Reflux Experimental Pilot Data and Simulated Results using Uwall values

Date		10/1/2015			
Case		A10ii			
Liquid Split, βL	fraction	0.25			
Data		pilot	sim, Uatm only	sim, Uwall 1	sim, Uwall 2
Uatm	W/m2K	-		14	14
Uwall	W/m2K	-	-	325	400
overhead reflux	mass flow rate, lb/hr	25.39	25.39	24.93	18.19
630A		13.01	16.43	16.17	12.81
630B		27.60	34.91	34.36	27.22
640B		30.55	29.95	37.45	30.02
ΔF, 640B-630B		2.95	(4.96)	3.09	2.80
ovhd LK	mass fraction	0.994	0.997	0.997	0.996
630 MK		0.707	0.799	0.801	0.770
sidedraw MK		0.989	0.998	0.999	0.998
bottoms HK		0.997	0.998	0.998	0.998
Qreboiler	kBtu/hr	41.16	33.74	33.87	30.80

Table 3. Summary of Uwall values for the Alcohol finite reflux cases

PF/MF	*Date*	*Name*	*Uwall #1*	*Uwall #2*
			U, W/m2K	
25/75	9/30/2015	A10i	250	350
25/75	10/1/2015	A10ii	325	400
50/50	10/9/2015	A13	320	340
75/25	10/7/2015	A12	120	140

Determining Impact of Uwall

Method

The experimental heat loss for the equimolar system was found to be 280 W/m2K. It was interesting to note that while the heat transfer did not significantly impact the product purities, it did impact the reboiler duty. To determine the impact of U_{wall} on the reported reboiler duty for the pilot column, simulations were performed by inputting a variety of Uwall values from 0-700 $\frac{W}{m^2K}$. The simulations were constrained to match the same reflux ratio and product purities of over 99 wt%.

In order to determine the impact of wall heat transfer on larger column diameters, the simulations was scaled-up from 0.5 feet diameter to 2 feet and 10 feet. HEEDS was used to scale-up the simulations by changing the feed flow rate and reboiler duty to match the experimental pilot column's overhead liquid reflux flow rate and the overhead liquid loading.

Results

Figure 5A shows the Plot of simulations at various U_{wall} values to determine the impact on the reboiler duty and the sidedraw product. The results show that U_{wall} impacts the reboiler duty, which impacts the overall energy savings. A U_{wall} value of approximately 340 W/m^2K results in a minimum energy duty, while a value of 700 W/m^2K results in a much higher reboiler duty. The U_{wall} range from 300-400 W/m^2K do not result in a significant change in reboiler duty savings. Figure 5B, Plot of the heat loss per stage of the reboiler duty at various U_{wall} values. It's interesting to note that the largest heat loss is at stage 12, which is the feed and product take-off stage. The U_{wall} value of 340 W/m^2K reaches a maximum heat loss at stage 12 in comparison to the other U_{wall} values.

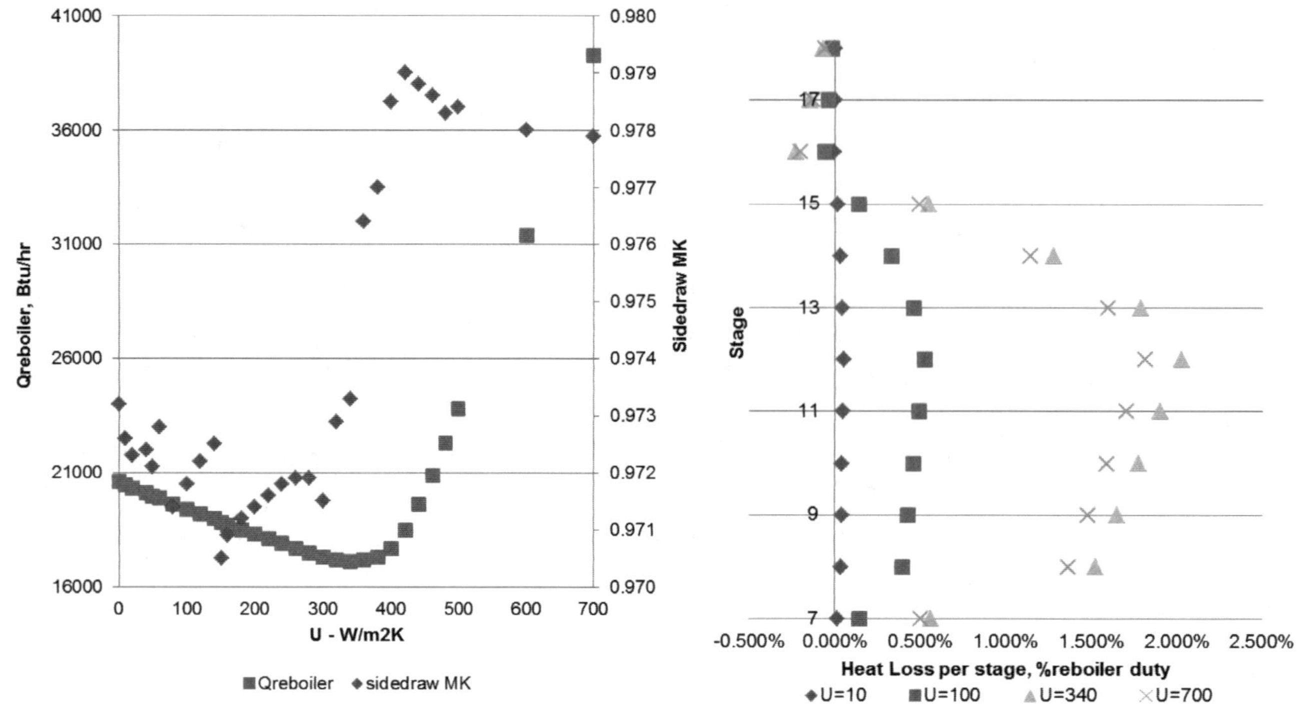

Figure 5. A) Left, Plot of simulations at various Uwall values to determine the impact on the reboiler duty and the sidedraw product; **B)** Right, Plot of the heat loss per stage at various Uwall values

The results of the column scale up are shown in Table 4 and Table 5. Table 4 scales up from a 0.5 ft. column to a 2 and 10 ft. by assuming an average Uatm of 14 W/m^{2}K and Uwall of 300 W/m^{2}K. The simulations show that at the 0.5 ft. diameter, the simulation is impacted by heat transfer across the wall, but as the simulation is scaled to 2 and 10 ft., the hydraulics are not as impacted by heat transfer.

Table 5 shows the results of a range of heat transfer across the wall for a 10 ft. diameter column. For U_{wall} values ranging from 300-1000 W/m^{2}K, no major impact was seen on the simulation results. This demonstrates that at a larger diameter column, heat transfer does not play as large of an impact in the hydraulics as for the smaller pilot column.

Table 4. Simulation Results from scaling up from a 0.5′ to a 2′ and 10′ diameter column using the pilot average Uwall, heat transfer coefficient across the dividing wall

		Units	Pilot	Simulation		
Column Diameter		feet	0.5	0.5	2.0	10.0
Uatm		W/m2K	?	14	14	14
Uwall		W/m2K	?	300	300	300
Reflux Ratio		unitless	**1.44**	**1.45**	**1.44**	**1.44**
Reboiler/Feed Ratio		(KBTU/hr)/(lb/hr)	0.57	0.49	0.26	0.28
Qloss/Qreboiler		%		47%	27%	21%
product purity	ovhd, LK	mass fraction	0.994	0.986	0.996	0.990
	sidedraw, MK		0.989	0.990	0.993	0.985
	bottom, HK		0.998	1.000	0.996	0.992
liquid loading	rectifying	gpm/ft2	**0.51**	**0.51**	**0.52**	**0.51**
	upper prefrac		0.48	0.48	0.44	0.42
	upper mainfrac		**1.52**	**1.53**	1.29	1.18
	lower mainfrac		1.55	1.54	0.78	0.59

Table 5. Simulation Results from scaling from a 0.5′ to 10′ diameter column at various Uwall, heat transfer coefficient across the dividing wall

		Units	Pilot	Simulation		
Column Diameter		feet	0.5	10.0	10.0	10.0
Uatm		W/m2K	?	14	14	14
Uwall		W/m2K	?	**300**	**1,000**	**3,000**
Reflux Ratio		unitless	**1.44**	**1.44**	**1.44**	**1.43**
Reboiler/Feed Ratio		(KBTU/hr)/(lb/hr)	0.57	0.28	0.29	0.28
Qloss/Qreboiler		%		21%	21%	21%
product purity	ovhd, LK	mass fraction	0.994	0.990	0.992	0.996
	sidedraw, MK		0.989	0.985	0.988	0.993
	bottom, HK		0.998	0.992	0.993	0.996
liquid loading	rectifying	gpm/ft2	**0.51**	**0.51**	**0.50**	**0.50**
	upper prefrac		0.48	0.42	0.40	0.37
	upper mainfrac		1.52	1.18	1.19	1.22
	lower mainfrac		1.55	0.59	0.62	0.71

Conclusions

In this research, an AspenPlus model is validated using experimental pilot plant data collected at the SRP. The model includes the heat loss to the atmosphere and heat transfer across the dividing wall. These calculations were performed in conjunction with the HEEDS optimization software.

Next, heat transfer across the dividing wall and heat loss to the atmosphere were calculated and implemented into each case and scale-ups from pilot scale to 2 ft and 10 ft diameter columns occurred to study any changes in design variables and observe the effects of heat transfer. In general, heat transfer had a larger effect on smaller diameter columns, mainly the pilot scale DWCs where in some cases products were thrown out of specification by heat transfer modeling. As the diameter increased, even though heat transfer and heat loss increased with the increasing heat transfer area, the effects on the column purities and hydraulics decreased to the point in the 10 ft diameter columns there was little to no change from the base case. The column hydraulics of each optimized case is studied and the pressure drops throughout the column are discussed.

References

1. Wright, O. R. Fractionation Apparatus. U.S. Patent 2.471.134, 1949.
2. Dejanovic, L., Matijasevic, Z. & Olujic, Z. DWC - a breakthrough towards sustainable distilling. Che. Eng. & Proc. 49, 559–580 (2010).
3. Niggemann, G., Hiller, C. & Fieg, G. Experimental and Theoretical Studies of a Dividing-Wall Column Used for the Recovery of High-Purity Products. Industrial and Chemical Engineering Research 49, 6566–6577 (2010).
4. Niggemann, Gerit, and Georg Fieg. 2012. "Validation of Dividing-Wall Columns Based on Experimental Data and Dynamic Simulations: Pilot-Plant and Production-Scale Columns." Industrial & Engineering Chemistry Research 51 (2): 931–43. doi:10.1021/ie201598q.
5. Niggemann, G., Hiller, C. & Fieg, G. Modeling and in-depth analysis of the start-up of dividing-wall columns. Chemical Engineering Science 66, 5268–5283 (2011).

2017 AIChE®

SPRING MEETING & 13TH GLOBAL CONGRESS ON PROCESS SAFETY

March 26–30, 2017
Henry B. Gonzalez
Convention Center
San Antonio, TX

17 AIChE Spring Meeting, San Antonio, TX

13TH GLOBAL CONGRESS ON PROCESS SAFETY

17 AIChE Spring Meeting, San Antonio, TX

13TH GLOBAL CONGRESS ON PROCESS SAFETY

Improve Energy Efficiency of DWCs

Manish Bhargava
Licensing Manager
GTC Technology

Norihito Suzuki
Staff Engineer
TonenGeneral Sekiyu K.K.

GTC Technology
Engineered to Innovate®

TonenGeneral

Concept of Dividing Wall Columns

17 AIChE Spring Meeting, San Antonio, TX
13TH GLOBAL CONGRESS ON PROCESS SAFETY
GTC Technology · Engineered to Innovate® TonenGeneral

Conventional 2-Column Fractionation Sequence For Separating Three Components

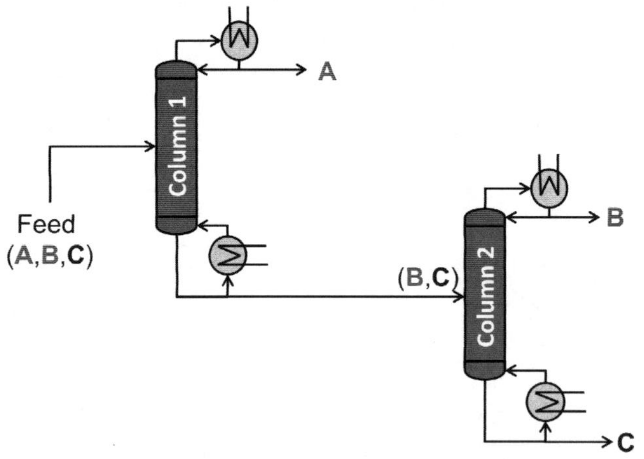

2017 AIChE® SPRING MEETING & 13TH GLOBAL CONGRESS ON PROCESS SAFETY

Thermodynamics Inefficiencies in Column 1

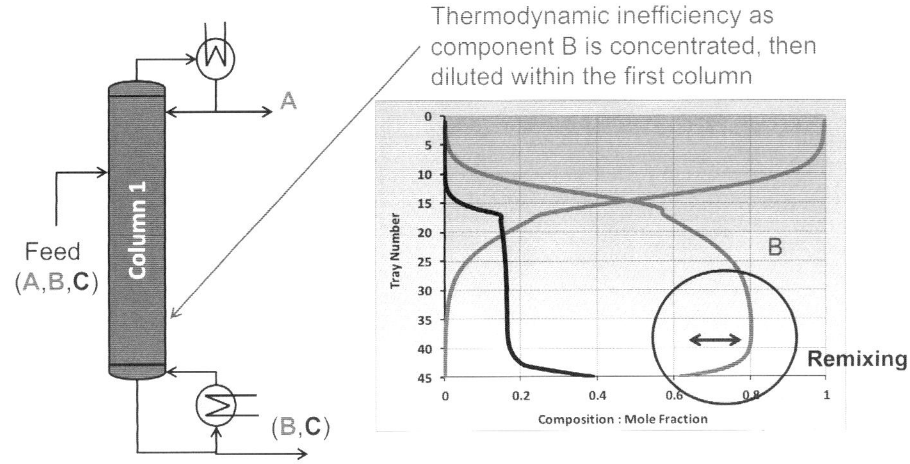

Pre-fractionator Main Column Arrangement

Composition Profile

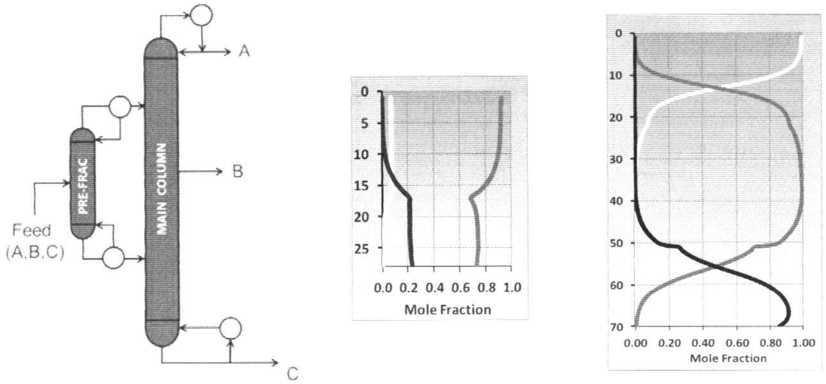

Main Column

Pre-fractionator Column

2017 AIChE® SPRING MEETING & 13TH GLOBAL CONGRESS ON PROCESS SAFETY

From Prefractionator Arrangement To Dividing Wall Column

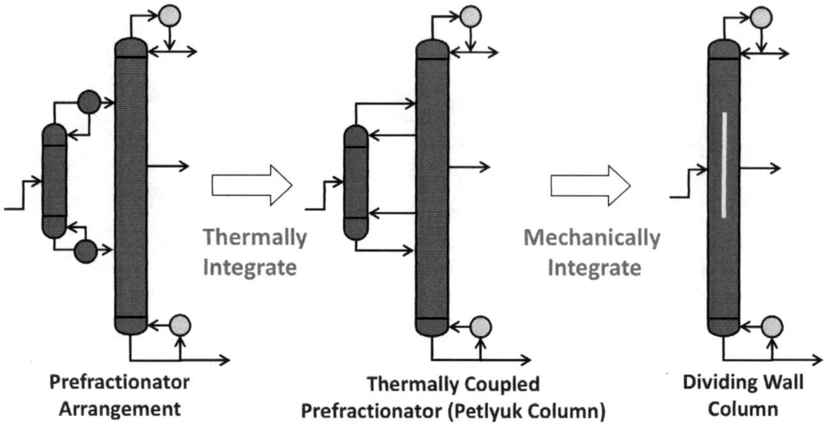

Typical Benefits of GT-DWC℠ (Dividing Wall Columns)

GT-DWC℠ CAPEX is lower by 20-30%

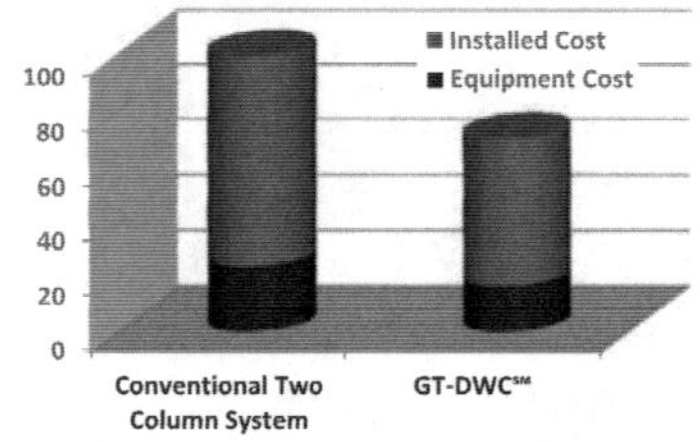

GT-DWC℠ OPEX is lower by 20-30%

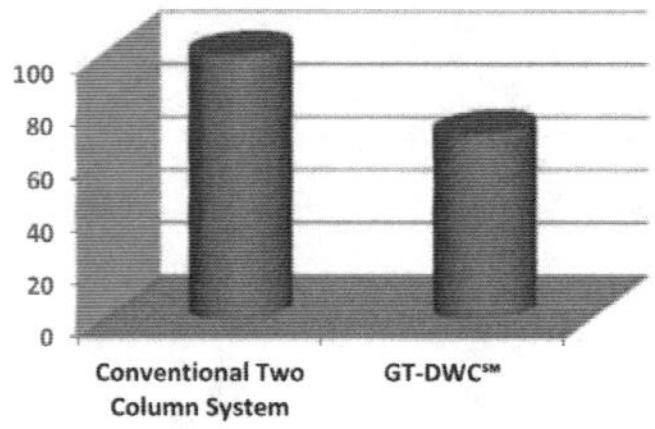

Equipment Count & Plot Space

	Conventional Two Column System	GT-DWC℠
Column	2	1
Reboiler	2	1
Condenser	2	1
Pump	4	3
Plot Space	100%	Approximately 30% less

2017 AIChE® SPRING MEETING & 13TH GLOBAL CONGRESS ON PROCESS SAFETY

Typical Dividing Wall Column Applications

- Naphtha Splitters
- Benzene/Toluene/Xylene/C9 Separation
- Debutanizer/Depropanizer
- Depropanizer/Deethanizer
- Reformate Splitters
- Naphtha Splitter/Deisohexanizer
- Depentanizer/Naphtha Splitter

17 AIChE Spring Meeting, San Antonio, TX
13TH GLOBAL CONGRESS ON PROCESS SAFETY
GTC Technology
TonenGeneral

Case Study 1 : Confidential Client in India
Two-column configuration modified to a single column

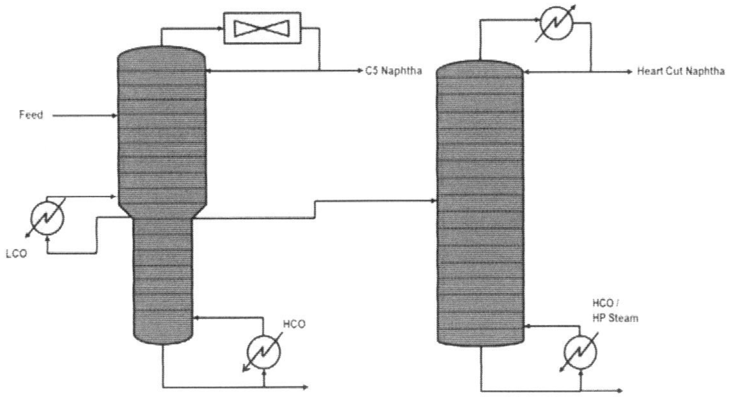

2017 AIChE® SPRING MEETING & 13TH GLOBAL CONGRESS ON PROCESS SAFETY

17 AIChE Spring Meeting, San Antonio, TX · 13TH GLOBAL CONGRESS ON PROCESS SAFETY · GTC Technology · Engineered to Innovate® · TonenGeneral

Objectives for Revamp

- Reduce utility consumption for the separation.
- Produce heart cut naphtha in Naphtha Splitter-1 column, while removing Naphtha Splitter-2 from service.

17 AIChE Spring Meeting, San Antonio, TX · 13TH GLOBAL CONGRESS ON PROCESS SAFETY · GTC Technology · Engineered to Innovate® · TonenGeneral

Bottlenecks in Column Design

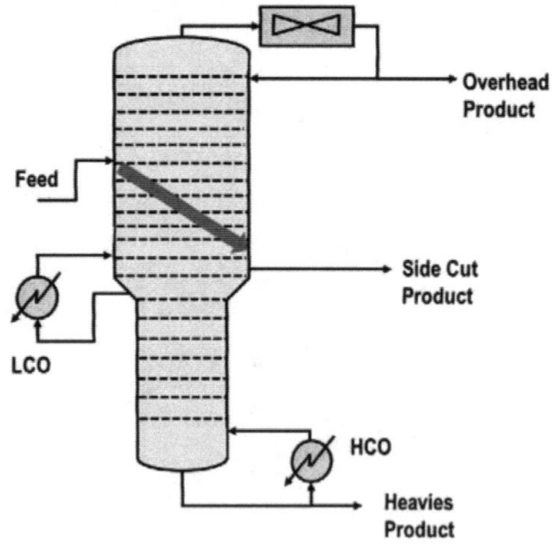

Intermixing of Feed with Side Cut

Lower Diameter Section Puts a restriction on Vapor Liquid Traffic

2017 AIChE® SPRING MEETING & 13TH GLOBAL CONGRESS ON PROCESS SAFETY

Revamp to GT-DWC℠

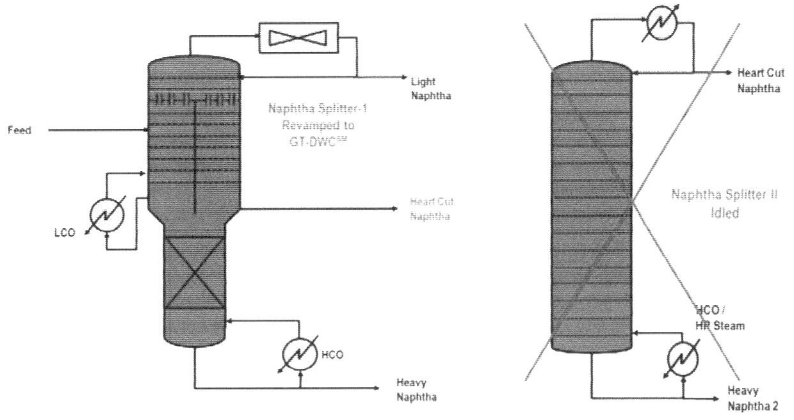

17 AIChE Spring Meeting, San Antonio, TX
13TH GLOBAL CONGRESS ON PROCESS SAFETY
GTC Technology TonenGeneral

Project Economics

Parameter	Units	Original Design (NS-I and NS-II)	Naphtha Splitter– I after Revamp to GT-DWC℠
Feed	TPH	308.0	308.0
Side Draw Flow Rate	TPH	165.0	165.0
D86 (IBP-FBP)	°C	110.6-170.0	110.0-170.0
Total Heating Duty	MMkcal/hr.	38.0	28.0

2017 AIChE® SPRING MEETING & 13TH GLOBAL CONGRESS ON PROCESS SAFETY

Benefits of GT-DWC℠

- Heating duty reduced by 26%.
- HP steam usage eliminated.
- Naphtha Splitter-II is idled.
- Heart Cut Naphtha obtained in Naphtha Splitter-I.

17 AIChE Spring Meeting, San Antonio, TX
13TH GLOBAL CONGRESS ON PROCESS SAFETY
GTC Technology · Engineered to Innovate®
TonenGeneral

Dividing Wall Column Trays

2017 AIChE® SPRING MEETING & 13TH GLOBAL CONGRESS ON PROCESS SAFETY

Top View of Dividing Wall

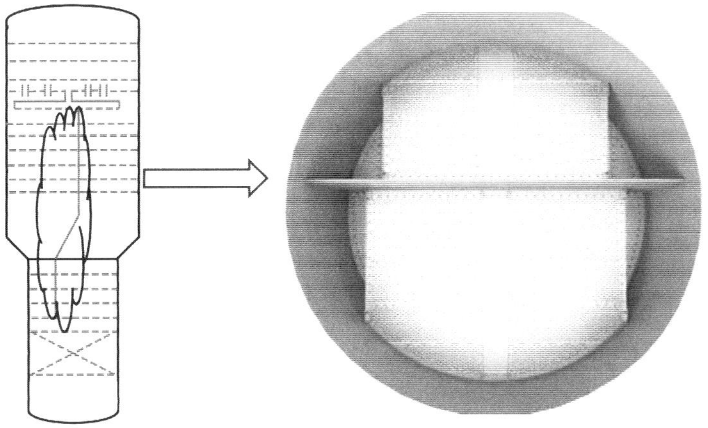

GTC Internal Liquid Split Distributer

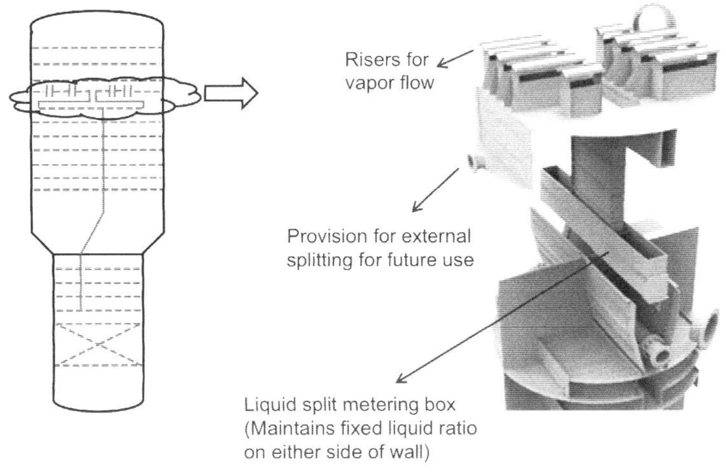

Case Study 2 : Grassroots Mixed Xylenes Recovery Unit at TonenGeneral (TG), Japan

- TG had an existing unit which produced C7+ product for gasoline blending.
- TG wanted to separate high purity petrochemicals (Toluene, Mixed Xylenes) from the feed.
- TG decided for a DWC solution against a two column configuration because of lower CAPEX and lack of plot space.

Case Study 2: Mixed Xylenes Recovery

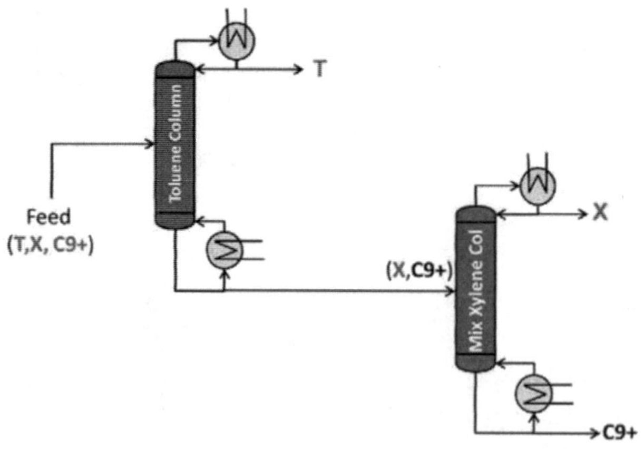

GT-DWC℠ for Mixed Xylenes Recovery

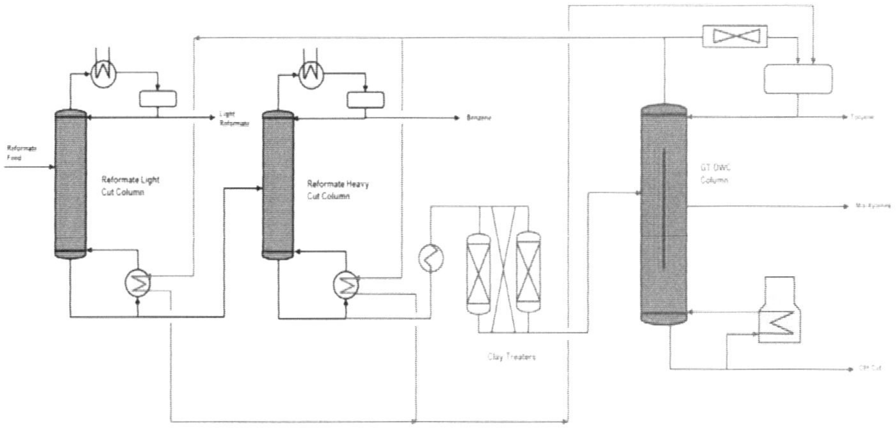

Comparison: Two Column System vs DWC

Parameter	Units	Two Column Configuration	GT-DWC℠ Configuration
Mix-xylenes product	Kg/hr	29,332	29,334
C8 aromatics	wt%.	99.2	99.3
Reboiler duties	MMkcal/hr.	21.6	17.2
Operating cost savings	%	-	20.0
Capital cost	US $ MM	26.0	21.0
Steam savings due to heat integration	TPH	-	20.0

Flexibility in Design

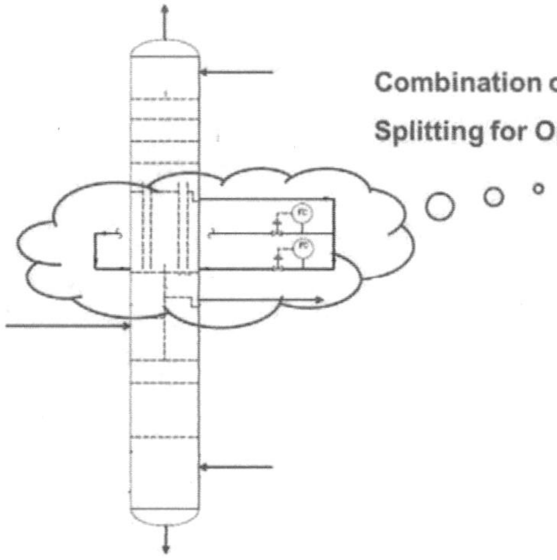

Combination of Internal and External Splitting for Operational Flexibility

17 AIChE Spring Meeting, San Antonio, TX
13TH GLOBAL CONGRESS ON PROCESS SAFETY
GTC Technology · Engineered to Innovate®
TonenGeneral

GT-DWC℠ Column Internals at TonenGeneral

Off Center Dividing Wall

2017 AIChE® SPRING MEETING & 13TH GLOBAL CONGRESS ON PROCESS SAFETY

TonenGeneral Mixed Xylenes Recovery Unit

Installation of GT-DWC℠ Column at Chiba Refinery, Japan

Column Operational since March 2016

17 AIChE Spring Meeting, San Antonio, TX
13TH GLOBAL CONGRESS ON PROCESS SAFETY
GTC Technology
TonenGeneral

GT-DWC℠ Internals Manufacturing - Korea

GT-DWC℠ Trays

Testing of GT-DWC℠ Internal Liquid Split Distributer

2017 AIChE® SPRING MEETING & 13TH GLOBAL CONGRESS ON PROCESS SAFETY

Heat Integrated GT-TDWC℠

- Conventional side draw column provides limited opportunity for heat integration
- Top dividing wall column gains the thermodynamic efficiency, plus affords meaningful heat integration.

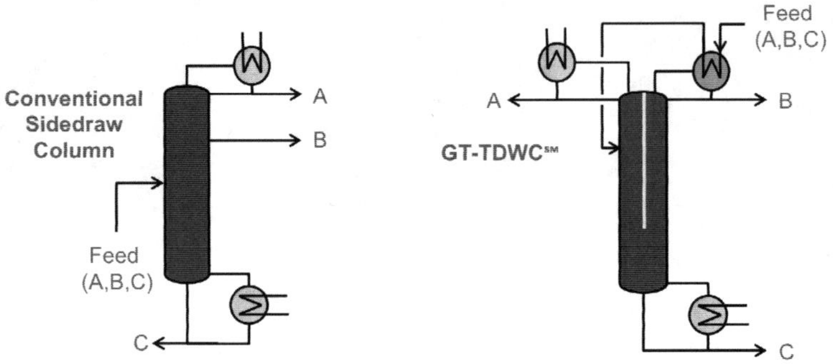

17 AIChE Spring Meeting, San Antonio, TX
13TH GLOBAL CONGRESS ON PROCESS SAFETY
GTC Technology · Engineered to Innovate®
TonenGeneral

Case Study 3: GT-TDWC℠ for Food Grade Hexane Production (FGH) for BPCL, India

FGH production requires a narrow cut of an intermediate product.

Options for application

1. Conventional 2 Column System : DIH + FGH column
2. Advanced Distillation Option : GT-TDWC

2017 AIChE® SPRING MEETING & 13TH GLOBAL CONGRESS ON PROCESS SAFETY

Conventional Option: DIH + New FGH Column

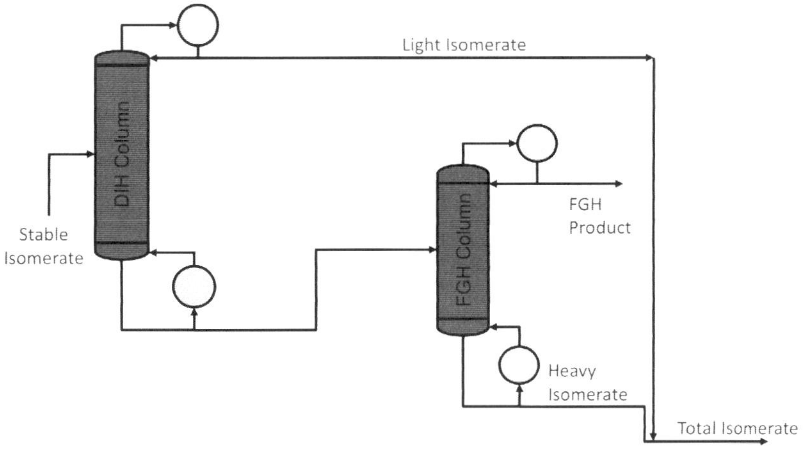

2017 AIChE® SPRING MEETING & 13TH GLOBAL CONGRESS ON PROCESS SAFETY

GT-TDWC℠ Column for Food Grade Hexane Production

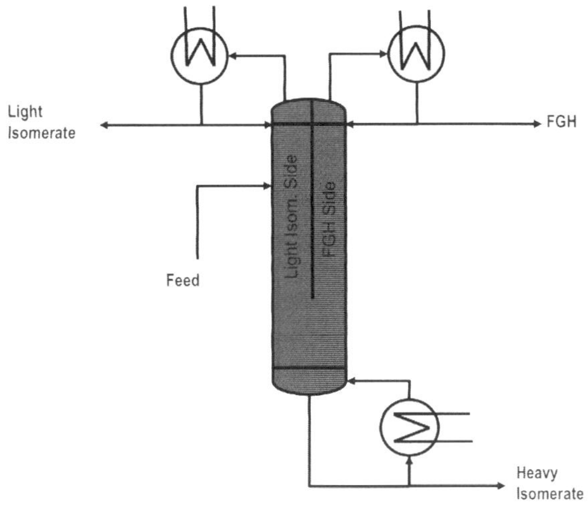

2017 AIChE® SPRING MEETING & 13TH GLOBAL CONGRESS ON PROCESS SAFETY

17 AIChE Spring Meeting, San Antonio, TX · 13TH GLOBAL CONGRESS ON PROCESS SAFETY · GTC Technology - Engineered to Innovate® · TonenGeneral

Comparison : GT-TDWC vs Conventional 2 Column System

	Conventional 2 Column System		GT-TDWC
	DIH Column	New FGH Column	
MP Steam Consumption (Tons/Hr)	26.20	8.50	26.46
	Total = 34.7		
Operating Cost (MM USD/year) Steam only	9.69		7.36
Savings (MM USD/year) Steam only	-		2.34

17 AIChE Spring Meeting, San Antonio, TX · 13TH GLOBAL CONGRESS ON PROCESS SAFETY · GTC Technology - Engineered to Innovate® · TonenGeneral

Top Dividing Wall Column Installation at BPCL, India

17 AIChE Spring Meeting — 13th Global Congress on Process Safety — GTC Technology — TonenGeneral

Case Study 4: GT-LPG Max℠ for LPG Recovery

Revamp of Existing LPG Recovery Unit in Texas to GT-LPG Max

Client Objectives:

- Propane recovery in LPG > 97%
- No Refrigeration

17 AIChE Spring Meeting — 13th Global Congress on Process Safety — GTC Technology — TonenGeneral

Case Study 4: GT-LPG Max℠ for LPG Recovery

Product Class	Product Specifications	Feed Composition: Component	Feed Composition: Wt%
Fuel Gas 3.8%	Minimize Propane	Hydrogen	0.0%
		Methane	0.3%
		Ethylene	0.0%
		Ethane	3.0%
		H_2S	0.5%
		CO_2	0.0%
LPG Product 11.8%	Propane > 90 wt% Ethane < 1 wt% Vapor Pressure < 208 psig @ 100°F	Propene	0.1%
		Propane	11.7%
I-C4 Product 8.3%	Propane < 3.0 wt% N-C4 < 4.0 wt%	i-Butane	8.2%
		i-Butene	0.1%
		1-Butene	0.1%
N-C4 Product 19.6%	Iso-Pentane < 3.0 wt% Iso-Butane < 5.3 wt%	n-Butane	19.6%
		Tr2-Butene	0.0%
Debutanizer Bottoms 56.5%		Heavies	56.6%

Lights are 3.7% of Total Feed

Considerable C5 and Heavy Components

17 AIChE Spring Meeting, San Antonio, TX | 13TH GLOBAL CONGRESS ON PROCESS SAFETY | GTC Technology · Engineered to Innovate® | TonenGeneral

Conventional Design of LPG Recovery Unit

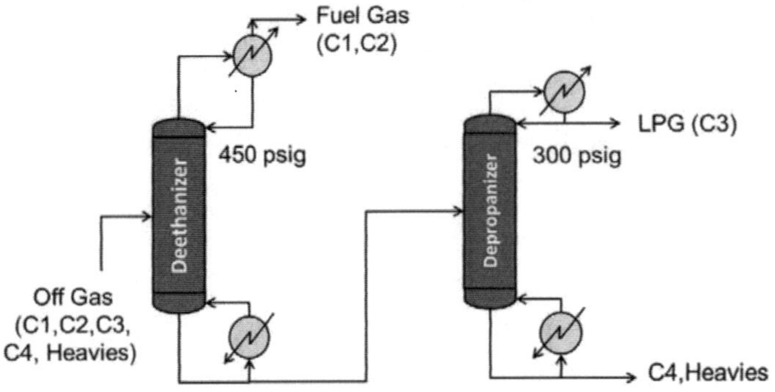

- LPG Recovery – 55%
- Energy – 22.0 MMBtu/hr

17 AIChE Spring Meeting, San Antonio, TX | 13TH GLOBAL CONGRESS ON PROCESS SAFETY | GTC Technology · Engineered to Innovate® | TonenGeneral

GT-LPG MAX℠ - Combined Absorption & Distillation in a Single Column

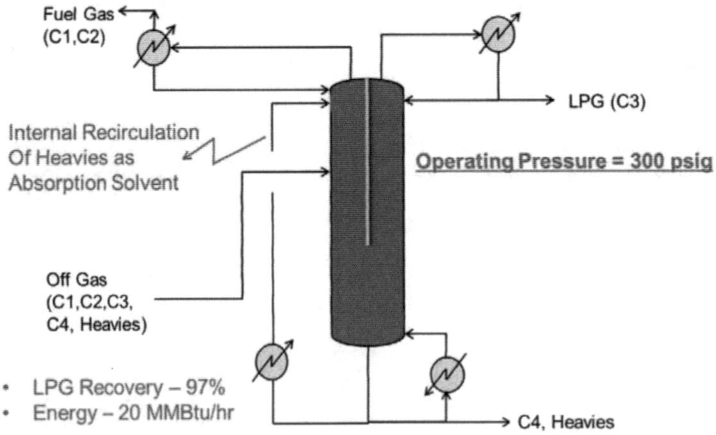

- LPG Recovery – 97%
- Energy – 20 MMBtu/hr

2017 AIChE® SPRING MEETING & 13TH GLOBAL CONGRESS ON PROCESS SAFETY

17 AIChE Spring Meeting, San Antonio, TX — 13TH GLOBAL CONGRESS ON PROCESS SAFETY — GTC Technology — TonenGeneral

Project Economics of GT-LPG MAX℠

Variables	Existing Configuration	GT-LPG Max℠
Overall Propane Recovery %	55 %	97 %
Total Duty MM kcal/hr	22.0 MMBtu/hr	20.0 MMBtu/hr

- Net Benefit/Year — $MM — 10.1
- Total Installed Cost — $MM — 15.0
- Simple Pay Back — Months — 18.0

2017 AIChE® SPRING MEETING & 13TH GLOBAL CONGRESS ON PROCESS SAFETY

17 AIChE Spring Meeting, San Antonio, TX — 13TH GLOBAL CONGRESS ON PROCESS SAFETY — GTC Technology — TonenGeneral

Summary

- Unconventional distillation applications are an overlooked means to reduce refinery energy consumption
  - GT-DWC reduces 20–30% OPEX through energy savings
  - GT-DWC reduces 20–30% CAPEX by requiring a single column for multi-component separation
- Advanced Distillation schemes can offer CAPEX and OPEX reduction of more than 40% by combining multiple unit operations and heat integration opportunities.

2017 AIChE® SPRING MEETING & 13TH GLOBAL CONGRESS ON PROCESS SAFETY

17 AIChE Spring Meeting, San Antonio, TX

13TH GLOBAL CONGRESS ON PROCESS SAFETY

Improve Energy Efficiency of DWCs With Overhead Heat Integration

Manish Bhargava
Licensing Manager
GTC Technology

GTC Technology · Engineered to Innovate®

Norihito Suzuki
Staff Engineer
TonenGeneral Sekiyu K.K.

TonenGeneral

2017 AIChE® SPRING MEETING & 13TH GLOBAL CONGRESS ON PROCESS SAFETY

Component Trapping with VLE Uncertainty: Principles and Troubleshooting

By

Paul M. Mathias,[1] *Henry Z. Kister,*[1] *Bruce Parker,*[2] *Lydia Narvaez,*[3] *Thomas Schafer,*[4] *Alan Erickson*[4]

[1] Fluor Corporation, 3 Polaris Way, Aliso Viejo, CA 92698

[2] Eli Lilly and Company, Lilly Corporate Center, Indianapolis, IN 46221

[3] Fluor Daniel Caribbean, Inc., Suite 500 St. 2 No. 14 Metro Office Park Guaynabo, Puerto Rico 00969

[4] Koch Modular Process Systems, LLC, 45 Eisenhower Drive, Suite 350, Paramus, NJ 07652

Presented at the Kister Distillation Symposium, AIChE Spring Meeting
San Antonio, Texas, March, 2017

Component Trapping with VLE Uncertainty:

Principles and Troubleshooting

Paul M. Mathias,[1] Henry Z. Kister,[1] Bruce Parker,[2] Lydia Narvaez,[3] Thomas Schafer,[4] Alan Erickson[4]

[1] Fluor Corporation, 3 Polaris Way, Aliso Viejo, CA 92698

[2] Eli Lilly and Company, Lilly Corporate Center, Indianapolis, IN 46221

[3] Fluor Daniel Caribbean, Inc., Suite 500 St. 2 No. 14 Metro Office Park Guaynabo, Puerto Rico 00969

[4] Koch Modular Process Systems, LLC, 45 Eisenhower Drive, Suite 350, Paramus, NJ 07652

Summary

Intermediate components tend to accumulate near the middle of a distillation column in many chemical and petroleum separations. Their accumulation may lead to off-spec products, corrosion, plugging, or periodic cycling. The most common method of removing these intermediate components is by taking them out in one or more a side draws. The addition of such side draws may not go far enough to achieve the desired product specs. If the side-draw approach fails, the expensive addition of a new tower may be required.

The question of whether a side draw is sufficient to remove the accumulating intermediate components, or whether an additional column is needed, depends on the relative volatilities of the intermediate components at tower conditions. Often, there is a high degree of uncertainty in the correlation of the volatilities of these components. The Margules Uncertainty Analysis method is valuable in analyzing these situations, and in guiding engineers to the most effective path.

The case analyzed here is an acetonitrile-water separation tower with t-crotonaldehyde and propionitrile as the intermediate components. There are uncertainties in the correlation of the volatilities of these intermediate components in this system. In the light of these uncertainties, our analysis addresses the question of whether the tight product specs can be achieved in one tower with a side draw or whether more than one tower is needed. The analysis shows that depending on the product specs and the tolerance for a high reflux ratio, there would be situations where one can be certain that one tower will work, while in other cases more than one tower is needed to assure achievement of the required product specs. In the present case, a single distillation tower is judged to be too risky, and hence the more conservative 3-column solution was chosen.

Introduction

This paper studies the options to purify an aqueous solution of ≈ 12 wt% acetonitrile (ACN) by distillation, and to recover about 90% of the ACN at the azeotropic composition (about 83 wt% at 1 atmosphere or 101.3 kPa). The separation is difficult due to intermediate components [propionitrile (PPN) and crotonaldehyde (T-CROTON] that build-up in the distillation column. The solution options are a single column with a side draw, or a more complex flowsheet with multiple columns. An example of a multiple-column solution is to remove all of the water from the ACN, PPN and T-CROTON via a pressure swing distillation, and then separate the dry ACN from the PPN and T-CROTON. The multiple-column solution is the safe solution, but costs more than the single column with a side draw in terms of both capital cost and operating costs. Here we quantify the risk of the single-column solution by introducing uncertainty analysis into the process modeling.

The problem of intermediate components in distillation is a general problem of component trapping that has been discussed extensively by Kister.[1,2] In some cases, the top temperature is too cold and the bottom temperature is too high to allow these components to leave the column at the same rate as they enter. Having nowhere to go, these components accumulate in the column, which causes problems like off-spec products, flooding, cycling slugging, and corrosion. Cycling on a variable time period (typically with frequencies between two hours and two weeks) is a common symptom indicating unsteady state component trapping. Diagnoses of the problem include studying time variation of temperature profiles, taking internal samples, shooting gamma scans, and simulation studies. Solutions comprise raising or lowering the temperature (usually at the expense of recovery), taking side draws, or removing the problem components from the feed. Diagnoses and solutions have been discussed in detail by Kister.[1,2] Here we apply uncertainty analysis to the problem caused by trapping of intermediate components in ACN purification.

Modeling of the distillation separations requires gaining quantitative understanding of the uncertainties in the vapor-liquid equilibrium (VLE) and applying these uncertainties to the problem at hand. While many technologists have highlighted the need to recognize the impact of uncertainties in property models on process design and plant operability[3,4,5,6,7] and several solutions have been suggested and evaluated in the literature,[8,9,10,11,12,13,14] uncertainty analysis is hardly used in chemical-engineering design practice. In fact, Kim et al.[15] observed, "Its practical implementation in a variety of scientific and engineering fields has typically seen less emphasis than it deserves." Mathias[16] recognized that liquid-phase nonideality is the major source of phase-equilibrium uncertainty, and developed an intuitive and easy-to-apply uncertainty-analysis method based upon treating the mixture, for the purpose of activity-coefficient perturbation, as a set of pseudobinaries described by the Margules equation. The Margules perturbation method has been applied to several case studies, including propylene-propane superfractionator, where small changes in relative volatility and product purity have a large effect on design;[16] a dehexanizer column that removes pentane and hexane from a heavier slate of components that are mostly aromatic and the system contains many components (≈ 60); [16] the "textbook problem" of acetone/benzene/chloroform separation, which revealed the relationship between uncertainty analysis and residue-curve analysis;[17] the butanol/water separation, which demonstrated the extension to liquid-liquid equilibrium (LLE) and is a case where the thermodynamic model is inadequate;[18] and ethyl acetate purification, which showed the importance of fitting multicomponent data for complex systems.[19]

The Margules uncertainty methodology used here follows that of the previous studies.[16,17,18,19] The first step is detailed data regression to obtain the best model fit in the region of application (atmospheric pressure, about 101.3 kPa), and quantitative estimation of the correlation uncertainties. This is a necessary first step. The data regression here is mainly applied to the water/ACN binary mixture, with PPN and T-CROTON at low concentration. In the second step, activity coefficients are varied within the process model to relate the design uncertainties to the confidence limits of the correlated phase equilibria.

Margules-Based Activity-Coefficient Perturbation Scheme

The perturbation scheme for the mixture nonideality has been published by Mathias,[16] and hence is only briefly presented here. The approach uses the approximation that the perturbed activity coefficients can be guided by the Margules equation.

$$\ln(\gamma_i) = \ln(\gamma_i^m) + \ln(\gamma_i^p) \tag{1}$$

where γ_i is the activity coefficient of component i after perturbation, γ_i^m is the activity coefficient calculated by the chosen model (the "best" model in the region of interest should be used), and γ_i^p represents the perturbation obtained through the estimated model uncertainty. It should be emphasized that γ_i^m is calculated from the "best" activity-coefficient model, and is not in any way constrained by the pseudo-binary approximation. γ_i^p is given by,

$$\ln(\gamma_i^p) = \delta_i (1-x_i)^2 \left[\frac{|\ln(\gamma_i^m)|}{(1-x_i)^2 + |\ln(\gamma_i^m)|} \right] \tag{2}$$

As previously noted,[16] Equation 2 is phenomenological, but has desired characteristics. The value of $\ln(\gamma_i^p)$ is small when γ_i^m is close to unity (i.e., low departures from ideal behavior), and also is low when the mole fraction of component i is close to unity. For components at concentrations close to infinite dilution and with activity coefficients significantly different from unity, the fractional change in γ_i resulting from the perturbation asymptotically approaches $\{exp(\delta_i) - 1\}$. In order to apply the perturbation method, the value of δ_i for each component must be related to the estimated uncertainty in its activity coefficient. This thorough data analysis is a necessary element of the perturbation scheme and is discussed in detail later in the paper.

Correlation of VLE in the Water/ACN/PPN/T-CROTON System

As in the previous uncertainty-analysis case studies,[16,17,18,19] the perturbation scheme has been implemented as a user activity-coefficient model in the NRTL-RK property option in Aspen Plus V8.8. The NRTL-RK property option uses the NRTL activity-coefficient model[20] and the Redlich-Kwong equation of state[21] for the vapor phase. The pure-component properties come from the Aspen Plus V8.8 pure-component database. This study is at a pressure of 1 atm (101.3 kPa), the temperatures of interest are approximately 70-100 °C, and at these temperatures the correlations for the vapor pressures of these four compounds from the Aspen Plus database are expected to be accurate to better than 1 %. The reason for this confidence is that the Aspen Plus pure-component properties come from the DIPPR 801 database,[22] and the correlated vapor pressures have been reported to be accurate to <1%. The Redlich-Kwong equation estimates the vapor-phase fugacity coefficient, but this thermodynamic quantity is expected to be close to unity and accurately predicted since the pressure of interest is low, ≈1 atm. For this application, the phase-equilibrium uncertainty is largely determined by the uncertainties in the component activity coefficients.

The NRTL model has two temperature-dependent binary parameters (α_{ij}^N and τ_{ij}). α_{ij}^N is a symmetric binary parameter and τ_{ij} is an asymmetric binary parameter,

$$\tau_{ij} = a_{ij} + b_{ij}/T + e_{ij}\ln(T) + f_{ij}T \tag{3}$$

$$\alpha_{ij}^N = c_{ij} + d_{ij}(T - 273.15) \tag{4}$$

where, T is the temperature in Kelvin.

The K-value of component i in a mixture (K_i) at vapor-liquid equilibrium (VLE) is defined to be the ratio of the vapor and liquid mole fractions at equilibrium, and the relative volatility between components i and j (α_{ij}) is the ratio of the K-values.

$$K_i \equiv \frac{x_i}{y_i} \tag{5}$$

$$\alpha_{ij} \equiv \frac{K_i}{K_j} \tag{6}$$

Note that the superscript "N" has been added to the NRTL parameter α_{ij}^N in order to distinguish it from the relative volatility, α_{ij}.

The Data Regression System (DRS) in Aspen Plus V8.8 has been used to obtain optimum values of the NRTL binary parameters at pressures at about 1 atm (101.3 kPa), which corresponds to the temperature range of ≈ 70-100°C. Phase-equilibrium data were obtained from the NIST-TDE[23] database, and key VLE data have been measured as part of this effort. The water/ACN binary is well-studied systems since a large number of datasets are available, however, the four binaries consisting of PPN and T-CROTON with water and ACN have only a few data. In fact, data for systems with PPN are very poorly covered, and hence this study has mainly focused on the effect of T-CROTON as the representative intermediate component.

Table 1 presents the 14 components considered in the Aspen Plus simulation together with their boiling points. The table also shows the estimated infinite-dilution activity coefficients in water and ACN at a representative temperature of 75°C; these correlations/estimations are discussed later in this paper. The last column of Table 1 designates components other than water and ACN as "light" or "heavy." The "light" components (1-4) have normal boiling points lower than that of ACN. These components will always have K-values greater than unity, and hence will split predominately to the distillate in the distillation column. The "heavy" components (10-14) have normal boiling points greater than that of water, and also are fairly ideal with water (infinite-dilution activity coefficients close to unity). These "heavy" components will always have K-values less than unity, and therefore will largely split to the bottoms stream of the distillation column. The "intermediate" components have normal boiling points higher than that of ACN, and form fairly ideal mixtures with ACN, consequently their K-values are likely to be below unity at the top of the column (in the vicinity of the water/ACN azeotrope). Further, the "intermediate" components have T_b values close to that of water and also water infinite-dilution activity coefficients significantly greater than unity (see Table 1), and therefore their K-values will be greater than unity at the bottom of the column. As a result, the intermediate components have K-values that cross unity and tend to build-up in the column.

Table 1. Components in the acetonitrile purification Aspen Plus model.

Component Name	T_b (°C)	Component ID	γ^∞ at 75°C in		Impurity Category
			Water	ACN	
1. Acetaldehyde	21.0	ACETALD	9.0	1.1	Light
2. Dichloromethane	39.8	DCM	235.9	1.2	Light
3. n-Propionaldehyde	49.0	PROPALD	32.2	1.3	Light
4. Acrolein	52.7	ACROLEIN	17.1	1.1	Light
5. Acetonitrile	81.6	ACN	11.7	1.0	-
6. cis-Crotonaldehyde	88.7	C-CROTON	32.5	1.4	Intermediate
7. Propionitrile	97.4	PPN	41.6	1.0	Intermediate
8. Water	100.0	WATER	1.0	6.1	-
9. trans-Crotonaldehyde	102.5	T-CROTON	32.5	1.4	Intermediate
10. Acetic Acid	117.9	ACETACID	1.3	1.9	Heavy
11. 2-Methoxyethanol	124.4	MEOXYETH	2.7	2.0	Heavy
12. Morpholine	128.0	MORPHOL	1.1	1.0	Heavy
13. Acetamide	221.2	ACETAMID	1.0	1.0	Heavy
14. 4-Ethylmorpholine	239.4	ACETMOR	1.1	1.0	Heavy

Crotonaldehyde has two isomers, *cis* and *trans*. The study of McGreer and Page[24] found that cis-crotonaldehyde is far less chemically favored than trans-crotonaldehyde, by about 1:50. Hence, although cis-crotonaldehyde is available in the simulation file, its properties are not considered reliable, and crotonaldehyde has been treated as entirely the trans isomer (T-CROTON)

It is clear that the system of interest to this study consists of 4 components: water, ACN, PPN and T-CROTON. For the other binary mixtures, the NRTL parameters were obtained by regressing data from NIST-TDE,[23] or were estimated using the UNIFAC method.[25] It is noted that the K-values of the other components (1-4 and 10-14 in Table 1) have negligible effect on this study.

Table 2. NRTL parameters for the water/ACN/PPN/T-CROTON system. See Eqs. 3 and 4 for a description of the parameters. Parameters e_{12}, e_{21}, f_{12}, f_{21} and d_{12} are always equal to zero.

Component 1	ACN	PPN	WATER	ACN	ACN	PPN
Component 2	WATER	WATER	T-CROTON	PPN	T-CROTON	T-CROTON
a_{12}	-2.34	-4.53	5.38	0	0	0
a_{21}	2.39	2.99	-2.34	0	0	0
b_{12}	1165.5	2041.1	-751.4	27.1	55.1	0
b_{21}	-199.2	-9.9	914.6	-30.0	55.1	0
c_{12}	0.45	0.41	0.3	0.3	0.3	0.3

Table 2 presents the NRTL parameters for the water/ACN/PPN/T-CROTON system. The water/ACN binary is well covered by experimental VLE data. The binaries of water with PPN and T-CROTON are also covered by VLE and LLE data; it is noted that these two binaries have regions of liquid-liquid immiscibility, but the concentrations of PPN and T-CROTON in the process are too low to exceed their solubility limit. No data were found in the literature for the other three binaries. Data for the water/ACN/T-CROTON ternary were measured as part of this project.[26] The NRTL parameters for the ACN/PPN are estimated based upon UNIFAC predictions.[25] Generally, UNIFAC predictions are not accurate, so future measurement was considered pending outcome of this study. The NRTL parameters for the PPN/T-CROTON binary have been defaulted to zero. UNIFAC estimations[25] indicate that this is a fairly ideal system (infinite-dilution activity coefficients ≤ 1.2). Minor components have a negligible effect on overall mixture nonideality, and it was decided that the assumption of ideality is the preferred choice.

Key comparisons of model results to experimental data are provided next in order to estimate the VLE uncertainties. Figure 1 compares the calculated relative volatility in the water/ACN binary to data from 9 sources available in NIST-TDE.[23] The model describes the data, including the azeotropic region, accurately, and there is no indication of model bias. The dashed lines show $\pm$ 20% from the model calculation. It is noted that that the water/ACN binary is not a significant cause of design uncertainty, and no uncertainty perturbation has been applied to the water/ACN binary. Figure 2 presents comparisons between model and data[26] for the K-value of T-CROTON at low concentration (effectively infinite dilution) in the water/ACN binary mixture at 1 atm (101.3 kPa). The dashed lines indicate $\pm$ 30% from the model results. Most of the data points fall within the dashed lines, and this may be considered to be the combined uncertainty (about 3 standard deviations) of the model K-values for T-CROTON. The vertical line in Figure 2 at 83 wt% shows the ACN/water azeotrope, which corresponds to the distillate composition from the column, and it is clear that both the model and experimental data indicate that the K-value of T-CROTON drops below unity before reaching the top of the column, which, of course, is the reason that T-CROTON is an intermediate component. Figure 3, which shows the K-value of PPN, is analogous to Figure 2. While the calculated results are equivalent for PPN and T-CROTON, there are no experimental data available for PPN, and hence the perturbation studies here are limited to T-CROTON, with the expectation that they will be approximately applicable to PPN.

Figure 4 provides a visual indication of the effect of $\delta_{T\text{-}CROTON}$ (hereafter simply referred to as δ, since the Margules perturbation in this work is only applied to T-CROTON). The bottoms concentration is nearly-pure water and the distillate concentration corresponds to the water/ACN azeotrope. Referring to Eq. 2 and Figure 2 it is clear that for a given value of δ, the perturbation at the bottom of the column (higher T-CROTON activity coefficient) will be higher than at the top of the column, and that a value of $\delta = 0.4$ is a "full" perturbation, corresponding effectively to the 30% uncertainty in Figure 2. Note that only positive values of δ are relevant to the study because increased values of δ will tend to increase the concentration of T-CROTON in the distillate, which makes it more difficult to meet the design goal of low T-CROTON concentration in the distillate.

Two design studies have been conducted to evaluate the effect of T-CROTON VLE uncertainty on the performance of the single column, and these are described next.

Design Study 1 – Single Column with Side Draw

Figure 5 presents the single column with side draw used for the first set of design studies. The feed flow is 770 kg/hr and is an aqueous mixture containing 12 wt% ACN, and with minor impurities. The impurity levels of PPN and T-CROTON in the feed are 100 ppmw and 20 ppmw, respectively. The column has 35 theoretical stages including a reboiler and total condenser. The distillate-to-feed ratio is set at 0.13, which corresponds to $\approx$ 91% ACN recovery at the azeotropic composition. The reflux ratio is varied to obtain a T-CROTON concentration in the distillate of 1 ppmw. The feed stage is varied from 12 to 24. The base case is the feed to stage 20, and for this case the calculated reflux ratio is 7.0. Figure 6 shows the vapor and liquid concentration profiles (weight fraction) for ACN, PPN and T-CROTON in the distillation column for the base case. The concentration of ACN rises with stages toward the condenser (here in decreasing stage number) and quickly reaches the water/ACN azeotropic concentration just a few stages above the feed stage. The concentration profiles of PPN and T-CROTON show the characteristics of intermediate components since they both have concentration peaks below the feed stage. Note that the concentration peak of T-CROTON is about three orders of magnitude above the distillate and bottoms concentrations. The distillate and bottoms concentrations of both PPN and T-CROTON are both small since most of these two components are removed in the side draw. Figure 7 emphasizes these points by showing the K-value profile of T-CROTON alongside its concentration profile. Note that the relative values of the vapor and liquid concentrations switch as the K-value moves from less than one at the top of the column (lower stage numbers) to greater than one at the bottom of the column.

Figure 8 studies how the reflux ratio needs to be adjusted to attain the spec of 1 ppmw T-CROTON in the distillate as the perturbation parameter (δ) and the number of rectification stages are varied. Increasing the number of rectification stages is helpful in reducing the required reflux ratio, and this can inferred from Figure 6 (and seen in Figure 8) because the concentrations of PPN and T-CROTON decrease monotonically in the rectification section of the column, but the effect is relatively weak. On the other hand, increasing δ (increased activity coefficient and therefore K-value of T-CROTON) results in a strong increase in the required reflux ratio. Very large reflux ratios cause problems, for example requiring large-diameter columns and significant increases in reboiler duty, and are usually not accepted in industrial process-design practice

Figure 9 studies the effect of the draw rate for the base case (Figure 5 with reflux ratio of 7.0 and feed to stage 20). It is clear that varying the draw rate has a strong effect on the concentration of T-CROTON in the bottoms (because it changes the total amount of T-CROTON removed in the side draw), but it has only a weak effect on the concentration of T-CROTON in the distillate. Since the main goal here is to reduce the concentration of PPN and T-CROTON in the distillate, the next study is done without the side draw.

Design Study 2 – Single Column with no Side Draw

Figure 10 presents the single column with no side draw used for the second set of design studies, and with the total number of stages increased from 35 to 40. Other design conditions are the same as in Figure 5. Figure 11 is qualitatively the same as Figure 8, which supports the conclusion that the side draw affects the bottoms concentration of T-CROTON, but has only a weak effect on the key result we desire, which is minimizing the concentration of T-CROTON in the distillate.

Figure 12 presents the results of a study where the reflux ratio has been fixed at 10.0, and the perturbation parameter (δ) and the number of rectification stages are varied. Similar to Figure 8, Figure 12 shows that it is beneficial to increase the number of rectification stages as this reduces the concentration of T-CROTON in the distillate. However, the key result is that the concentration of T-CROTON in the distillate is a very strong function of the uncertainty in the VLE. As δ increases not only does the concentration of T-CROTON in the distillate increase beyond the target value (1 ppmw), but it can also exceed the feed concentration (20 ppmw), in effect concentrating undesired impurities in the product!

The single-column solution was rejected as being too risky.

Conclusions

The purpose of this study was to apply the Margules uncertainty method to an industrially important component-trapping case, i.e., purification of acetonitrile used as a solvent in the pharmaceutical industry. Small amounts of T-CROTON and PPN turn a simple distillation into a very challenging problem. The study first focused on the VLE correlation, and determined that the key uncertainty is the K-values of the intermediate components, PPN and T-CROTON. Since K-value data are only available for T-CROTON, this component was chosen as the representative intermediate key component. The first set of studies indicated that the target of 1 ppmw T-CROTON in the distillate can be met, at least in the simulation model, but at a significant increase in reflux ratio if the T-Croton K-value is increased within the established uncertainty. The second set of studies concluded that increasing the K-value of T-CROTON within the established uncertainty caused a significant increase in the T-CROTON concentration in the distillate, and the product concentration could even exceed the feed concentration. The three-column solution was the eventual low-risk process solution selected.

List of Tables

List of Figures

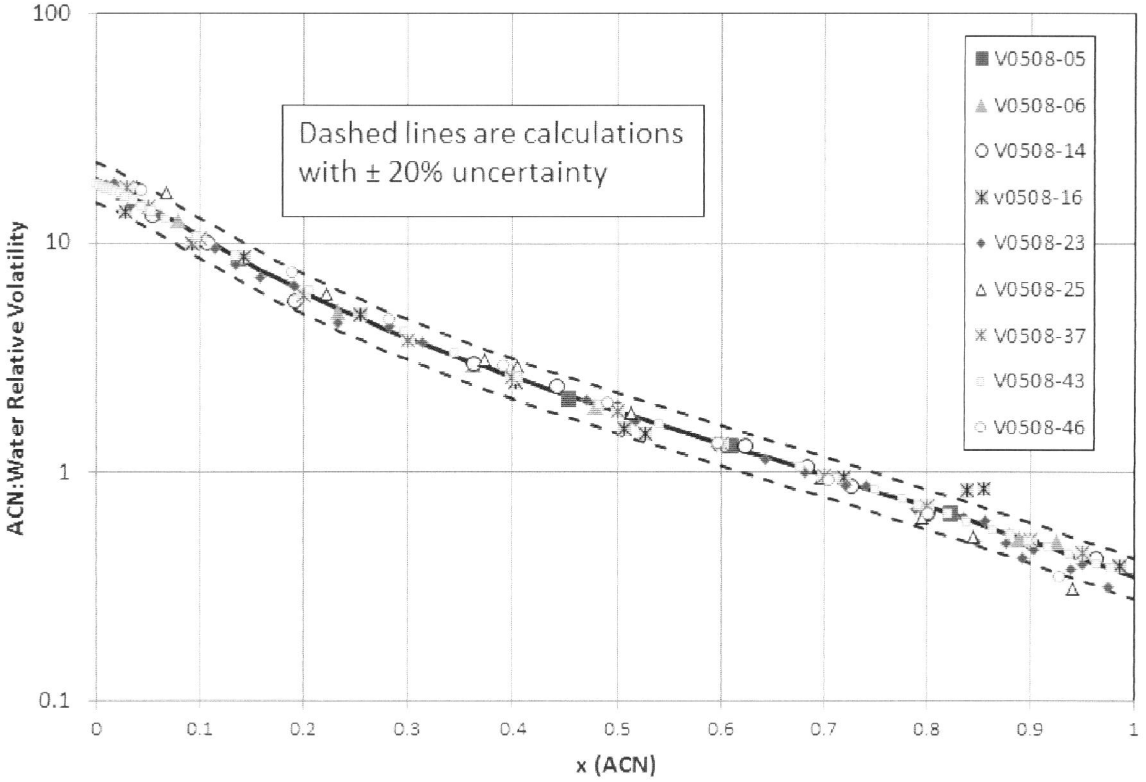

Figure 1. Water/ACN relative volatilities at 1 atm (101.3 kPa). Comparison of model calculations to 9 data sources from NIST-TDE.[23]

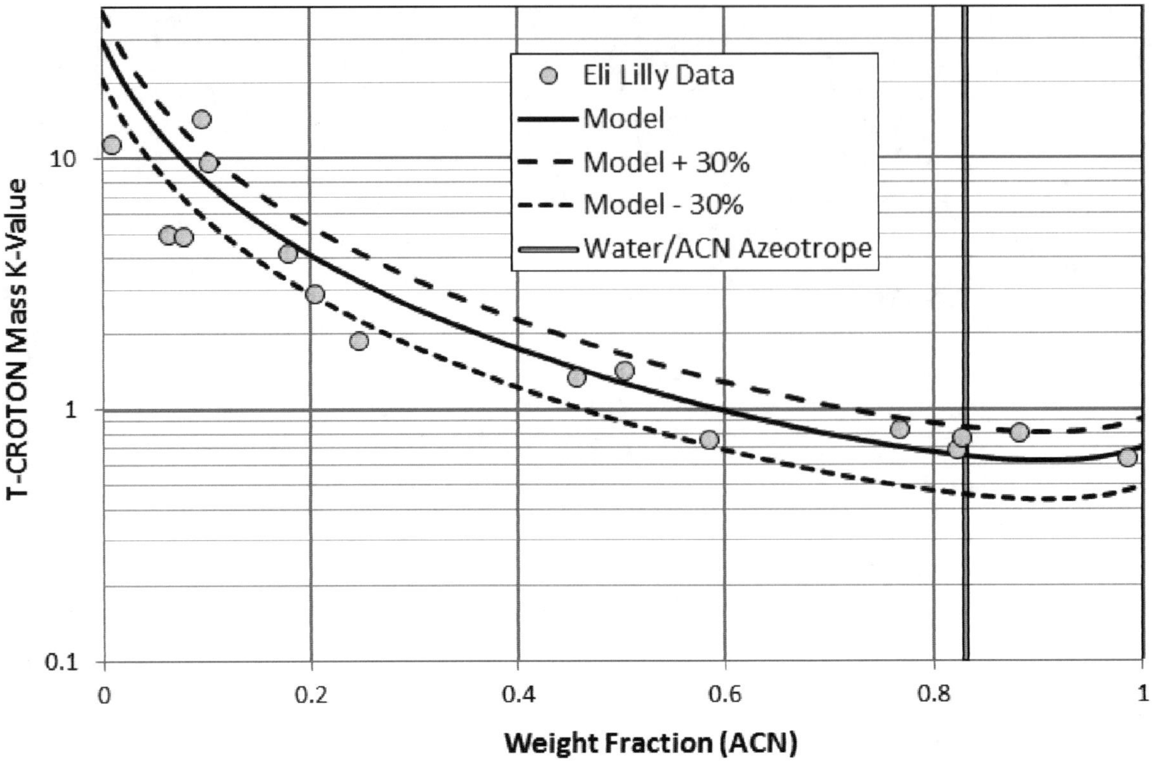

Figure 2. K-Value of T-CROTON at low concentration in water/ACN at 1 atm (101.3 kPa). Comparison of model calculations to Eli Lilly data.[26] The dashed lines show ± 30% from the model results. The vertical line indicates the water/ACN azeotrope.

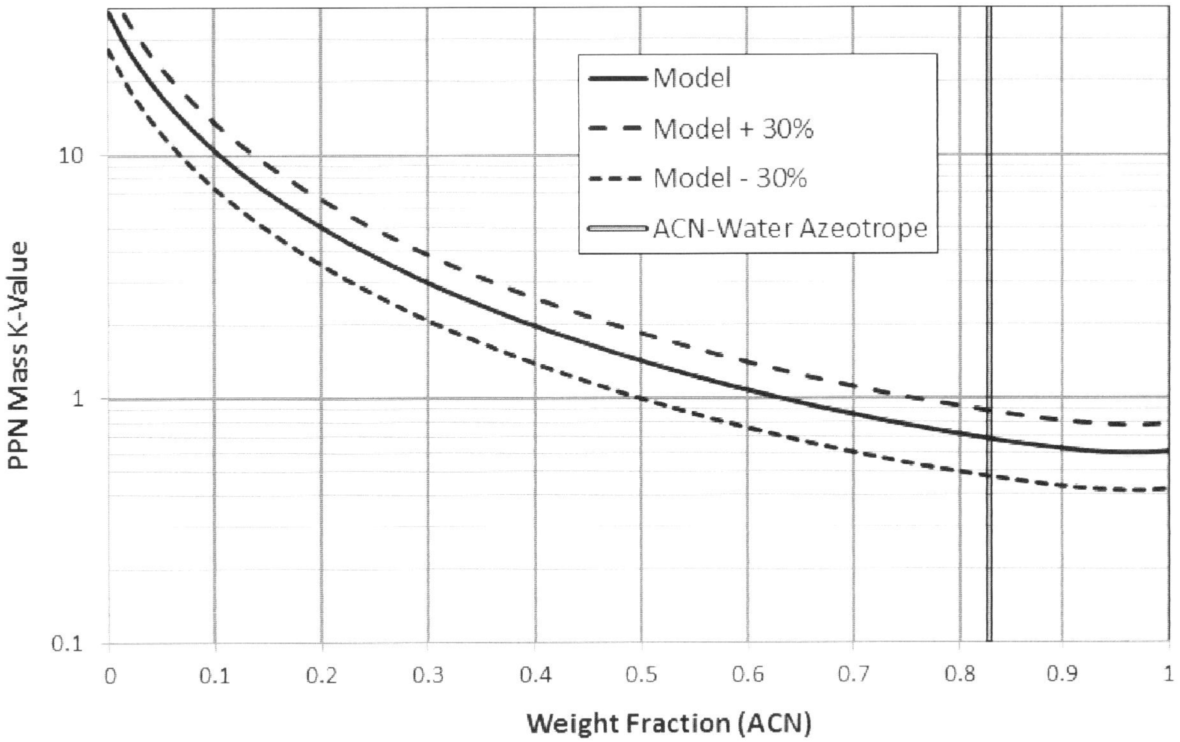

Figure 3. K-Value of PPN at low concentration in water/ACN at 1 atm (101.3 kPa). No experimental data are available. The dashed lines show ± 30% from the model results. The vertical line indicates the water/ACN azeotrope.

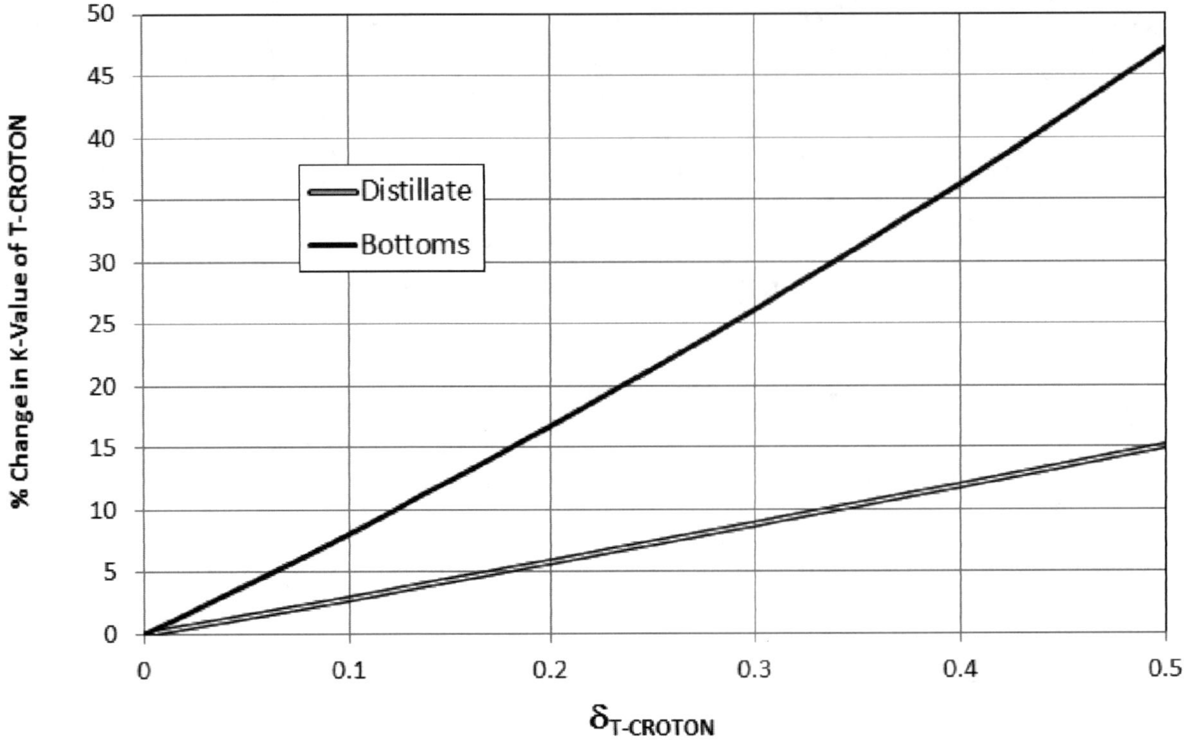

Figure 4. Effect of T-CROTON perturbation parameter (δ) on the percentage change in the K-value of T-CROTON. The bottoms concentration is nearly-pure water and the distillate concentration corresponds to the water/ACN azeotrope.

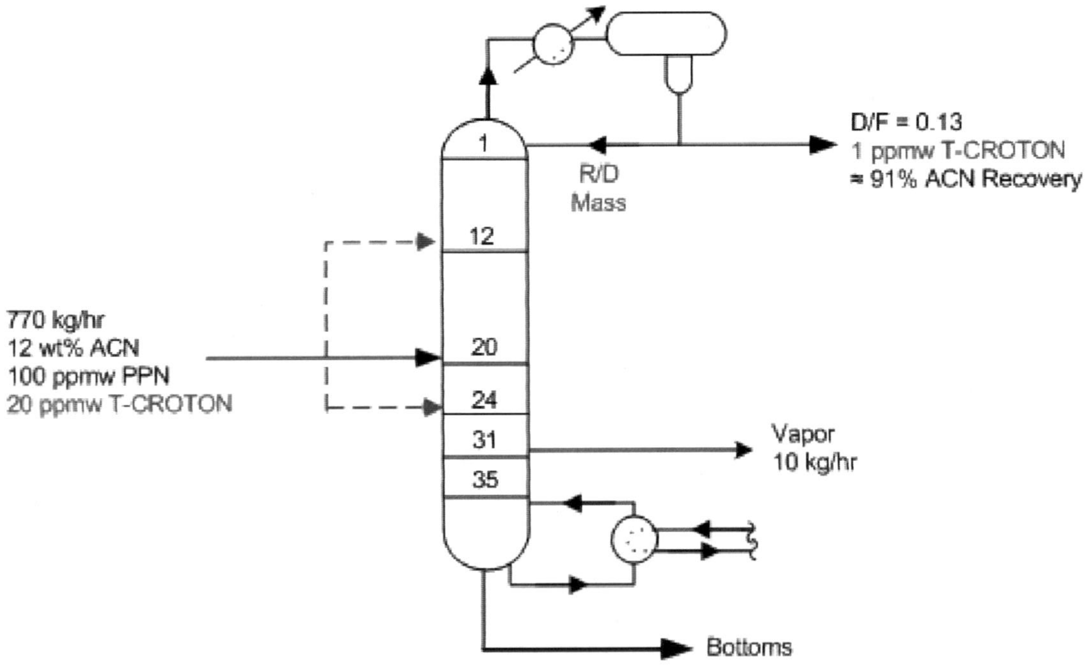

Figure 5. Single column with side draw used for perturbation studies-1. The column operates at 1 atm (101.3 kPa).

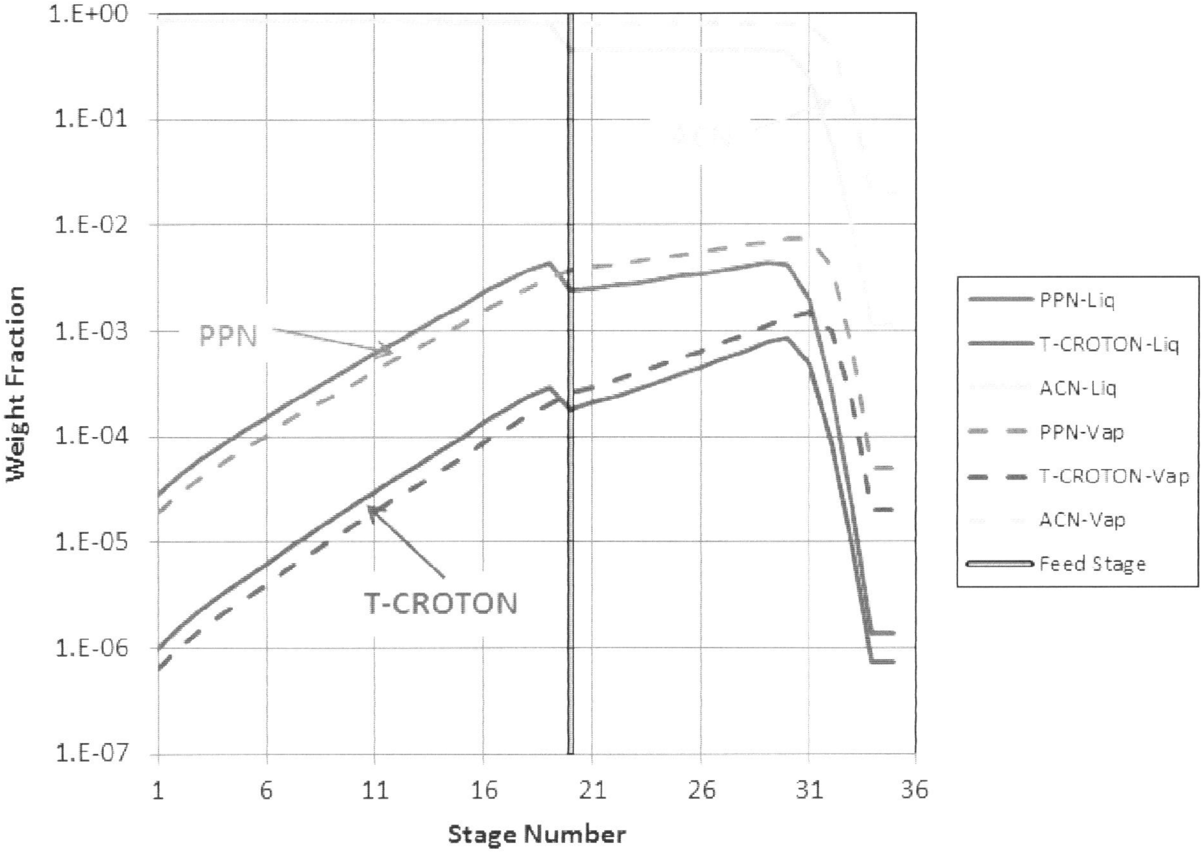

Figure 6. Vapor and liquid concentration profiles for ACN, PPN and T-CROTON in distillation column corresponding to Figure 5, and with feed to stage 20.

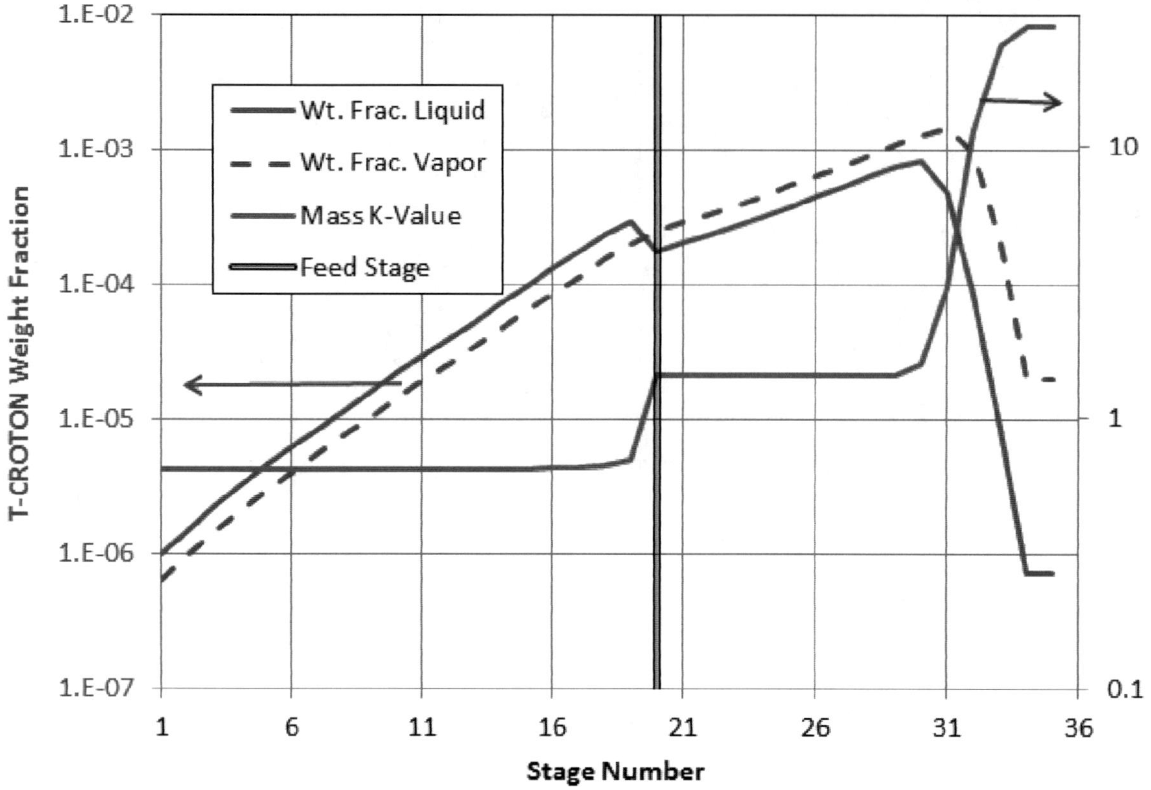

Figure 7. Mass-fraction and K-value profiles of T-CROTON in the base case. The specifications are the same as those in Figure 6.

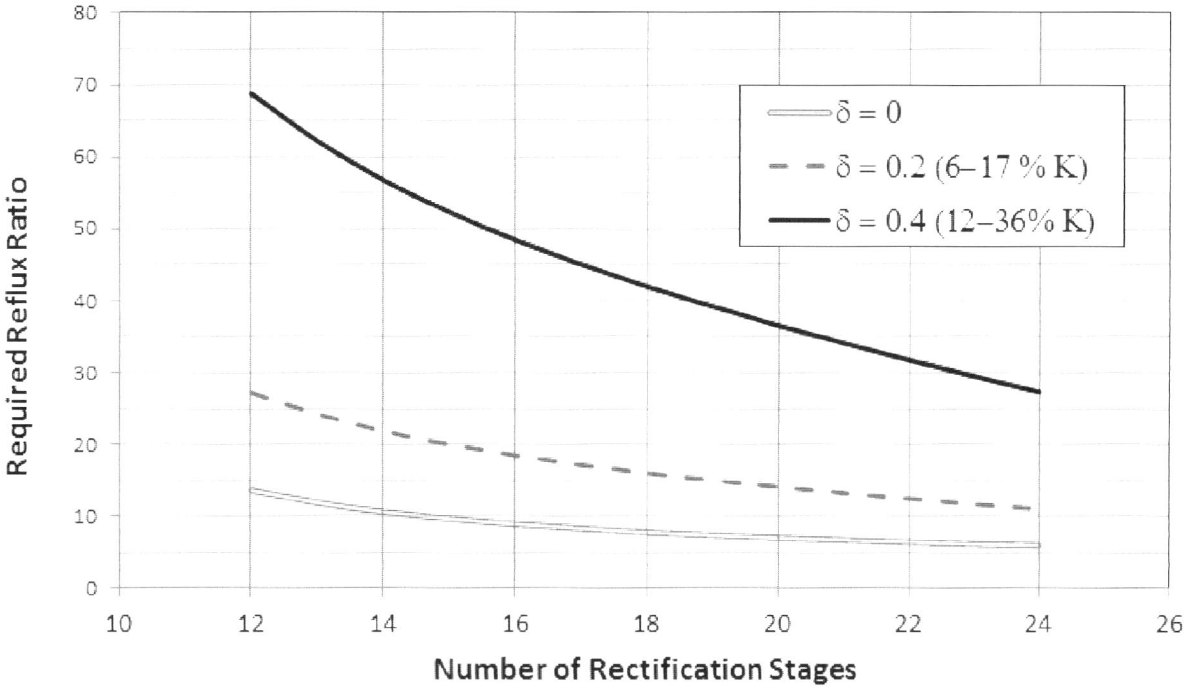

Figure 8. Effect of perturbation parameter (δ) and number of rectification stages on the reflux ratio required to meet the spec of 1 ppmw T-CROTON in the distillate. See Figure 5 for details.

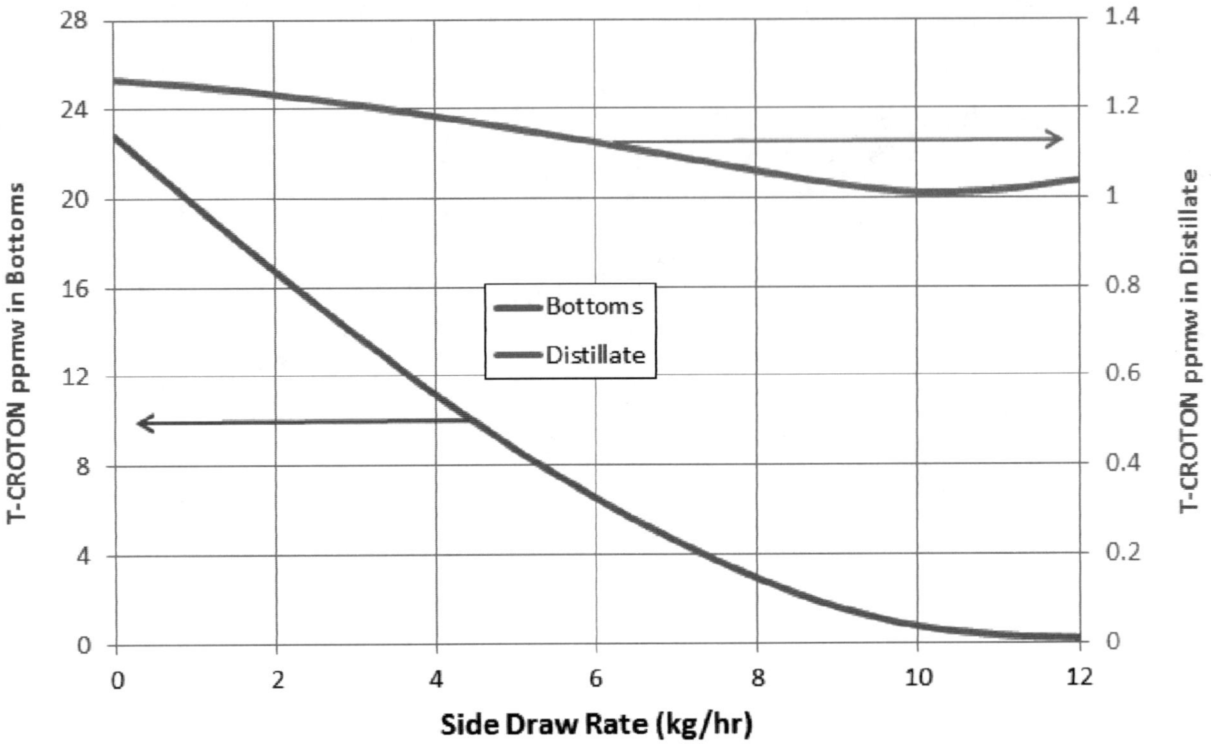

Figure 9. Effect of modifying side-draw rate. The feed is at stage 20 and the reflux ratio is fixed at 7.0.

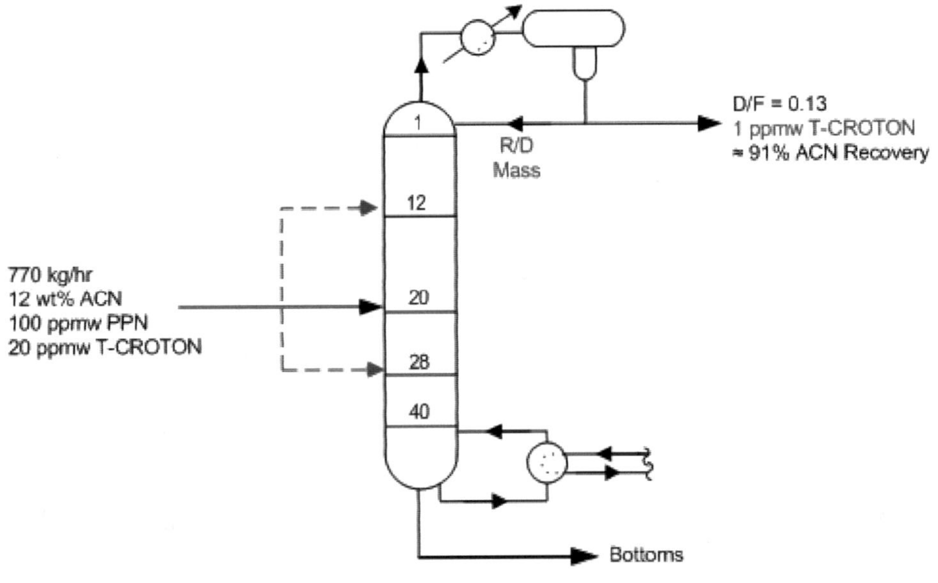

Figure 10. Single column with perturbation studies-2. The difference from Figure 5 is that the side draw has been eliminated.

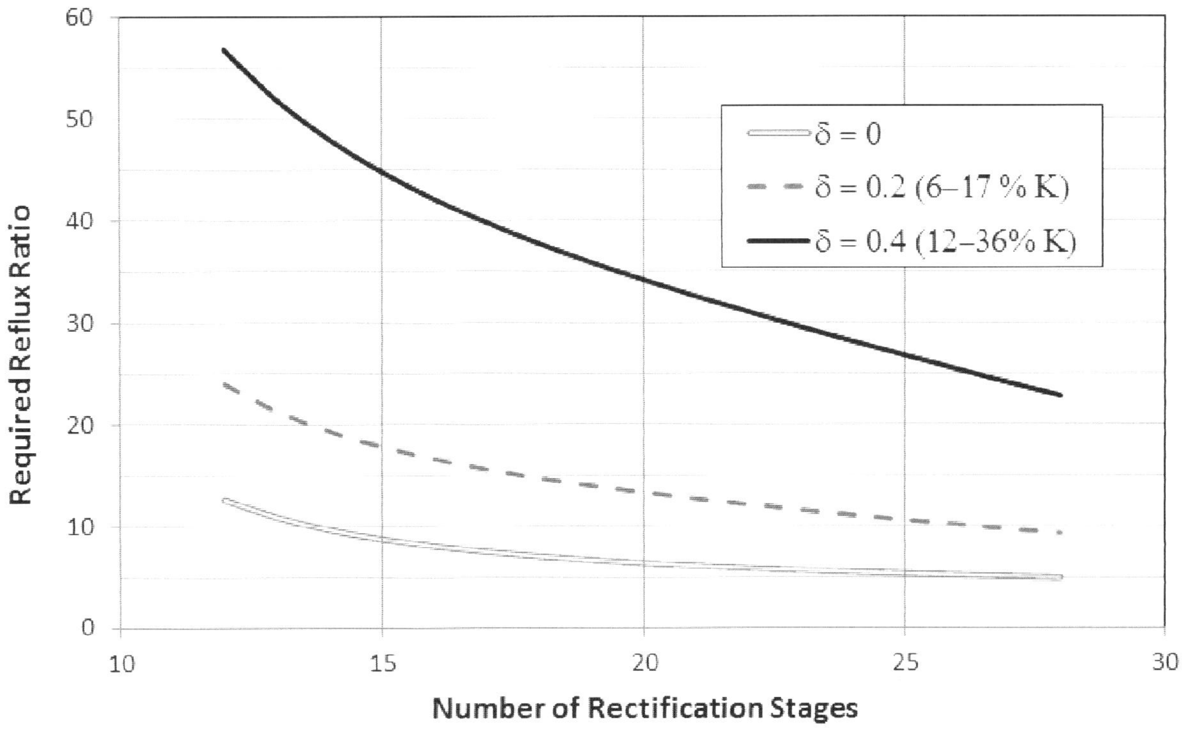

Figure 11. Repeat of study shown in Figure 8 with no side draw and total number of stages increased from 35 to 40 (See Figure 10).

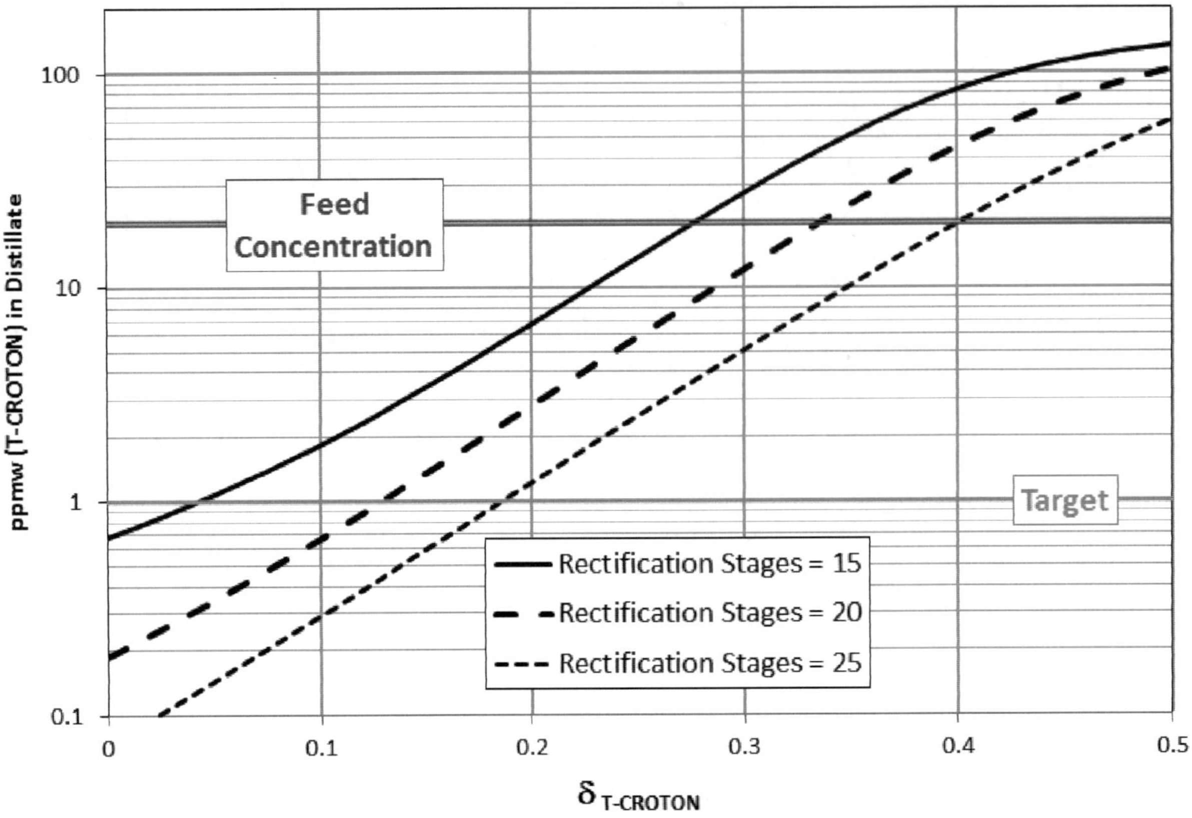

Figure 12. Effect of perturbation parameter (δ) and number of rectification stages on concentration of T-CROTON in distillate. Referring to Figure 10, the reflux ratio is fixed at 10.

Literature Cited

[1] Kister, H. Z. Components Trapping in Distillation Towers: Causes, Symptoms and Cures," *Chem. Eng. Prog.* **2004**, 100, 22-33.

[2] Kister, H. Z. *Distillation Troubleshooting*. John Wiley and Sons, Inc., Hoboken, NJ, **2006**.

[3] Streich, M.; Kistenmacher, H. Property Inaccuracies Influence Low Temperature Designs, *Hydrocarb. Proc.*, **1979**, 58, 237-241.

[4] Mah, R. S. H. Effects of Thermophysical Property Estimation on Process Design, *Computers Chem. Eng.*, **1977**, 1, 183-189.

[5] Zudkevitch, D. Imprecise Data Impacts Plant Design and Operation, *Hydrocarbon Process.* **1975**, 54, 97–103.

[6] Zudkevitch, D.; Gray, R. D. Impact of Fluid Properties on the Design of Equipment for Handling LNG, *Adv. Cryog. Eng.* **1975**, 20, 103–123.

[7] Larsen, A. H. Data Quality for Process Design, *Fluid Phase Equilib.*, **1986**, 29, 47–58.

[8] Zeck, S. Thermodynamics in Process Development in the Chemical Industry - Importance, Benefits, Current State and Future Development, *Fluid Phase Equilib.*, **1991**, 70, 125–140

[9] Macchietto, S.; Maduabeuke, G,; Szcepanski, R. Exact Determination of Process Sensitivity to Physical Properties, Fluid Phase Equilib., **1986**, 29, 59-67.

[10] Reed, M. E.; Whiting, W. B., Sensitivity and Uncertainty of Process Designs to Thermodynamic Model Parameters: A Monte Carlo Approach, *Chem. Eng. Commun.* **1993**, 124, 39–48.

[11] Whiting, W. B. Effects of Uncertainties in Thermodynamic Data and Models on Process Calculations, J. *Chem. Eng. Data*, **1996**, 41, 935–941.

[12] Whiting, W. B.; Vasquez, V. R.; Meerschaert, M. M. Techniques for Assessing the Effects of Uncertainties in Thermodynamic Models and Data, *Fluid Phase Equilib.* **1999**, 158–160, 627–641.

[13] Xin, Y.; Whiting, W. B. Case Studies of Computer-Aided Design Sensitivity to Thermodynamic Data and Models, *Ind. Eng. Chem. Res.* **2000**, 39, 2998–3006.

[14] Whiting, W. B.; Vasquez, V. R.; Meerschaert, M. M. Techniques for Assessing the Effects of Uncertainties in Thermodynamic Model and Data, *AIChE J.*, **1997**, 43, 440–447.

[15] Kim, S. H.; Kang, J. W.; Kroenlein, K.; Magee, J. W.; Diky, V.; Frenkel, M. Online Resources in Chemical Engineering Education: Impact of the Uncertainty Concept from Thermophysical Properties, *Chem. Eng. Educat.*, **2013**, 47, 48-57.

[16] Mathias, P. M., Sensitivity of Process Design to Phase Equilibrium – A New Perturbation Method Based Upon the Margules Equation. *J. Chem. Eng. Data*, **2014**, 59, 1006-1015.

[17] Mathias, P. M. Effect of VLE Uncertainties on the Design of Separation Sequences by Distillation – Study of the Benzene-Chloroform-Acetone System. *Fluid Phase Equilib.*, **2016**, 408, 265-272.

[18] Mathias, P. M. Effect of Phase-Equilibrium Uncertainties on the Separation of Heterogeneous Azeotropes – Application to the Water/1-Butanol System. *J. Chem. Eng. Data*, **2016**, 61, 4077-4084.

[19] Mathias, P. M.; Kister, H. Z. Effect of Phase-Equilibrium Uncertainties on Ethyl Acetate Purification, submitted for publication to *J. Chem. Eng. Data*, February **2017**.

[20] Renon, H.; Prausnitz, J. M. Local Compositions in Thermodynamic Excess Functions for Liquid Mixtures. *AIChE J.*, **1968**, 14, 135-144.

[21] Redlich, O.; Kwong, J. N. S. On the Thermodynamics of Solutions. V An Equation of State. Fugacities of Gaseous Solutions. *Chem. Rev.*, **1949**, 44, 233-244.

[22] Rowley, R. L.; Wilding, W. V.; Oscarson, J. L.; Yang, Y.; Zundel, N. A. DIPPR® *Data Compilation of Pure Chemical Properties, Design Institute for Physical Properties*, AIChE, New York, NY, **2010**.

[23] Frenkel, M.; Chirico, R. D.; Diky, V.; Yan, X.; Dong, Q.; Muzny, C. ThermoData Engine (TDE): Software Implementation of the Dynamic Data Evaluation Concept. *J. Chem. Inf. Model.*, **2005**, 45, 816-838.

[24] McGreer, D.E.; Page, B. D. cis-Crotonaldehyde and Related Compounds. *Can. J. Chem.*, **1969**, 47, 866-867.

[25] UNIFAC is an estimation method based upon group contributions. Several UNIFAC models are available in Aspen Plus V8.8. In the present study, since mainly approximate estimations are required, the base UNIFAC model has been used. Fredenslund, Aa.; Gmehling, J.; and Rasmussen, P. *Vapor-Liquid Equilibria Using UNIFAC, A group-Contribution Method*, 3rd Impression, Elsevier, Amsterdam, **1977**.

[26] Parker, B. Data measured in Eli Lilly Corporate Center Laboratory, August, **2016**.

创举塔器
CHUANGJU COLUMN

The Successful Application of a Partition Flow Spray Tray (PFST®) in C4 Extractive Distillation

Dr. Yan Zhuang, Advanced Column Technology
Enxia Shang, Tianjin ChuangJu Technology

ADVANCED COLUMN TECHNOLOGY PTY. LTD.
TIANJIN CHUANGJU TECHNOLOGY CO.,LTD.

Presentation Outline

1. Background for PFST Development
2. Introduction of PFST Spray Tray
3. Industrial Case Studies
4. Company Introduction

创举塔器
CHUANGJU COLUMN

C4 Down Stream Processing Market in China

Crude C4 production exceeded 30 million t/a in 2015

- C4 as Refinery by-product which accounts for the majority
- As Ethylene production by-product,
- As MTO by-product
- Further Crude C4 downstream processing becomes the trend.

One of the unit operations is to split butane from butene using extractive distillation.

创举塔器
CHUANGJU COLUMN

Current Issues with C4 Splitter using Extractive Distillation

- The butane and butene extractive distillation column uses a mixture of MEK and NFM as solvent.
- This is a typical system that operates under
  - High pressure,
  - Strong foaming,
  - High liquid to vapor ratio
  - Difficult to separate which requires a lot of stages (up to 205 stages).

创举塔器
CHUANGJU COLUMN

Failures of Conventional Valves Trays

- A variety of valve trays have been used such as, ADV float valve 、 DJ fixed valve、 guided trapzoidal floating valve、 and F1 float valve etc.
- The current issues are:
  - Fail to achieve design capacity;
  - Poor separation and recovery;
  - High column pressure drop;
  - High steam consumption.

创举塔器
CHUANGJU COLUMN

Why the conventional valve tray failed

- The liquid and vapor are contacted in a cross-flow pattern so there are hydraulic gradient and concentration gradient.
- The concentration gradient causes loss mass transfer driving force
- The hydraulic gradient leads to uneven distribution of vapor
- A lack of anti-foaming mechanism leads to premature flooding.
- mass transfer dead zones near the column wall with stagnant flow, side flow and circular flow.

创举塔器
CHUANGJU COLUMN

ChuangJu's Partition Flow Spray Tray (PFST®)

To overcome the difficulties ChuangJu developed the PFST Tray. The key innovations are:

- Total counter-current contact
- No liquid back mixing
- Provide a settling trough for vapor disengagement in the downcomer
- Use high performance spray tray

创举塔器
CHUANGJU COLUMN

How does a Spray Tray Work (Exhibitor Booth 622)

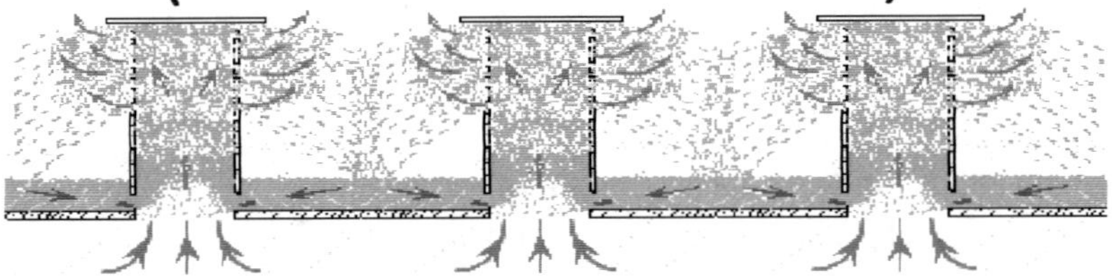

- Vapor rises through the opening in the tray; liquid enters through a gap at the bottom of the cap;
- Liquid is drawn up by the rising vapor as sheets, ligands, and broken into droplets, then impinges to the top of the cap;
- The liquid and vapor mixture is sprayed out through the holes.
- The vapor and liquid mixtures from the neighboring rows collide.
- The liquid droplets falling down to the tray floor and vapor rises to the tray above.

创举塔器
CHUANGJU COLUMN

Advantages of Spray Trays

- Higher mass transfer efficiency
- Higher Capacity
- Lower Pressure Drop
- Anti-fouling
- Reduce foaming

创举塔器
CHUANGJU COLUMN

Higher Mass Transfer Efficiency

- **Increased total areas of liquid-vapor interface** by dispersing liquid into a large amount of small droplets.
- **Longer phase contact time:** The liquid comes to contact with vapor by rising and mixing inside the cap, spraying out through the vertical sieve holes, then falling back to the tray;
- **Faster interface renewal** due to intensified phase contact inside the cap and outside of the cap.

The spray tray creates a **high impact**, more **turbulent** and liquid phase **highly dispersed** phase contact environment for mass transfer.

创举塔器
CHUANGJU COLUMN

F1 Valve Tray Air and Water System

创举塔器
CHUANGJU COLUMN

Spray Tray – Air and Water System

创举塔器
CHUANGJU COLUMN

Reduce Foaming

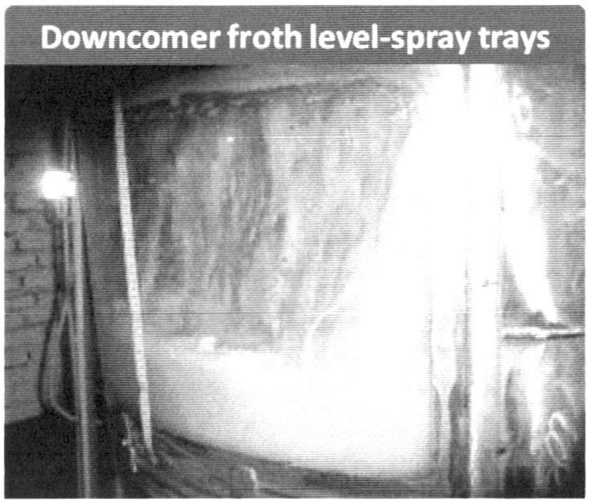

创举塔器
CHUANGJU COLUMN

How does a PFST Work

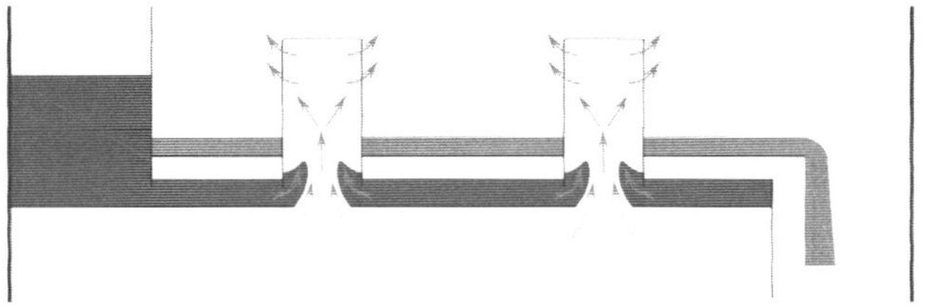

By using a partition plate to separate the portion of liquid that has been contacted with the vapor, from the liquid from the tray above, thus eliminates liquid phase back mixing:

- No hydraulic gradient
- No Concentration gradient
- No temperature gradient

which increases mass and energy transfer efficiency.

创举塔器
CHUANGJU COLUMN

Partition Flow Spray Tray (PFST®)

- Total counter-current contact
- No liquid back mixing
- Multiple Downcomer
- Provide a settling trough for vapor disengagement in the downcomer

创举塔器
CHUANGJU COLUMN

Multiple Downcomer with Settling Trough Design

To reduce vapor recycle, settling troughs have been incorporated in the downcomer design to extent the residence time for vapor to escape.

创举塔器
CHUANGJU COLUMN

PFST Tray at Work

创举塔器
CHUANGJU COLUMN

Case Study 1: Revamping three C4 Splitters for Ningbo Haode Chemical Industry Ltd.

- The original design used fixed valve trays
- Could only run at 70% design capacity
- The hydraulic gradient leads to uneven distribution of vapor
- After revamping, columns are running at above 30% design capacity.

创举塔器
CHUANGJU COLUMN

Ningbo Haode Chemical Industry Ltd. parameters before and after the revamp:

Parameters		Before Revamp	After Revamp	% of Change
Type of tray		Fixed Valves	PFST	
Feed rate, kg/h	Dist. Columns A/B	12,560	21,600	72%↑
	De-gassing Column	124,660	226,930	82%↑
Temperature℃	Dist. Columns A/B	142.8	140.3	Temp.↓
	De-gassing Column	156.8	154.4	Temp.↓
Steam Consumption kg/t feed	Dist. Columns A/B	520	295	43%↓
	De-gassing Column	53.8	37.5	30%↓
Solvent entrainment, ppm	Dist. Columns A/B	1,000	0	Reduced solvent loss
	De-gassing Column	1,000	0	Reduced solvent loss
Product Butane Content %		96%	96%	
Total olefin content%		90%	90%	
Scheduled maintenance time/a		4	1	Higher Plant availability

创举塔器
CHUANGJU COLUMN

Case Study 2: Design and Supply three C4 Splitters for Luoyang Petrochemical Hongda Industry Ltd

The unit reached the design capacity of 100,000 t/a C4 feed and met or exceeded all products specification within 65 hours of start-up with steady operation.

创举科技
CHUANGJU TECHNOLOGY

Luoyang Petrochemical Hongda Industry Ltd. parameters before and after the revamp

	Design	Achieved Results
Throughput, (t/a)	100,000	100,000
Extractive Distillation Column Temperature (℃)	132	118.8
De-gassing Column Temperature (℃)	172.9	165
Steam saving, t/d		48
Iso-butane product Content, %	≥96	98
Olefin Content, %	≥60%	66

创举塔器
CHUANGJU COLUMN

Case Study 3: Design and Supply three C4 Splitters for Hebei Haite Weiye Petrochemical Ltd.

The unit reached 150% of the design capacity and the tray pressure drop is averaged at 550Pa. After a successful start-up, the product quality all met or exceeded design specification at lower energy consumption.

创举科技
CHUANGJU TECHNOLOGY

Conclusion

The following points can be drawn from the industrial applications:

- The spray contactor intensifies the mass transfer by operating in **a stable and controlled spray regime**, without the risk of running into jet-flooding.
- The partition plate separates the fresh liquid and the liquid coming out of the spray contactors, eliminates **back mixing**, and ensures the maximum driving forces from **counter-current contact** of the liquid and vapor.
- The multiple down comer trough design gives more time for phase disengagement thus **reduces vapor recycle**, especially in a **foaming system**.

塔器
CHUANGJU COLUMN

Company Profile

Tianjin ChuangJu Technology Co., Ltd

Tianjin ChuangJu Technology Co., Ltd specializes in separation processes and equipment design, fabrication and installation, providing services to the chemical, oil and gas, pharmaceutical and environmental protection industries. ChuangJu developed its proprietary column internal designs and technologies with dozens of patents and know-hows, and successfully commercialised the spray tray technologies in nearly 1,800 industrial columns in 30 provinces of China and overseas. The company has its own fabrication plant covering 40,000m^2 areas with ASME U stamp certified pressure vessels manufacturing capabilities.

创举塔器
CHUANGJU COLUMN

Pilot & Testing Capabilities

Ambient Temperature Column Testing Rig

Swinging Column Testing Unit Simulating a Floating LNG Rig

Column Testing Rig with Temperature Control

创举科技
CHUANGJU TECHNOLOGY

Company held design and Manufacturer Licenses and Certificate

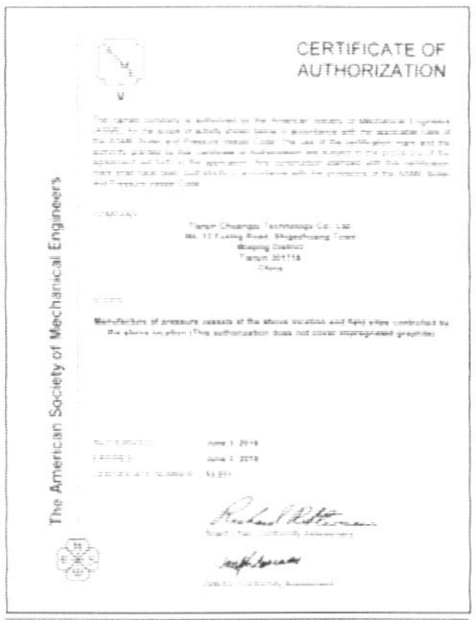

ASME Pressure Vessel Manufacture Accreditation U Stamp

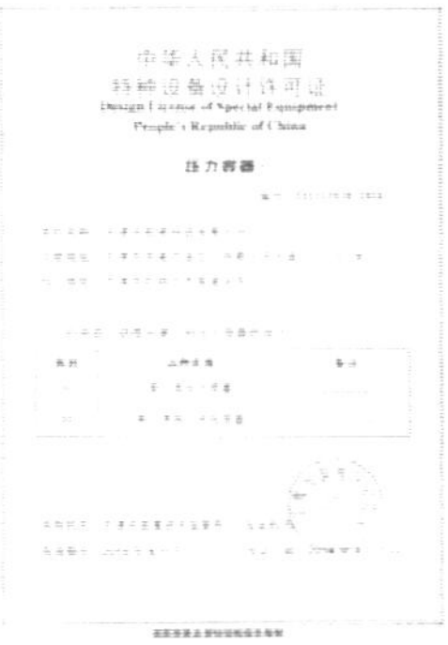

Chinese Special Equipment Design Licenses

Chinese Special Equipment Manufacture Licenses

创举塔器
CHUANGJU COLUMN

Manufacturing Capabilities

Manufacturing Workshop

创举科技
CHUANGJU TECHNOLOGY

Manufacturing Capabilities

Production Equipment

Numerical controlled welding machine

OTC Welding Robot

Messer CNC numerical controlled cutting machine

Automatic welding machine for pipe and flange

创举塔器
CHUANGJU COLUMN

Manufacturing Capabilities

Production Equipment

Automatic submerged arc welding machine

CNC plasma flame cutting machine

Automatic molding equipment of packing

创举塔器
CHUANGJU COLUMN

Manufacturing Capabilities

Products

创举塔器
CHUANGJU COLUMN

Manufacturing Capabilities

Products

创举塔器
CHUANGJU COLUMN

Thanks !

创举塔器 科技造就

Chuangju Column Technology Innovate

创举塔器
CHUANGJU COLUMN

AIChE Paper 122A

TRAY OPERATING WINDOW – DESCRIPTION AND APPLICATIONS

André Bernard
Process Design Engineer
NOVA Chemicals (Canada) Ltd.

Prepared for Presentation at the 2017 Spring National Meeting
San Antonio, TX, March 26th – March 30th, 2017

AIChE Paper 118A

TRAY OPERATING WINDOW – DESCRIPTION AND APPLICATIONS

André Bernard
Process Design Engineer
NOVA Chemicals (Canada) Ltd.

Abstract

Stable and effective operation of distillation trays are characterized by vapor and liquid loads falling within a number of limitations imposed by the tray geometry. Flooding limits capacity of trays and can be induced by vapor and/ or liquid rates. At the low operating end, weeping, dumping and/ or flow regime transitions can severely affect the ability of a tray in achieving the desired separation. Significant amount of research has been done over the years for correlating those limitations to operating conditions, physical properties of fluids involved and tray geometry. Tray designers rely strongly on those correlations for optimizing the design of a tray based on a given set of process loads and physical constraints. Those correlations can be used to plot tray operating limits as a function of liquid and vapor loads for a given tray geometry and are often referred to as tray operating windows or stability diagrams. Those diagram are extremely useful to process engineer in setting up operating conditions or making process adjustments that ensure effective fractionation in tray towers. They are also a valuable tool for a trouble shooter when assessing poor performance of distillation towers. This paper describes the different limits plotted on typical tray operating windows and their applications in process design and operation through a few practical examples.

Introduction

Whether developing a new design or troubleshooting an existing process unit, process engineers must understand the process loads imposed on equipment's as well as limitation of that equipment. A basic example is the evaluation of a piping system, where the engineer would evaluate the flow of fluid using fluid mechanics correlations and establish a system curve. The characteristics of the equipment, in this case a pump, would be represented by a head curve. The junction of the system and head curve would determine the expected flow rate. To get more or less flow a valve would be introduced to shift the system curve such that the intersection with the pump head curve provides the desired flow. The pump head curve also provides the limitation of the equipment such as minimum, maximum flow and NPSH requirements. Rather than using a head curve, complex hydrodynamic calculations can be performed on the pump. However, frequently that information may be unavailable and proprietary and adds on the complexity of the evaluation. A simple observation of the head and system curves simplifies and speeds up the evaluation.

A similar approach can be used for evaluating tray towers where process loads and tray limitations can be plotted as a function of vapor and liquid loads. The process loads can be determined by process simulations and/ or plant data. The tray limitations could be either calculated from known correlations, rating programs or obtained from vendors. Similar to the above pump example, an operating line can be plotted over the tray operating limits and quickly provide the expected operating range of a tray. For a given feed rate, the process load profile will quickly indicates the need for varying the tray design in the different section of a tower. This graphical method for evaluating tower performance or new design is relatively simple and allow a quick identification of potential pinch points.

Trays Operation and Limitations

Trays are normally design to meet two primary objectives: maximizing mass transfer (efficiency) and meeting the vapor/ liquid traffic (capacity) required for achieving a desired separation. Efficient tray operation is defined by an operating range where limitations are characterized by equipment configuration and properties of the fluid involved. Operation outside the tray limitations can have significant impacts on fractionation efficiency and in some case lead to poor or no fractionation at all. Figure 1 illustrates a conventional single pass tray where liquids contact rising vapor as it flows across the tray deck. The tray deck is often referred as active area as it is the location where mass transfer occurs between vapor and liquid. With proper design and conditions, the vapor/ liquid mixture on the active area form a froth. The height of the froth is driven by vapor velocity across the active area. With sufficient tray spacing, the froth would transition to a spray followed by a clear vapor layer. Liquid travels downward from tray to tray through downcomers. The fluid inside the downcomer is characterized by three layers, the frothy liquid leaving the tray, an aerated liquid layer where vapor disengages from the liquid phase and a clear liquid layer at the bottom of the downcomer.

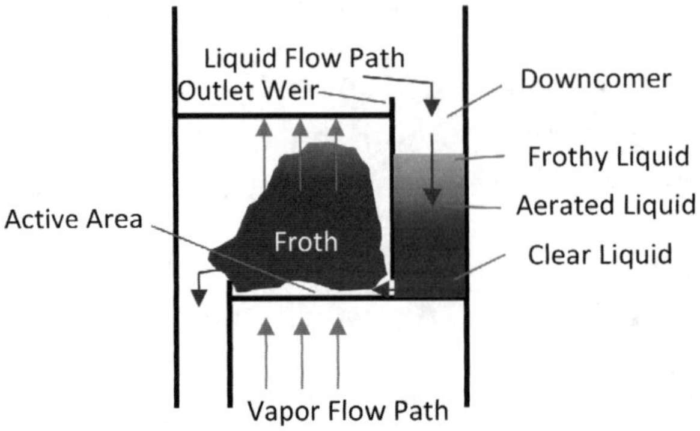

Figure 1: Conventional Single Pass Tray

System Limit

Conventional tray technology is based on gravitational forces which drive the liquid to flow downward in the tower through downcomers. The tray deck, while exhibiting a cross flow path still requires the liquid to eventually drop down in the downcomer. However, as vapor velocity increases it reaches a limit where large liquid droplets can no longer flow downward but are entrained by rising vapors. This condition is referred as the system limit. There are tray technologies that are design to operate above the system limit. These technologies[2,3,4] operates differently than conventional trays where vapor and liquid flow co-currently in an active area followed by a vapor/ liquid disengagement section. Disengagement is typically accomplished through centrifugal force or mechanical mist elimination.

Jet Flood

Jet flooding is induced by high vapor velocities causing liquid entrainment on the tray above. As previously stated, the fluid on the tray deck can be characterized by three layers: froth, spray and clear vapor. As vapor velocity increases, the spray layer reaches the tray above and causes accumulation of liquid on the tray above. This condition is the most commonly known form of jet flooding and is also referred as spray entrainment [1]. At short tray spacing (less than 18"), vapor velocity can be such that the top of the froth layer approaches the tray above causing liquid accumulation on the tray above. This condition is known as froth entrainment flooding [1].

Downcomer Backup

Downcomer backup occurs when the aerated liquid in the downcomer accumulates and backs up on the tray above. It is caused by the combination of tray pressure drop, liquid height on the tray and pressure drop across the downcomer apron that causes the aerated liquid height in the downcomer to exceed the tray spacing. Downcomer backup is a function of both liquid and vapor rates.

Downcomer Choke

Unlike downcomer backup, downcomer choke is mostly dependent on the liquid rate and has little if no dependency with vapor rates. Downcomer choke is characterized by the aerated liquid velocity at which friction losses in the downcomer and downcomer entrance exceeds the capability of the downcomer in moving the liquid to the tray below.

Minimum Liquid Rate

There are 3 elements that can define the minimum liquid rate. The first two are governed by the weir loadings. At low weir load liquid maldistribution can occur, which impacts the tray efficiency. That limit would be independent from the vapor velocity. A second element would involve a flow regime transition from froth to spray as the vapor load increases on a tray operating at low weir load. This limit would be a function of both liquid and vapor rate and would be associated with an increase in liquid entrainment. A third element that can govern the minimum liquid rate applies to downcomer design with dynamic seal. As shown on Figure 2, downcomer with dynamic seals are intended to increase the active area by reclaiming the area occupied by the downcomer outlets. On conventional downcomers, the downcomer remains sealed at all times when the outlet weir is higher than the downcomer clearance. With a dynamic seal, a minimum liquid load must be provided to ensure a minimum liquid head in the downcomer. Below that minimum, vapor blows through the downcomer and severely impact fractionation.

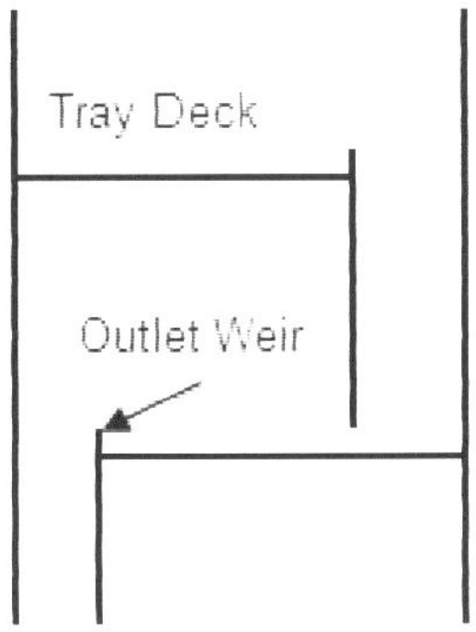

Conventional Downcomer with Outlet Weir

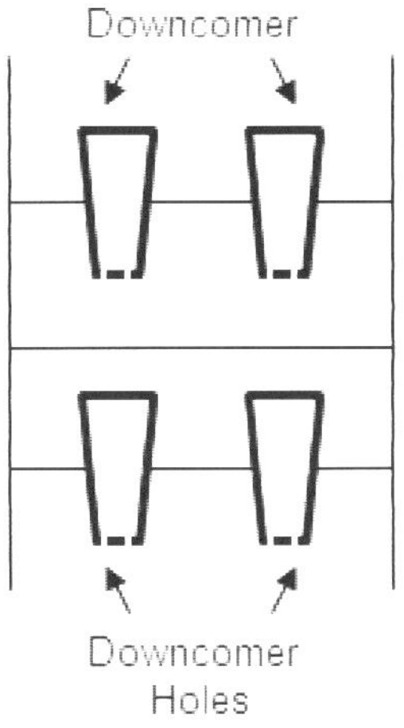

Downcomer with Dynamic Seal

Figure 2: Comparison Between Conventional and Downcomers with Dynamic Seal

Minimum Vapor Rate (Weeping)

Weeping is define by liquids falling to the tray below through perforations on the tray deck rather than following the intended path through downcomers. It short-circuits the normal path through the active area and depending on the extent affects the efficiency. Normal tray operation exhibits a minor fraction of weeping, however, as the vapor rate is reduced the dry pressure drop across the tray becomes deficient in keeping liquid on the tray deck and the extent of weeping increases and become associated with a loss of efficiency. Weeping is mostly dependent on the vapor load but has a small dependency on the liquid at high liquid load.

Tray Deck Seal Point (Dumping)

Dumping is an extension of weeping where the vapor rate is reduced to a point where all the liquid can now flow through the tray deck. Under these conditions, the downcomers could likely become unseal and vapor could easily short circuit the active area by flowing through the downcomers. Those conditions would be associated with a sever loss of fractionation. Dumping is a function of both liquid and vapor rates.

Downcomer Critical Vapor Rate

The downcomer critical vapor rate is a condition that can occur during startup or following a significant tower upset. It's a condition defined by the vapor velocity in the downcomer where the liquid cannot descend freely but is entrained by the vapor. Towers that are started with a vapor feed are at risk of encountering this condition. In the initial startup phase, the downcomer are unseal. A vapor initially introduced in the tower would first pressurize the tower then after condensing in the overhead condenser, liquid would flow down the tower from the reflux drum. At high vapor rate, the liquid will not be able to seal the downcomer and could accumulate on higher trays and/ or weep through the tray deck. The tower will not fractionate until the downcomer are sealed. Once the downcomers are sealed this limit no longer exists.

Tray Operating Windows

The tray limitations discussed above are a function of tray geometry, vapor/ liquid rates and fluid properties such as densities and liquid surface tension. For a given tray geometry and assuming relatively constant fluid properties, those limitations can be expressed as a function of vapor/ liquid rates and plotted into a diagram referred as tray operating window or stability diagram. Such a diagram, as illustrated on Figure 3, provides the characteristics of the equipment. Complementing this diagram with an operating line that represents the loads imposed by the process allows a quick evaluation of the expected operating range. Although illustrated on Figure 3, the system limit, downcomer critical vapor rate and dumping limits are rarely illustrated on tray operating windows intended to define the normal operating envelope. However, their understanding may become practical when evaluating abnormal situations, startup conditions or potential capacity revamps.

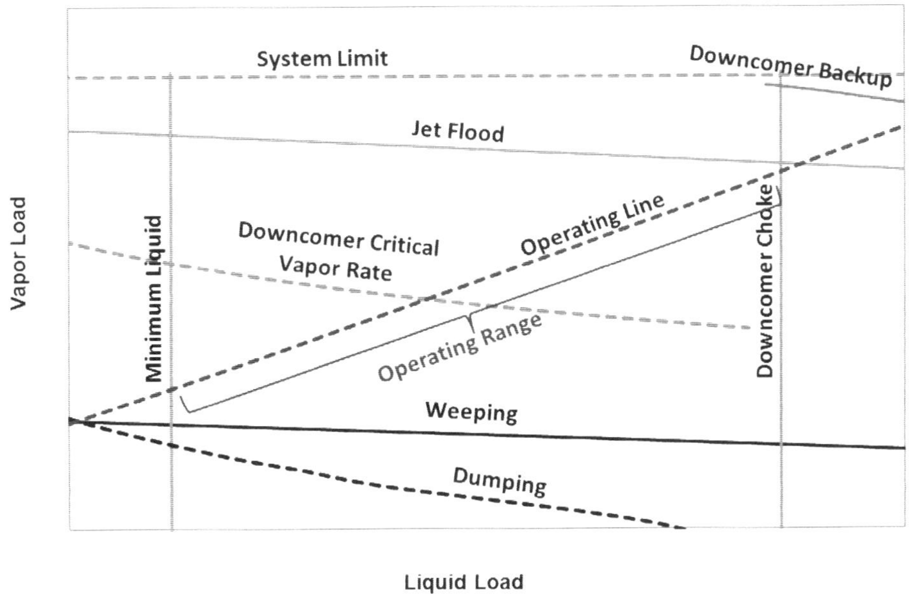

Figure 3: Tray Operating Window

Unlike all the limitations illustrated on Figure 3, the downcomer critical vapor rate is only present when the downcomers are unsealed. This condition is typical of startup or any upset that can cause the downcomer to unseal. Once the downcomer gets sealed, this tray limitation disappears and the operating point can progress upward towards the next limit.

The tray operating window illustrated on Figure 3 suggests a relatively well optimized tray as it becomes limited at the junction of jet flood and downcomer choke. This suggests that the active area and downcomer area are fully utilized. However, as shown on Figure 4, process loads reflecting "Operating Line 2" would suggest a deficiency in active area and larger than required downcomer area. Intuitively, increasing the active area by reducing the downcomer area may shift jet flood upward while moving the downcomer choke line to the left in this case.

The tray operating windows illustrated on Figure 3 and 4 shows a downcomer backup limit being somewhat irrelevant as the tray is limited by downcomer choke and jet flood. At large tray spacing, the downcomer backup limit tends to be above the jet flood limit. However, at short tray spacing it is not unusual to see downcomer back up becoming the controlling flood mechanism.

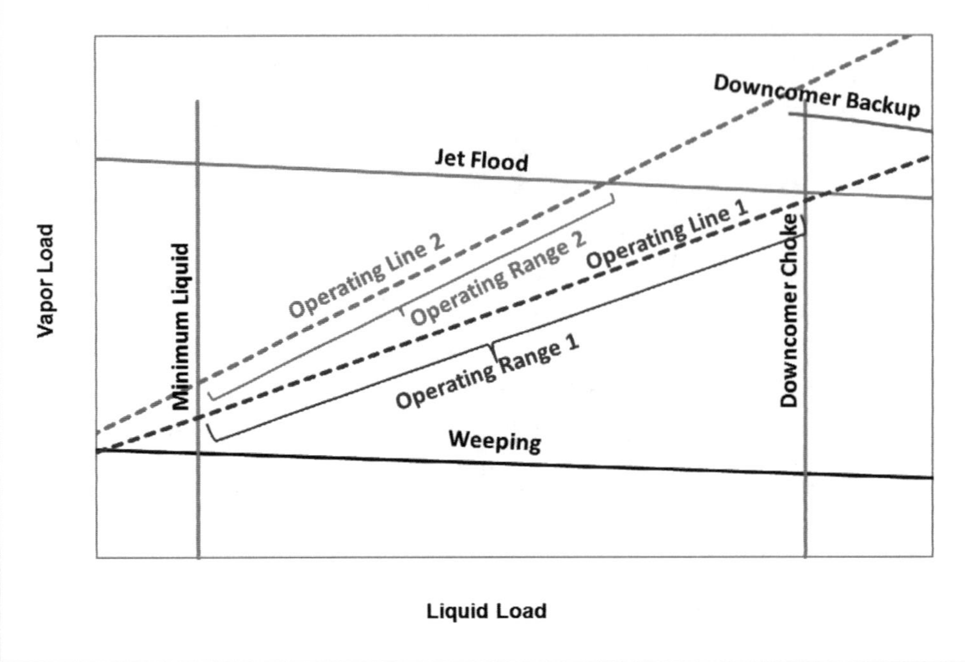

Figure 4: Operating Line vs Optimum Tray Design

A common practice is to express the vapor load as volumetric rate normalized by the square root of vapor and liquid density ratio as described by equation (1).

$$VLOAD = V_{rate}\sqrt{\frac{\rho_V}{(\rho_L-\rho_V)}} \tag{1}$$

Or in term of velocity by the C-factor.

$$C_{S=}\ \frac{VLOAD}{A} = \frac{V_{rate}}{A}\sqrt{\frac{\rho_V}{(\rho_L-\rho_V)}} \tag{2}$$

The C-factor can be calculated based on the tray active area or the tower cross section. In most cases it might be advantageous to express the C-factor in terms of the tower cross section. For one thing, tray test data is frequently reported as function of C-factor based on tower cross section. Process loads demonstrated on similar services can be use as reference for towers having different diameters. It also provides a reference relative to the tower system limit and may help selecting the appropriate tray technology on revamps.

Factor Affecting Tray Operating Windows

Feedstock & Operating Conditions

In recent years many ethylene plants have transitioned to larger volume of ethane cracking, resulting in significant changes in tower feed composition. While it is understood that variation in feed composition affects the operating line due to differences in reflux ratio and reboiler duty, the validity of a tray operating window originally developed for heavier feedstock may be questioned for evaluating the impact of such a feedstock conversion. Tower operating pressure is another parameter that can vary from the tray operating window original basis. Although many towers exhibit relatively small variation in operating pressure due to their integration with refrigeration systems, warmer service based on cooling water may operate over a wider range given the dependency of cooling water temperature with atmospheric conditions. Summers[5,6] suggest that once a tray operating window is established, expressed on a volumetric basis, it can be used for different conditions and different services providing changes in surface tension are small. The impact of feedstock changes and operating pressures have been reviewed by comparing densities and surface tensions at different conditions. Table 1 provides a comparison between the bottom section of a Deethanizer relative to 3 different feed compositions and the top section of a C3 Splitter operating at 3 different operating pressures. The Deethanizer condenser cooling is provided by propylene refrigerant while the C3 Splitter (chemical grade) uses cooling water.

Table 1: Impact of Feedstock and Operating Pressure on Properties

	Deethanizer vs Feed			C3 Splitter vs Operating Pressure		
	Feed 1	Feed 2	Feed 3	225 psig	250 psig	270 psig
ρ_V (lb/ft^3)	3.5	3.5	3.4	2.3	2.6	2.8
ρ_L (lb/ft^3)	28.9	28.8	28.1	29.4	28.8	28.5
σ (Dyne/cm)	3.4	3.3	2.8	5.0	4.5	4.2

The data in Table 1 shows that the pressure operating range on the C3 Splitter causes larger differences in properties than the one caused by a feedstock transition on the Deethanizer. The impact on a tray operating window has been evaluated on a hypothetical two-pass sieve tray with the geometry summarized on Table 2:

Table 2: Hypothetical Two-Pass Sieve Tray Geometry

Tower Diameter (ft)	7.5
Hole diameter (in)	0.5
Percent Hole Area (%)	10
Weir Length (in)	122
Active Area (ft^2)	30

A tray operating window has been developed for the C3 Splitter operating at 225 and 270 psig. The vapor load has been represented by the C-factor based on the tower cross-section and the liquid load expressed as volumetric rate. As shown on Figure 5, except for the system limit the change in properties have minor impact on the tray operating limits. The correlation developed by Stupin and Kister[7] for the system limit seems more sensitive to the change in surface tension than the other tray limitations evaluated. Note that the system limit is not a tray limitation per se, but a reference regarding the ultimate capacity of cross-flow tray technology.

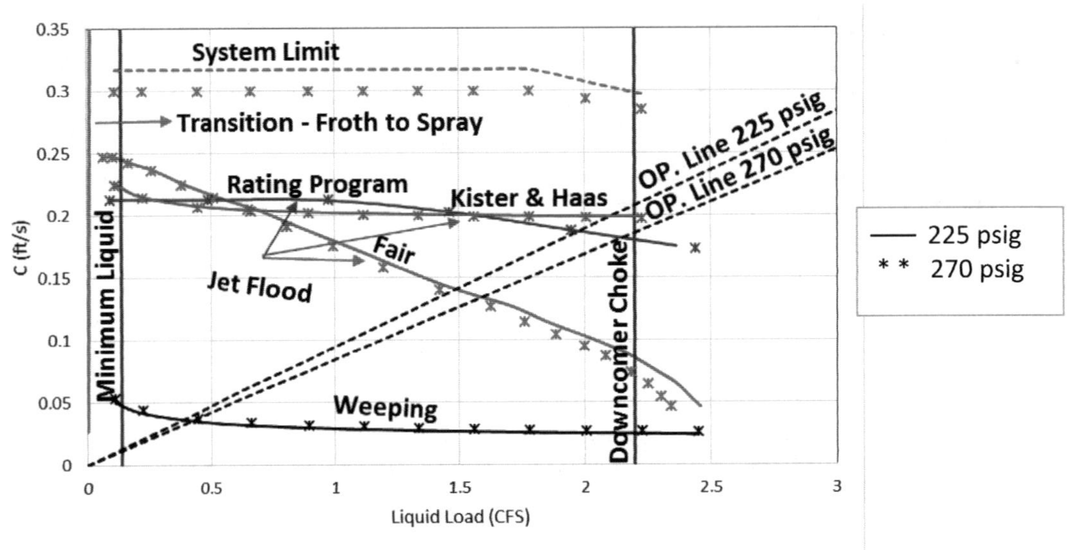

Figure 5: Impact of Operating Pressure on Tray Operating Window

The jet flood limit has been evaluated at 85% of flood by three different methods; a rating program, the Fair correlation and the Kister and Haas correlation[1]. Figure 5 suggest no apparent differences for the rating program and the Kister and Haas correlations between the two operating pressures evaluated. The difference is in fact small where the jet flood C-factor at 225 psig is approximately 0.5% higher than at 270 psig. The Fair correlation exhibits a more noticeable difference between the two operating pressures. The difference tends to increase at liquid loads beyond the predicted downcomer choke. However, the jet flood C-factor predicted for both conditions remains relatively similar between the minimum liquid and downcomer choke limits. Figure 5 also highlights the level of conservatism of the Fair correlation when compared to the rating program and the Kister and Haas correlations.

Downcomer choke can be expressed based on a maximum allowable weir loading or on downcomer residence time. Depending on the correlation use, downcomer choke would be set based on a fixed liquid volumetric rate and therefore be insensitive to change in operating conditions. Some correlations are based on rates and densities, but the variation considered in the present case would not cause significant differences.

The weeping limit was calculated with a 10% weeping fraction using the Lockett and Banik[1] correlation with the clear liquid height calculated by the Colwell[1] correlation. As shown on Figure 5, the weeping limit prediction at 270 psig is slightly higher than the one predicted for 225 psig. On a relative basis, there is 6 to 7% differences in C-factor between the two conditions evaluated. However given the relatively small C-factor values, those differences looks relatively insignificant on the tray operating window. Note that the correlation used for evaluating weeping does not depend on surface tension but on vapor/ liquid rates, densities and tray geometry.

The minimum liquid load is either based on minimum weir loading or transition from froth to spray. Since this example does not consider dynamic seal downcomers, liquid sealing the downcomers is not a criteria for defining the minimum liquid load. Figure 5 shows the minimum liquid load based on a minimum weir load of 0.5 gpm/in of weir[6]. The transition from froth to spray as evaluated by the Porter and Jenkins[1] correlation is also plotted and demonstrates that it is not the controlling mechanism for defining the minimum liquid load on this tray. Since the minimum liquid load is defined based on volumetric liquid rate it is independent from variations in properties.

This example supports the affirmation by Summers[5,6] that once a tray operating window is developed it can be used for different conditions and services providing changes in surface tensions are small. In this case it has been demonstrated that varying the operating pressure between 225 and 270 psig did not cause sufficient changes in properties to affect the tray operating window. Although tray operating windows are typically developed for specific systems, it can be further demonstrated that with the same tray geometry small variations in jet flood and weeping limitations would be observed for the properties reported on Table 1 for the Deethanizer service. The minimum liquid load and downcomer choke would remain unchanged as they are based on liquid volumetric rates.

Fouling

Many distillation services are known to foul throughout the plant run length. The accumulation of fouling material on a tray causes alteration to its geometry and depending on the extent, could be detrimental to performance and capacity. Figure 6 shows how fouling material can reduce the hole area on a tray deck. In the case of what looks like a mild loss of open area, a layer of material accumulated around the holes reduces on average the hole diameter by about 1/16 in. This would be equivalent to a 45% reduction in open area on the tray layout used as example.

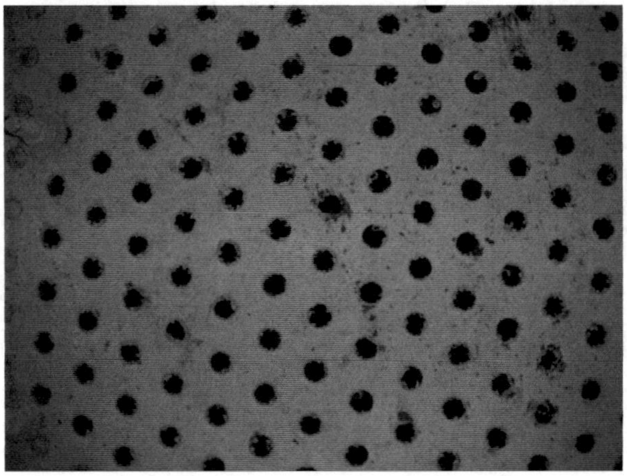

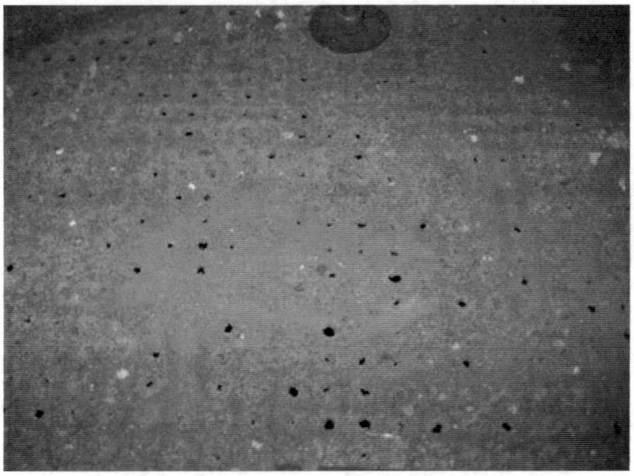

Mild Loss of Open Area **Severe Loss of Open Area**

Figure 6: Impact of Fouling on the Tray Deck

The impact of this reduction on the tray window has been evaluated with the Kister and Haas correlation for jet flood and the Lockett and Banik correlation for weeping. The hole area reduction causes the jet flood and weeping limits to shift down. As shown on Figure 7, this results in a 16% reduction of capacity when assuming no change in tray efficiency. A perfectly uniform reduction of hole area, although affecting capacity may provide benefits on efficiency. However, fouling rarely occurs uniformly and will likely creates dead zones and maldistribution, which will negatively impact the tray efficiency. The loss of efficiency compounded by the reduction in tray capacity will further decrease the capacity of the tower.

The minimum liquid load might be unaffected assuming no alteration on the weir. Again if the mouth of the downcomer is not affected the downcomer choke limit would remain the same. Although not shown on Figure 7, the downcomer backup limit will shift down due to an increase in tray pressure drop. Downcomer backup would only affect the tray capacity if it becomes the controlling flood mechanism.

To summarize the above observations, premature flooding relative to the tray operating window along with a reduction in efficiency are symptomatic of fouling. In practice, this can be demonstrated by trending the offset in reflux ratio relative to a clean system and plotting historical process loads on the tray operating window.

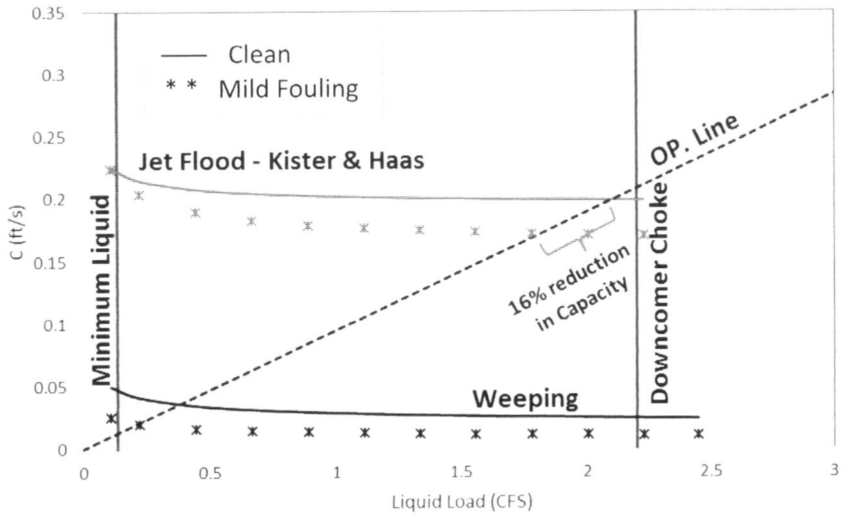

Figure 7: Impact of Fouling on Tray Operating Window

The severe fouling case illustrated in Figure 6 results in alterations that are beyond the capability of some of the correlation used for developing the tray operating window. The hole area is severely reduced and will likely cause high tray pressure drops. Liquid accumulation on the tray due to downcomer backup will likely become the predominant flood mechanism. Vapor maldistribution due to blockage of many holes will severely affect fractionation and will likely required an outage for cleaning up the tower.

Unless fouling becomes excessive and makes a tower inoperable, tray operating windows along with process simulation can be used as a tool for evaluating the extent and progression of fouling in a tower.

Applications

C2 Splitter – Feedstock Conversion

The capacity of a C2 Splitter in an ethylene plant needed to be evaluated in the early stage of a feedstock conversion project. The C2 Splitter system had been designed based on the plant processing a mix feed of naphtha and NGL. The NGL volume was predominantly propane and butane, ethane cracking being limited to the recycle stream from the C2 Splitter. The new plant feed slate would be based on 100% NGL with ethane becoming the predominant feed. The impact of the new plant feed slate caused the ethane feed composition in the C2 Splitter to increase from 15 to 32 mole%. The change in feed composition translates in a 24 % increase in reflux ratio. Process loads, hence tower capacity would therefore be affected by this change.

The C2 Splitter system is illustrated on Figure 8. It is a high pressure system that uses -40 °C propylene refrigerant for condensing overhead vapors. Condensation of propylene refrigerant vapor provides heat to the bottom reboiler. A side reboiler allows heat integration between the tower and cracked gas at the front end of the chilling train. The tower is equipped with 123 high capacity trays, which were installed during a plant revamp in 2005. Prior to the feed conversion project, the plant was producing a dilute ethylene product stream, which caused a reduction in process loads in the upper section of the tower. In the post feed conversion project, the dilute ethylene stream was to be eliminated. This, along with the expected higher reflux ratio would further increase the load on the tower.

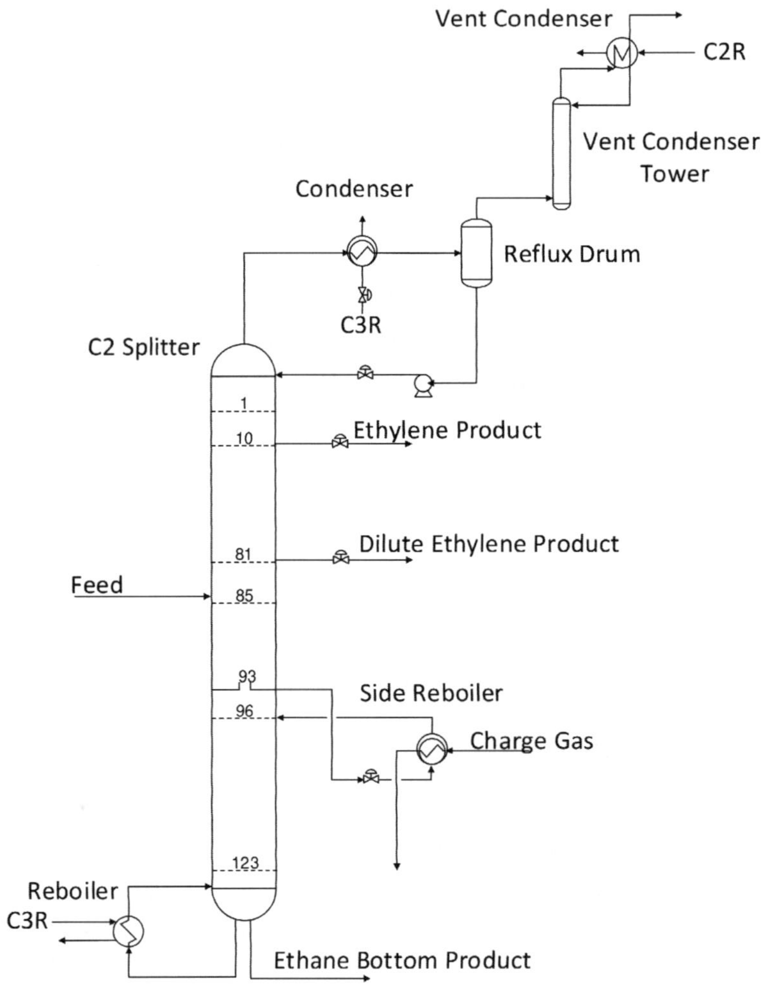

Figure 8: C2 Splitter System

Tray operating windows for the rectifying section (trays 10-84) and the bottom section (trays 98-123) are provided on Figure 9. The main differences between the two sections lies in the controlling flood mechanism. Plotting the operating line would show that the trays in the rectifying section would cross the junction between jet flood and downcomer backup, which also occurs close to the downcomer choke limit. The tray section between the side and bottom reboiler (bottom section) is quite different as it is only limited by downcomer backup. It also has a relatively narrow operating range between weeping and downcomer backup. The difference between the two tower sections is mainly explained by the difference in tray spacing. A 20-inch tray spacing is used in the rectifying section vs 12-inch in the bottom. The operating point in the bottom section is very sensitive to the side/ bottom reboiler operation. High side reboiler duty reduces tray load in the bottom below the weep point, while insufficient side reboiler requires the bottom reboiler to provide more duty and causes the bottom trays to flood due to downcomer backup.

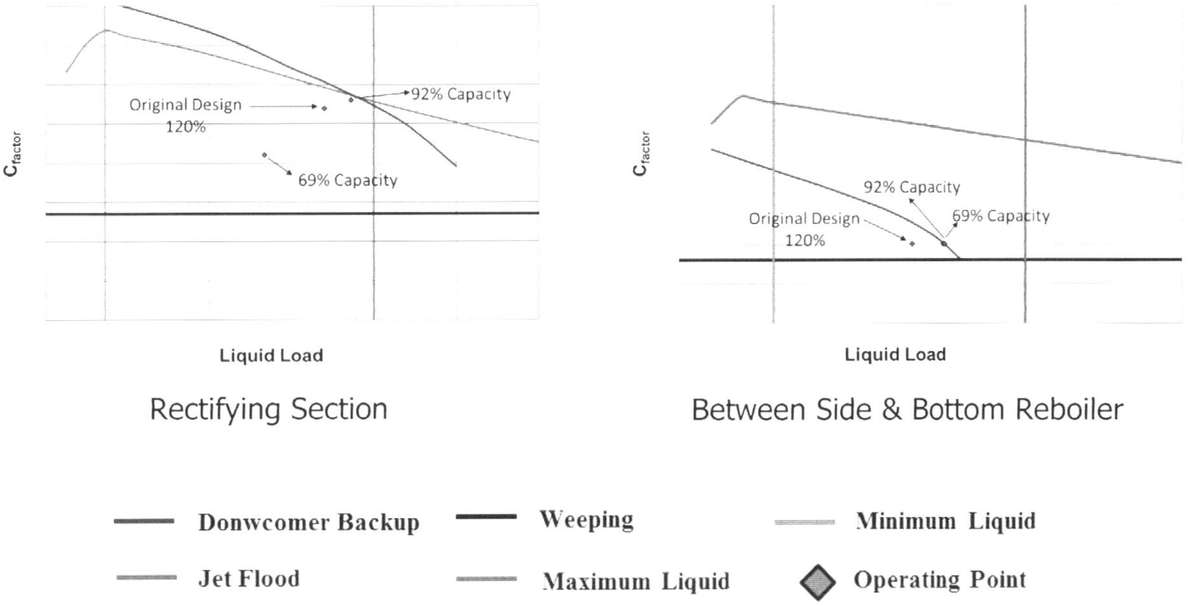

Figure 9: Operating Points vs Tray Operating Limits

The feedstock conversion affects operation of the side reboiler in two ways. The higher volume of ethane cracking causes a reduction in cracked gas flow through the side reboiler. The higher ethane content in the C2 Splitter feed causes the temperature profile to warm up in the tower and reduces the temperature difference between the C2 Splitter side reboiler stream and cracked gas. With the heat duty available from the side reboiler, the tower would become limited by downcomer backup in the bottom section at 69% of the desired production rates. The operating point for 69% capacity in the rectifying section shows a lot of remaining available capacity. Removing the side reboiler constraint by providing a parallel side reboiler, which uses the same level of propylene refrigerant as the bottom reboiler would allow pushing the tower to the next constraint. As shown on Figure 9, the side reboiler duty is manipulated to maintain the operating point in the bottom section at the downcomer backup limit while allowing the operating point in the rectifying section to reach the flood limit. By doing so, the tower capacity can be increased from 69 to 92% capacity.

Still short of meeting the target production and based on other considerations, a parallel C2 Splitter was installed[8]. However, this example highlights the effectiveness of using tray operating windows for understanding limitations and potential debottleneck options. Furthermore, the tray operating windows for this C2 Splitter proved valuable following the 2005 re-tray in terms of understanding the relationship between the bottom tray operation and side/ bottom reboiler balance.

Depropanizer – Unsealed Downcomers

A detailed evaluation of a Depropanizer performing poorly during a plant startup has been covered by the author in a previous paper[9]. Unsealed downcomers in the bottom section of the tower were suspected to cause the problem, however, the conditions preventing the downcomer from sealing were based on speculation at the time. Luckily a few changes in process conditions were successful in bringing the tower back to normal operation. A post event analysis demonstrated that the loss of downcomer was caused initially by operation at rates lower than the minimum liquid limit followed by a transition above the downcomer critical vapor rate. This represent an interesting case involving the downcomer critical vapor rate. For one thing, the depropanizer feed had a relatively low vapor fraction (13% mole) and second how could the downcomer be unsealed, while it should have sealed once the operating point passed the minimum liquid limit.

As illustrated on Figure 10, the Depropanizer has a total of 62 trays. Trays 1 to 33 being conventional two-pass sieve trays and trays 34 to 62 high capacity trays. On startup, the tower receives feed from the Deethanizer bottom. The feed is normally split between tray 34 and tray 39 due to limitations of feed nozzles and distributors. Under normal conditions the tower can be fed entirely from the Deethanizer bottom or from the Deethanizer and a Condensate Stripper, which provides pre-fractionation to the Depropanizer. External streams and/ or off-specification material can be fed to the tower at tray 39. The high capacity tray downcomers have dynamic seals, which in this case defines the minimum liquid rate on the tray.

Soon after the tower was started, high C3 content in the Depropanizer bottom streams was noticed. The tower was reported responsive to reflux in the top section but irresponsive to reboil rates in the bottom. Review of operating data suggested many operating moves in a relatively short period of time. A gamma scan done on the tower indicated no signs of mechanical anomalies or flooding. The scan suggested a significant reduction in froth height relative to previous scan conducted at normal plant rates. A loss of downcomer seal was suspected but did not fully align with the operating point location on the tray operating window.

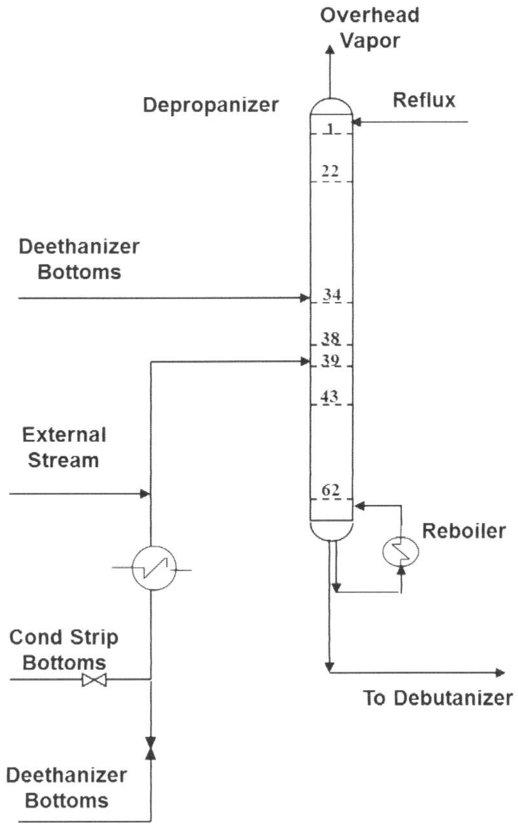

Figure 10: Depropanizer

In a post event analysis, operating data for the startup period was extracted from the plant data historian. The bottom tray load transition was calculated from plant data and plotted against the low end of the tray operating window. As shown on Figure 11, the calculated tray loads were found dispersed on each side of the expected operating line. However, many of the operating points were found to fall within the stable area of the operating window.

The operating data were then sorted relative to the C3 composition transition in the tower bottom stream. Conditions for which the C3 composition were trending up was plotted against the tray operating window as shown on Figure 12. The downcomer critical vapor rate was first calculated based on a clean tray (red dashed line). However, this service is known to foul especially in the bottom of the tower, and sensitivity analysis suggested that the tower behavior would line up with a 20% reduction in hole area, which shift the downcomer critical vapor rate limit and the tray deck seal point downward. By doing so, operating points with an upward trend in C3 composition are grouped on each side of the minimum liquid load and downcomer critical vapor rate limit.

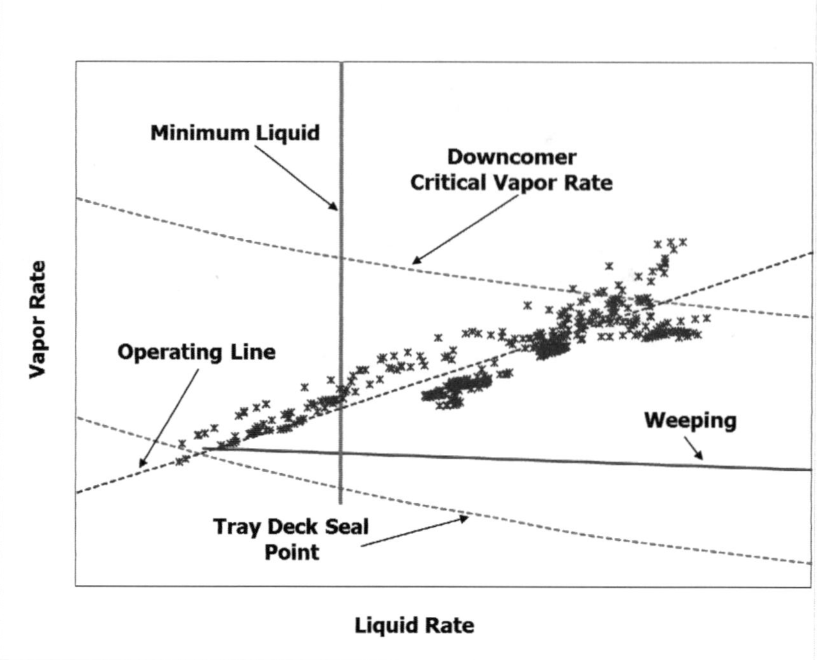

Figure 11: Operating Point Transition

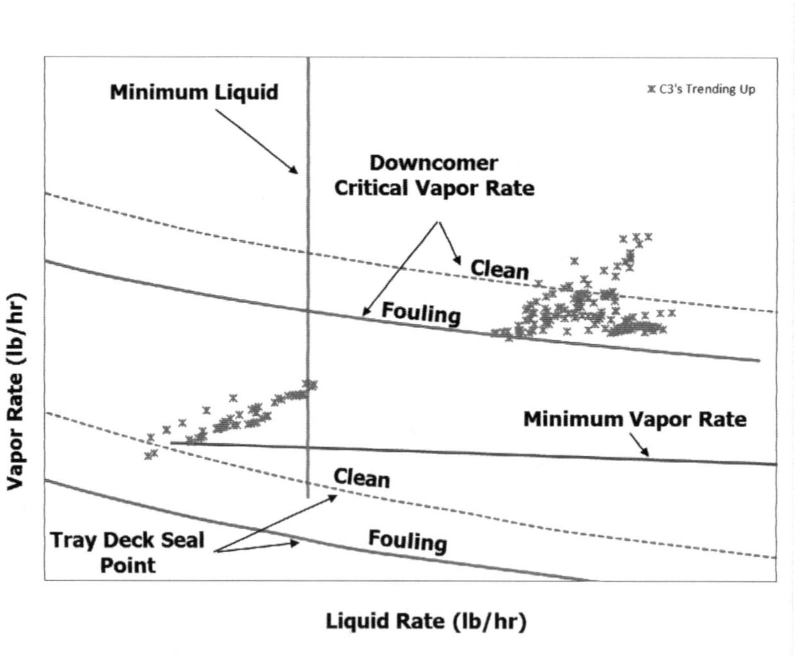

Figure 12: Operating Point with C3 Composition Trending Up

The key question to resolve at this point was how could exceeding the downcomer critical vapor rate cause the downcomer to unseal. As shown on Figure 11, the operating point had transitioned by more or less following the operating line. The downcomer would be expected to seal once it passed the minimum liquid limit. As indicated earlier, once the downcomers are sealed, the downcomer critical vapor rate limit no longer exists. Although not meeting product specification, many of the operating points shown on Figure 11 were associated with the C3 composition trending down. This mystery was resolved by reviewing the many operating changes done on the tower in a relatively short period of time. Per normal practice the Depropanizer was fed entirely from the Deethanizer. The loads were increased which brought the operating point within the stable area of the tray operating window. It is assumed that the downcomers were sealed at that time. However, as shown on Figure 11; most of the operating points inside the stable area of the tray operating window are not lining up with the operating line. This suggests deviations in operating conditions for meeting the expected bottom product quality. At one point the feed to the Depropanizer was changed by commissioning the Condensate Stripper. The transition caused an upset in the Depropanizer as the feed from the Deethanizer was all shifted to feed tray 34. The feed transition occurred before the Condensate Stripper could relieve the load from the Deethanizer. The feed at tray 34 then exceeded the capability of the feed distributor. It is suspected that the reduction in cross section due to the feed distributor caused a high vapor velocity that exceeded the system limit. This caused severe liquid entrainment and starved the tray below from liquid, hence resulting in a loss of downcomer seal. This theory was supported by a drop in bottom liquid level and a spike of the overhead product flow, which is being regulated by a reflux drum level/ flow cascade control loop.

The loss of downcomer seal was suspected during the startup but was believed to be caused by a liquid load deficiency. The liquid load was then increased by feeding off-specification material. The reboil rate was also reduced to help liquid seal the downcomers. The post event evaluation demonstrated that the reduction in vapor load was the key move that allowed the tower to regain stability. A reduction in liquid load in alignment with the vapor load and operating line would have been more effective than the increase in load which was made to re-establish stable operation.

Figure 13 shows the transition of the operating point relative to the operating line and tray operating window on the following startup. As shown, dispersion of the operating point relative to the operating line occurs below the minimum liquid limit. A good alignment is then observed once the operating point reaches the stable area of the tray operating window. In this normal startup, the downcomer critical vapor rate was no longer a constraint once the operating point passed the minimum liquid limit.

This case is a good example on how the downcomer critical vapor rate, typically unexpected, affects operation of trays. As demonstrated above, attributing the loss of downcomer seal on a liquid rate deficiency would suggest an increase in loads, while the remedy required a load reduction. It also highlights the impacts and importance of understanding how fouling can change the expected tray limitations. The learning from this event allowed revision of procedures to prevent future re-occurrence.

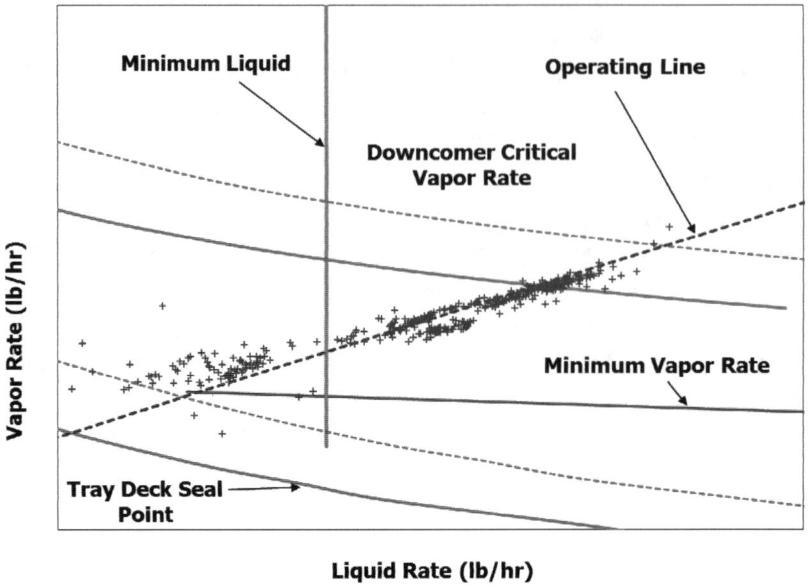

Figure 13: Normal Startup

Conclusion

As stated earlier, whether designing or trouble shooting; process engineers must understand the characteristics of equipment and loads imposed by the process. In distillation, the characteristics of tray tower can be effectively described by tray operating windows. The relationships between equipment/ process requirements can be evaluated by plotting an operating line on the tray window. The tray operating window/ operating line combination defines the expected operating range of a tray and also defines the limitations mechanism.

As demonstrated, the change in properties expected by change in feedstocks and operating conditions are expected to have negligible impact on tray windows developed for a specific applications. Tray operating windows developed for a given service might be used as reference for evaluating other services providing that change in properties do not cause major shift on some of the limitations.

Fouling would typically cause a reduction in hole area and may restrict the downcomer clearance. This will affect a tray operating window by shifting the jet flood, downcomer backup and weep limits downward. Minimum liquid and downcomer choke are expected to be less sensitive to fouling. As demonstrated in the Depropanizer example, the extent of fouling can be estimated by aligning the tray limitations with plant data.

The C2 Splitter example demonstrates the advantage of visualizing the tray constraints. Although well balanced for the original design case, a change in feedstock caused the bottom section to be prematurely constrained, while the rectifying section still had a lot of unused capacity. Understanding the location of the pinch point, flooding mechanism and relationship with side reboiler operation allowed identifying a solution that would increase capacity and properly balance each section of the tower.

Due to the graphical nature of operating window, a multitude of operating points gathered from plant data can be overlaid and analyzed on a single diagram. As done in the Depropanizer example, sorting the data based on specific tower behavior helps in identifying potential tray issues. A tray operating window will not replace the need for tools such has gamma scan but it provides additional information for diagnosing distillation problems.

Nomenclature

A:	Tower Cross Section
C2R :	Ethylene Refrigerant
C3R :	Propylene Refrigerant
C_S:	C-factor based on tower cross section
VLOAD:	Vapor Load as defined by eq. 1
V_{rate}:	Volumetric Vapor Rate
ρ_L:	Liquid Density
ρ_V:	Vapor Density
σ:	Liquid Surface Tension

References

1. H. Z. Kister, "Distillation Design", McGraw-Hill.
2. J. Penciak, I. Nieuwoudt, G. Spencer, "High-Performance Trays : Getting The Best Capacity and Efficiency", Symposium Series No. 152, 2006 IChem
3. D. Summers, "Ultra-High-Capacity ConSep Trays", Sulzer Technical Review 3+ 4/2005
4. Z. P. Xu, B. J. Nowak, K. J. Richardson, B. Miller, "SimulFlow Device Capacity Beyond System Limit", 2007 AIChE Spring Meeting, Houston, TX
5. D. Summers, "Performance Diagrams All Your Tray Hydraulics In One Place", 2004 AIChE Annual Meeting, Austin, TX
6. D. Summers, "Distillation Tray Design Through A Better Understanding Of Operating Window", 2016 AIChE Annual Meeting, San Francisco, CA
7. W. J. Stupin, H. Z. Kister, "System Limit The Ultimate Capacity of Fractionators", IChem, Vol 81, Part A. January 2003
8. A. Bernard, Design, "Operation and Startup Strategy of a New Parallel C2 Splitter", Ethylene Producer Conference, 2016 AIChE Spring Meeting, Houston, TX

9. A. Bernard, "Distillation: Avoid Problems During Tower Startup", Chemical Engineering, June 2012
10. H. Z. Kister, "When Tower Startup Has Problems", Hydrocarbon Processing, February 1979
11. M. J. Lockett, "Distillation Tray Fundamentals", Cambridge University Press

Hidden ties - An explanation for O'Connell's success

Markus Duss
Sulzer Chemtech Ltd.
Winterthur, Switzerland

and

Ross Taylor
Clarkson University
Potsdam, NY, USA

Presented at the Distillation Topical Conference,
AIChE Spring Meeting, San Antonio, Texas, March, 2017

SULZER
Clarkson
UNIVERSITY

Hidden ties – An explanation for O'Connell's success

Markus Duss, Sulzer Chemtech Ltd, Winterthur, Switzerland
Ross Taylor, Clarkson University, NY, USA

Summary:
When predicting tray efficiencies, O'Connell's simple method has survived all attempts at being replaced by more rigorous models. O'Connell's method is an empirically obtained correlation to determine section efficiencies that has been widely applied in industry and has demonstrated applicability to many distillation processes. This paper presents a simple theoretical but fundamentally sound approach that achieves similar results to O'Connell's correlation. It is proposed to modify the α-term in order to have a close to perfect match between the modified O'Connell correlation and the derived model.

Introduction

Tray efficiency models are still lacking the accuracy that would allow process engineers to rely on their outcome; this despite considerable effort to improve their predictive power. Lewis[1] has presented a fundamental approach to obtain the section efficiency by correlating the Murphree point efficiency (usually based on the number of transfer units) and to apply it to predict the performance of cross flow trays by integrating over the flow path length. Horizontal liquid back-mixing, entrainment and weeping might be included and the final result yields then the Murphree tray efficiency. The section efficiency is then determined as a function of the tray efficiency and the stripping factor.

A supposedly rigorous tray efficiency model requires a large amount of input data:

- detailed tray geometry
- physical properties of the liquid and vapor phase
- operating conditions
- thermodynamic information

From such information, mass transfer coefficients, interfacial area, froth height, bubble regimes, horizontal back mixing, entrainment, weeping etc. can be derived, all of which are required to calculate the Murphree tray efficiency in a rigorous manner.

In sharp contrast to the considerable effort required to obtain all of this information collected, the well-known O'Connell correlation[2] to estimate the section efficiency has been used quite successfully. Perhaps due to its simplicity, it is the most often applied method.

O'Connell published in 1946 a graph showing section efficiencies for bubble cap trays which is represented by Eqn. (1).

$$\eta_{Section} = 0.503 \cdot (\mu_L \cdot \alpha)^{-0.226} \tag{1}$$

This paper will reveal how this correlation is tied to a more fundamental mass transfer approach to determine tray efficiencies and what the assumptions need to be.

Attempts to Explain O'Connell's Success

O'Connell's correlation is widely used in industry to estimate section efficiencies and, therefore, several attempts had been undertaken to find theoretical explanations for its success.

King[3] noted that "*the overall efficiency decreases with increasing feed viscosity, presumably reflecting a poorer dispersion for higher viscosity feeds*" and interpreted that the higher relative volatility increases liquid side mass transfer resistance.

Lockett[4] noted that the increase in liquid viscosity is associated with a decrease in liquid diffusivity and that "*...there was little justification for including α in the correlation.*"

Chen and Chuang[5] derived their explanation by assuming that the vapor side mass transfer resistance is negligible and assumed the concentration of the light boiling component to be 29 mol-% in order to find the slope of the equilibrium line to be approximately proportional to $\alpha^{-0.25}$ for α-values between 1 and 2. They further imposed the restriction that the number of transfer units, N_{OG} is small so the point efficiency equals N_{OG}. Furthermore, they assumed horizontally well-mixed liquid to make the tray efficiency equal to the point efficiency and, finally, it was assumed that the relative volatility is close to unity so that the section efficiency equals the tray efficiency. By employing such contradictory assumptions, they could show that O'Connell's correlation yielded similar results to their model. However, none of the assumptions made by Chen and Chuang hold true in general for any system!

It will be shown that no such restrictive assumptions are required in order to find an explanation to justify O'Connell's correlation. A simplified – albeit traditional – approach to predict section efficiencies for trays will be derived, matching well with the outcome of the O'Connell model.

Traditional Approach

The section efficiency is defined as the number of theoretical stages found in a section divided by the number of actual trays:

$$\eta_{Section} \equiv \frac{n_{th}}{n_{trays}} \tag{2}$$

The traditional approach to model section efficiencies based on mass transfer considerations is via use of the point efficiency, which can be determined by the number of transfer units. The number of transfer units overall gas, N_{OG}, can be expressed as follows[4].

$$\frac{1}{N_{OG}} = \frac{1}{N_G} + \frac{\lambda}{N_L} \tag{3}$$

with

$$\lambda = \frac{m}{L/G} \qquad N_G \equiv \frac{k_G \cdot a_I \cdot h_f}{G} \qquad N_L \equiv \frac{k_L \cdot a_I \cdot h_f}{L} \tag{4}$$

Rigorous models place considerable effort on determining the number of transfer units, which includes correlations for mass transfer coefficients, interfacial area and froth height. In addition, the slope of the equilibrium line is required in order to properly sum the separate liquid and vapor side mass transfer resistances as in Eqns. (3) and (4). Most of this effort can be avoided, simply by assuming that for a given system the section efficiency remains independent on the throughput when columns are operated within their hydraulic limits at given L/G ratio.

The approach to equilibrium which can be achieved with a given number of transfer units depends on the driving force for mass transfer, i.e. on the flow regimes of vapor and liquid phase. Most flow models for cross flow trays assume vapor flowing vertically in plug flow and liquid being vertically well-mixed. With these assumptions, the Murphree point efficiency is given by[4]:

$$\eta_{Point} = 1 - e^{-N_{OG}} \tag{5}$$

From the point efficiency, the Murphree vapor tray efficiency can be determined. To achieve a mathematically simple solution, the following assumptions are made:

- the liquid travels in plug flow across the tray
- the point efficiency is constant across the tray
- there is an infinite number of flashes in flow path, each approaching equilibrium as per the assumed point efficiency
- uniform vapor concentration at the inlet to the tray (Lewis case 1)

Based on the above assumptions, the Murphree vapor tray efficiency becomes:

$$\eta_{Tray} = \frac{e^{\lambda \cdot \eta_{Point}} - 1}{\lambda} \tag{6}$$

The Murphree vapor tray efficiency defines the fractional approach to equilibrium for a single cross flow tray based on the vapor concentrations on the operating and equilibrium lines. In order to retrieve the section efficiency, the change in the driving force along the operating line must be considered; this can be done using the stripping factor:

$$\eta_{Section} = \frac{\ln[1 + \eta_{Tray} \cdot (\lambda - 1)]}{\ln(\lambda)} \tag{7}$$

When the slope of the equilibrium line is steeper than the slope of the operating line, as typically found in stripping sections, the stripping factor is greater than unity and the section efficiency becomes smaller than the tray efficiency. When the stripping factor is smaller than unity, the section efficiency is higher than the tray efficiency, which typically applies to rectifying sections. When the stripping factor is exactly unity then section and tray efficiencies are equal.

Eqns. (6) and (7) were derived by Lewis[1] in 1936.

Eqns. (3), (5), (6) and (7) can be combined to give:

$$\eta_{Section} = \frac{\ln[1+\frac{e^{\lambda\cdot\left(1-e^{-\left(\frac{1}{N_G}+\frac{\lambda}{N_L}\right)^{-1}}\right)}-1}{\lambda}\cdot(\lambda-1)]}{\ln(\lambda)} \tag{8}$$

Eqn. (8) is the basis for our approach tying O'Connell's empirical correlations to a theoretical more fundamental mass transfer model.

Postulating the Ties

There seems to be a discrepancy between the simple O'Connell correlation and the traditional model given by Eqn. (8). Should there be any hidden ties, then the following approximation must hold:

$$\eta_{Section} = \frac{\ln[1+\frac{e^{\lambda\cdot\left(1-e^{-\left(\frac{1}{N_G}+\frac{\lambda}{N_L}\right)^{-1}}\right)}-1}{\lambda}\cdot(\lambda-1)]}{\ln(\lambda)} \approx 0.503\cdot(\mu_L\cdot\alpha)^{-0.226} \tag{9}$$

Whereas O'Connell uses the relative volatility, the traditional approach requires the use of the stripping factor instead. But whereas α might remain constant and is always greater than unity for the whole concentration range (for non-azeotropic systems), the stripping factor is changing with concentration and is smaller than unity for high concentrations of the light boiler and greater than unity for low concentrations. Figure 1 represents the conditions for a binary system assuming $\alpha = 4$ and for the case of total reflux. The slope of the equilibrium line at zero light boiler concentration (x→0) becomes α, whereas the slope for pure light boiler (x→1) becomes $1/\alpha$. L/G is unity (total reflux) for the case shown, and thus the stripping factor becomes α and $1/\alpha$, respectively at the respective concentrations.

In order to establish the conditions where O'Connell's correlation can be tied to the traditional model, the stripping factor must be used instead of α and N_G and N_L must be expressed using a single variable, namely the liquid viscosity. As per O'Connell's approach, one constant of proportionality and one fixed exponent shall be used to correlate the number of transfer units.

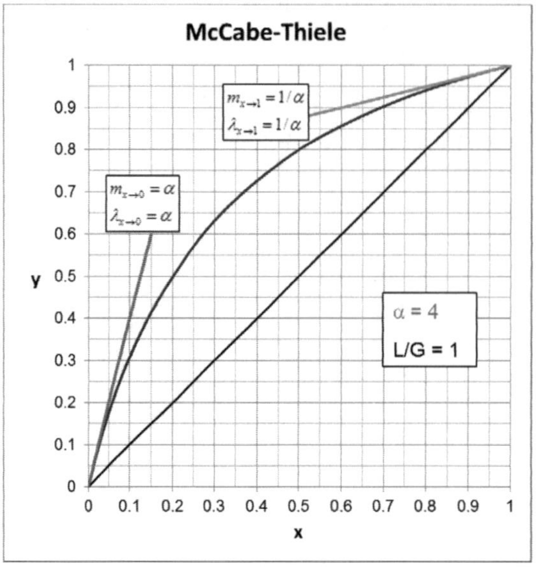

Figure 1: McCabe-Thiele diagram for a binary system with $\alpha = 4$ and $L/G = 1$. The red line indicates the slope of the equilibrium line for x→0 and the green line for x→1.

In order to be in the position to correlate N_G and N_L with the few given parameters, the following assumption was made:

$$N_G = N_L = c_1 \cdot \mu_L^{x_1} \tag{10}$$

The number of overall gas transfer units becomes based on Eqns. (3) and (10):

$$N_{OG} = \frac{c_1 \cdot \mu_L^{x_1}}{1+\lambda} \tag{11}$$

c_1 and x_1 need now to be adjusted in order to minimize the difference of the derived model and O'Connell's correlation, as indicated in Eqn. (9).

Revealing the Ties

Minimizing the difference for the term on the left hand side and O'Connell as given with Eqn. (9) results for the constant of proportionality in $c_1 = 0.936$ and for the exponent in $x_1 = -0.25$. Thus, Eqn. (10) results in:

$$N_G = N_L = 0.936 \cdot \mu_L^{-0.25} \tag{12}$$

Figure 2 shows the parity plot of the section efficiency based on O'Connell's correlation and the derived model using Eqns. (9) and (12). The relative volatility was varied in the range of 1 to 5 and the liquid viscosity between 0.08 and 5 mPa·s. This covers the majority of distillation applications.

Figure 2 clearly reveals ties of the two approaches predicting section efficiencies. The deviation in the investigated range is less than 10%.

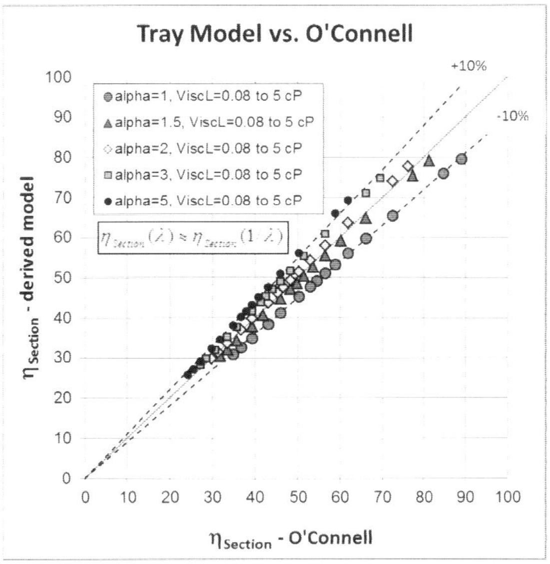

Figure 2: Parity plot showing section efficiencies derived from O'Connell's correlations and equation (9) assuming $N_G = N_L = 0.936 \cdot \mu_L^{0.25}$

The question which arises is this: which concentrations were used to calculate the slope of the equilibrium line to retrieve the section efficiencies with the derived model? - The answer is: Fig. 2 shows the section efficiencies derived with the traditional model approach for x=0 and x=1! The astonishing outcome is that Eqn. (9) using Eqn. (12) achieves close to identical results for λ and $1/\lambda$.

This outcome needs to be understood in more depth since this will undergird the explanation why using the relative volatility instead of the stripping factor can represent relevant behaviour.

From Figure 2 it can be further interpreted that the dependency on the liquid viscosity agrees very closely for the two correlations. However, the dependency on α or λ, respectively shows that O'Connell over estimates the impact of α on the section efficiency, compared to the model derived based on the traditional approach.

Understanding the Ties

In order to understand why Eqn. (9) using Eqn. (12) achieves an almost identical section efficiency for a given value of the stripping factor and its reciprocal value, Fig. 3 is given here. It shows the calculated section efficiencies for 3 different ratios of N_G and N_L, i.e. for the ratios 1:1, 1:3 and 3:1. In order to achieve the same maximum section efficiency, N_G was adjusted accordingly and the stripping factor was varied. The purpose of Fig. 3 is not to compare absolute values but to reveal apparent symmetries.

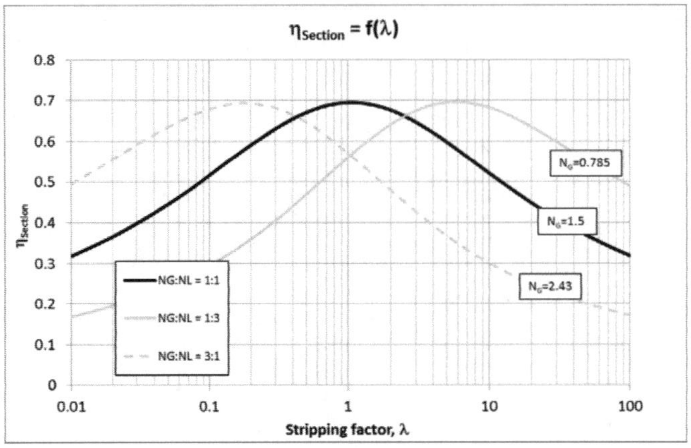

Figure 3: Section efficiencies for varying stripping factors obtained using equation 9. The following ratios of N_G and N_L were assumed: black line 1:1; green solid line 1:3; green dotted line 3:1. Values of N_G are adjusted to achieve the same maximum section efficiency.

As can be seen from Fig. 3, an apparent symmetry about $\lambda=1$ resulted for the case $N_G = N_L$. Such symmetry is not achieved for the other two cases. However, the black curve is not strictly symmetrical about $\lambda=1$, but achieves slightly different values for λ and $1/\lambda$ for values other than unity.

Assuming $N_G = N_L$ does not imply that the liquid and vapor side resistances to mass transfer are equal since the slope of the equilibrium line will influence how the resistances distribute across the two phases as given with Eqn.(13).

$$\frac{1}{k_{OG}} = \frac{1}{k_G} + \frac{m}{k_L} \tag{13}$$

The liquid phase resistance, LPR, is defined by Eqn. (14).

$$LPR \equiv \frac{m \cdot k_{OG}}{k_L} = \frac{\lambda \cdot N_{OG}}{N_L} \tag{14}$$

Figure 4 shows the dependency of LPR on the stripping factors: the liquid phase resistance increases with increasing stripping factor and for small values of λ, the liquid phase resistance is negligible whereas for high values of λ, LPR will become dominant.
It can now be explained, based on Fig. 4, why, for increasing values of λ above unity, the section efficiency is reduced: the increase in liquid side resistance dominates. But what is then the explanation for the decreasing efficiency with small λ when the liquid side resistance tends to be negligible? To understand this behaviour it is important to quantify the impact of the stripping factor when converting the Murphree tray efficiency to the section efficiency.

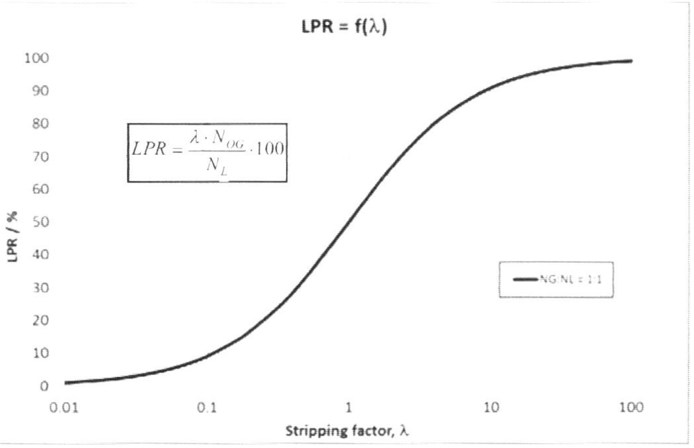

Figure 4: Liquid phase resistance depending on the stripping factor for the case $N_G=N_L=1$.

Figure 5 shows the graphs of the 4 basic Eqns. (3), (5), (6) and (7), which were the basis of the derived model. The dotted black line indicates the section efficiency depending on the stripping factor, assuming the Murphree tray efficiency to be constant, in our case close to 0.5. Based on Eqn. (7) the section efficiency decreases with decreasing λ. This is the explanation for the drop in the section efficiency when the stripping factor becomes small, despite the fact that the Murphree tray efficiency increases. And this explains that O'Connell's correlation can use α instead of λ and is still achieving meaningful results, when the curve is symmetrical about 1.

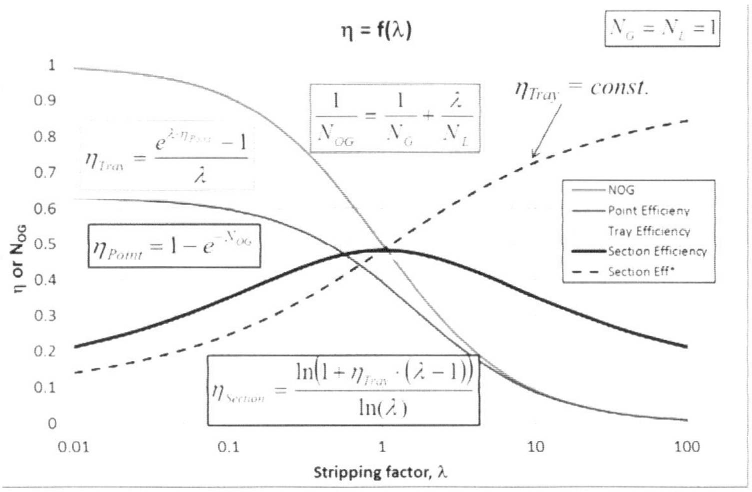

Figure 5: Graphs of Eqns. (3), (5), (6) and (7) for varying stripping factors and for $N_G=N_L=1$. The dotted line shows the section efficiency assuming a constant tray efficiency of 0.5.

Figure 6 summarizes the outcome: for increasing λ the liquid phase resistance becomes dominant and makes the section efficiency poor, whereas for decreasing λ the driving force for mass transfer increases and the translation of the tray efficiency to section efficiency is the reason for the decrease. Such behaviour is known when converting transfer units to theoretical stages in counter-current systems using packing.

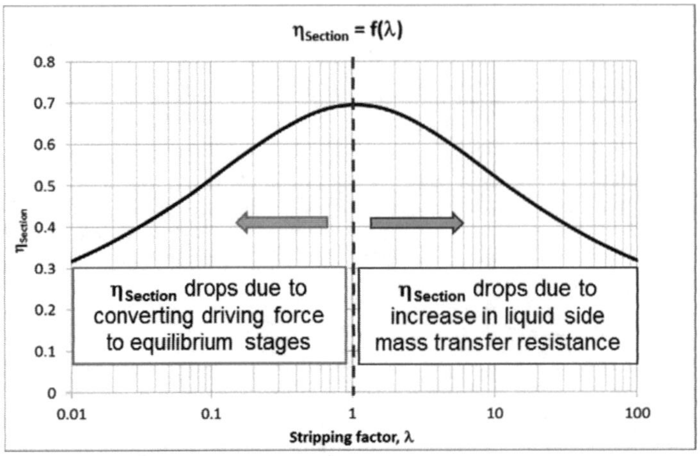

Figure 6: Section efficiency based on equation (9) assuming $N_G=N_L$, indicating the reasons for the drop in efficiency when $\lambda \neq 1$.

Modified O'Connell

As already was concluded before, O'Connell's correlation is over estimating the impact of α compared to the retrieved model (see Fig. 2). Therefore, it is proposed to adjust the exponent of α individually to match the theoretical outcome. The relevant range for α between 1 and 5 was chosen to adjust the exponent. Using an exponent of -0.08 achieved a close match in the chosen range with the traditionally derived model, as shown in Fig. 7.

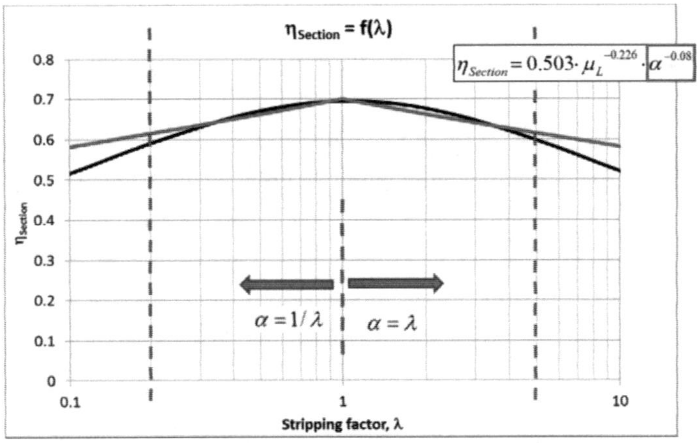

Figure 7: Adjustment of the exponent of α to match the theoretical dependency according to equation (9) assuming $N_G=N_L$.

The modified O'Connell correlation becomes:

$$\eta_{Section} = 0.503 \cdot \mu_L^{-0.226} \cdot \alpha^{-0.08} \quad (15)$$

The constant c_1 in Eqn. (12) needs then to be refit in order to match O'Connell's correlation best and the following value was retrieved:

$$N_G = N_L = 1.05 \cdot \mu_L^{-0.25} \quad (16)$$

As the parity plot in Fig. 8 shows, Eqn. (9), now using Eqn. (16), achieves an almost perfect fit with the modified O'Connell correlation.

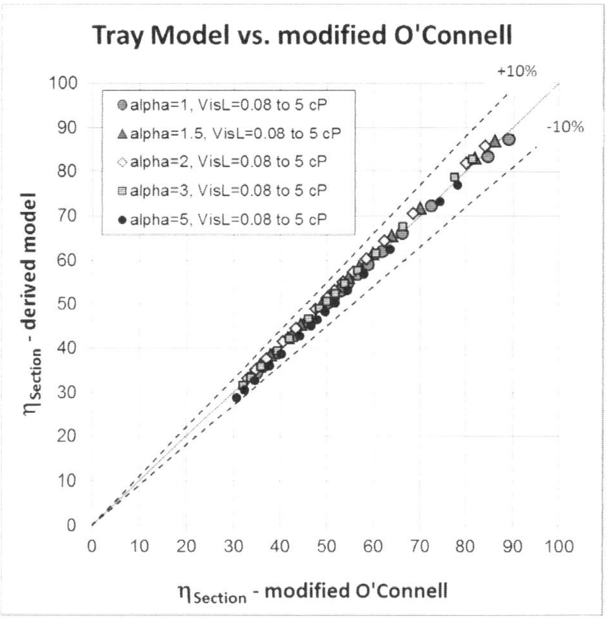

Figure 8: Parity plot showing section efficiencies derived from the modified O'Connell correlation equation (15) and equation (9) assuming $N_G = N_L = 1.05 \cdot \mu_L^{-0.25}$

Based on the gained insight it is concluded that using the stripping factor or its reciprocal value, respectively, is preferred over using the α -value from a theoretical point of view. The reason is that when using α, it is not possible to distinguish between total and partial reflux conditions; the use of the stripping factor accounts for this. Therefore, when using the modified O'Connell correlation, we recommend using the following equation:

$$\eta_{Section} = 0.503 \cdot \mu_L^{-0.226} \cdot \bar{\lambda}^{-0.08} \quad (17)$$

where

If $\lambda > 1$: $\bar{\lambda} = \lambda$ (17a)

If $\lambda < 1$: $\bar{\lambda} = \dfrac{1}{\lambda}$ (17b)

The reciprocal value of the stripping factor is often referred to as the absorption factor. The use of Eqn. (17) is expected to achieve a better match with plant data.

Conclusions / Recommendations

It was shown that O'Connell's empirical correlation has ties to the traditional tray efficiency model approach when assuming that the number of transfer units is equal for the liquid and vapor phase. Though this assumption might not be very accurate in all cases, it is an acceptable assumption for numerous applications.

A modification to O'Connell's correlation was proposed by adjusting the exponent of the α-term to match the theoretically expected dependency on the stripping factor. The adjustment was made for the range of α between 1 and 5. Based on the gained insight, it was recommended to use $\overline{\lambda}$, i.e. the stripping factor or its reciprocal value the absorption factor, respectively instead of the α-value.

Reviewing earlier attempts trying to explain O'Connell's success requires adding some comments:

King[3] was partly right that with increasing relative volatility the liquid phase resistance increases but this holds true only for increasing stripping factors when $\lambda > 1$. However, decreasing stripping factors will reduce the liquid side resistance but still a drop in section efficiency is expected (and postulated by O'Connell's correlation when using α). The reason for such behavior is the translation of driving force to theoretical stages which was not included in King's interpretation.

Lockett[4] was right when stating that there was little justification to use α. Based on the derived model, α has an impact on the efficiency, which, however, is small and the stripping or the absorption factor, $\overline{\lambda}$ should be preferably used, instead.

The arguments used by Chen and Chuang[5] must not be followed. Assuming a system to be liquid side controlled excludes most systems. Assuming a specific concentration where the slope of the equilibrium is according to the desired value would make it valid only in a very small range. And the requirement that $N_{OG}=\eta_{Point}= \eta_{Tray}= \eta_{Section}$ must hold in order to postulate ties to mass transfer considerations, are far from realistic.

In contrast, the assumption $N_G = N_L$ to show the ties of O'Connell's empirical approach matches reality much better.

Finally, it can be concluded that the insight gained might help to learn about limitations of O'Connell's method. However, maybe the most important aspect of revealing ties to a more fundamental model is that those using O'Connell can now more easily justify its use.

Acknowledgements

This paper is based in part on work carried out during the sabbatical of Markus Duss at Clarkson University in 2016.

SULZER Clarkson UNIVERSITY

Symbols

a_I	(m^2/m^3)	Specific vapour-liquid interfacial area
c_I	(-)	Constant of proportionality
G	$(kmol/(m^2 s))$	Gas throughput per cross sectional bubbling area
h_f	(m)	Froth height
k_G	$(kmol/(m^2 s))$	Gas side mass transfer coefficient
k_L	$(kmol/(m^2 s))$	Liquid side mass transfer coefficient
k_{OG}	$(kmol/(m^2 s))$	Overall gas side mass transfer coefficient
L	$(kmol/(m^2 s))$	Liquid throughput per cross sectional bubbling area
LPR	(-)	Liquid phase resistance see eq. (13)
m	(-)	Slope of the equilibrium line
N_G	(-)	Number of transfer units in gas phase
N_L	(-)	Number of transfer units in liquid phase
N_{OG}	(-)	Number of transfer units overall gas
n_{th}	(-)	Number of theoretical stages
n_{tray}	(-)	Number of actual trays
x_I	(-)	Exponent to be applied to the liquid viscosity

Greek

α	(-)	Relative volatility
η_{Point}	(-)	Murphree vapor point efficiency
η_{Tray}	(-)	Murphree vapor tray efficiency
$\eta_{Section}$	(-)	Section efficiency or overall efficiency
μ	(mPa s) or (cP)	Viscosity
λ	(-)	Stripping factor = $m / (L/G)$
$\overline{\lambda}$	(-)	Stripping or absorption factor in Eqn. (17): if $\lambda>1: \overline{\lambda}=\lambda$; if $\lambda<1: \overline{\lambda}=1/\lambda$

Indices

G	Vapor or Gas
L	Liquid

Literature

[1] Lewis, W.K., Rectification of Binary mixtures, *Ind. Eng. Chem.*, **1936**, *28* (4), pp 399

[2] O'Connell, H.E., Plate Efficiency of Fractionating Columns and Absorbers, *Trans. Am. Inst. Chem. Eng.* , **1946.** *42*, pp 741

[3] King, C.J., *Separation Processes*, 2nd ed., McGraw-Hill, New-York, **1980**

[4] Lockett, M.J., *Distillation tray fundamentals*, Cambridge University, New York, NY, **1986**

[5] Chen, G.X and Chuang, K.T, Liquid-Phase Resistance to Mass Transfer on Distillation Trays, *Ind. Eng. Chem. Res.*, **1995**, *34* (9), pp 3078

BUBBLE CAP TRAY VAPOR TURNDOWN

By

Daniel R. Summers, P.E.
SULZER CHEMTECH USA, Inc.

.

Presented at the AIChE Spring National Meeting

Topical 8:

Kister Distillation Symposium

Paper 122c

March 28, 2017
San Antonio, TX

Abstract

The Bubble Cap tray has been in the market place as a distillation device since the early 1800's[1]. It has been used extensively on distillation tray equipment worldwide as a highly efficient vapor/liquid-contacting device. Many people have examined the maximum capacity of Bubble Caps[2,3], but very few have studied turndown. The author of this paper was a coauthor of a US Patent in 1987[4] that endeavored to enhance the turndown capability of Bubble Cap trays to as high a turndown ratio value of 15:1. It is the intent of this paper to establish a turndown criterion for bubble cap trays such that the tray designer can maintain efficient operation at the lowest loads. A relationship is established based on operational data that shows what the minimum efficient vapor load can be on Bubble Cap trays.

Discussion

The author wrote a paper on the stability of sieve trays, movable valve trays and fixed valve trays in 2010[5]. The theory presented is that there is a relationship between the upwards force of the vapor to a tray that will balance against the gravitational force of the liquid (froth) depth and maintain good efficient vapor distribution onto the tray. Maintaining good vapor distribution should enable good tray performance. A "stability" term was introduced which was basically a modified Froude number. Bubble cap trays were excluded from the discussion at that time because the liquid is not balanced against the vapor on such trays. The liquid is prevented from weeping through a bubble cap tray by the physical presence of the bubble cap risers. However, the author has frequently been queried about the turndown capability of bubble cap trays. Normally the topic of turndown is not an issue since trays that have a loss of efficiency (and theoretical stages) at turndown can easily be compensated by an increase in reflux ratio (heat duty) to the tower. However, there are times (in highly heat integrated units) when there is little or no extra heat available to overdrive the tower when operated at turndown. In these cases, the issue of efficient turndown of bubble cap trays becomes extremely important.

With the typical sieve or valve tray, the prevalence of weeping can be one way to determine the minimum operating vapor load to such a device. As mentioned above, the author has offered an alternate idea to weeping on how to determine the turndown of such typical trays with the introduction of the "stability" concept.

$$\eta = (\Delta P_{DRY} / H_{CL})^{0.5} \quad \text{(Eq. 1)}$$

Where,

η = stability factor
ΔP_{DRY} = inches of liquid
H_{CL} = Hydrostatic Head of Liquid on the tray, inches of liquid

This stability concept should, in theory, be applicable to bubble cap trays with some modification. The theory is that to maintain good efficient operation of a bubble cap tray, the vapor must be flowing through a high percentage of the bubble cap's risers.

One modification to the concept of equation 1 would be that the hydrostatic head of liquid has to have the height of the slots on the bubble caps removed. The vapor that is introduced to the liquid on a bubble cap tray does not have to "fight" against the full depth of liquid (or froth) on the tray to be able to flow through the cap. The other modification to equation 1 is the definition of dry tray pressure drop itself. Dry tray pressure drop is a simple theory which has been explored by the author numerous times[(6,7,8)].

$$\Delta P_{DRY} = 12\ \rho_V\ (V_H/C_V)^2\ /\ (2\ g_C\ \rho_L) \qquad \text{(Eq. 2)}$$

Where,

ΔP_{DRY} = inches of Liquid
V_H = Hole Velocity, ft/sec
ρ_V = Vapor Density, lb/ft^3
C_V = Orifice Coefficient
ρ_L = Water Density, lb/ft^3
g_C = Acceleration of Gravity = 32.174 ft/sec^2

If one assumes that the 4 critical areas of a bubble cap are equal (riser area, reverse area, annular area and slot area) or are at least limited by the riser area, then the dry tray pressure drop through the riser can be used to establish the ΔP_{DRY} in equation 2 above. The C_V for smooth bore risers can be assumed to be 0.43. With this, equation 2 reduces to;

$$\Delta P_{DRY} = 1.009\ (\rho_V/\rho_L)\ V_R^2 \qquad \text{(Eq. 3)}$$

Where,

V_R = Riser Velocity, ft/sec

The data to determine if this theory works was established by Fractionation Research Inc. (FRI) in Alhambra, CA between the years 1956 and 1968[(9)]. The data was taken in an industrial size tower (48 inch diameter) at total reflux and shows the tray efficiency vs. vapor load. There were 125 data sets taken during this time period of with numerous compounds operating from deep vacuum to 500 psia. Figure 1 shows an example of one such data set with cyclohexane/n-heptane operating at 24 psia pressure. One can see that where the tray efficiency drops from about 80% to less than 50% as the CB falls below 0.1 ft/s, that this would be an indication of minimum efficient operation.

Through the 386 pages of FRI data, the author looked for turndown data such as this. He also looked for data that showed no signs of weeping. The data sets from FRI included an indication of weeping from the trays. Normally a bubble cap tray should not weep because most are either welded in or fully gasketed. The FRI trays were neither since they have to change out the trays frequently to be able test more trays quickly. The weeping observed was present at the tray seams and at the tray support rings in the FRI column. Recent work at FRI has shown that even minor wall weeping/leakage can cause a significant loss of tray efficiency at low vapor rates. Therefore, the author needed to avoid any data that exhibited signs of weeping.

Most of the tray data was for 4” size bubble caps. Of the 4” caps reported, there were two styles; slotted and “tea cup.” There was a comprehensive set of data from the “tea cup” style caps that covered a wide set of pressures. In addition, by examining the “tea cup” cap which is the industry standard today, all subjectivity with regards to slot shape and size is eliminated from the discussion.

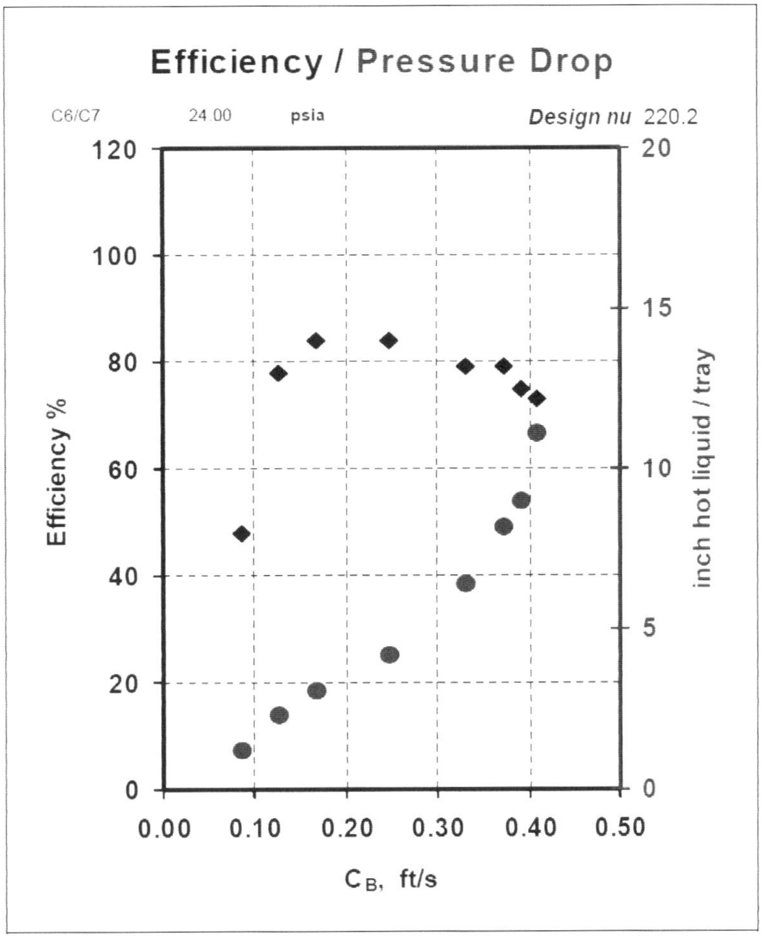

Figure 1 – Typical Bubble Cap Operation at 24 psia

Examination of the tremendous amount of data did lead to some subjectivities. Namely, what is a tolerable reduction in tray efficiency at turndown? Figure 2 shows what was picked as the turndown limit of efficient operation (which was about a 10% loss in tray efficiency) for a set of data at 165 psia. The tray geometry examined here is reported thoroughly in Reference 9. The diameter is 47.75 inches ID, distance to outlet weir from tray centerline is 15 inches and flow path length is 30 inches. Outlet weir height is 2 inches. There are 37 bubble caps per tray each being 4" in size and having a ¼" gap under the cap. The downcomers are sloped and there is a recessed inlet pan. The outlet weir length is 37 inches. Table 1 shows the examination of the 2 data points surrounding this choice of minimum tolerable operation in Figure 2. The value of the stability between these two data points is about 0.2.

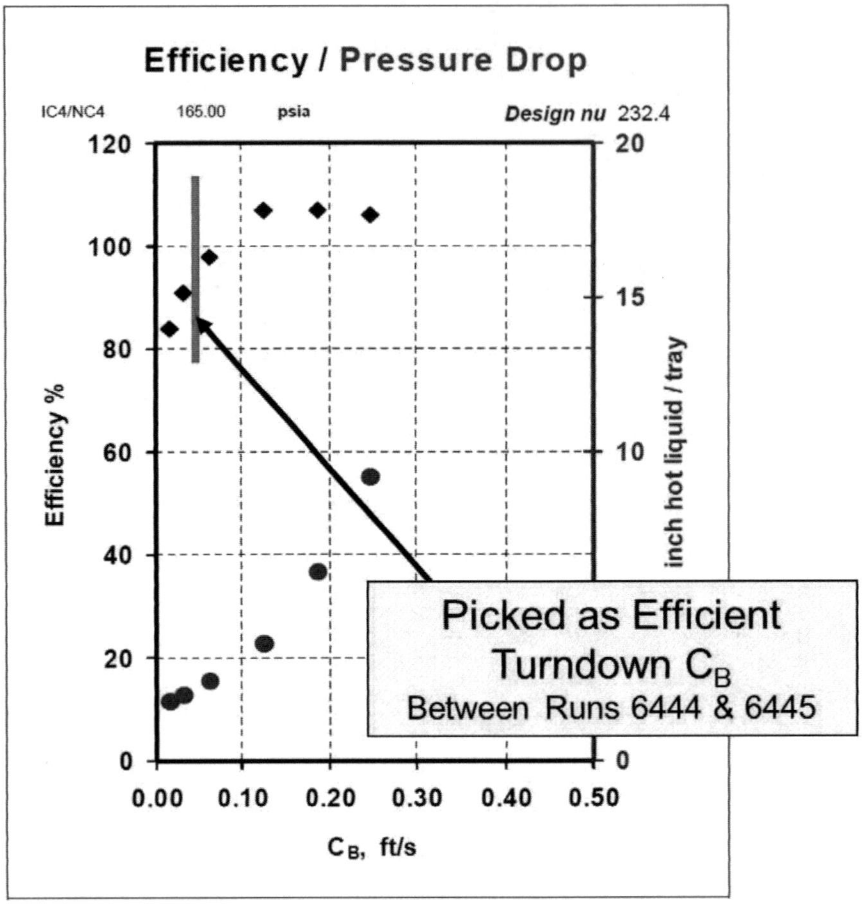

Figure 2 – Turndown Determination of C_B

Table 1 – Data from 2 Data Points in Figure 2 at Turndown

Run	Riser Area	Vapor Den	Liquid Den	V-rate	CFS	Vr	Dry Drop	HS	Stability
	ft2	lb/ft3	lb/ft3	lb/hr		ft/s	in. Liq	in. Liq	
6444	1.39055358	1.74	30.7	14720	2.34993614	1.68992851	0.16325136	2.34	0.26413167
6445	1.39055358	1.74	30.7	7370	1.1765645	0.8461123	0.04092382	2.06	0.14094656

The exact same trays were also operated in C_6/C_7 service at 24 psia, see Figure 3. Here the 10% loss of efficiency is very close to the data point at run #6489. Table 2 shows the examination of the 2 data points surrounding this choice of minimum tolerable operation in Figure 3. The value of the stability between these two data points, and close to run 6489, is about 0.2.

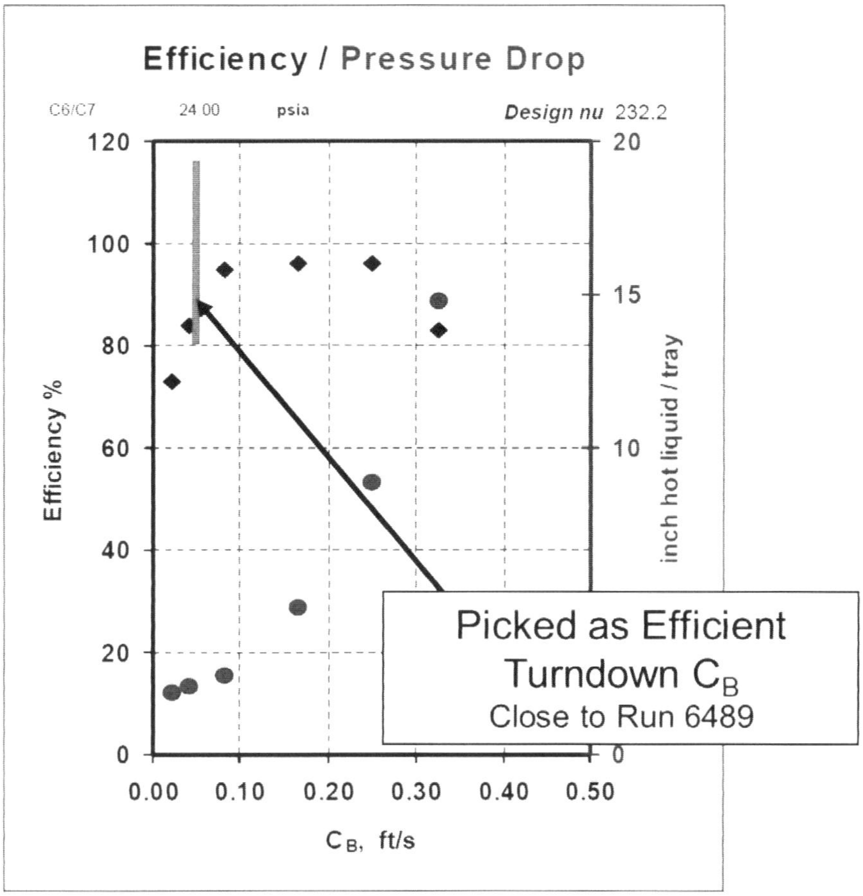

Figure 3 – Turndown Determination of C_B

Table 2 – Data from 2 Data Points in Figure 3 at Turndown

Run	Riser Area	Vapor Density	Liquid Den	V-rate	CFS	Vr	Dry Drop	HS	Stability
	ft2	lb/ft3	lb/ft3	lb/hr		ft/s	in. Liq	in. Liq	
6488	1.39055358	0.305	42.7	9850	8.9708561	6.45128401	0.29982895	1.99	0.38815952
6489	1.39055358	0.302	42.5	4940	4.54378219	3.26760669	0.07652209	2.02	0.19463357

Repeating this procedure for all the other "Tea Cup" data sets yields an interesting set of results, see Table 3. It appears that there is a constant stability factor number for turndown of 4 inch "Tea Cup" bubble cap trays. This value is approximately 0.2 which is considerably smaller than other trays as reported earlier. Reference 5 shows that most other trays need to stay above a stability number of 0.6 to remain efficient. It makes sense that bubble caps trays can use a smaller stability number since they should not weep.

Table 3 – Tea Cup Bubble Cap Stability Results

System	Pressure psia	Run Numbers	Stability
i/n C4	500	8307-8313	0.19
i/n C4	400	8297-8303	0.18
i/n C4	300	8280-8286	0.20
i/n C4	165	6441-6446	0.20
C6/C7	24	6485-6490	0.20
C6/C7	5	6517-6520	0.21

Keep in mind that all tray data examined thus far used only a 2" high outlet weir. The question remains, "What happens when a different outlet weir height is chosen?" There was one set of data examined at FRI using very tall 6" high outlet weirs. The bubble caps for this data set were slotted, not tea cup. This data set (Run #4094-4101 operating at 24 psia) was thus examined to see if a stability of 0.2 was reasonable for this unusually tall outlet weir height. Figure 4 shows this data set and subsequently Table 4 shows the examination of the 2 data points surrounding the choice of minimum tolerable operation. Table 4 results did employ a reduction of the hydrostatic head due to the elevation of the 1.75" tall slots on the bubble caps. The value of the stability between these two data points is very close to 0.2. Therefore a conclusion can be drawn that bubble cap trays can employ a minimum stability factor of 0.2 and that the correlation above can apply to both "tea cup" and slotted caps.

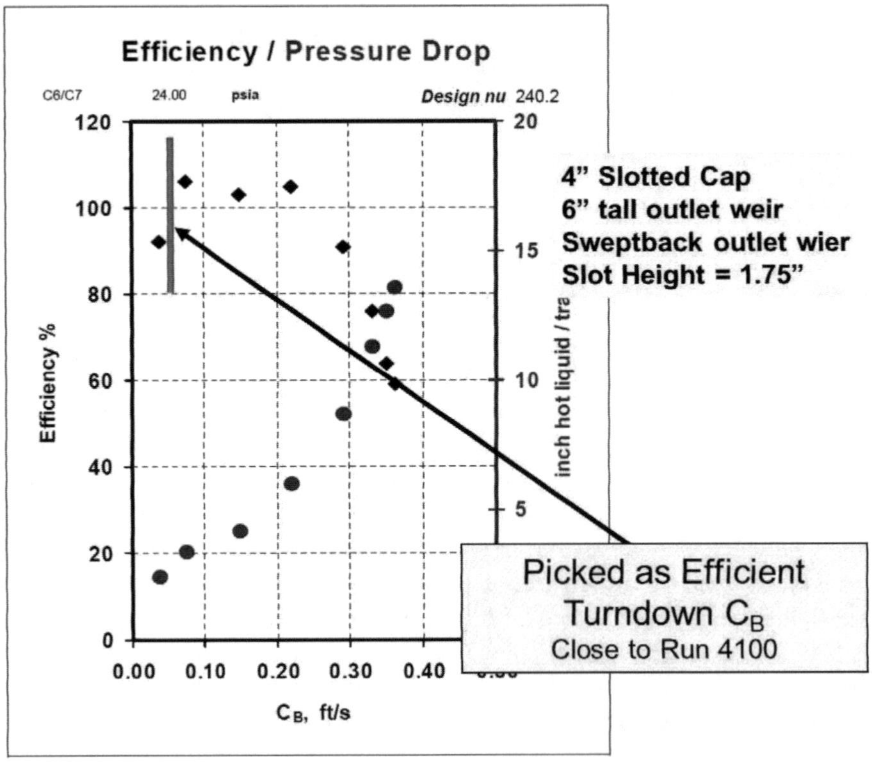

Figure 4 – Turndown Determination with 6" outlet Weir

Table 4 – Data from 2 Data Points in Figure 4 at Turndown

Run	Riser Area	Vapor Density	Liquid Den	V-rate	CFS	Vr	Dry Drop	HS	Stability
	ft2	lb/ft3	lb/ft3	lb/hr		ft/s	in. Liq	in. Liq	
4100	1.18730113	0.31	41.9	9080	8.13620072	6.85268506	0.35041162	5.38	0.25521025
4101	1.18730113	0.31	41.9	4600	4.1218638	3.47162459	0.0899337	5.65	0.12616445

Summary

The concept of tray Stability does apply to Bubble Caps and can be used to determine their minimum efficient capacity of these devices. A new stability correlation for Bubble Cap Trays has been developed and checked against FRI data. A turndown guideline has also been established such that the tray stability correlation result needs to remain above a value of 0.2 for bubble cap trays.

References

(1) Forbes, R. J., "Short History of the Art of Distillation", E. J. Brill Publisher, Leiden, NL (1948), pp 308-309

(2) Bolles, W. L., "Optimum Bubble Cap Tray Design", 4 part series, McGraw Hill Publishing (1956)

(3) Distillation Subcommittee of the Research Committee, "Bubble-Tray Design Manual", AIChE, NY (1958)

(4) Lockett, M. J., Summers, D. R., Smith, V. C., Upchurch, J. C., "High Turndown Bubble Cap Tray", U.S. Patent 4,711,745, December 8, 1987.

(5) Summers, D. R., Spiegel, L., and Kolesnikov, E., "Tray Stability at Low Vapor Load", Conference Proceedings of "Distillation and Absorption 2010", p.611, Eindhoven, NL September 12-15, 2010.

(6) Summers, D. R., van Sinderen, A., "Dry Tray Pressure Drop of Rectangular Float Valve and V-Grid Trays", in Distillation 2001 Topical Conference Proceedings, AIChE 2001 Spring National Meeting, Houston, TX, April 25, 2001

(7) Summers, D.R., "Dry Tray Pressure Drop of Sieve Trays", Chemical Engineering, June 2009, pp 36-39.

(8) Summers, D.R., Cai, T. J. "Dry Tray Pressure Drop of Sieve Trays Revisited", AIChE Spring Meeting 2017 Topical 8 Conference Proceedings, March 27, 2017

(9) FRI Bubble Cap Tray Data Base – Public Domain files, Oklahoma State University Special Collections and University Archives

Marinization of Mass Transfer Columns for FLNG Applications

Michael Schultes[1], Jeremy Brauer[2], Pengfei Chen[2], Shain Doong[2]
(1) Raschig GmbH, Germany
(2) Honeywell UOP, USA

2017 AICHE Spring Meeting & 13th Global Congress on Process Safety, Distillation Symposium, San Antonio, TX March 26-30

Test facility
at
Raschig Ludwigshafen

Raschig Rocking Column Test Facility

Test facility at Raschig Ludwigshafen

D = 450 mm = 1,5 ft	
H = 1.6 m = 5,3 ft	
Constant tilt	**Period of motion**
$0^o, 1^o$	6 sec
$2^o, 5^o$	12 sec
10°	30 sec
u_L = 2 - 250 m3/m2h = 0.8 - 100 gpm/ft^2	

Raschig Rocking Column Test Program

7 different Structured Packings

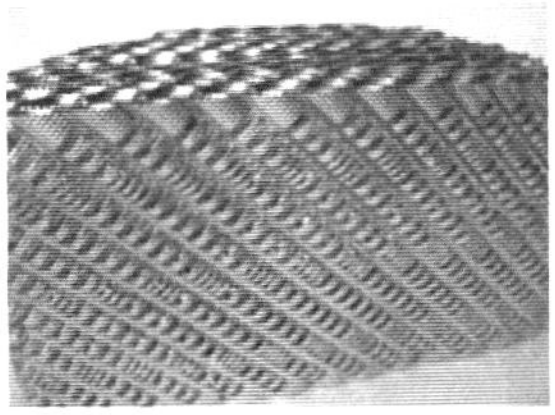

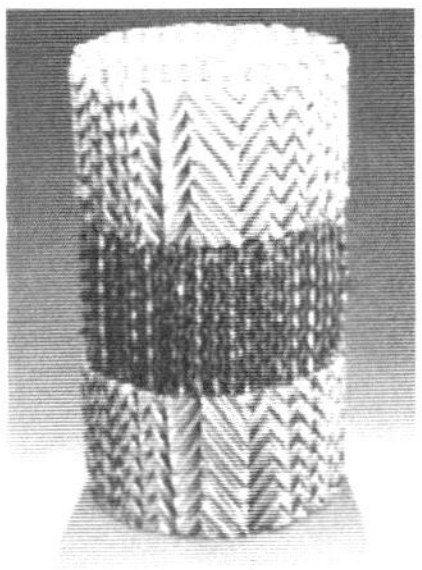

6 different Random Packings

Honeywell UOP

RASCHIG

Evaluation of Fluiddynamic Test Result

Honeywell UOP

RASCHIG

No tilt
Maldistribution Factor $M_{V,0}$

Raschig Experimental Data:

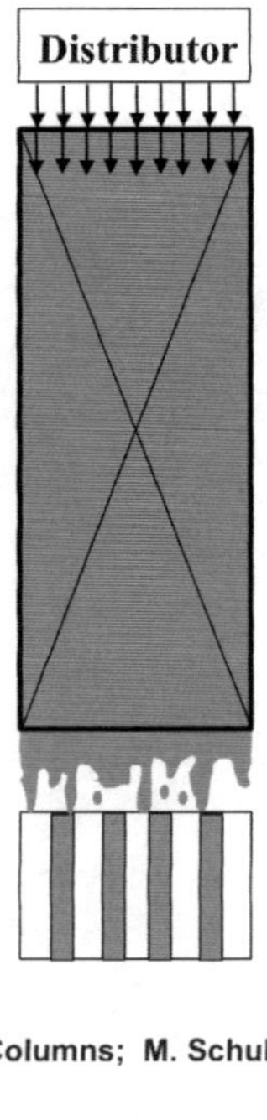

Column top $M_{V,0} = 0$ %

D = 450 mm = 1,5 ft

H = 1.6 m = 5,3 ft

H/D = 3.6

Raschig Super-Ring No. 2

$M_{V,0} = 28$ %

Ref.: Influence of Liquid Redistributors on the Mass-Transfer Efficiency of Packed Columns; M. Schultes, *Ind. Eng. Chem. Res.* 2000, *39*, 1381-1389

Honeywell UOP

Maldistribution Factor $M_{V,0}$

RASCHIG

Packing	Material	Maldistribution Factor	Data Base
25 mm Ralu-Ring	Plastic	0.55/0.50	Stikkelman/Author
38 mm Ralu-Ring	Plastic	0.69	Stikkelman
50 mm Ralu-Ring	Plastic	0.46	Author
25 mm Pall-Ring	Metal	0.81/1.1	Stikkelman/Author
50 mm Pall-Ring	Metal	0.50	Author
25 mm IMTP-Ring	Metal	0.57	Stikkelman
Raschig Super-Ring No. 2	Metal	0.28	Author
Ralu-Pac 250 YC	Metal	0.21	Author
Mellapak 250 Y	Metal	0.24	Stikkelman
Mellapak 500 Y	Metal	0.21	Stikkelman
Mellapak BX	Plastic	0.22	Stikkelman
Montz B1-250	Metal	0.23	Stikkelman
Montz BS-450	Metal	0.44	Stikkelman

Ref.: Influence of Liquid Redistributors on the Mass-Transfer Efficiency of Packed Columns; *Ind. Eng. Chem. Res.* 2000, *39*, 1381-1389

Honeywell UOP

With tilt Maldistribution Factor M_V

RASCHIG

Raschig Experimental Data:

Column top: $M_V = 0$ %

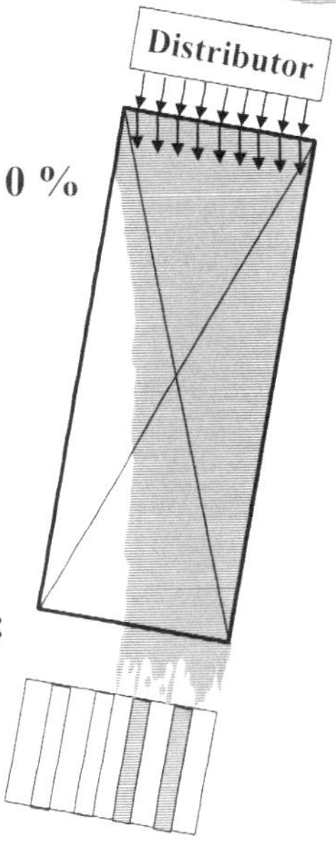

D = 450 mm = 1,5 ft

H = 1.6 m = 5,3 ft

H/D = 3.6

5° constant tilt

Raschig Super-Ring No. 2:

$M_V = 77$ %

Honeywell UOP

RASCHIG

Raschig Rocking Column Test Facility

RASCHIG

Honeywell UOP

RASCHIG

Relative Standard-Deviation $C_V/C_{V,0}$

Initiated by column oscillation or constant column tilt

Cv/Cv,0 over liquid load

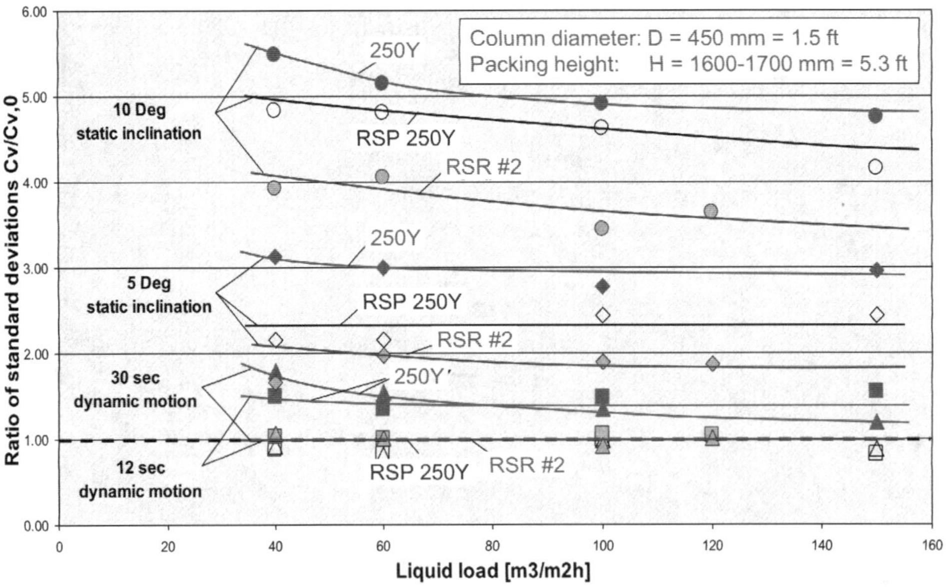

Honeywell UOP

RASCHIG

Relative Standard-Deviation $C_V/C_{V,0}$

Initiated by column oscillation or constant column tilt

Cv/Cv,0 over liquid load

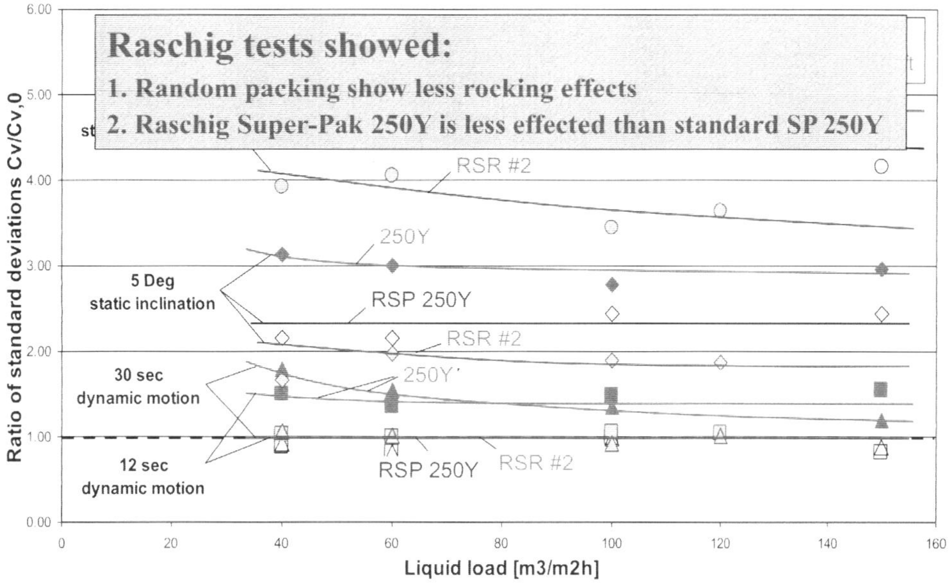

Honeywell UOP

RASCHIG

Evaluation of Mass Transfer Test Result

Deterioration in Mass Transfer
Oxygen Absorption into Water

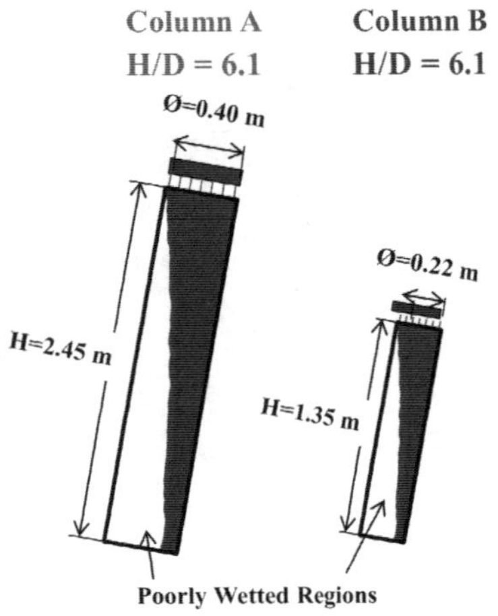

Packing: 25 mm Pall-Ring 16 mm Pall-Ring

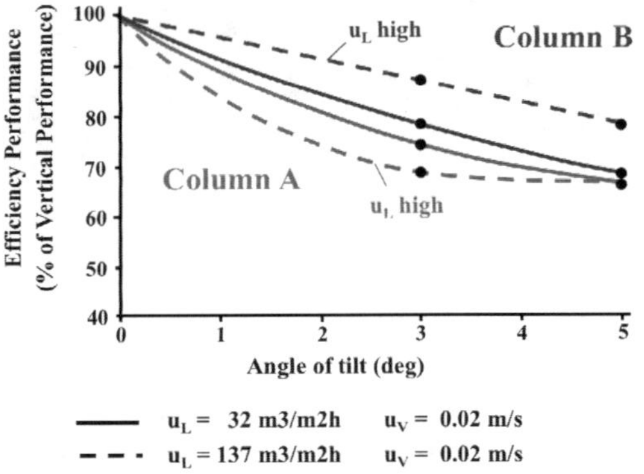

Conclusion: Tall beds are more sensitive to loss in efficiency
Tall beds are more effected by tilt as liquid load increases

Ref:
Tanner, Baker, Waldie; Heriot-Watt University; Distillation and Absorption; IChemE Symposium Series No. 128, 1992

Deterioration in Mass Transfer
Oxygen Absorption into Water

Effect of H/D ratio

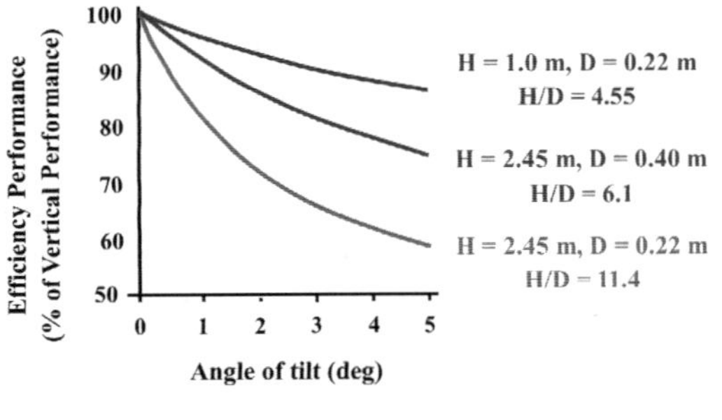

Packing: 16 mm Pall-Ring

u_L = 137 m3/m2h u_V = 0.01 m/s

Effect of motion

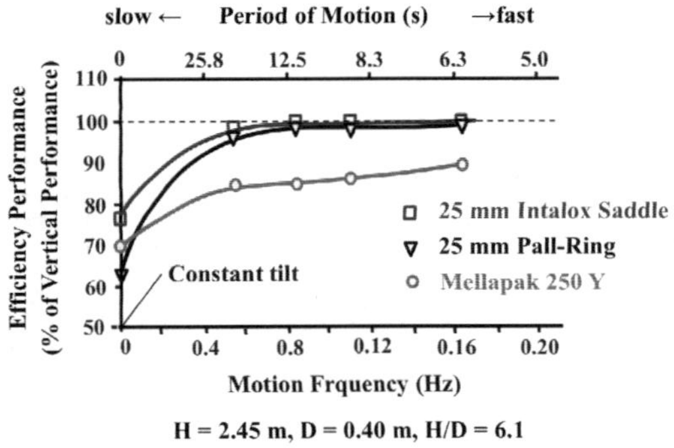

H = 2.45 m, D = 0.40 m, H/D = 6.1

u_L = 80 m3/m2h, u_V = 0.01 m/s

3° Maximum Tilt

Conclusion: High H/D ratios are more sensitive to loss in efficiency
Structured packings are more effected by motion than random packings

Ref:
Tanner, Baker, Waldie; Heriot-Watt University; Distillation and Absorption; IChemE Symposium Series No. 128, 1992

Honeywell UOP RASCHIG

CFD Analysis

Honeywell UOP RASCHIG

Mal-Distribution Factor Calculated by CFD Model Matches Testing Results

Test

5° tilt

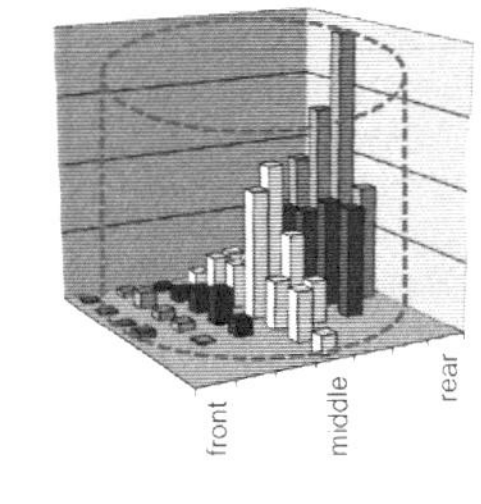

Mal-distribution Factor
(Tests and CFD results)

CFD

Liquid mass flux on bottom

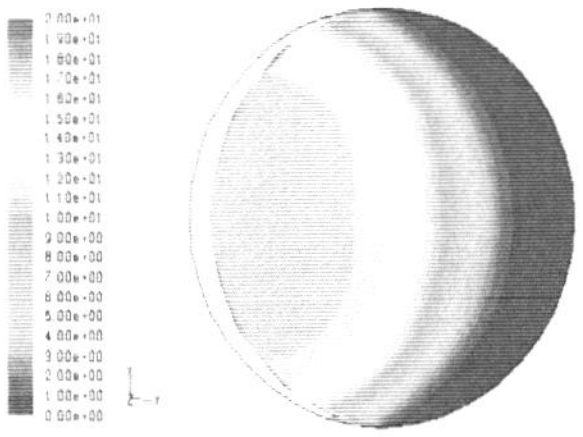

10° tilt

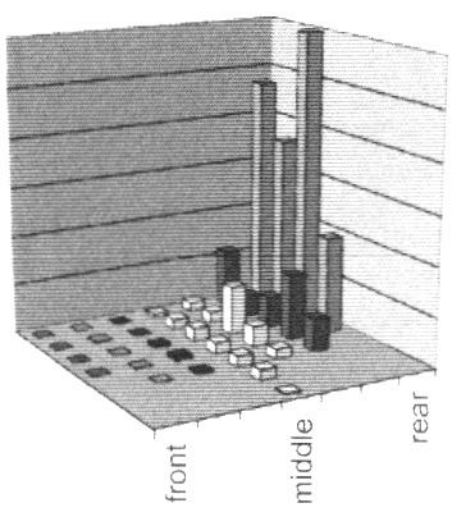

	$C_v/C_{v,O}$
1° tilt	1.21
5° tilt	1.96
10° tilt	4.01

Liquid mass flux on bottom

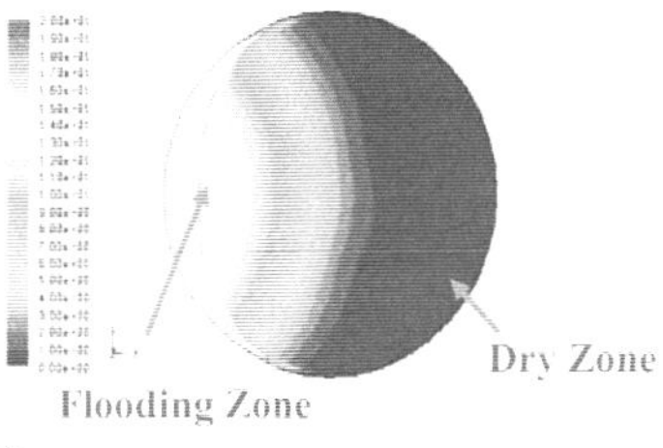

RSR2, D=0.45 m, H=1.6 m, liquid: 60m3/m2/h

Honeywell UOP

Oscillation show less Impact on Mal-Distribution

RASCHIG

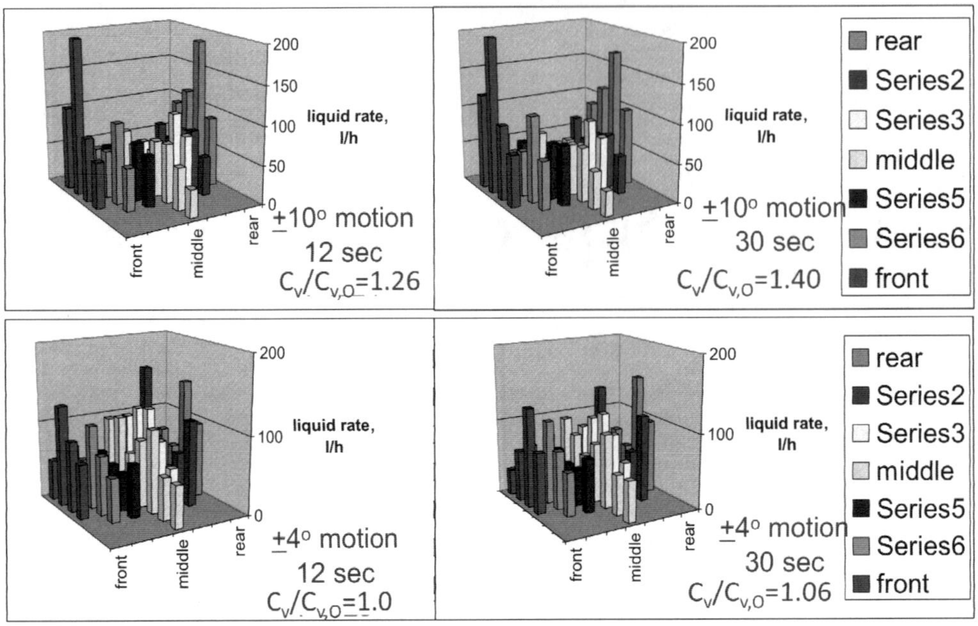

RSR2, D=0.45 m, H=1.6 m, liquid: 60m3/m2/h

UOP 5549A-17

Honeywell UOP

RASCHIG

Rocking Columns: Tests *vs.* CFD

Dynamic Maldistribution: ±10° motion, 12 sec period

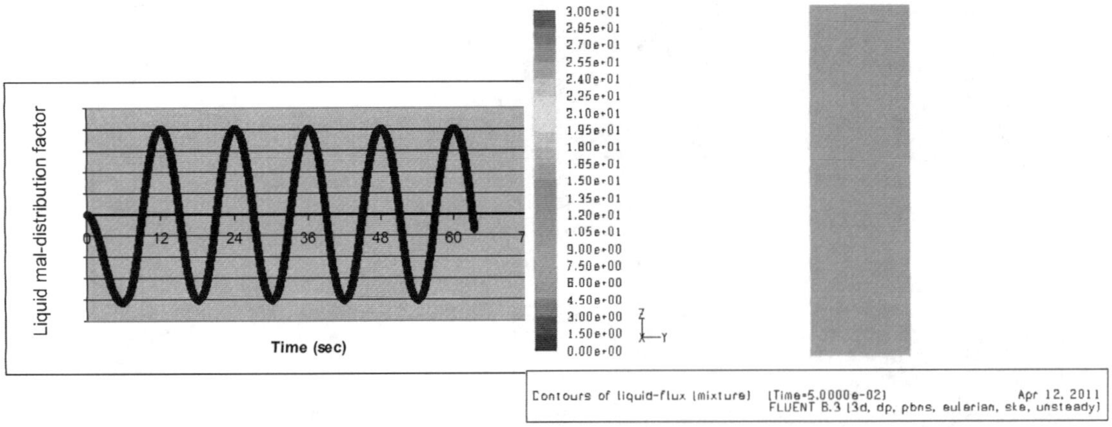

Raschig column: D=0.45 m, H=1.6 m

Honeywell UOP

Real vessel motion and bed elevation

RASCHIG

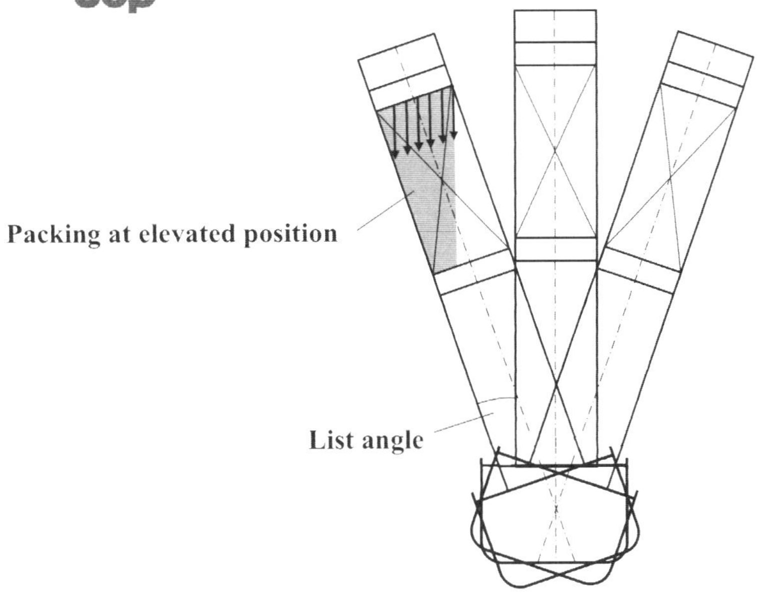

CFD modeling helps to predict maldistribution at elevated position

Honeywell UOP

Transfer of CFD results into Floating Column Model/ Process Simulation

RASCHIG

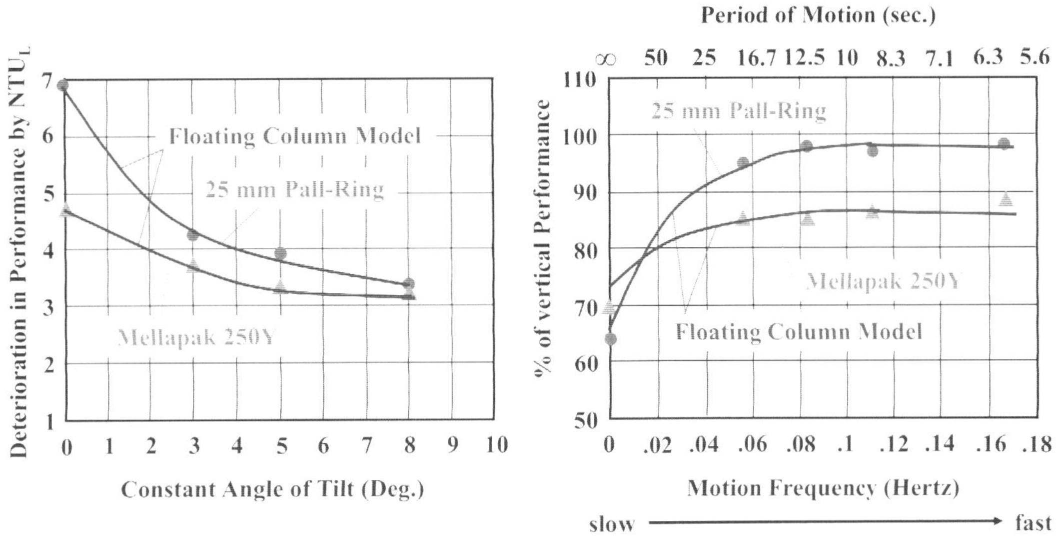

Ref: Baker, Tanner, Waldie; Heriot-Watt University; Institution of Chemical Engineering, 2005

Honeywell UOP

RASCHIG

Conclusions for FLNGs

- Random and Structured Packings can be applied if motional conditions are properly considered
- Random packings show less hydraulic and efficiency effect at floating conditions than structured packings
- Structured Packings should be applied if high effective surface area at low liquid rate is required
- Random Packings should be applied if high liquid rates are required
- Random Packings should be applied in high pressure distillation services
- Random Packings can handle foaming liquids more convenient
- Random Packings can handle dirty liquids more convenient

Honeywell UOP

End of Presentation

RASCHIG

Floating Columns

Questions ?

Fractionation Research, Inc.

RESEARCH - KEY TO BETTER DESIGN

Flow-Induced Tray Vibrations and Prevention

Tony Cai
Fractionation research Inc., Stillwater, OK
T.Y. Lin
Chang-Chun & Dairen Group, Taiwan

Kister Distillation Symposium
2017 AIChE Spring Meeting

San Antonio, TX

28 March 2017

What is Flow-induced Vibration?

FRI

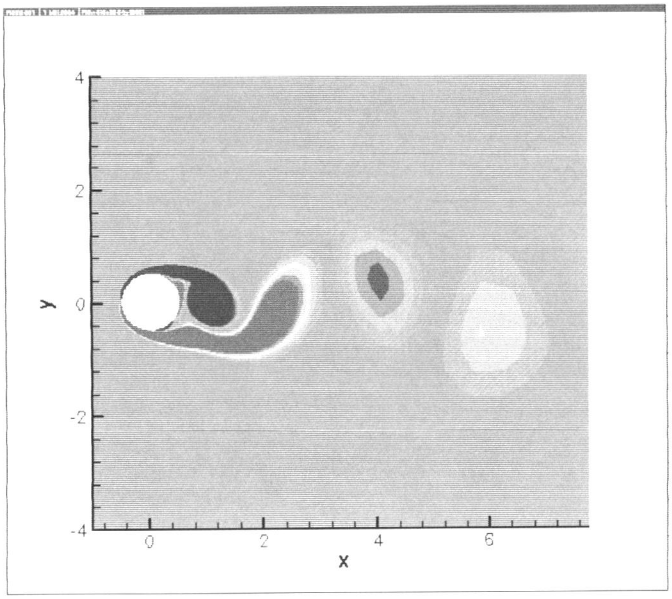

3 Distillation Symposium -AIChE Spring 2017 Tray Vibrations

Tacoma Narrows Bridge Collapse

https://www.youtube.com/watch?v=nFzu6CNtqec

Collapsed due to a oscillation caused by interactions between the bridge and the winds passing through its structure

4 Distillation Symposium -AIChE Spring 2017 Tray Vibrations

Flow-induced Vibration of Tray

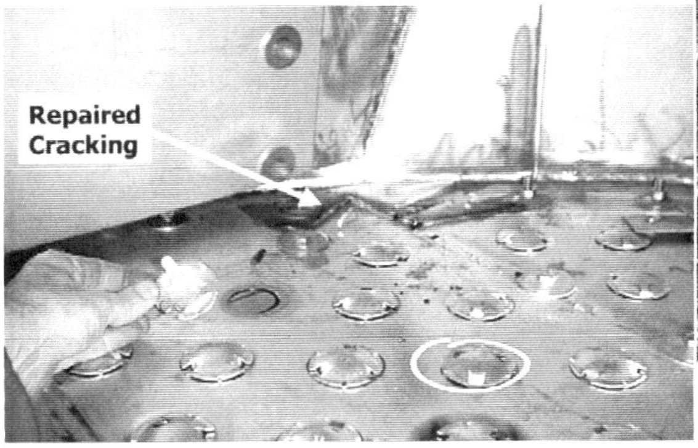

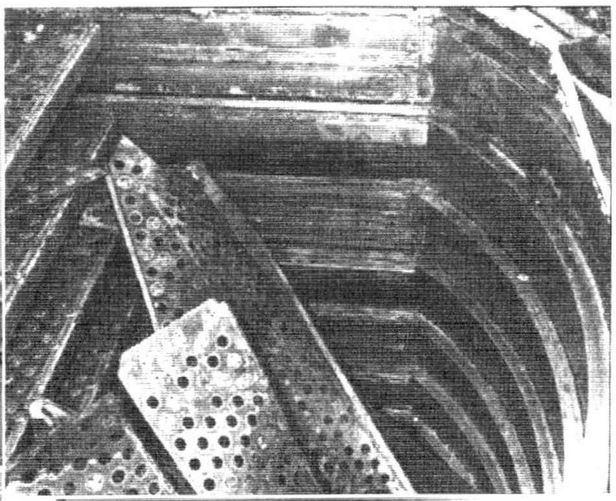

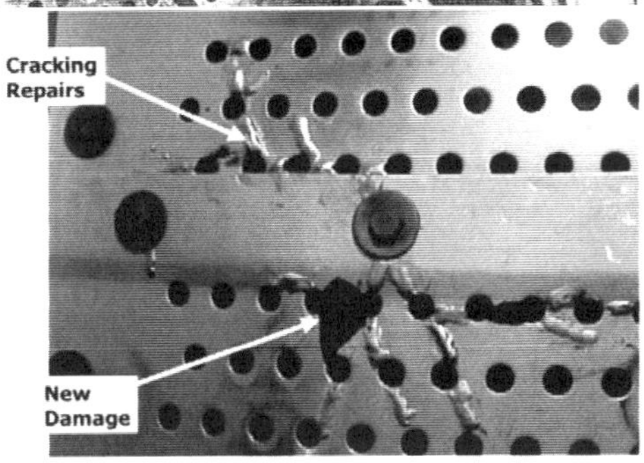

Tray vibration can cause severe damage.

1 & 3, DR Summers; 2, JR Winter

Simple Forced Vibrations

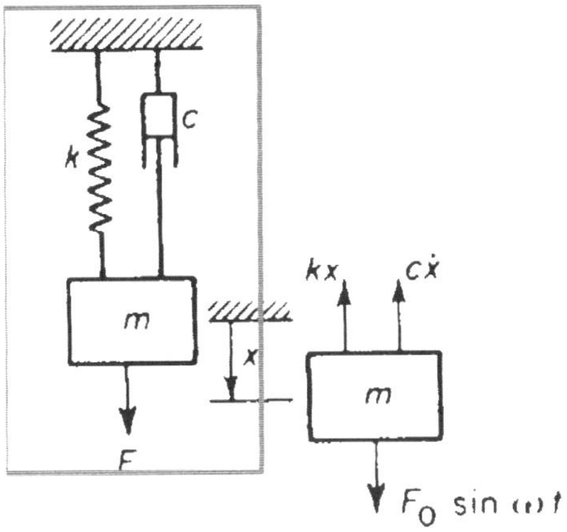

1. Excitation force, F
2. Natural frequencies, f_n
3. Force/excitation frequency: f ($\omega=2\pi f$)
4. Damping factor C (ξ)

Resonance

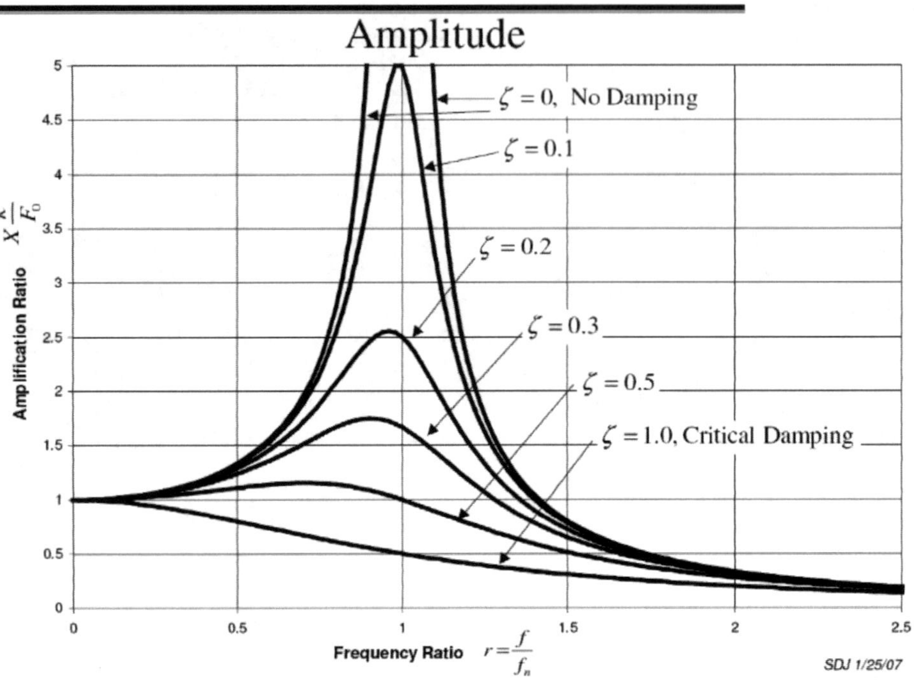

Two Type Tray Vibrations

- When a tray oscillating at or near-resonance condition, the tray can be damaged very quickly
- When tray vibration is not at or near a resonance condition, the tray may still be damaged by the fatigue caused by long period of vibration

Natural Frequencies of Tray

- Natural frequencies of a tray depend on:
  - Materials
  - Tray thickness
  - Tray supports (beam, truss, clamps)
  - Open area
  - Hole/valve sizes
  - Mass (Trays +Liquid holdups)

Where causes the excitation?

- There are two schools of thought when it comes to what causes the excitation:
  - Negative Resistance Theory
  - Pressure Pulsations

Negative Flow Resistance Theory

Waddington, 1973; Brierley et al., 1979

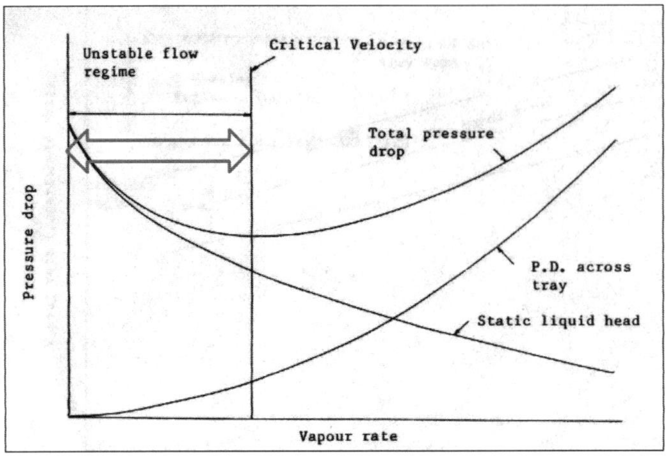

Pressure Pulsations/Fluctuations

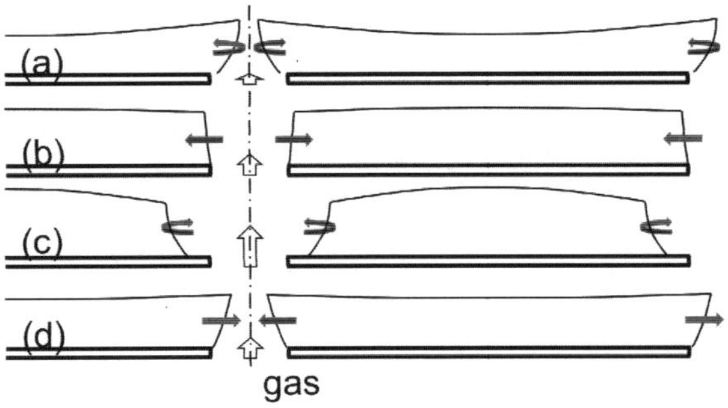

1. Oscillation initiated at the holes at the deck
2. Caused by:
   - Gas-liquid interactions;
   - Bubble formation and break;
   - Pulsating jet;
   - Synchronous bubbling or jetting (oscillation)

Courtesy by Zhang, Taylor and Darton, D& A 2014

Excitation Frequency

- The frequencies depend on:
  - Hole velocity
  - Liquid head/Holdup
  - Tray geometry (Hole size & pitch)
  - Physical properties (density, surface tensions etc.)
- The volume of inter-tray spacing does not affect the frequencies

Predicting Excitation Frequencies

Zhang et al., 10th International Symposium on Distillation and Absorption, Friedrichshafen, 2014. Correlates data of Priestman et al., r^2=0.974

$$\frac{fd}{U} = 0.005\left(1+\sqrt{\frac{\rho_L g d^2}{\sigma}}\right) + 0.46\left(\frac{d}{s-d}\right)\sqrt{\frac{gH_L}{U^2}}$$

Strouhal number fd/U is ratio of time scales: gas flow through hole, to pulsation

Bond number $\rho_L g d^2/\sigma$ relates gravity to surface tension forces. Affects shape of waves, droplets, jets

Froude number $U/\surd(gH_L)$ relates hole velocity to speed of surface wave

Potential Risks of Tray Vibrations?

- Trays may operate much lower capacities than designed
- Vibrations generally occurred at the gas rate close to weep point
- If plant throughput is reduced, check all trayed columns for flow-induced vibration.
- Trays with long flow path, low liquid head, low hole velocity

How to Avoid Tray Vibrations?

- Estimate the natural frequencies f_n, at least two modes, of the tray including the deck liquid holdup
- Estimate the excitation frequency, f
- Ensure **f/fn <0.8 or f/fn>1.2**

Forced Vibrations with Damping

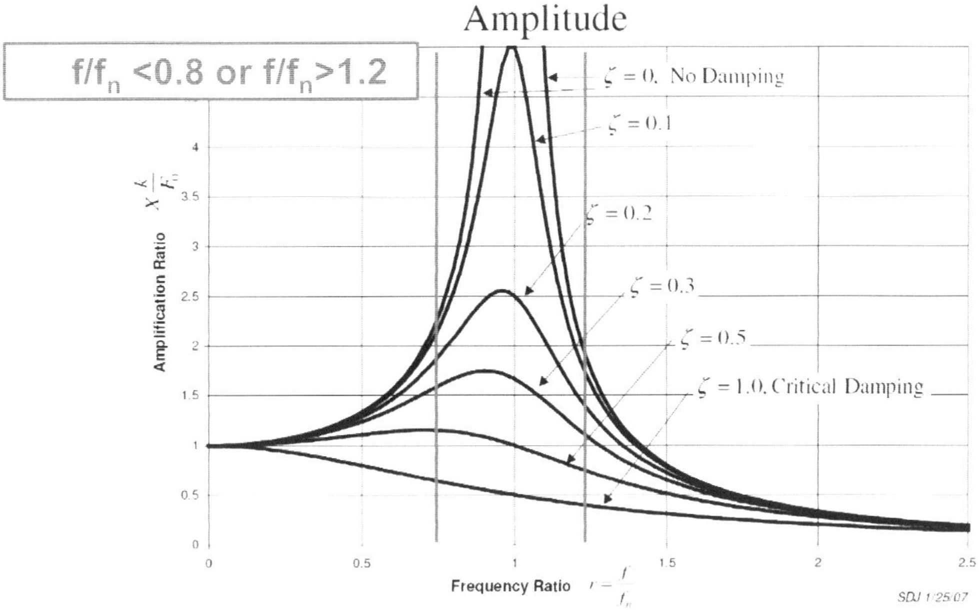

Design Considerations

- If a tray vibrates or may vibrate,
  - it is very effective to avoid the problem by reducing the open area
    - Open area reduced, actual hole velocity increases, excitation frequency increases
  - Or increasing the f_n by increasing the stiffness of the support beams
  - Or changing the operation conditions away from the critical condition

Industrial Cases (Column T2-531)

Industrial Cases (Column T2-531)

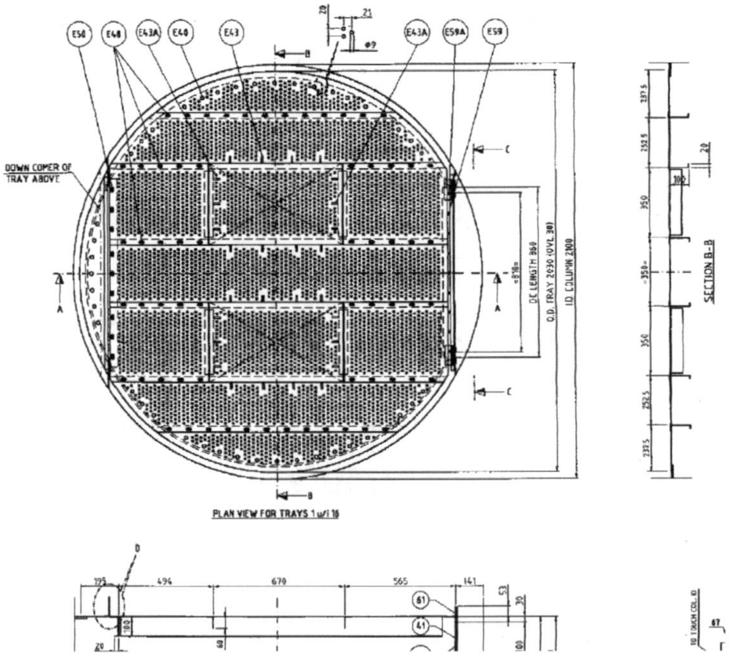

Details of Damaged Trays

Equipment No. : Column T-531 Dehydration 3rd Tower

Tray No. (from top)	Trays Damaged					
	Clamp			Sieve Tray Deck		
	Clamp broke	Clamp loosed	Bolt	Deck broke	Location (Panel No.)	Bolt loosed
1	0	0	0	Yes	5, 6, 8	0
2	1	0	0	Yes	1, 4, 6, 7, 8	2
3	11	1	0	Yes	4, 6, 8	2
4	3	0	0	Yes	1, 5, 6, 8	1
5	5	0	0	Yes	1, 6, 7, 8	2
6	1	0	0	Yes	1, 6, 7	3
7	8	0	0	Yes	1, 6, 7	2
8	9	0	0	Yes	1, 5, 6, 7, 8	4
9	7	0	0	Yes	1, 7	2
10	4	0	0	Yes	1, 6	1
11	4	0	0	No	-	2
12	5	0	0	No	-	1
13	4	0	0	Yes	1, 6	4
14	0	0	0	Yes	4, 5, 6, 7, 8	6
15	0	0	0	No	-	5
16	5	0	0	Yes	5, 6	0
17	0	0	0	No	-	0
18	0	0	0	No	-	0
19	0	0	0	No	-	0
20	0	0	0	No	-	0

Column ID	mm	2,100
Tray Spacing	mm	550
Upper Section		**Tray #1 - #16 [1-Pass]**
Downcomer Top Width	mm	141
Downcomer Btm Width	mm	141
Downcomer Clearance	mm	15
Outlet Weir Height	mm	30 + 20 (Notch)
Liquid Surface Tension (s)	dyne/cm	52.2
Tray Type / Size (d)	mm	9.0
Hole Pitch - Trans./Longi. (v)	mm	20.0 / 21.0
Lower Section		**Tray #17 - #20 [1-Pass]**
Downcomer Top Width	mm	355
Downcomer Btm Width	mm	355
Downcomer Clearance	mm	20
Outlet Weir Height	mm	25
Liquid Surface Tension (s)	dyne/cm	49.3
Tray Type / Size (d)	mm	Sieve Tray / 6.0
Hole Pitch - Trans./Longi. (v)	mm	13.0 / 15.0

Photos of Tray Damaged

Photos of Tray Damaged

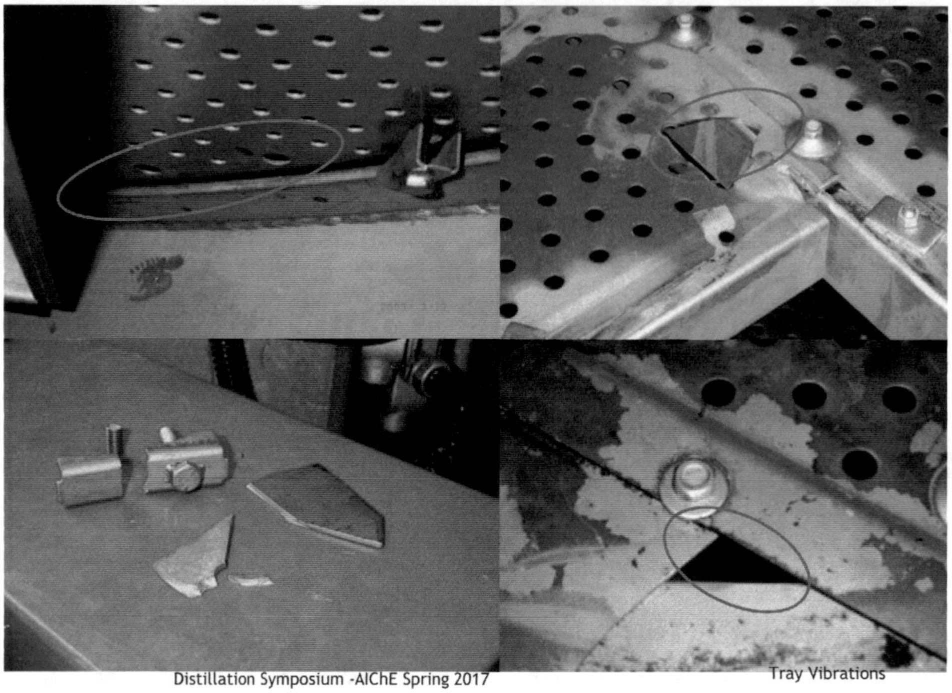

Diagnosis of Tray Vibrations

- To avoid tray vibration problems:
  - $f/f_n < 0.8$ or $f/f_n > 1.2$
- If $0.8 < f/f_n < 1.2$, the trays will likely have vibration problems
- To diagnose, it is important to estimate the excitation frequency f and natural frequency fn

Natural Frequency Estimation

- For most of trays, following assumptions were used to calculate the natural frequencies:
  - All edges of the tray deck/panel are supported/fixed
  - Uniform load per unit area including its own weight
  - Deck/panel are uniformly perforated

Natural Frequency (Cont'd)

- The natural frequencies can be calculated from following equation*:
  - $f_n = K_n\sqrt{\frac{Dg}{\omega a^4}}$ where $K_n = \pi^2\left[m_a^2 + \left(\frac{a}{b}\right)^2 m_b^2\right]$
  - a: length of short edge, m
  - b: length of long edge, m
  - $D = \frac{Et^3}{12(1-\mu^2)}$: flexural stiffness, N.m
    - E: Young's modulus, Pa μ: Poisson's ratio, t:deck thickness
  - For the first order natural frequency $m_a=m_b=1$

*Roark's formulas for stress and strain, 7th Edition

Natural Frequency (Cont'd)

- E: Young's modulus, and μ: Poisson's ratio are dependent of the deck open

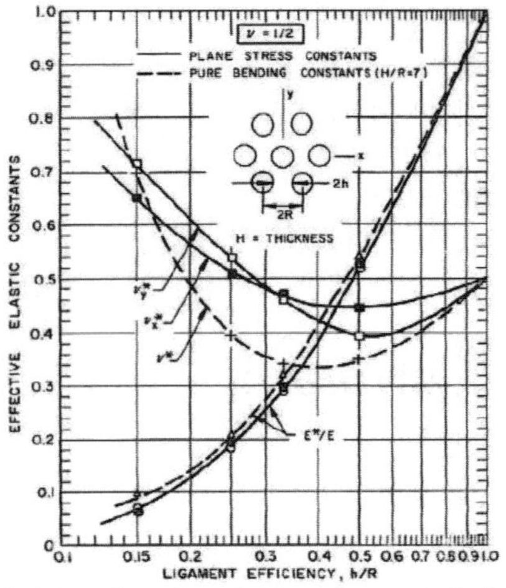

Fig. 3 Comparison of Sampson effective elastic constants for bending and plane stress

Source: Design of Perforated Plates
O'Donnell W.J & Langer B.F.
Journal of Engineering for Industry
Trans ASME, VOL. 84, August 1962

Predicting Excitation Frequencies

Zhang et al., 10th International Symposium on Distillation and Absorption, Friedrichshafen, 2014. Correlates data of Priestman et al., r^2=0.974

$$\frac{fd}{U} = 0.005\left(1+\sqrt{\frac{\rho_L g d^2}{\sigma}}\right) + 0.46\left(\frac{d}{s-d}\right)\sqrt{\frac{gH_L}{U^2}}$$

Strouhal number fd/U is ratio of time scales: gas flow through hole, to pulsation

Bond number $\rho_L g d^2/\sigma$ relates gravity to surface tension forces. Affects shape of waves, droplets, jets

Froude number $U/\sqrt{(gH_L)}$ relates hole velocity to speed of surface wave

Liquid Head, H_L, was calculated using FRI Sieve Tray Liquid Head Model

Results for Column T2-531

FRI

- The calculated natural frequency
  - fn=93 Hz
- The estimated excitation frequency
  - f=82 Hz

  - $0.8 < f/fn = 0.88 < 1.2$

✓ So the tray would indeed have the problem of flow-induced vibration

Solutions

FRI

- Retrofit with fixed valve trays / support beams, butterfly clamps, shear clips to increase the natural frequency
- In operation of load-up or load-down, shorten the period of low-load near weeping point
- No tray cracking or damage has been found since 2012

Conclusions

- Trays will have vibration problems if the excitation frequency and tray natural frequency are too close
- Operating conditions affect both frequencies so a well designed tray may still vibrate if the conditions change
- To avoid the problem, those two frequencies must be separated apart **f/fn <0.8 or f/fn>1.2**
- It is crucial to estimate both frequencies accurately
- The perforated trays will affect mechanical properties

FRI Tray Vibration Test Project

Acknowledgements

- Authors would like to extend their great appreciations to:
  - Chang-Chun & Dairen Group for sharing their experiences and information of tray vibrations
  - Henry, Mike and Nick for their great work on this and all previous symposiums

Fractionation Research, Inc.

RESEARCH - KEY TO BETTER DESIGN

Flow-Induced Tray Vibrations and Prevention

Thank You for Your Attention

Kister Distillation Symposium

2017 AIChE Spring Meeting

San Antonio, TX

28 March 2017

High Performance Trays for a Fractionator in an Aromatics Plant

By

Lee Siang Hua & Ang Chew Peng
Sulzer Singapore Pte. Ltd.

Senthil Krishnamoorthy
Sulzer Chemtech USA

Mario Roza
Sulzer Chemtech AG

Paper 4f

For presentation at the Session 4, Distillation Symposium
AIChE Spring Meeting, San Antonio, TX, March, 2017

Introduction

Distillation is a very important separation technology in refineries and downstream chemical processing. Mass transfer internals in distillation columns can be equipped with trays, structured packing or random packing. Trays are generally the preferred option in columns with high liquid rates or in services with high fouling tendencies. After the acquisition of Nutter Engineering in 1998, Sulzer Chemtech expanded its tray product portfolio with inclusion of the acclaimed BDH™ rectangular float valves and the V-Grid™ fixed valves. The technological alliance with Shell Global Solutions since 2000 has brought Shell high performance trays into Sulzer Chemtech's range of tray products. This case study highlights an integrated refinery complex in the East Asia, where various tray options were evaluated and Shell tray technology was eventually applied for a new xylene rerun column in an aromatic unit.

Project Background

Increased demand for synthetic fibre, partly fuelled by the rise in price of natural cotton, has led to a proliferation of world scale aromatics plants in recent years. Para-xylene is the key raw material in production of purified terepthalic acid (PTA) which is the main ingredient in polyester manufacturing. A typical aromatic unit has more than 30 columns of various sizes, among them small strippers of diameters less than 1 meter and large super-fractionators of diameters spanning more than 10 meters. The biggest columns in an aromatic unit include xylene rerun column, xylene splitters in the xylene fractionation unit as well as extract and raffinate columns in the para-xylene adsorption unit. A simplified process flow diagram of a typical aromatic unit is shown in **Figure 1**.

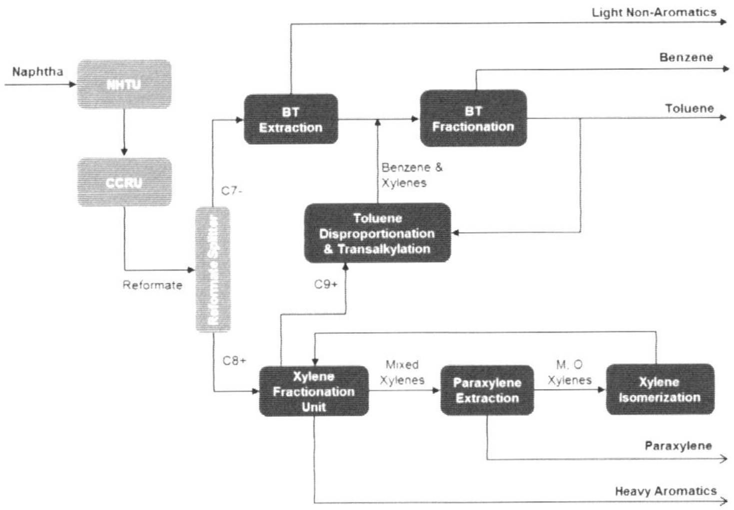

Figure 1: Simplified process flow diagram of an aromatic unit

An integrated refinery complex in the East Asia planned for an expansion of its aromatic unit to increase aromatics production, to meet market demand. A process licensor was engaged for the project study. Based on the detailed evaluation and analysis, it was concluded that a new xylene rerun column was required to handle the new, increased, production loads. The xylene rerun column, in the xylene fractionation unit, is one of the heaviest loaded columns in the aromatic plant. It usually takes the bottom stream of a deheptanizer, together with the effluent from the transalkylation unit. This column separates the mixed xylenes (ortho-xylene, meta-xylene and para-xylene) from the heavies, and the mixed xylenes are then sent to the xylene splitters for further separation.

The new xylene rerun column T-4001 was simulated with 88 theoretical stages including condenser and reboiler. It was designed to handle 2 feeds, with the top feed entering onto Stage 22 from the top of the tower and the bottom feed entering onto Stage 61. The column had a specification of the C9 aromatics and heavies content in the distillate not to be more than 300ppm by weight and the recovery of ortho-xylene to be more than 99% by weight. **Figure 2** shows the simplified column sketch of T-4001.

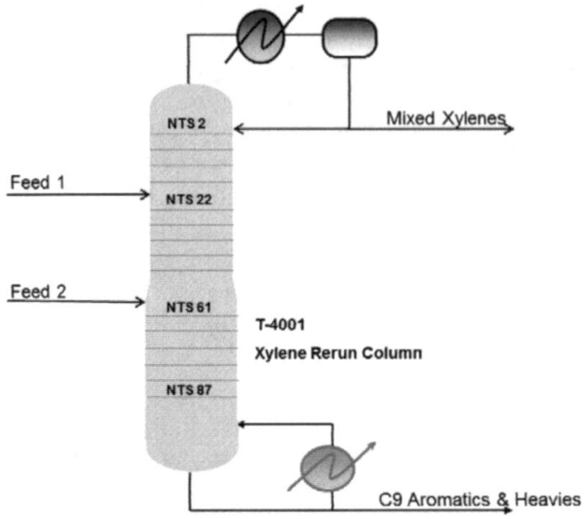

Figure 2: Simplified column sketch of T-4001

Tray Design Options

The selection of the most appropriate tray type requires detailed evaluation. The two main factors are tray performance and economics. Over the last decade, there has been a trend to equip new large fractionators with high performance trays. Use of advanced tray designs with high capacity valves and enhanced downcomers to handle the high vapor and liquid internal loadings inside the columns, can result in a substantial reduction in column sizes which translates into immediate savings in capital investment for plant owners.

For the new xylene rerun column T-4001, Sulzer Chemtech evaluated three tray designs.

- Design A: Conventional Trays
- Design B: High Performance Trays with Chordal Downcomers
- Design C: High Performance Trays with Multi-Downcomers

For the base tray Design A, BDH conventional valves with straight downcomers were considered. BDH valves are Sulzer Chemtech's proprietary rectangular float valves, the lateral vapor release of these valves minimize the tendency of liquid weeping from upstream edge of the valves. In the process of tray design optimization, it was apparent that high weir loading was the limiting factor for these chordal downcomer trays. Weir loading is defined as the liquid volumetric flowrate [in m^3/hr] per unit weir length [in m]. It is a common design practice not to design trays with weir loading more than 120m^3/(hr.m) [13.4 gpm/inch] and it would be ideal to keep the value less than 90m^3/(hr.m) [10 gpm/inch] for increased safety margins in the tray design. The most effective method to reduce the weir loading is to increase the tray weir length by increasing the number of liquid passes on the tray. Hydraulic evaluation with the above weir loadings suggested that the diameters of these BDH conventional trays would be approximately 12 meters if they were designed with 4 liquid passes per tray. By increasing the number of liquid passes per tray to 6, the column diameter could be reduced to slightly over 10 meters.

Design B is for high performance trays with chordal downcomers. It was considered to employ proprietary MVG fixed valves with sloped downcomers, for an increased active area. MVG is the most established member of the V-Grid family which comprises of MMVG^{TM}, MVG^{TM}, SVG^{TM} and XVG^{TM} in the order of increasing valve sizes. V-Grid fixed valves are punched out from the tray decks and they release vapor laterally from openings at the sides. The MVG, being smaller than conventional valves, offers higher vapor handling capacity due to improved distribution of vapor from the tray deck which reduces the froth height. The combination of MVG fixed valves with sloped downcomers resulted in a reduction in column diameter of more than 1 meter from the conventional tray design with 6 liquid passes per tray.

Design C uses Shell HiFi™ high performance trays. These trays are equipped with multiple envelope downcomers which are oriented offset to the tray's center line. This layout allows for the placement of an increased number of downcomers. The long weir length of these downcomers reduces the clear liquid height on the tray which results in a higher vapor handling capacity. Shell HiFi™ Plus trays are an enhanced version featuring high capacity valves (such as MVG, MMVG) to further boost the tray deck performance.

Figure 3: 3D model of the Shell HiFi™ tray

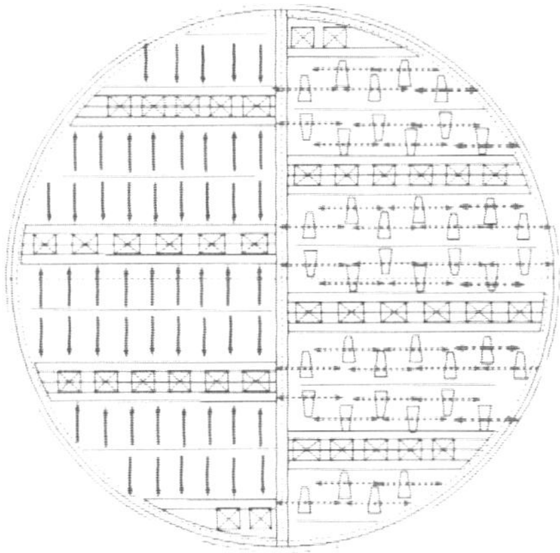

Figure 4: Flow direction on the Shell HiFi™ Plus tray

The direction of vapor and liquid flows on a Shell HiFi tray is illustrated in **Figure 4**. The unique orientation of the downcomers allows uniform liquid distribution on each active tray deck (blue arrows) which then enables the vapor to get distributed uniformly underneath the tray deck and released laterally through the MVG valves perpendicular to the liquid flow (red arrows). The liquid flow path lengths are the same for all liquid passes, which bring about equal mass transfer on each active panel. With no obstruction between the various compartments of the trays, liquid and vapor are naturally self-balanced all over the tray for the most uniform mixing, contacting and highest separation efficiency achievable for multi-downcomer trays.

With the use of Shell HiFi Plus trays, the column diameter of xylene rerun column T-4001 was reduced to 8,300mm [27.25 feet] for the top and middle sections and 8,600mm [28.25 feet] for the bottom section. Compared to the conventional tray design, there was a reduction of more than 2 meters in column diameters and 32% in estimated column weight including column internals. **Table 1** shows side-by-side comparisons of all options.

Table 1: Summary of different tray design options

Tray Options		Design A	Design B	Design C
Tray Type		Conventional Trays	High Performance Trays	Shell HiFi™ Plus Trays
Deck Type		BDH™ Float Valve	High Capacity MVG™ Fixed Valves	High Capacity MVG™ Fixed Valves
Downcomer Type		Straight Downcomers	Sloped Downcomers	Shell HiFi™ Downcomers
Column Diameter [mm]	Top Section	10,500	9,400	8,300
	Middle Section	10,500	9,400	8,300
	Bottom Section	11,000	9,700	8,600
Estimated Total Column Weight including Trays [kg]		Base	20% Less	32% Less

The xylene rerun column T-4001 was eventually designed with 117 Shell HiFi Plus trays with 80 trays in the top and middle sections (of diameter 8,300mm), and 37 trays in the bottom section (of diameter 8,600mm). As the trays in the bottom section had to handle much larger liquid loads, the HiFi Plus trays in the bottom section were designed with 12 downcomers per tray whereas the each HiFi Plus tray in the top and middle section was designed with 10 downcomers. Hence, a redistribution tray was required for the transition, to ensure that the liquid from the last tray of the middle section was properly distributed to the first tray of the bottom section. Sulzer Chemtech also added a collector tray below the last HiFi Plus tray. The collector tray was built to collect and direct the liquid from multiple downcomers to the column sump. Use of the collector tray also prevented the potential re-entrainment of the liquid from the last tray by the vapor from the reboiler return.

Column Design – Feed Distribution

Even a perfectly designed tray will not perform to its maximum potential without a sound feed distribution system. A good distribution, for the liquid feeds and reflux, is especially important for Shell HiFi trays with multiple downcomers. The ideal liquid distributor is one which discharges the liquid proportionally to the active area, for a constant V/L mass transfer on all panels. A badly designed distributor can have an adverse impact on column performance such as loss in tray efficiency, due to unequal mixing on the panels and hydraulic issues such as weeping and vapor channeling if there is a substantial difference in clear liquid height on the panels. For this application, with less than 15 vol% vapor, an appropriately designed spider distributing the liquid proportionally to each active area was used.

Figure 5 shows the layout of HiFi Plus tray in the top section with 10 downcomers and **Figure 6** shows the feed pipe design for first tray of the top section. Both Figures have labels on one half of the equipment design, to illustrate the distribution of liquid from the spider pipe. Branch Pipe I, positioned above Downcomer A, had a row of side orifices which projected liquid feed onto the panel a, while branch Pipe II, positioned above Downcomer B, had 2 rows of side orifices which discharged the feed onto the centre-line of active panels b and c. Branch Pipe III, positioned above Downcomer C, served panels d and e by applying the same working philosophy.

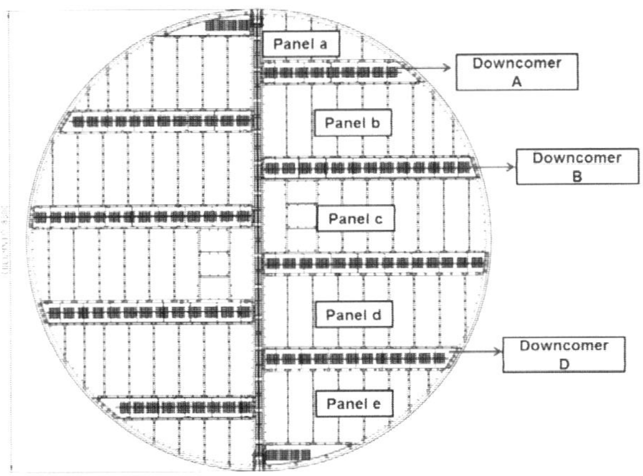

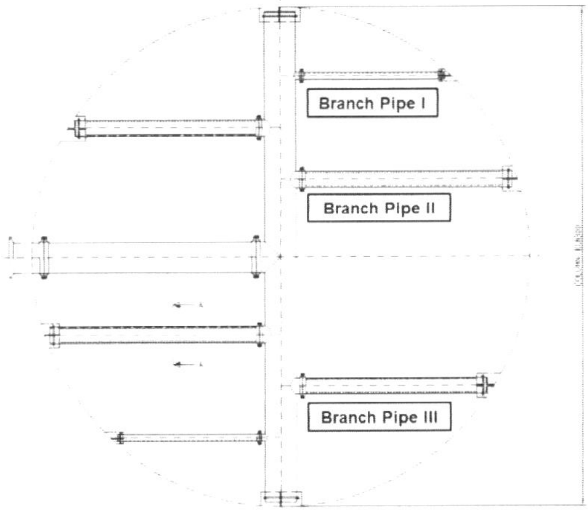

Figure 5: Layout of Shell HiFi™ Plus trays with 10 downcomers
Figure 6: Feedpipe design for Shell HiFi™ Trays

The vapor distribution from the reboiler return nozzles is as important as the liquid distribution from the feed. A well designed vapor feed inlet ensures that the vapor is evenly distributed before the vapor passes through the column internals. There are several factors which affect the selection of vapor feed inlet, which include the column diameter, vapor space above nozzle, nozzle orientation, nozzle size and amount of liquid in feed. For mixed phase feeds which may contain some amount of liquid, vane type feed inlets are preferred as they act as impingement baffles that facilitate the bulk separation of liquid from mixed phase feeds. The Shell Schoepentoeter™ is probably the most commonly used vane inlet device for introducing gas/liquid mixtures into these columns. For xylene rerun column T-4001, the reboiler return was designed with 4 Schoepentoeters positioned at right angle to one another. Owing to the large column diameter and huge amount of vapor in the reboiler return, the use of 4 Schoepentoeters from 4 radial reboiler return nozzles provided the most uniform vapor distribution below the collector tray.

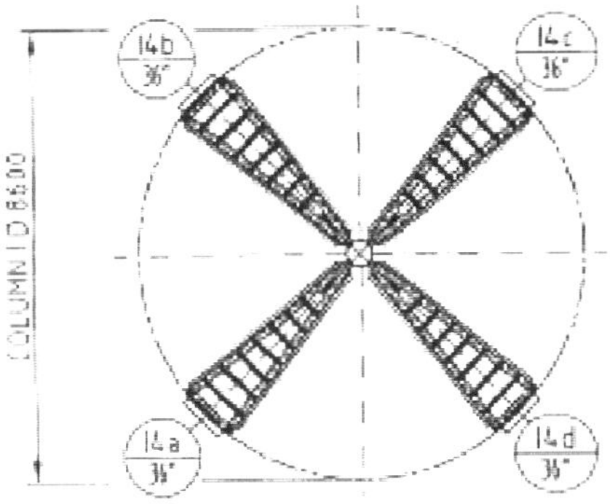

Figure 7: Picture of a large Shell Schoepentoeter™
Figure 8: Orientation layout of 4 Schoepentoeters™ for T-4001

Column Design – Support System

Shell HiFi trays are high performance trays and are often designed with low tray spacing, especially in tall fractionators with more than 100 trays. As a result, the design of the support system for these HiFi trays with low tray spacing and large column diameters has become a challenging task. For xylene rerun column T-4001 with tray spacing as low as 400mm [15.75 inches], lattice beams were used. Every set of lattice beam supported 2 trays – the tray above and below it. These beams, which were constructed of angle bars, offered an open structure which allowed free vapor communication on the tray deck and provided the necessary mechanical strength to support these trays of diameters over 8 meters.

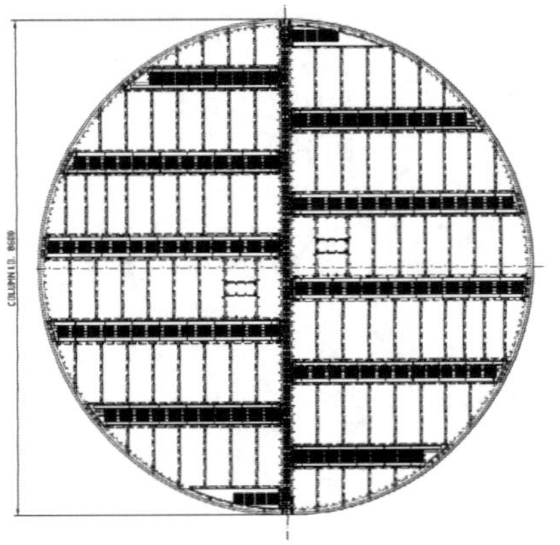

Figure 9: Shell HiFi™ Tray Downcomer Layout, with lattice beams
Figure 10: Installation picture of one set of lattice beams between two HiFi trays

Column Performance

After the installation of the Shell HiFi Plus Trays and the subsequent column commissioning, the xylene rerun column T-4001 has been operating successfully for the past few years. The trays perform well, providing the required production capacity while meeting the process requirements. The column performance exceeds expectations with an ortho-xylene recovery of 99.2% which is higher than the design recovery of 99.0%. C9+ aromatics in distillate are around 200 ppm by weight which is lower than design value of 300 ppm by weight.

The following **Table 2** shows a comparison of design values and actual plant operating data.

Table 2: Comparison of Plant Data and Design

Parameters	Units	Design Max	Plant Data
Feed 1 Flowrate*	Sm^3/hr	100	101.5
Feed 2 Flowrate*	Sm^3/hr	61	58
Top Pressure	MPa(g)	0.971	0.980
Pressure at Feed 1	MPa(g)	0.994	0.984
Pressure at Feed 2	MPa(g)	1.043	1.049
Bottom Pressure	MPa(g)	1.081	1.080
Feed 1 Temperature	°C	250	259
Feed 2 Temperature	°C	240	255
Top Temperature	°C	256	257
Bottoms Temperature	°C	299	303
Reflux Temperature	°C	234	229
Distillate Flowrate*	Sm^3/hr	133	136.5
Bottoms Flowrate*	Sm^3/hr	28	23
Reflux Flowrate*	Sm^3/hr	294	246

Reflux/Feed Ratio (Vol Ratio)	-	1.82	1.54
Condenser Duty**	M*Watt	32	-
Reboiler Duty**	M*Watt	32	-
C9 Aromatics in Feed	wt%	13.0	12.7
C9 Aromatics in Distillate	wt%	0.03	0.02
O-Xylene in Bottoms	wt%	1.05	1.02
O-Xylene Recovery	%	≥ 99	99.2

* All flows normalized to a 100 Sm3/h feed as true column capacity is confidential
** All duties adjusted to a 100 Sm3/h feed basis

There are 4 localized pressure indicators on this column, one on the overhead line, one at the feed 1, one at the feed 2 and one below the bottom tray. The observed pressure drop across the top 27 trays was only 4 kPa; which seems too low. An explanation could be due a faulty reading from the pressure indicator at the feed 1 location. After eliminating this reading, the observed pressure drop agreed well with the expected pressure drop. The detailed hydraulic results and pressure drop data are tabulated in **Table 3.** The temperature of both feed streams were higher than design value, this could explained the lower pressure drop in the bottom section.

Table 3: Comparison of Tray Hydraulic Parameters

Sections	Units	Design Max	Plant Data
Top (27# trays)			DP ~ 69 kPa
- Max. useful capacity	%	81	
- Weir loading	m^3/mh	37.2	
- Tray pressure drop*	kPa/tray	0.610	
- Section pressure drop**	kPa	21.1	
Middle (53# trays)			
- Max. useful capacity	%	82	
- Weir loading	m^3/mh	50.4	
- Tray pressure drop*	kPa/tray	0.653	
- Section pressure drop**	kPa	44.2	
Bottom (37# trays)			DP ~ 31 kPa
- Max. useful capacity	%	89	
- Weir loading	m^3/mh	54.0	
- Tray pressure drop*	kPa/tray	0.690	
- Section pressure drop**	kPa	34.2	

* Exclude vapor head
** Include vapor head

Conclusion

Shell HiFi Plus trays are one of the highest capacity trays in the market and have proven to be the optimal tray design for the new fractionator in the aromatic unit developed for the subject of this case study. The ideal combination of the Shell HiFi multi-downcomer design to handle the high liquid loads, and the Sulzer MVG high performance valves to handle the high vapor load, is an advanced tray design which delivers proven performance with high capacity whilst maintaining good tray efficiency. Well-designed column internals including feed distributors, collectors and support beams complement these trays, enabling the fractionator to achieve its maximum potential.

Bibliography

1. Ang, C.P. and M. Pilling., Tray Design for High Liquid Loads Applications, Digital Refining, July 2014
2. Summers, D.R., and M. Pilling, D. Wiesman, *Mega Tower Design Considerations*, Distillation Topical Conference, 2014 AIChE Spring Meeting, New Orleans, March 2014
3. DeVilliers, W., and P. Wilkinson and E. Vos, *Performance Data for Shell HiFi High Capacity Trays,* 4th AIChE Regional Process Technology Conference, Houston, October 2012
4. Pilling, M., Ensure Proper Design and Operation of Multi-Pass Trays, Chemical Engineering Progress, June 2005, pp. 22-27
5. Shell HiFi, Shell HiFi Plus, Shell Schoepentoeter, are Trademarks of Shell
6. Sulzer MVG is Trademark of Sulzer

创举塔器
CHUANGJU COLUMN

ADVANCE IN HIGH PERFORMANCE SPRAY TRAY DEVELOPMENT AND THEIR INDUSTRIAL APPLICATIONS

Dr. Yan Zhuang
Advanced Column Technology

ADVANCED COLUMN TECHNOLOGY PTY. LTD.

Presentation Outline

1. Comparison of Flow Regimes
2. How does a Spray Tray Work
3. The Development of Spray Trays
4. The Industrial Application of CJST Spray Trays

创举塔器
CHUANGJU COLUMN

Vapor and Liquid Contact Equipment

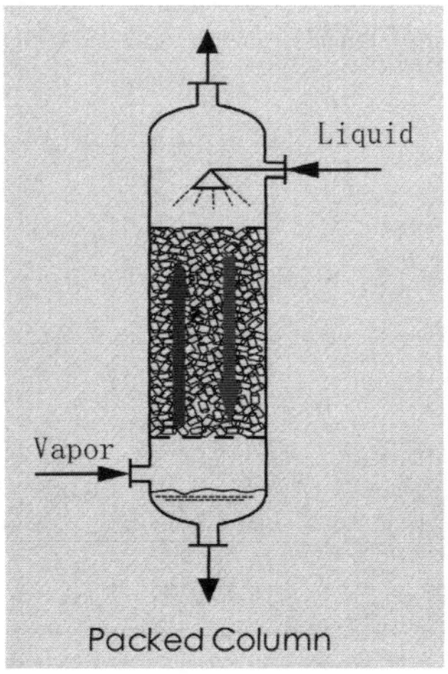

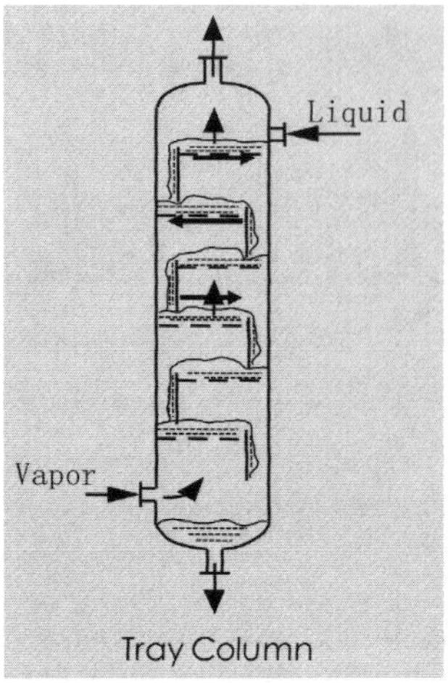

Packed columns and Tray columns are the two main types of equipment used for distillation.

The hydraulics ie. flow regimes dictate the efficiency of the mass transfer.

创举塔器 CHUANGJU COLUMN

Common Operating Regimes on a Conventional Tray

Froth Regime:

- The most common operating regime in distillation
- Liquid as the continuous phase and vapor the dispersed

Spray Regime:

- Exists at relatively low liquid load and high vapor load
- Liquid as the dispersed phase and vapor the continuous

Emulsion Regime:

- Exists at high liquid load and relatively low vapor load

创举塔器 CHUANGJU COLUMN

Froth Regime vs Spray Regime on a Conventional Tray

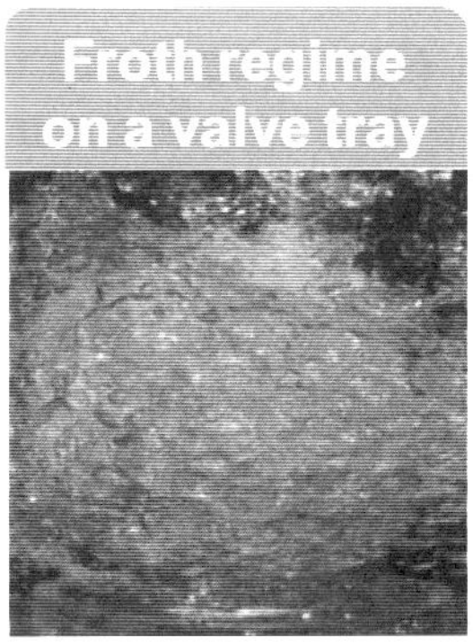

VS

Froth Regime: On a valve tray, the vapor is dispersed in the liquid

Spray Regime: On a sieve tray, the liquid phase is dispersed in the vapor phase.

创举塔器
CHUANGJU COLUMN

Transition of Flow Regimes

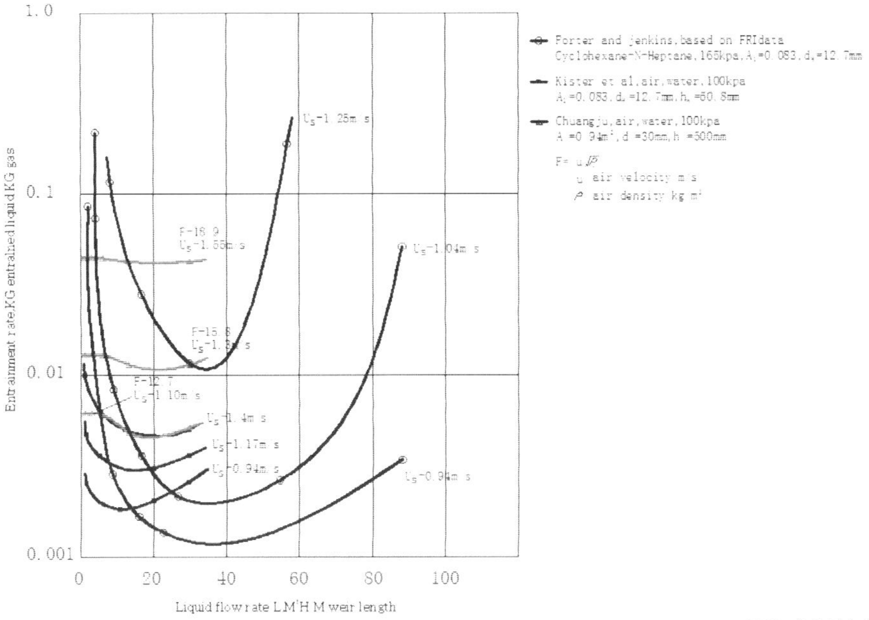

创举塔器
CHUANGJU COLUMN

How does a Spray Tray Work
(Exhibitor Booth 622)

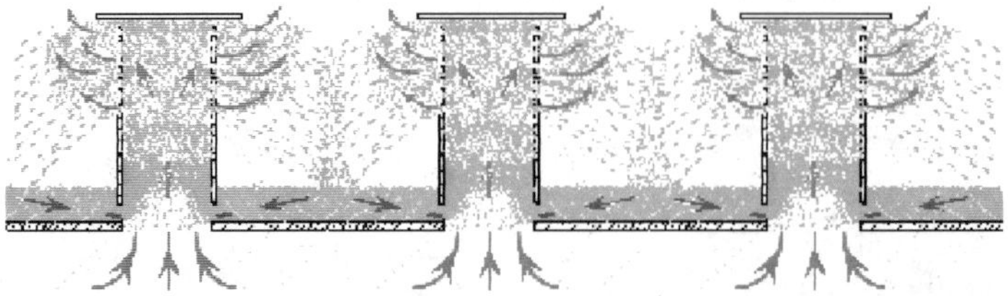

- Vapor rises through the opening in the tray; liquid enters through a gap at the bottom of the cap;
- Liquid is drawn up by the rising vapor as sheets, ligands, and broken into droplets, then impinges to the top of the cap;
- The liquid and vapor mixture is sprayed out through the holes.
- The vapor and liquid mixtures from the neighboring rows collide.
- The liquid droplets falling down to the tray floor and vapor rises to the tray above.

创举塔器
CHUANGJU COLUMN

Spray Tray at Work with different Vapor Load

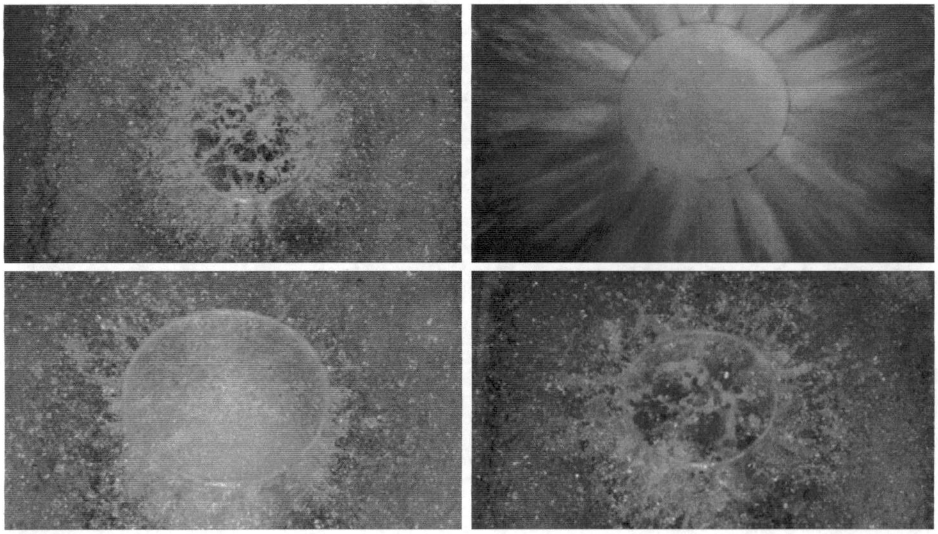

创举塔器
CHUANGJU COLUMN

创举塔器
CHUANGJU COLUMN

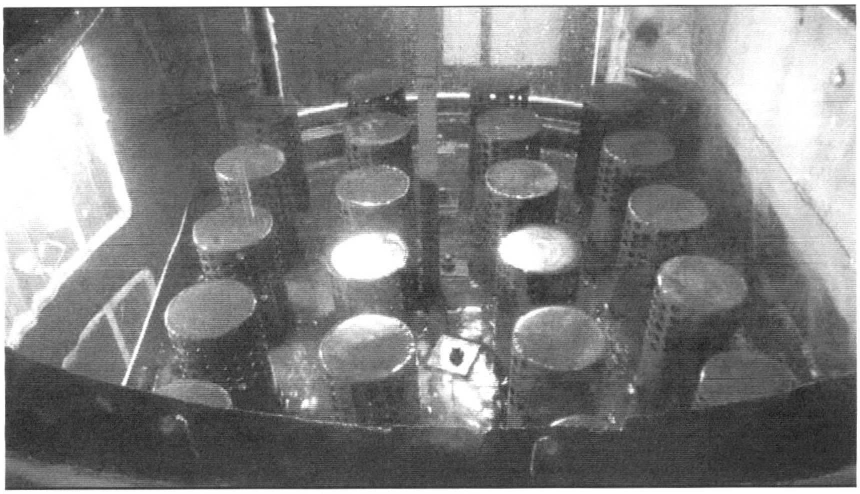

创举塔器
CHUANGJU COLUMN

Spray Regime vs. Spray Tray

Spray Regime on a Conventional Tray	Spray Tray
Common characteristics: Liquid dispersed & vapor continuous **Differences:** Exists at a relatively low liquid load & high vapor load Requires bigger tray spacing, preferably >24 inches Unstable when close to jet flooding	**Common characteristics:** Liquid dispersed & vapor continuous **Differences:** Much wider liquid and vapor load range (with liquid load up to 35gpm/ft^2 and hole F factor 8 - 15 in vacuum ambient pressure or up to 30 in a pressure system) Normal tray spacing, 12 to 20 inches Very stable spray operation in the entire operating range

创举塔器
CHUANGJU COLUMN

Advantages of Spray Trays

1	**Higher mass transfer efficiency**
2	**Higher Capacity**
3	**Lower Pressure Drop**
4	**Anti-fouling**
5	**Reduce foaming**

创举塔器
CHUANGJU COLUMN

Higher Mass Transfer Efficiency

- **Increased total areas of liquid-vapor interface** by dispersing liquid into a large amount of small droplets.
- **Longer phase contact time**: The liquid comes to contact with vapor by rising and mixing inside the cap, spraying out through the vertical sieve holes, then falling back to the tray;
- **Faster interface renewal** due to intensified phase contact inside the cap and outside of the cap.

The spray tray creates a high impact, more turbulent and liquid phase highly dispersed phase contact environment for mass transfer.

创举塔器 CHUANGJU COLUMN

Efficiency Comparison
Spray Tray vs F1 Valve Tray

- Compared with F1 valve tray, the CJST tray has a 10% plus higher column efficiency
- Industrial data showed 10% to 40% steam saving after revamping using spray trays.

Project name	Before Revamp	After Revamp	Percent of saving
Jiantao Stage 2 Methanol Project Revamp	1.25t(steam)/t (product)	1.0-1.1t(steam)/t (product)	12-20%
Natural Gas Treatment Plants of Changqing Oil Field of PetroChina	1.08t steam/10,000 m^3 Natural Gas	0.86t steam/10,000m^3 Natural Gas	20%
Qilu Petrochemical Technology's Wastewater Stripping Columns			10-15%
Ningbo Haode Chemical Industry's Distillation Columns A/B	520 kg steam /t feed	295 kg steam /t feed	43%
Ningbo Haode Chemical Industry's De-gassing Column	53.8 kg steam /t feed	37.5 kg steam /t feed	30%

创举塔器 CHUANGJU COLUMN

Higher Vapor Handling Capacity

- The sealed cap forces the rising vapor and liquid change direction and spray out horizontally, the entrainment is much lower.

- The spray design hole F factor is normally 8 to 15, compared with a valve tray's 8 to 12.

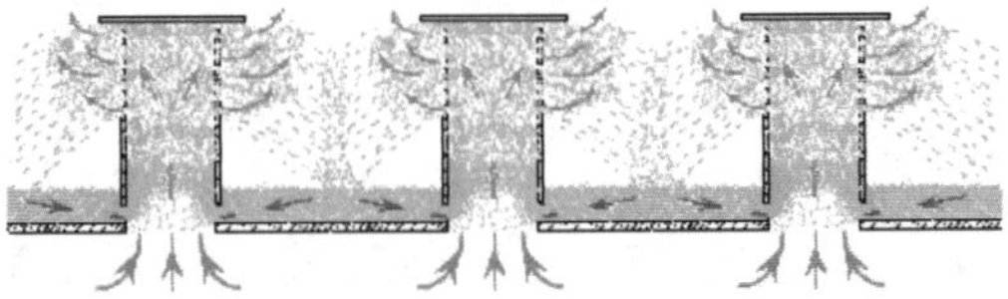

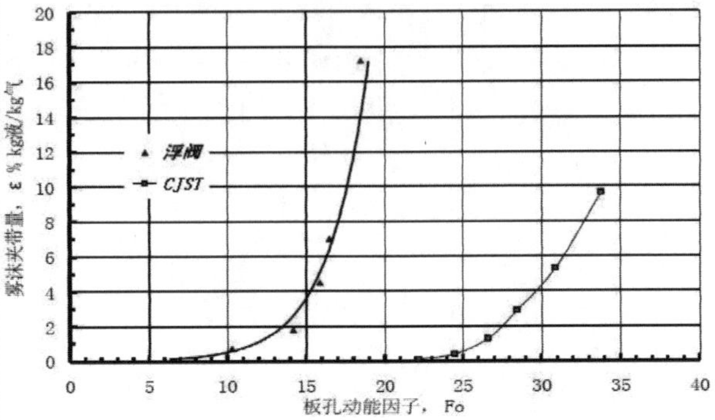

Hole F-factor

创举塔器
CHUANGJU COLUMN

Higher Liquid Handling Capacity

On a spray tray the vapor does not pass through the liquid layer, so the liquid is basically clear.

The down comer liquid is also not aerated, so the tendency for down comer back up flooding and choke flooding is greatly reduced.

The spray tray reduces foaming tendency, thus can handle foaming systems such as amine much better than conventional trays.

创举塔器
CHUANGJU COLUMN

Lower Pressure Drop

- The spray tray weir height is normally 30 to 50 mm
- The traditional tray's weir height is normally 40 to 80 mm to facilitate longer mass transfer time.

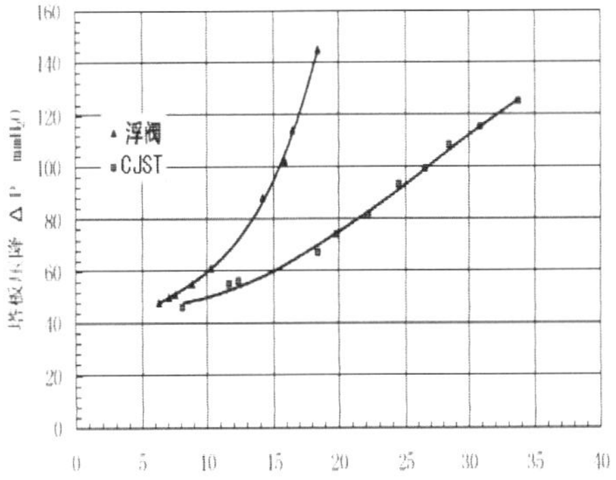

CJST与F1塔板压降比较图，开孔率11%

创举塔器
CHUANGJU COLUMN

Anti-Fouling Capability

The valves in the Waste Water Stripping Column was blocked in one month - Sinopec Qilu Petrochemicals After revamped using CJMP the column can run 12 month without blocking.

创举塔器
CHUANGJU COLUMN

Reduce Foaming

Downcomer froth level-spray trays

Downcomer froth level-Valve trays

创举塔器
CHUANGJU COLUMN

The Development of CJST Series Spray Trays

- ★ New-VST 1968
- ★ Radial Side-guiding Spray Tray （CJST®） 2006
- ★ Low Resistance Membrane Spray Tray （CJMP Ⅰ®） 2006
- ★ Membrane Spray without Back-mixing Tray （CJMP Ⅱ®） 2009
- ★ Partitioned Flow Spray Tray （CJPFST®） 2012
- ★ Low Back-mixing Spray Tray （CJLBST®） 2015

创举塔器
CHUANGJU COLUMN

New-VST

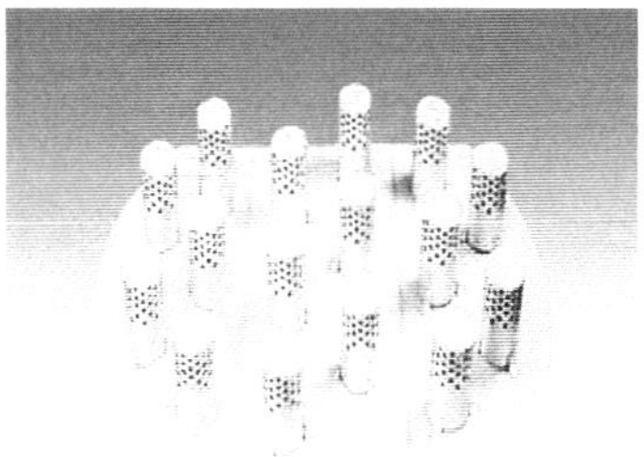

- Developed by Japan's Mitsui Engineering & Shipbuilding (MES) in 1963-1968.
- Since the 1980s, the New-VST was studied extensively by Chinese researchers and consequently commercialized in industrial applications in China successfully.

创举塔器
CHUANGJU COLUMN

Radial Side-guiding Spray Tray （CJST®)

- The liquid is sprayed to the direction of liquid flow on the tray.
- Reduced liquid back mixing.
- Can handle large liquid load.

创举科技
CHUANGJU TECHNOLOGY

Low Resistance Membrane Spray Tray （CJMP Ⅰ®）

- Lower pressure drop compared with a CJST®
- Can incorporated cooling coils on the tray

创举塔器
CHUANGJU COLUMN

Membrane Spray without Back-mixing Tray （CJMP Ⅱ®）

- No liquid back mixing
- No liquid concentration gradient on the tray
- Uniform mass transfer driving force on the tray
- Anti-fouling
- Can fit in heat/cooling coils easily

创举塔器
CHUANGJU COLUMN

Partitioned Flow Spray Tray（PFST®）

- No liquid back mixing
- Vapor and liquid contact counter-currently
- Multiple downcomer design
- Extended downcomer residence time for vapor disengagement

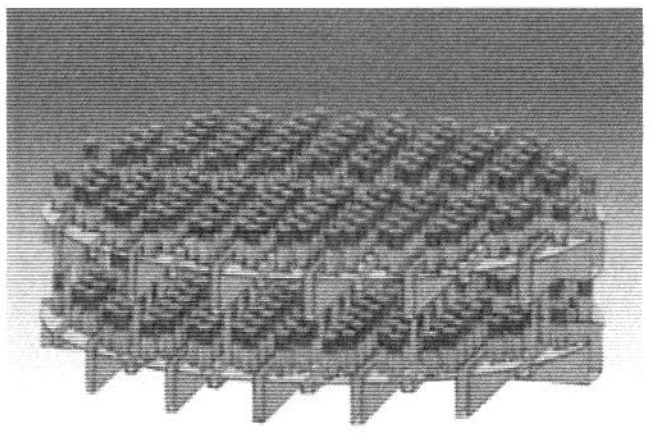

创举塔器
CHUANGJU COLUMN

Low Back-mixing Spray Tray （LBST®）

•The down comer design has a liquid distribution trough to feed fresh liquid to each individual vertical sieve caps on the tray, so each cap will have the same mass transfer driving force to achieve a uniform efficiency.

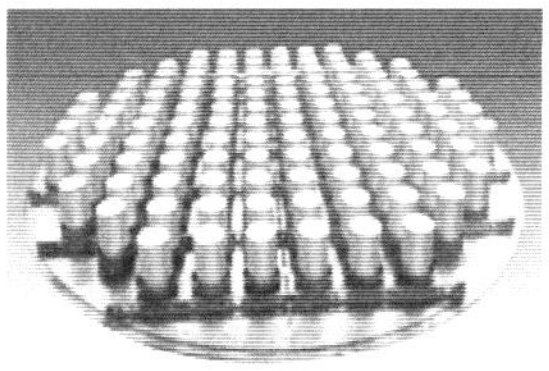

创举塔器
CHUANGJU COLUMN

CJST Spray Tray Industrial Applications

Coking Industry	
Benzol recovery	151
Ammonia recovery	195
Benezen, toluene and xylene	51
Naphthalene recovery	9
Tar oil	24
Sub-total	430
Other Industries	
Fertiliser ammonia stills	157
Methanol distillation columns	96
Phamerceutical solvent recovery	21
C4 splitter distillation	27
CTL, CTG waste water NH3 or phenol recovery	108
Petrochemical and refineries	62
Sub-total	471
Total Distillation Columns	901

Applied Industries:
Fertilizer
Coking
Coal to Oil
Coal to Gas
Fine Chemicals
Oil and Gas
Petrochemicals
Environmental protection
Pharmaceutical

创举塔器 CHUANGJU COLUMN

Case Study 1: Revamping Sinopec Qilu Petrochemical's Wastewater Stripping Column

- The valve tray was blocked in a month time
- ChuangJu Revamped using CJMP trays
- The column is running for a year without blockage

	Before Revamp	After Revamp
Tray Type	Floating Valve	CJMP II
Processing Throughput	110m^3/h	140m^3/h
Ammonia Content	300 -400mg/L	80mg/L
Steam Consumption saving		10-15%
Maintenance shutdown	Once per month	Once per 12 month

创举塔器 CHUANGJU COLUMN

Case Study 2:Revamping ENN's 600,000 t/a Coal to Methanol Pressure Distillation Column

The flooded rectification stage became the bottle-neck for the pressure distillation column.

ChuangJu replaced the 10 valve trays with 10 LBST trays and solved the problem.

	Before Revamp	After Revamp
Tray Type	Valve	LBST®
Pressure distillation column feed rate	90m³/h	130 m³/h
Ethanol content in Methanol Product	600ppm	<100ppm

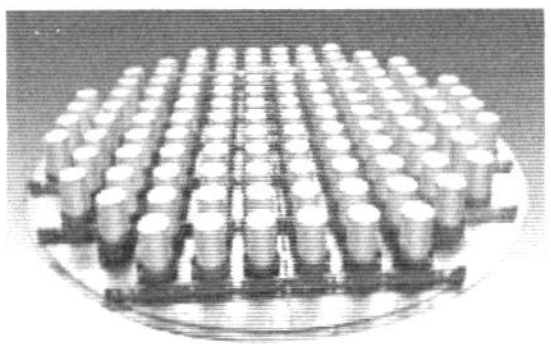

低返混均效喷射态塔板

创举塔器 CHUANGJU COLUMN

Case Study 3:Revamping Natural Gas Treatment Plants of Changqing Oil Field of PetroChina

Gas Sweeting Column	Before Revamp	After Revamp
Column tray design	Kohn-Glitch floating valve	CJST® spray tray
Processing throughput	1.5 million-2.3million m³ /d	3million m³/d
Single Tray Pressure Drop	1700Pa	750-900Pa
Steam Consumption	1.08t steam/10,000 m³	0.86t steam/10,000m³
Flooding occurrence/month	3 times	0
Anti-foaming Agent Dosage	1000-3000mL/d	0

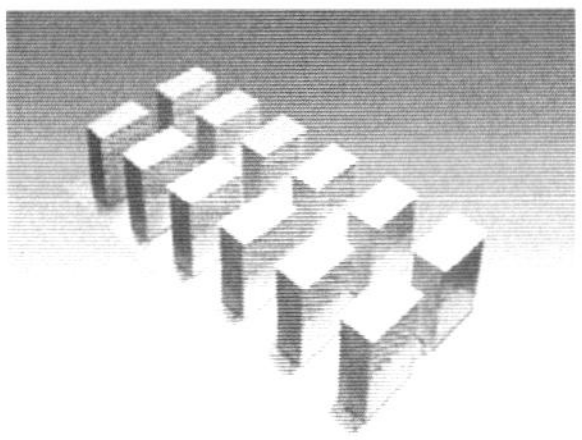

径向侧导喷射塔板（CJST）

From 2007, ChuangJu revamped all the columns in the three gas sweetening plants with its CJST® trays with successful results.

创举塔器 CHUANGJU COLUMN

Application of Spray Tray Technology

Revamping	New Projects
• Capacity Expansion • Better separation efficiency • Fouling Resistance • Anti-foaming • Can tailor design spray trays for your specific requirements	• Lower capex due to high efficiency and capacity • Lower Opex due to energy savings • Provide free equipment selection and costing during feasibility study stage • Can provide equipment design and fabrication

创举塔器
CHUANGJU COLUMN

Company Profile

Tianjin ChuangJu Technology Co., Ltd

Tianjin ChuangJu Technology Co., Ltd specializes in separation processes and equipment design, fabrication and installation, providing services to the chemical, oil and gas, pharmaceutical and environmental protection industries. ChuangJu developed its proprietary column internal designs and technologies with dozens of patents and know-hows, and successfully commercialised the spray tray technologies in nearly 1,800 industrial columns in 30 provinces of China and overseas. The company has its own fabrication plant covering 40,000m^2 areas with ASME U stamp certified pressure vessels manufacturing capabilities.

创举塔器
CHUANGJU COLUMN

Pilot & Testing Capabilities

Ambient Temperature Column Testing Rig	Swinging Column Testing Unit Simulating a Floating LNG Rig	Column Testing Rig with Temperature Control

创举科技
CHUANGJU TECHNOLOGY

Company held design and Manufacturer Licenses and Certificate

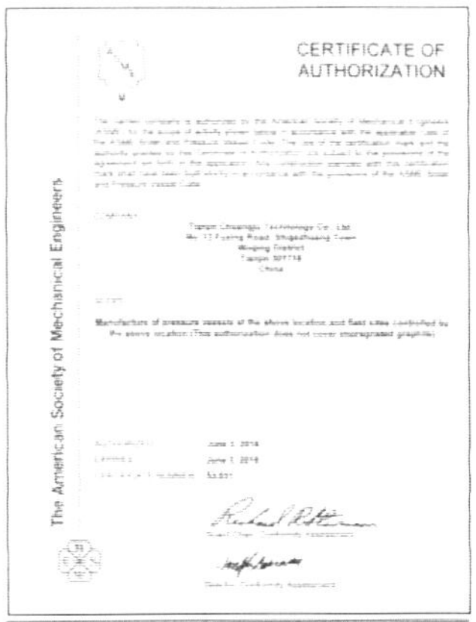

ASME Pressure Vessel Manufacture Accreditation U Stamp

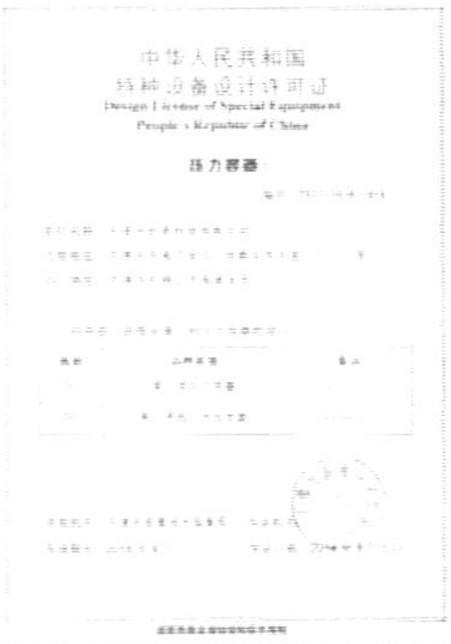

Chinese Special Equipment Design Licenses

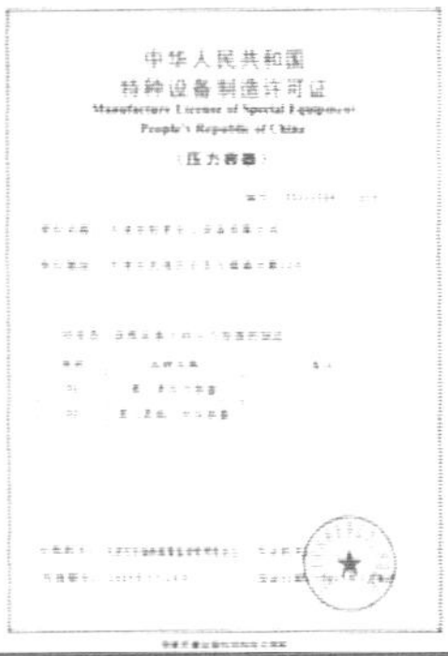

Chinese Special Equipment Manufacture Licenses

创举塔器
CHUANGJU COLUMN

Manufacturing Capabilities

Manufacturing Workshop

创举科技
CHUANGJU TECHNOLOGY

Manufacturing Capabilities

Production Equipment

Numerical controlled welding machine

OTC Welding Robot

Messer CNC numerical controlled cutting machine

Automatic welding machine for pipe and flange

创举塔器
CHUANGJU COLUMN

Manufacturing Capabilities

Production Equipment

Automatic submerged arc welding machine

CNC plasma flame cutting machine

Automatic molding equipment of packing

创举塔器
CHUANGJU COLUMN

Manufacturing Capabilities

Products

创举塔器
CHUANGJU COLUMN

Manufacturing Capabilities

Products

创举塔器
CHUANGJU COLUMN

Thanks

创举塔器 科技造就

Chuangju Column Technology Innovate

创举塔器
CHUANGJU COLUMN

Johnson Matthey

Packed Bed Performance Analytics Based on Gamma Scans

Lowell Pless

Business Development Manager

Tracerco

Tracerco

What do you learn from Scanning Packed Columns?

Johnson Matthey

Objective Measurements

- Condition of packed beds
- Base Liquid Level
- Chimney Tray and Distributor Liquid Levels
  - Damage?
  - Overflowing?
- Liquid Maldistribution?
- Flooding?
- Foaming?
- Excess Liquid Holdup?

Subjective Measurements

- Degree of Liquid Maldistribution
- Liquid or Vapor Maldistribution

Tracerco

A.I.Ch.E. 2017 Spring Meeting – Henry Kister Distillation Symposium

THE Equation

Johnson Matthey

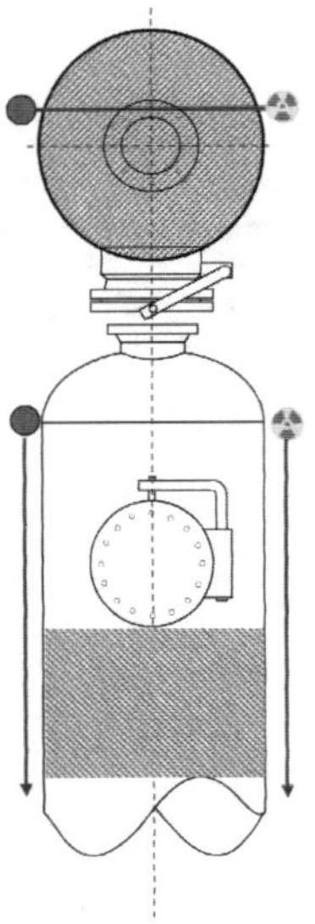

$$I = I_o\, e^{-\rho\mu\chi}$$

I_o is the initial gamma ray intensity measured at a given distance with no material interfering with the radiation transmission

I is radiation intensity measurements taken through a material

μ is the absorption coefficient specific to the material being measured

ρ is the material density

χ is the material thickness

Tracerco

A.I.Ch.E. 2017 Spring Meeting – Henry Kister Distillation Symposium

Gamma Scanning Procedure

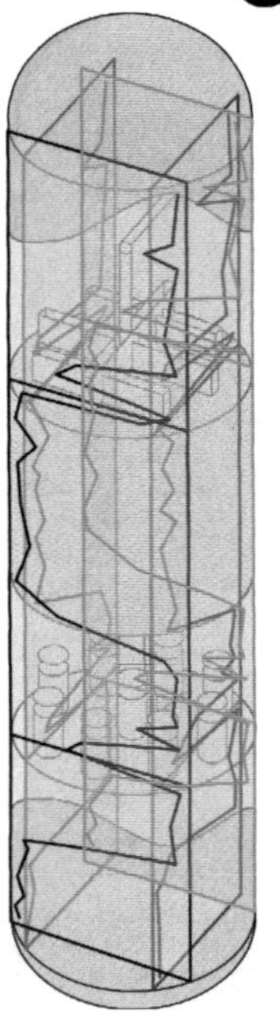

$$I = I_o\, e^{-\rho\mu\chi}$$

Objective:
Keep the various scan lines equal length so χ stays constant.
Then comparing responses between the scan lines shows differences in overall density.

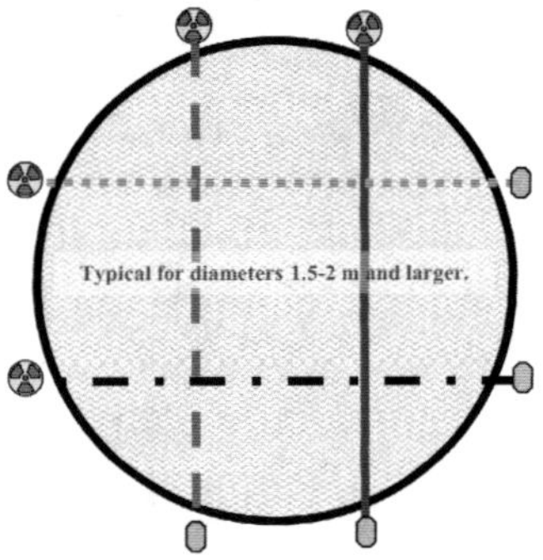

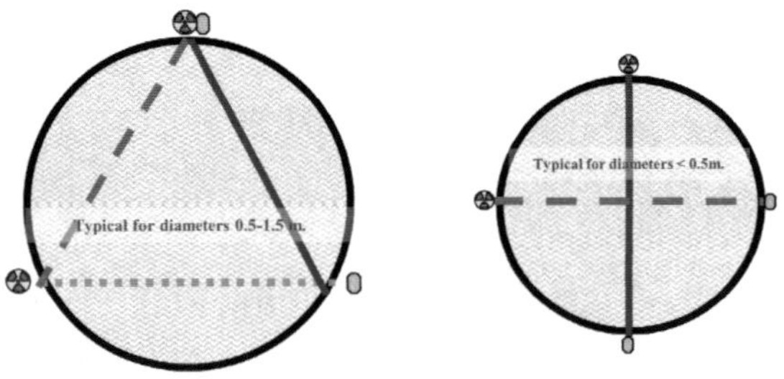

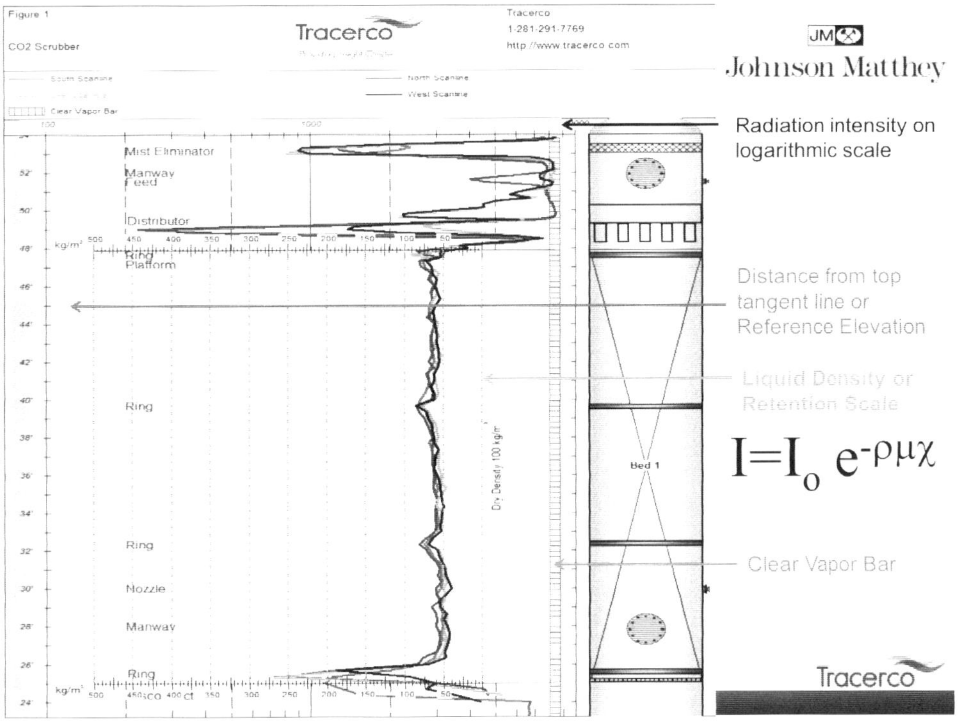

Example of Good Liquid Distribution

- **Scans show the trough-n-lateral distributor in place**

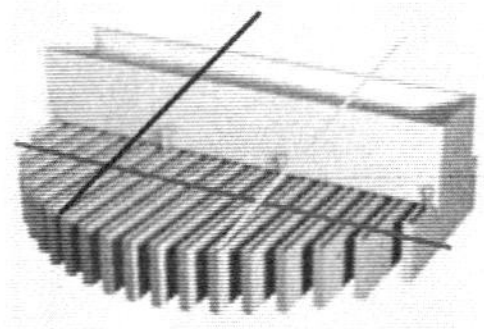

Courtesy Sulzer Chemtech

- **Overlaying scanlines indicates good liquid distribution**

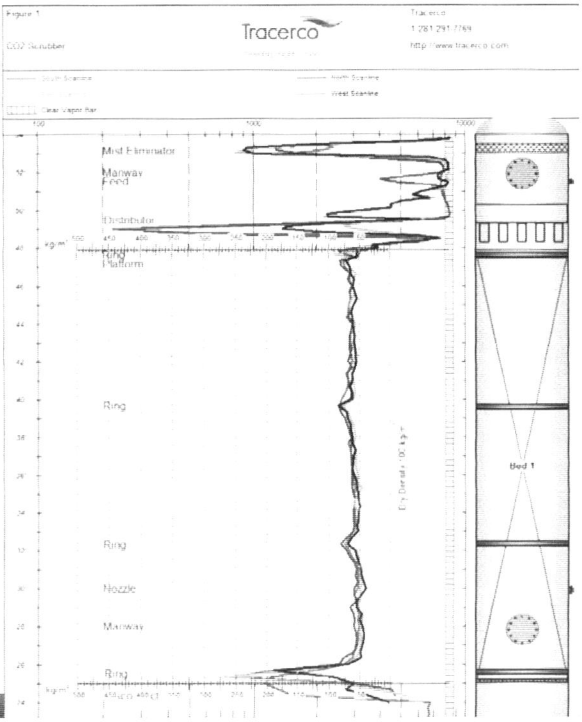

Example of Bad Liquid Distribution

- Scanlines do not overlay with each other indicating poor liquid distribution
- The spread in bulk density through the bed varies from 40 – 150 kg/m³

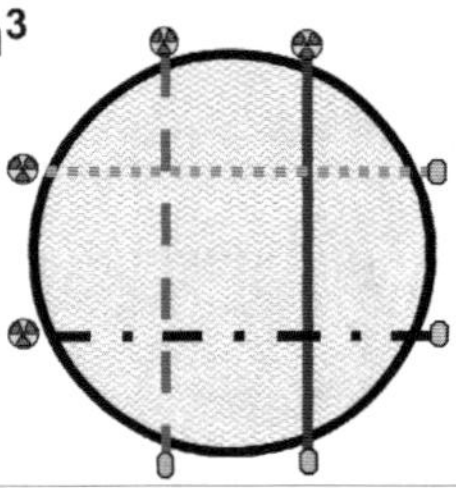

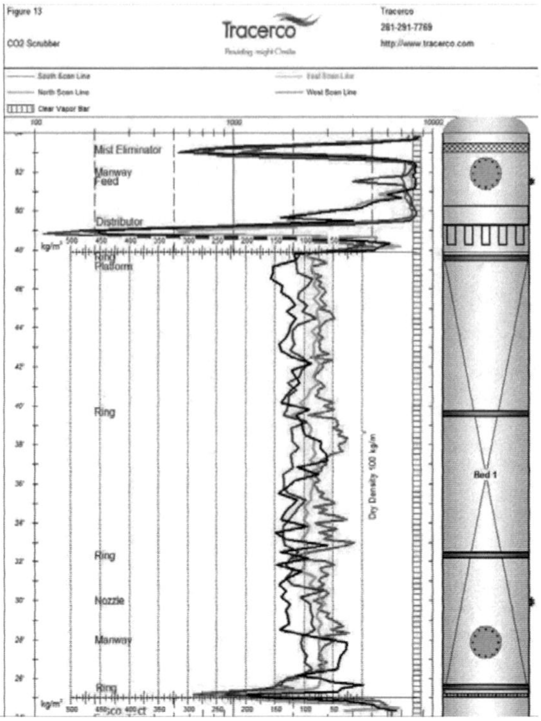

PackView™

- Scans show a density response (count reduction from clear vapor counts) for packing at correct elevation for top of bed.
- Lower half of packing more dense, presumably with liquid retention.

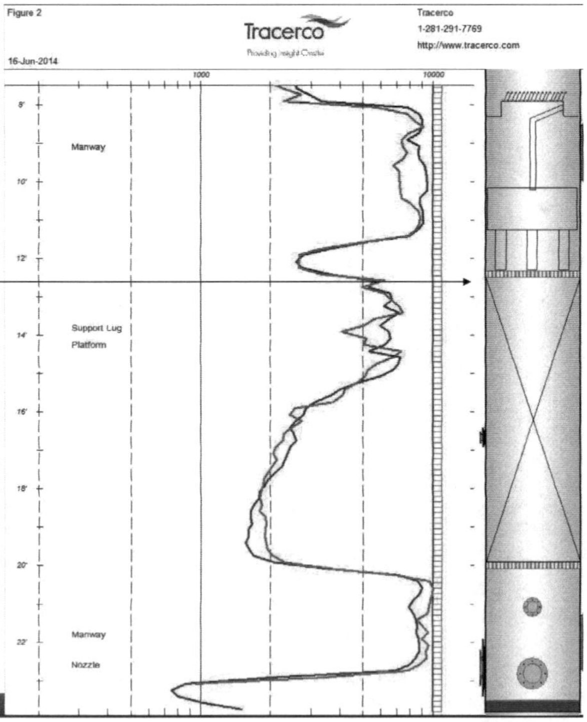

PackView™

- **Response at top of bed represents density for packing <u>and</u> liquid.**
- **However, packing density alone = 160 kg/m^3. Therefore, something seems to be missing.**

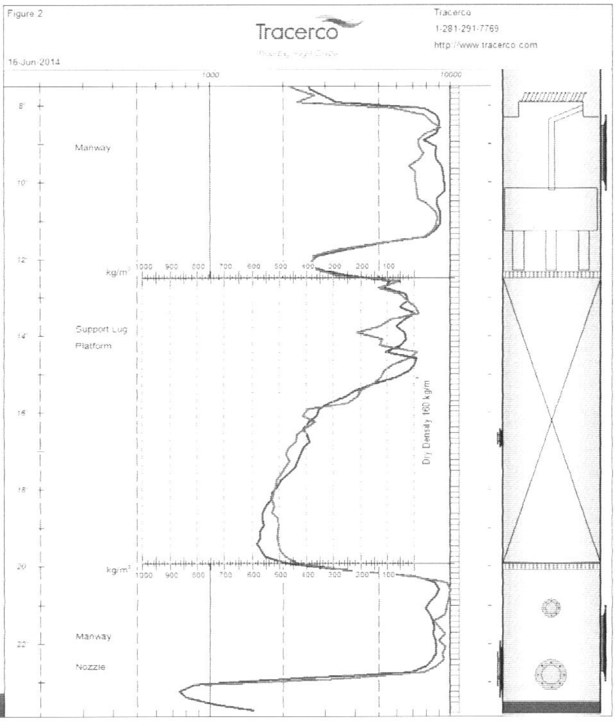

PackView™

∴ Instead of packing at top of bed being OK and only excess liquid retained in bottom of bed, PackView™ analysis showed packing damage. Random packing had partially collapsed leaving voids in packing at top of bed and crushed packing / liquid holdup in bottom of bed.

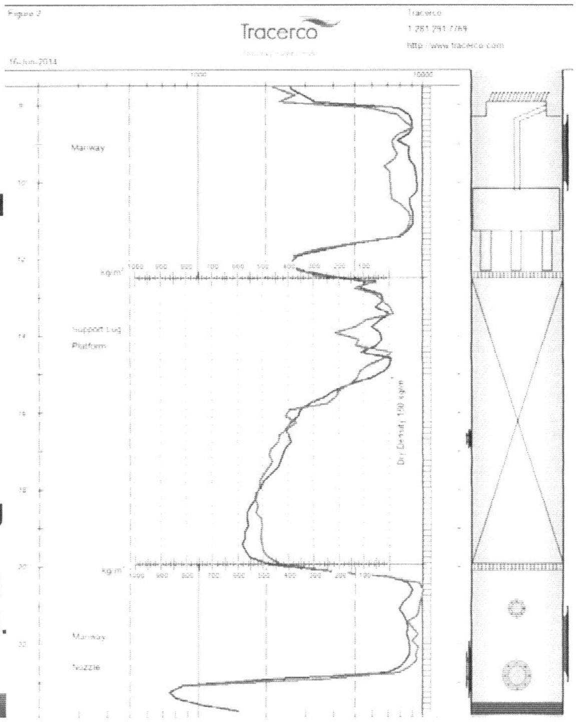

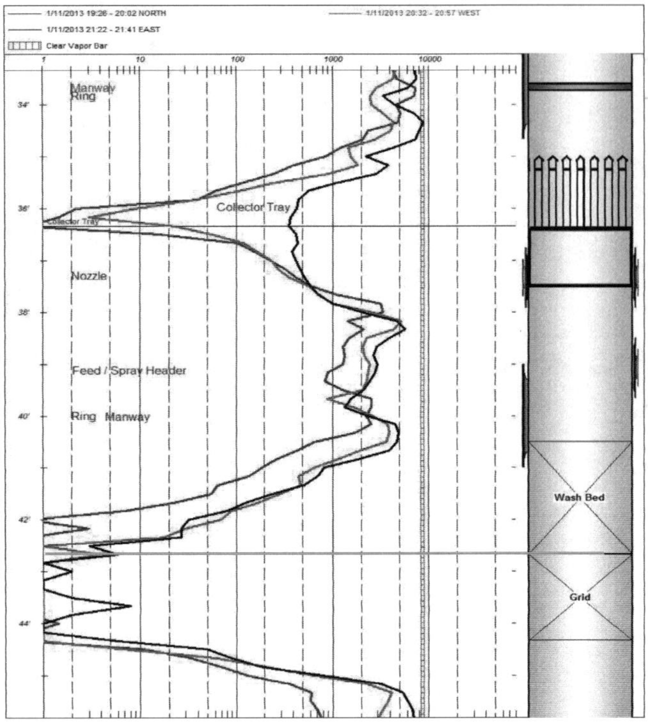

Johnson Matthey

Grid packing tends to be high density, typically 240 – 320 kg/m^3

Radiation counts through grid section and packed section very low, <10 which is at or near background levels

So was source size adequate or what was density of the packing?

Tracerco

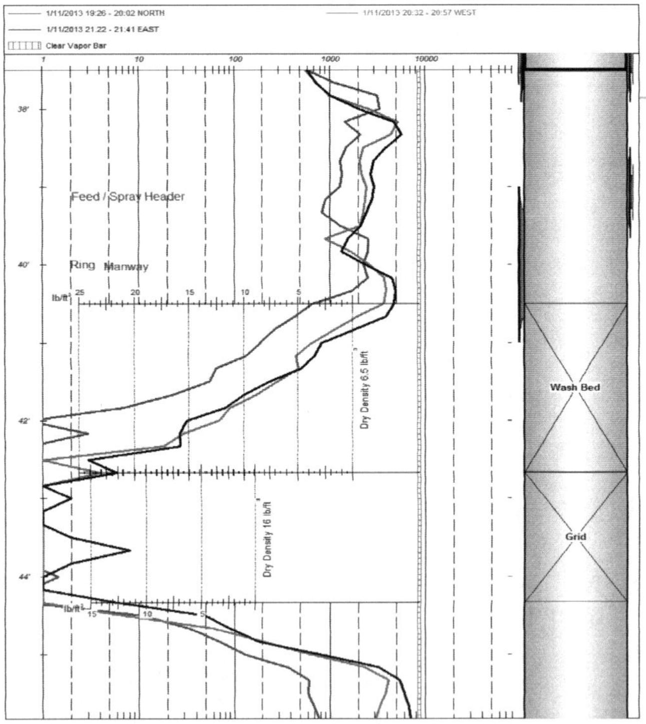

Johnson Matthey

PackView™ analysis showed grid and packed sections had liquid retention > 20 lb/ft^3 (> 320 kg/m^3)

This is much too high for a vacuum column wash bed; therefore, source size was more than adequate and grid/packing was coked up and retaining excessive liquid, i.e. flooding!

Tracerco

Johnson Matthey

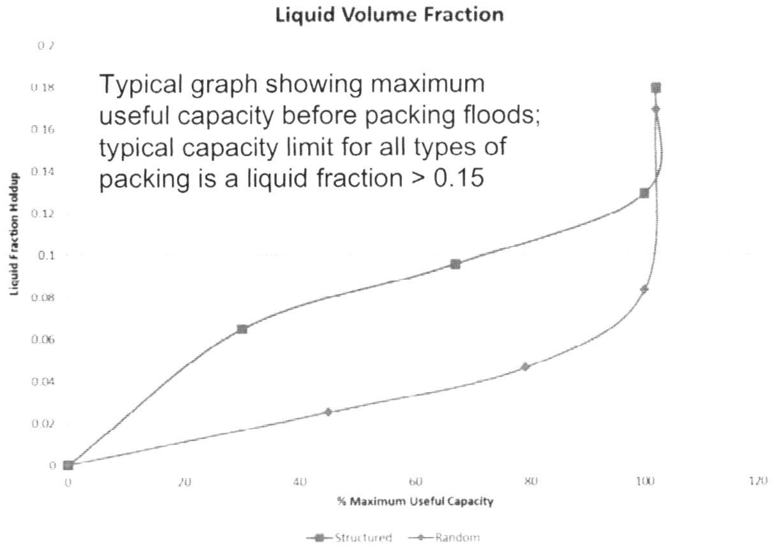

Tracerco

A.I.Ch.E. 2017 Spring Meeting – Henry Kister Distillation Symposium

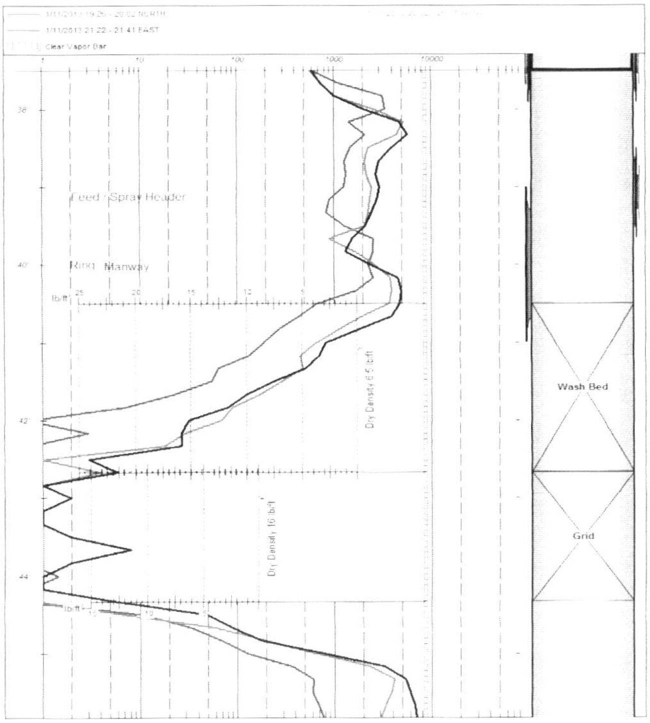

Johnson Matthey

Liquid Volume Fraction:

$$\frac{\text{Retained Liquid Density}}{\text{Liquid Density}}$$

Liquid Volume Fraction through packing section ≈ 0.56

Liquid Volume Fraction through grid section ≈ 0.4

Both grid and packing section way beyond their maximum useful capacity

Tracerco

Johnson Matthey

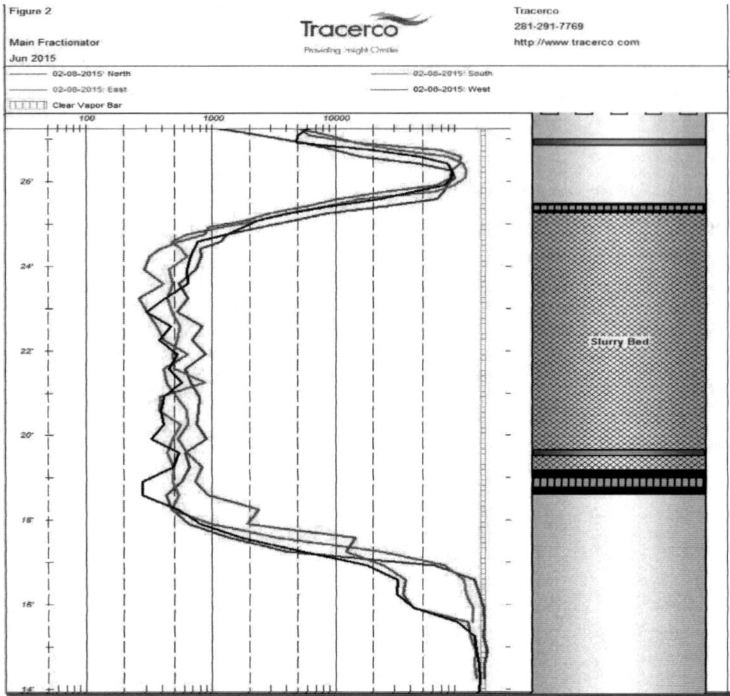

Tower 1:

Δ Counts = **300** to **870**

Δ Density = 5.1 lbs/ft³ (80 kg/m³)

So, would a 570 count difference always equate to a 5.1 lbs/ft³ or 80 kg/m³ density difference?

Tracerco

A.I.Ch.E. 2017 Spring Meeting – Henry Kister Distillation Symposium

Johnson Matthey

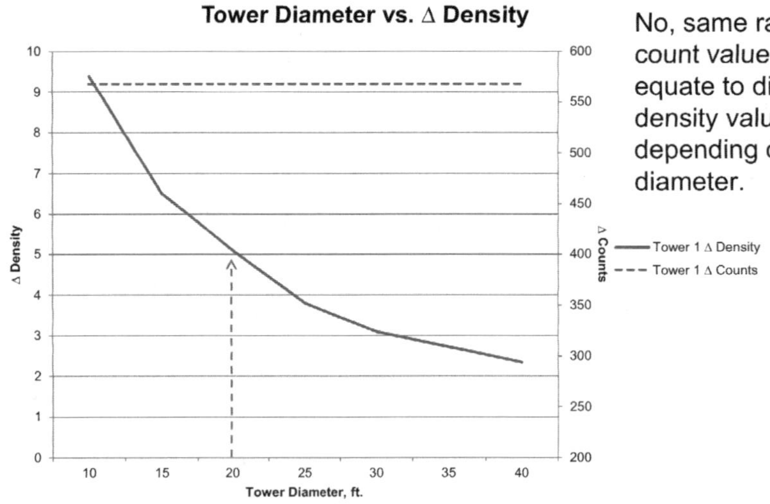

No, same radiation count value would equate to different density value depending on tower diameter.

Tracerco

A.I.Ch.E. 2017 Spring Meeting – Henry Kister Distillation Symposium

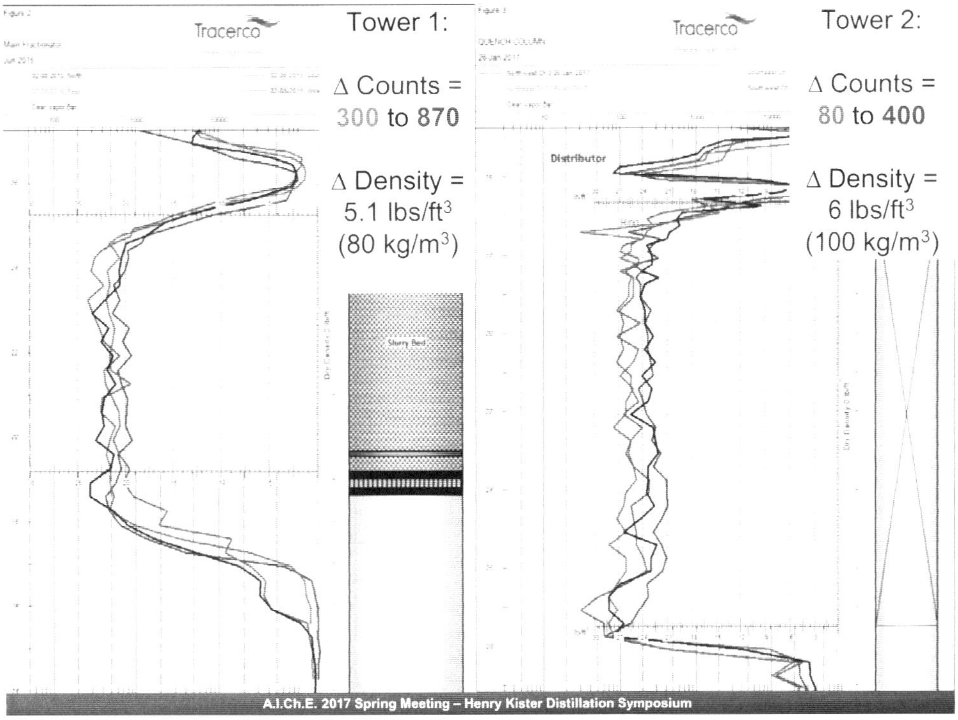
Tracerco
Tower 1:
Δ Counts =
300 to 870
Δ Density =
5.1 lbs/ft3
(80 kg/m3)
Slurry Bed
Tracerco
Tower 2:
Δ Counts =
80 to 400
Δ Density =
6 lbs/ft3
(100 kg/m3)
Distributor
A.I.Ch.E. 2017 Spring Meeting – Henry Kister Distillation Symposium

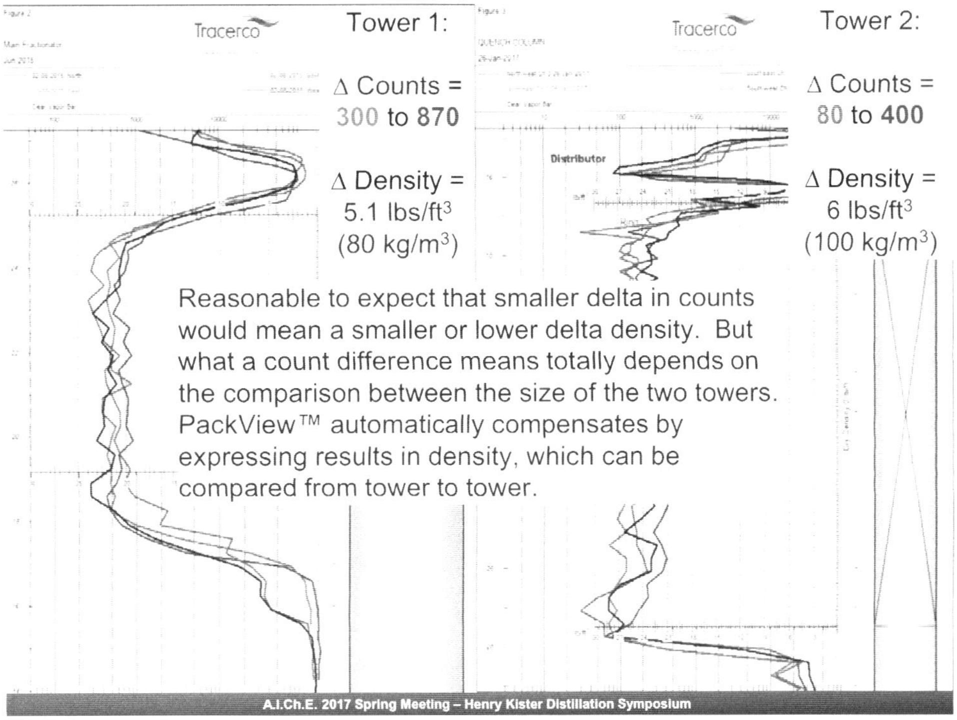
Tracerco
Tower 1:
Δ Counts =
300 to 870
Δ Density =
5.1 lbs/ft3
(80 kg/m3)
Tracerco
Tower 2:
Δ Counts =
80 to 400
Δ Density =
6 lbs/ft3
(100 kg/m3)
Distributor
Reasonable to expect that smaller delta in counts would mean a smaller or lower delta density. But what a count difference means totally depends on the comparison between the size of the two towers. PackView™ automatically compensates by expressing results in density, which can be compared from tower to tower.
A.I.Ch.E. 2017 Spring Meeting – Henry Kister Distillation Symposium

PackView™ allows direct comparison of scan results, e.g. comparing scan results from two different sized crude vacuum columns.

JM Johnson Matthey

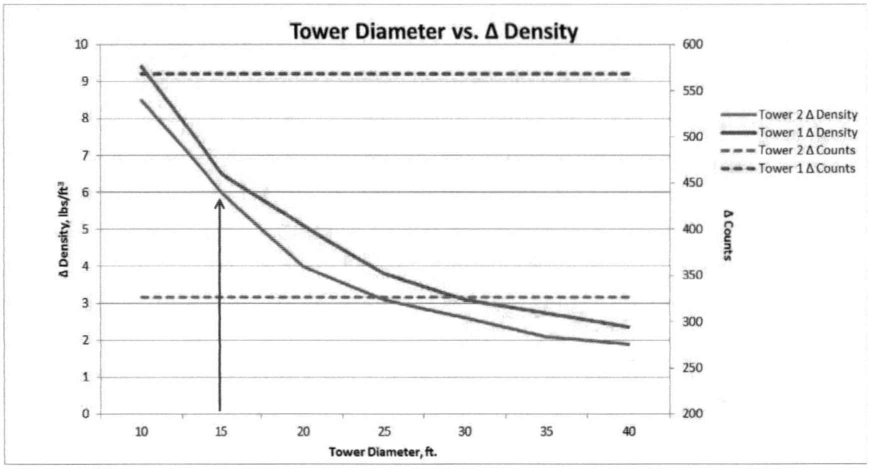

Tracerco

JM Johnson Matthey

Liquid Distribution Quality

General "Rules of thumb":

- **Good Liquid Distribution –**
  Δ = 0 – 2 lbs/ft^3 (0 – 50 kg/m^3)
- **Fair Liquid Distribution –**
  Δ = 2 – 5 lbs/ft^3 (30 – 100 kg/m^3)
- **Worrisome (Poor) Liquid Distribution –**
  Δ > 5 lbs/ft^3 (> 100 kg/m^3)

Tracerco

Conclusions – *PackView™ Analytics*

- Gamma scans measure (quantify) liquid mal-distribution through a packed bed in terms of liquid density.
- Gamma scans measure the liquid retention in packing, perhaps signaling excess liquid holdup caused by fouling.
- PackView™ compensates for tower size differences, scanline path length differences, radiation source differences, etc. so analysis can be done on a consistent density basis.
- If the actual liquid density is known, a simple calculation can determine the % liquid fraction, a measure of the Maximum Useful Capacity of the packing.

Rectifying a Reactive Distillation Column

By

Greg Theunick

The Dow Chemical Company, Midland, Michigan

Presented at the Distillation Topical Conference

AIChE Spring Meeting, San Antonio, Texas, March, 2017

Rectifying a Reactive Distillation Column

By Greg Theunick, The Dow Chemical Company, Midland, MI

ABSTRACT

Reactive distillation technology is used to produce Material A silane. Since the inception of making this material at a commercial scale, the process has had issues challenging stable operation. A troubleshooting investigation solved long term issues of pressure instability, steam surges and achieving product specifications within a limited startup timeframe. This paper captures the diagnosis and resolution for each issue.

BACKGROUND

"Material A" silane is made using reactive distillation technology. The reactive column is a shared asset which gets utilized for several reactive distillation products. Since Material A has been made at a commercial scale it has had issues that have made production a challenge. Other products made on this asset have experienced similar issues but Material A has been the most impacted.

Material A is produced by reacting Material B and C (see equation 1 below). A noncondensible coproduct is also formed. This is an equilibrium limited reaction so the products are quickly separated to keep the reaction favoring product formation. Some of Material B is in the liquid phase and co-fed to the column with Material C. This ensures good mixing of reactant species before the material enters the column. A fraction of Material B is vaporized and introduced lower in the column to reduce the amount of unreacted material leaving in the bottoms product. Material B is a low boiler relative to the bottoms product so it quickly migrates up the column and reacts with any unreacted material it encounters along the way. Only noncondensibles vent out, all other materials come out in the bottoms product. This reaction has the potential to form undesired side products. These are impurities that require additional processing downstream before the final product can be sold.

Eq. 1: *Material B + Material C* ⟷ *Material A + Noncondensible*

Figure 1 is a high level schematic of the distillation asset. The process equipment includes:

- Column 1 – This is the reaction tower and contains structured packing internals. Liquid feeds are introduced at the top of the column and the vapor feed is introduced above the bottom packed bed. The reboiler operates as a thermosiphon with steam as the heat source for the exchanger. On the overhead there are two condensers. Reflux gravity drains from the second condenser (column 2) to column 1.

- Condenser and Vapor/Liquid Separator - The first condenser is a vertical shell and tube unit connected to a vertical vapor/liquid separator. Process material flows through the shell side and a refrigerated liquid is on the tube side. In the condenser the refrigerated material does not contact the process material. As the refrigerated material vaporizes, it provides cooling to the process. The tube side is directed to the vapor/liquid separator which ensures only vapor is returned to the refrigerated material supply process. The shell side outlet is a two phase stream that continues into the next condenser (column 2).

- Column 2 - This packed column functions as the second condenser. In this column the same refrigerated liquid is introduced as was used on the condenser but in this vessel the liquid contacts the vapor rising through the column. This creates the functionality of a contact condenser.

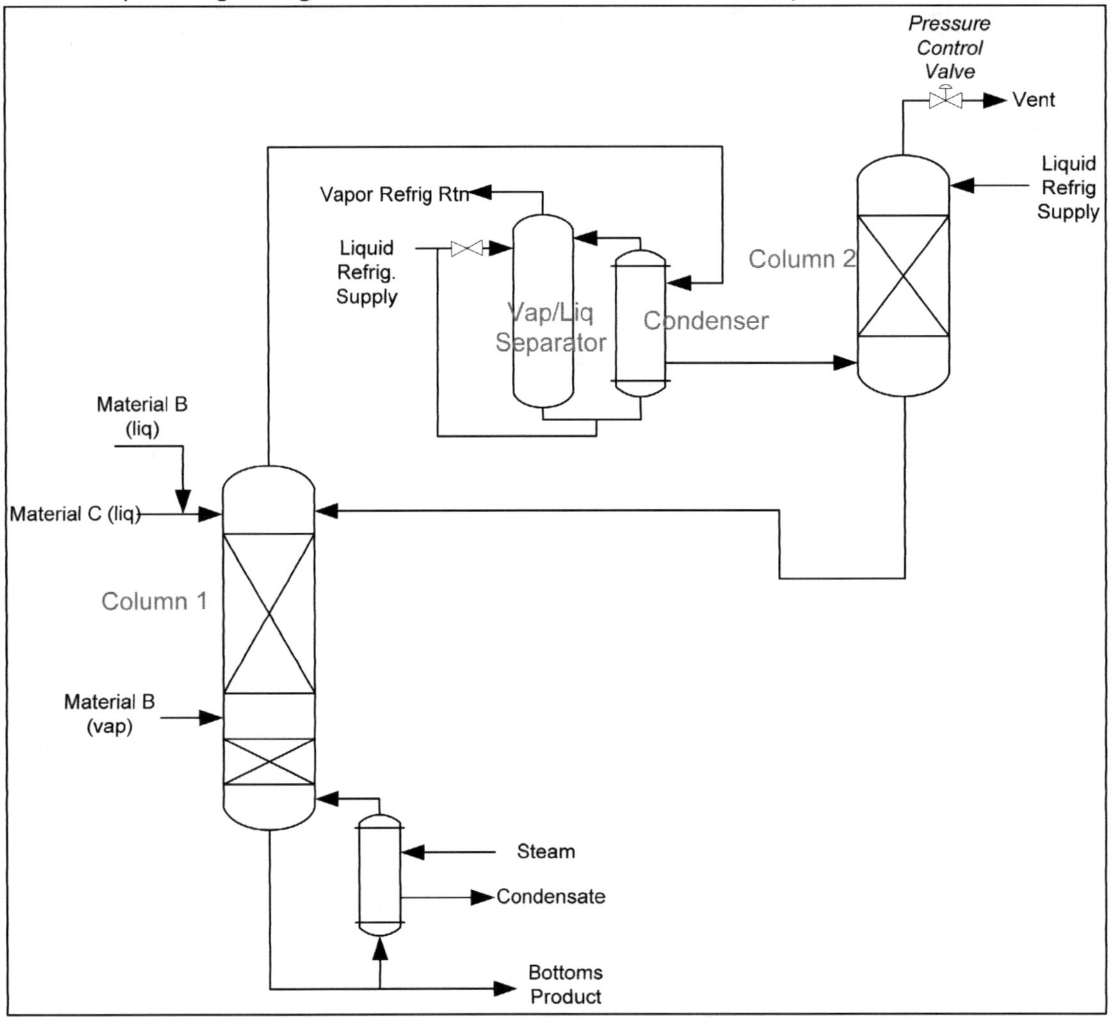

Figure 1: Process Diagram for reactive distillation asset and supporting equipment.

There are three main challenges impacting Material A production; unstable column pressure, steam flow surges and an inability to achieve product specifications in a short startup window. The following sections provide details on the troubleshooting and resolution for each of these issues.

PRESSURE INSTABILITY

Stable column pressure is an essential part of stable distillation operation. Pressure impacts column temperatures which in this application are critical as a way to monitor column performance and reaction conditions. Pressure also impacts hydraulic loading, vaporization, and condensation.

Often while producing Material A (and some other products), the column pressure would cycle and not be controllable with the pressure control scheme. Figure 2 shows plant operating data during the pressure cycling event. The cycling does not occur every campaign but when it does the cycle has a consistent repeating pattern with peaks approximately every two minutes. With this cycling the column temperatures would also cycle and become unreliable to use as process performance indicators. At times the cycling was so extreme that the pressure would drop low enough to allow the sump to flash, rapidly shifting column inventory and loading up the packing. These were significant upsets and it would take time to get the column lined back out, sometimes even resulting in a shutdown.

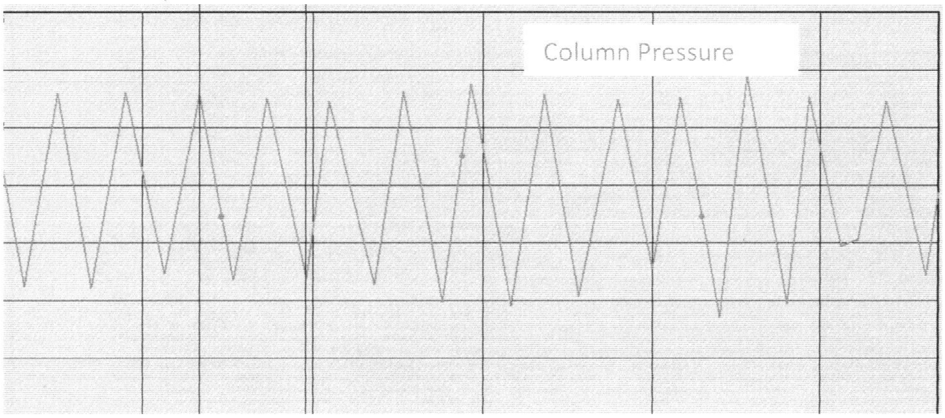

Figure 2: Column overhead pressure readings during cycling period. Peaks are approximately every 2 minutes, consistently, swings 20% of operating pressure

As this problem has persisted for several years many theories have been developed as to why this pressure cycling occurs. The most credible theories were investigated further:

- Tuning Problems – theory was that poor tuning was causing pressure instability. Improper tuning of control loops has been known to result in similar problems controlling pressure. The team conducted trials to develop a new set of tuning parameters for the column pressure control. However the pressure oscillations persisted after the tuning updates were made so tuning was not believed to be cause of the problem. The pressure control valve was even put in manual for a period of time which had no impact on the pressure cycling, which was also a clue something else was dominating the pressure control.

- Mid-boiler Accumulation – Other reactive distillation columns on site experience mid-boiler accumulation with the extreme temperature difference between the condenser and the column bottoms. This tends to trap in mid-boiling species that accumulate over time and have been known to cause column issues. However the problems with this column appear from the start, prior to allowing a mid-boiler to accumulate to a sufficient quantity to cause issues. This theory was dismissed.

- Column 2 Inlet Flooding – The reflux return to Column 1 is approximately 2ft elevation lower than the inlet to the second condenser (Column 2). The theory was the pressure drop in the overheads was causing a higher liquid head in Column 2 in order to gravity drain reflux back to Column 1. If the inlet to Column 2 was flooded, vapor would bubble through the liquid causing erratic differential pressure in Column 2 as the vapor carries liquid up into the packing. This would impact the overall system pressure. Column 2 and the gravity drain line were gamma scanned during stable operation and during a pressure upset. In both cases Column 2 inlet was not flooded. Therefore this theory was dismissed as the possible cause.

- Two phase flow in lines (surging) – The outlet of the condenser has both liquid and vapor flowing in a horizontal section of pipe. The theory was the line was experiencing slugging flow which was causing periodic surging and pressure instability. After investigating further it was found that the line was adequately sized for this two phase flow. Plus the pressure oscillations showed an even amount of time building pressure and then losing pressure. In a surging event, it would have been expected to see a more rapid loss of pressure and a slower increase in pressure. This theory was dismissed as well.

Fixing the Pressure Problem

Working with the operators and reviewing the historical plant operating data, it was discovered that when the temperature setpoint between the condenser and column 2 was lowered, the pressure oscillation stopped. Figure 3 shows plant data and how the pressure steadies as the temperature is lowered. The methodology works every time, but the setpoint needed varied each time. This worked for the other products that were also experiencing the same pressure instability.

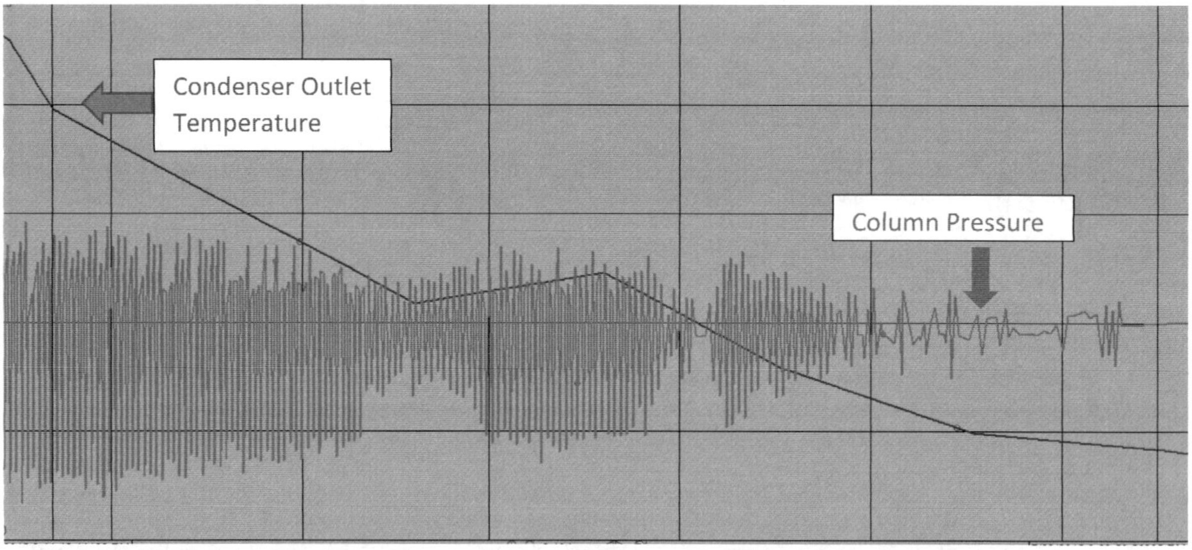

Figure 3: Plant data showing as the condenser outlet is cooled (blue) the column pressure becomes stable (red).

What Was Fixed?

A methodology was developed that solved the problem but the root cause was still unknown. The process seemed sensitive to the condenser outlet temperature but what was changing? Was this even the right variable? To validate that this was the correct variable a trial was conducted going the opposite direction. If cooling down the temperature stabilized the pressure, then warming it up could destabilize things. The condenser outlet temperature was warmed and as predicted the column pressure cycling returned as shown in Figure 4.

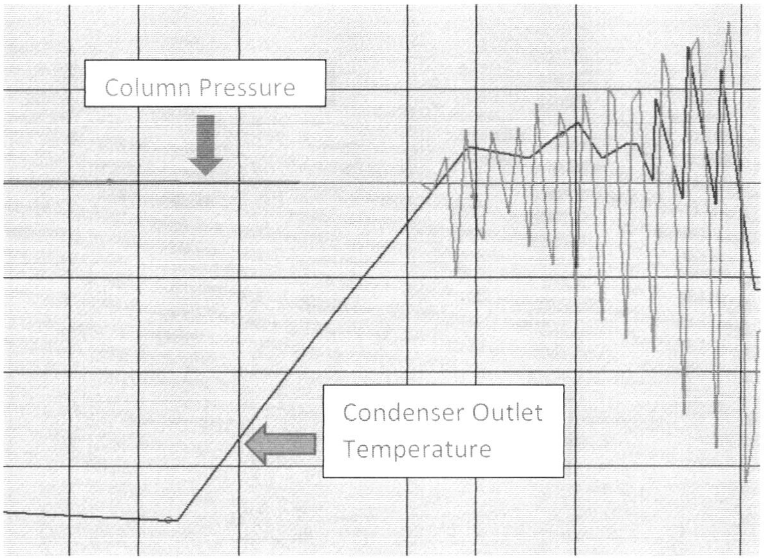

Figure 4: Plant data showing as the condenser outlet temperature increases (blue) the column pressure becomes unstable (red).

Further investigation revealed that the pressure oscillations occur only when condenser was used. Some products bypass the condenser and only use Column 2 to condense the overheads. This caused a more focused look at the condenser and its companion vessel. It was concluded that condenser was meant to operate as a thermosiphon evaporator. A thermosiphon is commonly used as a reboiler with process material on the tube side and steam or some other heat source on the shell side. For this reactive tower, the process is on the shell side and refrigerated liquid is on the tube side. The vapor/liquid separator functions similar to the sump of a column, providing vapor/liquid separation, residence time and ability to provide static head to drive the thermosiphon.

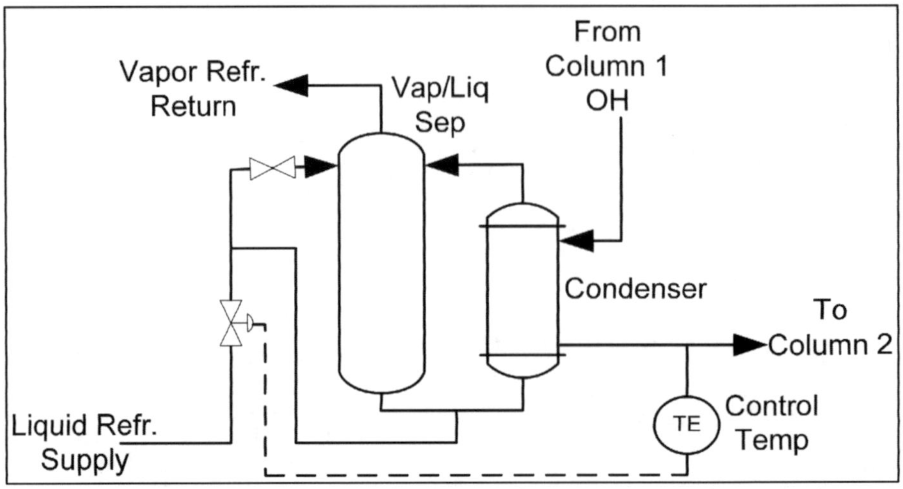

Figure 5: Condenser and Vapor/Separator functioning as a thermosiphon.

With the condenser functionality discovery a connection was still needed explaining how a thermosiphon could be related to the pressure cycling. Leveraging a troubleshooting effort from a separate process, a thermosiphon was unstable and as a result the shell side pressure was also very erratic (see figure 6). Once the thermosiphon vaporization was stabilized the shell side pressure also becomes stable. This thermosiphon was fixed by increasing the sump level so there was sufficient head to drive the thermosiphon circulation.

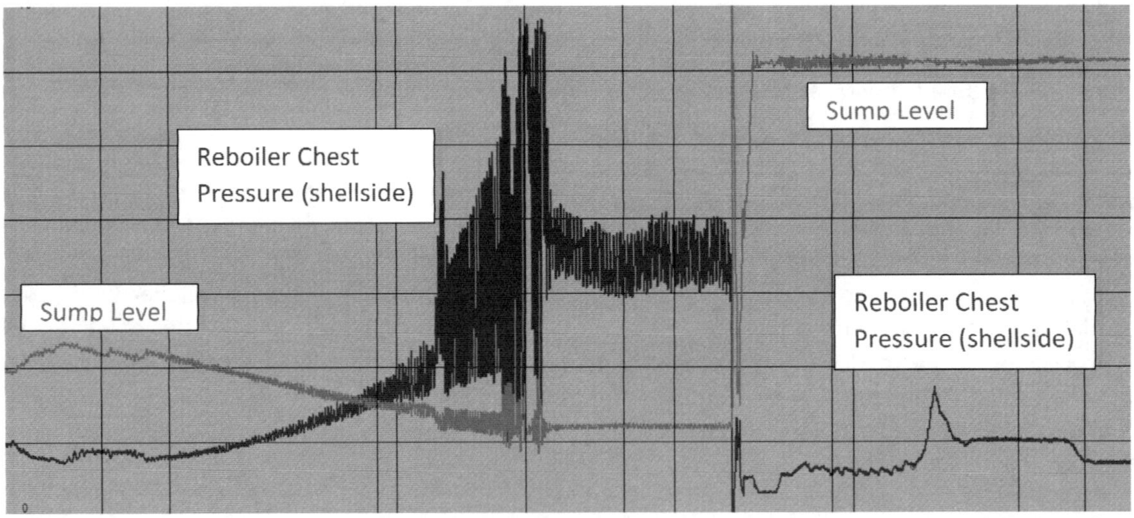

Figure 6: Plant data from a typical steam heated reboiler on a column. Chest pressure (shell side) is shown in blue, sump level is in red. Pressure oscillations stop once level is increased.

The previous example helps put the pieces together to explain why the methodology used to fix the pressure actually works. For Material A production, when the operators implement the solution of decreasing the temperature setpoint on the condenser outlet, the control system responds by increasing refrigerant flow. With more flow the level will increase in the vapor/liquid separator (which acts like a column sump in this installation). This increased level provides head pressure allowing for a more stable thermosiphon and the result is stable column pressure which is on the shell side of the unit. Just like the separate process example shared above, more level in the "sump" fixed the problem and steadied the shell side pressure, in this case the column pressure. Gamma scans on the vapor liquid separator show the level increase in the unit as the temperature setpoint is lowered, which helps validate the theory.

With this new understanding of what causes the pressure instability and the already established methodology to stabilize the column pressure; the process has been able to successfully run with a controllable pressure.

STEAM SURGES

Steam is supplied to the reboiler on the column as the heating medium. At times, the steam flow to the reboiler experienced sudden surges in flow up to 150% of the control setpoint. This would be followed by a loss in steam flow to the reboiler. In the process history steam surges have caused a number of upsets and have affected Material A production.

The issue wasn't unique to Material A, other products made on the column also had issues related to steam. While making another product, the process could only run a few days before the steam valve would open to 100% and flow would drop off. Precedence from previous troubleshooting experience led to the theory that this was an indication of the reboiler flooding with condensate.

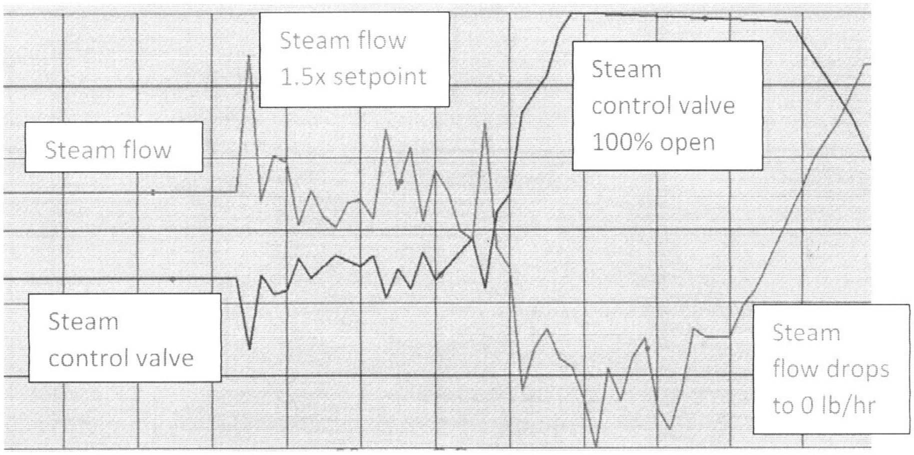

Figure 7: Plant Data showing steam flow (red) and steam control valve position (blue). Steam flow surged above setpoint. Then flow dropped off while steam flow valve went 100% open.

The steam line from the highline to the reboiler was walked down. It was found that there were areas of the existing installation that were not best practice. The main takeoff from the highline came off the bottom of the header, which made the branch vulnerable to condensate from the highline. Then as the piping worked its way through the control valve station and out to the reboiler, there were two large sections of pipe with low points and no way to drain any condensate. These pockets could collect condensate, which could eventually slug to the reboiler flooding in the reboiler shell. Even around the control valve there was a pocket, which could collect condensate and create a false flow measurement. Figure 8 shows a representation of the condensate collection points in the piping. This theory matched up well with all the evidence and historical plant data.

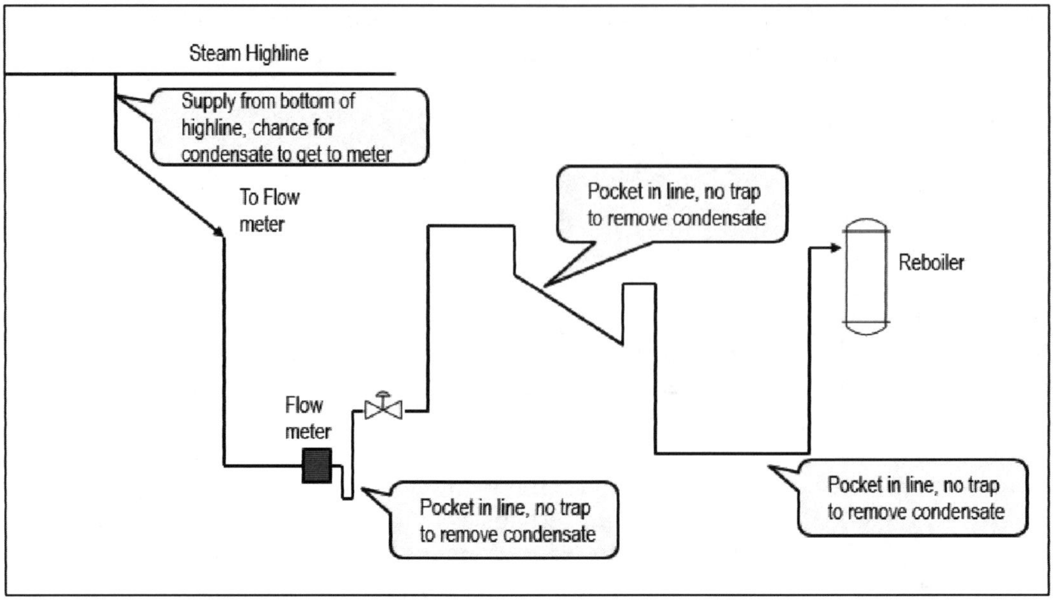

Figure 8: Orthographic representation of steam piping from highline to the reboiler.

Once trapped condensate was identified as the likely root cause, a temporary solution was put in place by adding drip traps at the low points in the steam piping. This gave any condensate a way out without impacting the reboiler. Later a capital project repiped the steam line to the reboiler to eliminate the pockets. Since the solutions have been put in place, there have been no reoccurrences of the steam surges.

OUT OF SPEC STARTUP MATERIAL

At startup all out of spec material leaving the column bottoms is segregated to a collection tank, which has limited capacity. Once the tank is full if the column bottoms is still out of spec the process has to shutdown. Shutdown is followed by a series of cleanout steps to essentially "reset" the column for another try at making Material A. This generates a lot of waste and consumes valuable raw materials. The last remaining challenge impacting successful operation relates to getting product that meets specification before exceeding the startup tank volume.

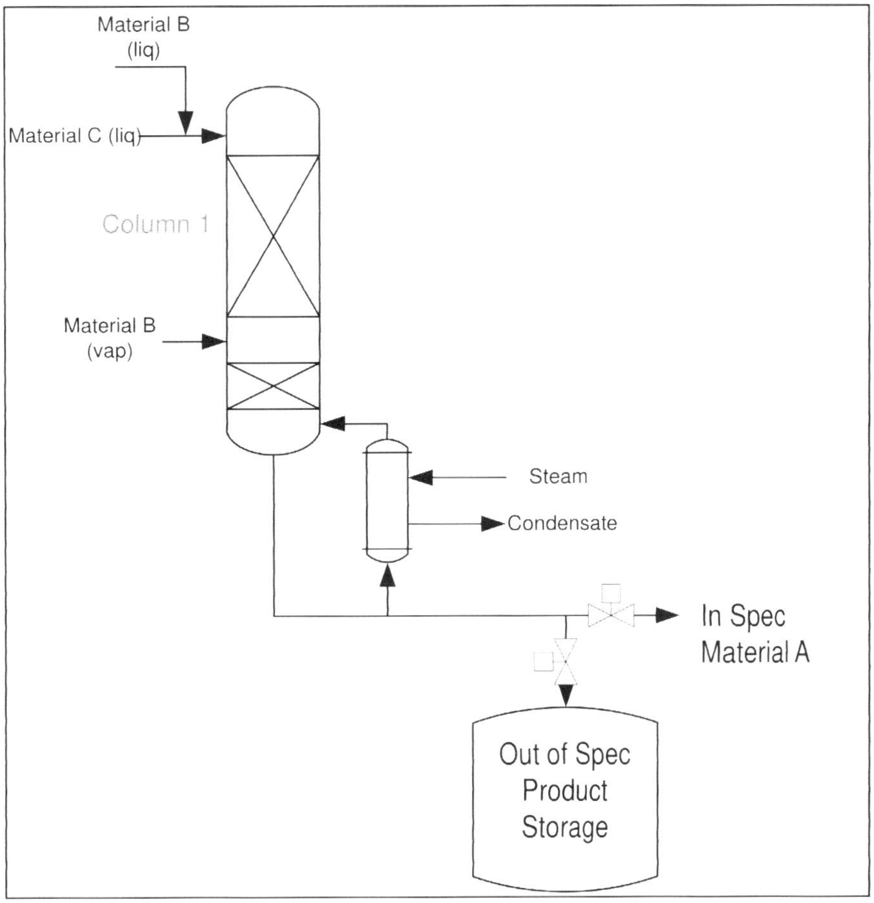

Figure 9: Bottoms Product diverted to holding tank until it meets specification

The startup material is out of spec due to the way the process is intentionally started. To avoid creation of unwanted byproducts, the process is started with less than the stoichiometric amount of Material B needed to complete the reaction. As a result, Material C is only partially reacted which causes the bottoms to be out of spec high on this material. While this is effective at managing side product formation it creates an inventory of out of spec material that must be corrected.

Historically to recover from this condition, more Material B is added in small slow increments above what is needed for the stoichiometric ratio. This recovery period takes most of the startup duration. Sometimes the product cannot be brought within spec by the time the out of spec tank fills.

Since the current solution is to slowly add more material B, a theory was developed to add Material B at a faster rate to see if the process would recover more quickly. Plant trials were completed and it was found that Material A came into spec much faster with this approach. The addition rate was optimized to the point where a 50% reduction in the startup duration was achieved. This addressed the problem and resulted in financial benefits for the process.

EPILOGUE

With the solutions implemented the plant has successfully run several dozen stable campaigns over the last year without any reoccurrence of the previous issues. Better still is the added process control and stability has allowed for extended campaigns achieving higher production volumes.

This experience sheds light on the unique application of a thermosiphon exchanger functioning as an overhead condenser. The paper has shown the challenges that can occur if this type of exchanger is not functioning properly and the impact that can have on the process. For years the production facility operated without understanding how the condenser was designed to operate and as a result went through several unsuccessful attempts to resolve the pressure instability with no effect. However once the true intent of the exchanger was understood and a methodology for control was established, the process was able to develop a robust pressure control strategy.

This study also emphasized the value in walking down a process to look for installations that are out of the ordinary or not best practice that could be causes for process problems. A simple walk down revealed the cause of the steam issue. This is a valuable lesson that valuable troubleshooting knowledge can be gained by a trip out to the process.

Finally as encouragement to future troubleshooters, don't be afraid to step up to a challenge. It doesn't matter how many have come before you and have been unsuccessful, all it takes is the right set of experiences to know how to solve a problem.

REFERENCES

1. Kister, H.Z., "Distillation Operation", McGraw-Hill, NY, 1990

2. Kister, H.Z., "Distillation Design", McGraw-Hill, NY, 1992

Distillation Control – Still an Issue

Sigifredo Nino, P.Eng.

Summa Control Solutions Inc.

Montreal, QC, CANADA

SNino@SummaControlSolutions.com

Presented at the 2017 Distillation Symposium, AIChE Spring Meeting, San Antonio, Texas, March 26 – 30, 2017

DISTILLATION CONTROL - STILL AN ISSUE

Sigifredo Nino, Summa Control Solutions Inc., Quebec, Canada
Email: SNino@SummaControlSolutions.com

Distillation control is and will always be a challenging application for the process control engineer. While not a subject of active research, its relevance remains. Moreover, being a multivariable control application does not make it a perfect fit for Model Predictive Control either. The distillation columns are typically delivered to the end user with the conventional LV controls where reflux (L) and boilup (V) control top and bottom composition respectively. The bigger the reflux ratio, the harder it becomes to control the overhead drum level; it will take an enthalpy change for this dual-composition controls array to interact. That is to say, a change in the reflux flow and the subsequent change in reboiler heat input will result in an oscillation that will only be stopped by operating a single composition in closed loop and leaving the other manipulated variable for the operator, to keep the product composition specifications in check and providing the slack to deal with unmeasured disturbances. The end result is the over purification of the column main product by using more energy than required, and the missed opportunity to optimize product recovery.

When it comes to distillation control strategies, there is no "one size fits all" approach. A clear understanding of the column's behaviour and purpose are critical to the design of a control strategy that can respond to specific requirements. For this reason, distillation control remains, and will likely always be, an issue. Due to the disappearing understanding of Advanced Regulatory Control (ARC), the use of Model Predictive Control (MPC) is currently being favoured, according to McMillan et al. [4]. However, as stated in this article, in which the author collaborated, how can the deficient understanding of a process be superseded by the use of an MPC tool that is not properly comprehended?

None of this should come as news. As the pendulum swings back to remind the industry that it is time to optimize production units, the purpose of this paper is to demonstrate through case studies how material balance controls have been implemented to improve/stabilize several distillation columns. These improvements have invariably resulted in quantifiable economic benefits, with an atypically short timeline for Return On Investment (ROI) measured in days, not months.

Acknowledging MPC's place in process control and understanding that Process Control technologies are meant to complement each other and not to be in competition, one of the presented cases studies is specific on the selection of ARC over MPC.

For this article, a couple of notes are in order: First, since temperature sensors located within the column can be used as composition surrogates, the text may refer to temperatures when mentioning a product composition. Secondly, the distillation column models described in this paper are only suitable for process analysis and control strategy design.

CASE 1: Xylene Tower: Single composition control means one degree of freedom available for optimization

This column is the last in a BTX aromatics train at an oil refinery. It was designed to separate the petrochemical feedstock exiting as distillate product from the ortho-xylene to be removed as vapour sidedraw below the feed tray and the bottom stream directed to the gasoline pool. The bottom sidestream was no longer in use when the ARC optimization was implemented. In consequence, the ortho-xylene distributed along the tower exits with both distillate and bottoms product.

The petrochemical feedstock coming out as distillate has more value than the bottom product. Since there was no specification for the bottom product, it was an opportunity to optimize the distillation operation profitability by bringing the top product non-aromatics impurities to the allowed maximum.

In the original control strategy, top composition was not controlled and bottom composition control was attempted by using the fired heater reboiler Coil Outlet Temperature (COT) with little success, as this temperature is not a reflection of the components in the tower bottoms. The reflux drum level was controlled simultaneously by the reflux and the distillate flow by using a reflux ratio (RR) set up by the operator.

Shinskey's four-component model operating lines and relative gains methodology [8] was used to analyze the possible selections of a single-end composition control pairing, under the scenarios of the predominant operating conditions, and to compare them to the optimization mode that would maximize bottom product recovery into the distillate, as it is the most valuable stream. For completeness sake, that methodology is briefly outlined herein; interested readers will find the details in [7].

The starting point of the methodology is the use of relative gain (RG) as the quantification of the interaction amongst control loops in a multivariable system. Relative Gain metric was proposed by Bristol [1], and its application to a Two Inputs Two Outputs (TITO) multivariable system is a perfect fit for its use in the selection of distillation control configuration.

In a two-loops interacting process control system, there are 2 controlled variables (CV1 and CV2) and 2 manipulated variables (MV1 and MV2). If we designate the difference of a variable between final and initial steady state values by Δ, and the relative gain by λ, the latter is calculated by the ratio of the change in CV1 caused by a change in MV1, when both loops are in manual, over the change in CV1 over the MV1 change when the second control loop is in automatic, as shown in Eq. 1

$$\lambda = \frac{(\Delta CV_1/\Delta MV_1)|_{both\ loops\ manual}}{(\Delta CV_1/\Delta MV_1)|_{loop\ 2\ in\ automatic}} \qquad (Eq. 1)$$

A typical distillation tower has 5 CVs and 5 MVs; the system matrix is reduced to a 2x2 by assuming that pressures and levels are in control and that only the top and bottom composition, y and x respectively, are to be controlled by using 2 MVs. This simplification is possible based on the dynamic decoupling with the other CVs.

The procedure begins by having all the species in the feed distillate and bottoms assigned into a maximum of four components; two of them, Heavy Key (HK) in the Distillate, and Light Key (LK) in the Bottom product becoming the key components to be controlled. The other two are to group the lighter-than-light and heavier-than-heavy keys, LLK and HHK respectively. The theoretical number of trays (N) and the reflux ratio (RR=L/D) are the other two pieces of information required to calculate the relative gains that are presented in a tabular format to facilitate comparison among the possible pre-selected pairings. By combining material balance, Separation factor (S) and the Douglas et al [2] Separation model, the operating curves can be traced to demonstrate graphically how the product compositions change in the neighbourhood of the column products impurities specification, when the independent manipulated variable indicated is kept constant. All curve slopes around the impurity specification visually display what is calculated by the relative gains.

The Separation factor (S) – shown as Eq.2- for control purposes [7] is an extension of the Fenske-Underwood equation that calculates the minimum number of plates in a distillation column operating at total reflux. In this case, it is used to evaluate the split or, feed separation, into its components across the entire column as indicated by the ratio of the light key (y_{LK}) to the heavy key (y_{HK}) in the distillate (D) product composition specification, compared to the same quotient in the bottom (B) stream (x_{LK} and x_{HK}) when the column operates at other reflux conditions

$$S = \frac{y_{LK}/y_{HK}}{x_{LK}/x_{HK}} \qquad (Eq. 2)$$

The Douglas Separation model shown in Eq.3, allows the calculation of RR once the compositions are known.

$$1/RR = \left[\left(\frac{\alpha}{S^{1/N}}\right)^2 - 1\right] \cdot z \qquad (Eq.3)$$

Where α is relative volatility, N the theoretical number of trays calculated by multiplying the actual number of trays in the column by the tray efficiency (N=n×E). Finally, z is the mol-fraction of the light keys in the feed. Efficiency of 70% has been the typical value used in the calculations.

Continuing with the case at hand, the species were distributed into four components. Keep in mind that the model requires that none of the LLK leaves through the bottom, that the LK's concentrate in the D, that the HK's concentrate is in the B, and that neither the HHK's leave through the top nor the LLK's through the bottom.

Accordingly:

- LLK is Toluene
- LK is the mix of Meta-xylene, Para-xylene and Ethyl-Benzene
- HK is Cumene
- HHK is C9+

In a variation in the use of the method, the ortho-xylene was excluded altogether from the distribution amongst the four components considered as it would appear in similar concentrations in the LK and HK.

The Feed, Bottoms and Distillate current compositions in %-mol, number of theoretical trays and RR complete the information entered as shown in the top table on the right hand side of Figure 1. The cells in green are entries and the yellow cells are calculated values. The table below that shows the calculated Relative Gains (RG) with the grey cell indicating that the impurities in the top product (y_{HK}) should be controlled by D/F, and that if the bottom composition was going to be controlled at the same time, then Boilup Ratio (VR) would be the first choice as the calculated RG, is closest to one. However, that would have required the tower bottom level to be controlled by manipulating the fuel gas to the fired heater. There are two quite important drawbacks to this: one, inverse response and secondly the limited operating range in the manipulated variable intended to control inventory.

The Feed, Bottoms and Distillate current compositions in %-mol, number of theoretical trays and RR complete the information entered as shown in the top table on the right hand side of the Figure 1; the cells in green are entries and the yellow are calculated values. The table underneath shows the calculated Relative Gains (RG) with the grey cell indicating that the impurities in the top product, y_{HK}, should be controlled by D/F, and that if the bottom composition was going to be controlled at the same time then Boilup Ratio (VR) would be the first choice as the calculated RG is the closest to one; that however would have required the tower bottom level control to be done by manipulating the fuel gas to the fired heater, against two quite important drawbacks: inverse response and limited operating range in the manipulated variable intended to control inventory.

The D/F ratio is 0.852, is in agreement with the intended operation of the tower to remove heavy keys from the petrochemical feedstock that exits from the top of the tower. The relative volatility does not have much practical meaning in this particular case, due to the multiple components of the feed and the exclusion of the ortho-xylene. This value can be used to run other scenarios such as the changes in product specifications and its implications in RR and energy requirements.

Figure 1 also shows that the DV strategy can be used to control the separation of the heavy components from the petrochemical feedstock. The nearly horizontal D/F means that by keeping this ratio constant, the top composition will stay very close to the specifications. The lack of specifications for the bottom product ruled out the otherwise severe interaction associated with dual-composition controls should the conventional LV pairing was used, as can be realized by the calculated RG of 156.

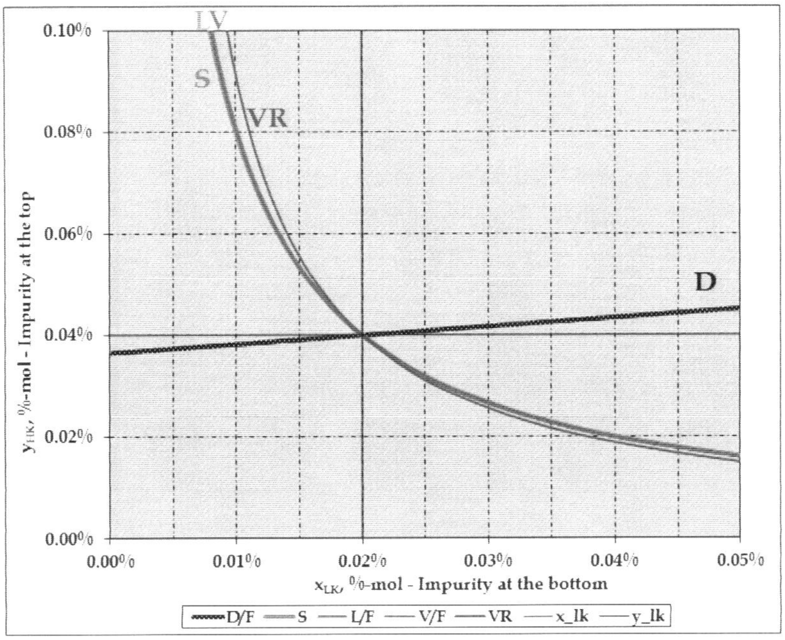

KEY		Feed	Distillate	Bottoms
		z%	y%	x%
Lighter	LLK	0.200	0.235	
Light	LK	85.000	99.725	0.020
Heavy	HK	14.000	0.040	94.563
Heavier	HHK	0.800		5.417
Theoretical Stages		100	D/F =	0.852
Reflux Ratio (RR)		1.5	α =	1.571

Relative Gains		MV2→y		
		D [D/F]	L [L/F]	S [L/D]
MV1→ x	B [B/F]		0.079	0.080
	V [V/F]	0.921	156.011	62.957
	VR [V/B]	0.928	10.186	9.366

Figure 1 - Xylene tower - Specifications, RG and Operating Curves

The single-end composition control applied to the bottom stream released one degree of freedom and opened the door for what Shinskey [8] calls "local optimization" opportunities in distillation. When addressing this issue, he proposed an objective function that combines product losses and utility cost per unit feed. This shows that objective function minimization reduces the loss of the most valuable product and minimizes the heat input cost.

Extracted from the numerical application of this objective function as used in the Example 11 [Ref. 8, p.323], the relevant result for the purposes of this writing, has been plotted in Figure 2 to graphically demonstrate how the low Reflux/Feed ratio (L/F) sensitivity to feed composition changes can be used in the maximization of impurities in the top product. Explicitly: Once the optimum L/F ratio is found and maintained constant, it will keep the top composition nearly constant despite the feed composition variability. The arrows indicate the direction of different variables increase.

Because the numerical values of the example are not applicable here, the direction the different variables increase has been indicated by arrows.

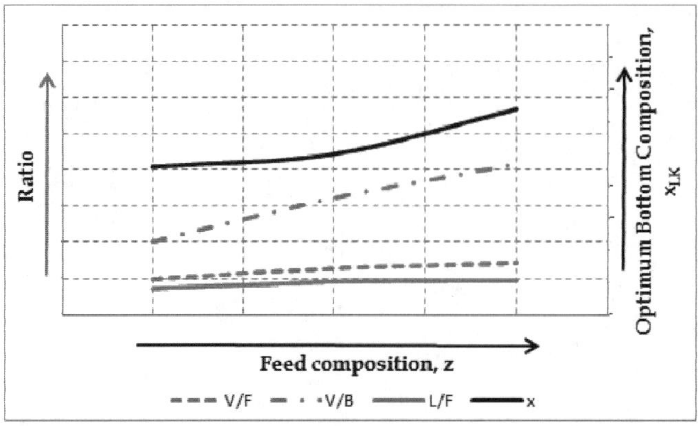

Figure 2 - Typical ratios Sensitivity to feed composition changes in a distillation column

The final control strategy, combining the top composition control and the optimization strategy is depicted in Figure 3. It includes two distinct dynamic compensation blocks: one in the direction of the top composition control and the second towards the reflux optimization loop. The high and low fuel gas valve openings are guaranteed by limiting the operating span of the reflux optimization controller (indicated by "Opt" in Figure 3), and a switch is available for the operator, should the fired heater operation controlling the coil outlet temperature is required, e.g. Cold Start-up. Although not shown in the drawing, there is a low oxygen control override to prevent the formation of fuel rich atmosphere in the fired heater; pass balance control was not required for the furnace in the application due to the inherent uniformity of temperatures and flows in the different passes.

Better control of the top of the column allowed the use of the optimization strategy to bring impurities from 0.04 up to 1.6, causing the D/F to increase. This additional volume produced at the top, multiplied by the difference in value between the petrochemical feedstock and the gasoline pool grade bottom product, accounts for about 70% of the total benefits. The remaining 30% of the profit is coming from the lesser V/F ratio resulting from the reduction the L/D ratio, directly reflected in fuel gas consumption reduction in fired heater reboiler. The total savings are in excess of 2 MM$/yr. for an ROI of 12 days. It can be reported that 5 years after implementation, the strategies are still in use, requiring only the routine field devices maintenance activities.

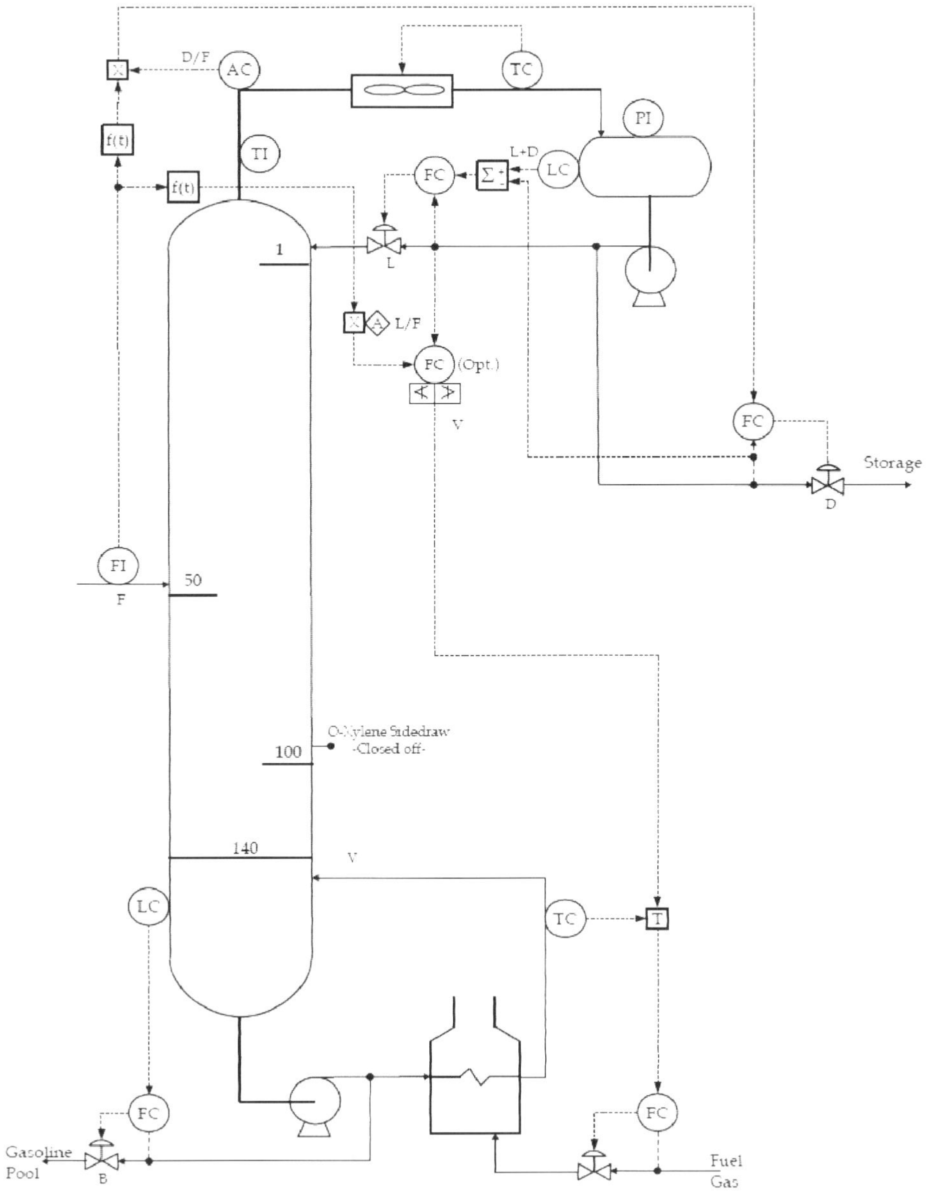

Figure 3 - Xylene tower - Control Strategies

CASE 2: Benzene-Toluene Heat Integrated columns – Why was ARC chosen over MPC?

A paper detailing the implementation of ARC-based dual-composition material balance control in a high-purity distillation heat integrated Benzene-Toluene tower application was published by this author in 2011[5]. The untold story, thus far, is that the ARC option was presented to the customer by a seasoned MPC engineer whose experience has been that in high purity distillation and other columns with long dynamics, it is quite difficult to implement model-based multivariable control and would still not perform as well as with ARC.

Dual-composition control in distillation columns has always been a challenging application. However, once the strategies are properly formulated based on the understanding of the interaction between material and energy balance in distillation, as well as the interaction associated with the CV-MV pairing, it is possible to select an array of control strategies to minimize interaction without sacrificing either robustness or performance of the operation.

There were multiple challenges in this application. Despite requiring a double-end composition, being heat integrated and having a ternary sidestream, amongst others, all difficulties were successfully overcome. After a recent visit to this installation, it can be reported that the application is still fully functional with the same control arrangement and tuning of the PID and dynamic compensators used in the commissioning over six years ago.

The pairing analysis done on the benzene tower by using Relative Gain (RG) showed the conventional LV – or energy balance to be the worst with an RG of 23.7 – 1.0 being optimal; compared to the material balance SV strategy that had a value of 0.998. The energy balance strategy instability driven by the LV based double-ended composition interaction was not affecting the benzene, but also the toluene tower through the heat integration.

One of the most counter-intuitive loops configured in the benzene tower was the control of the LK impurities in a tray below the feed tray. This configuration required the use of lead-lag compensation and the use of a considerable derivative time in order to improve the performance of the interacting-lag dominated dynamic response of the composition variable. This fact, along with the closure of the reboiler/condenser bypass valve highlights the importance of the thorough understanding of the process that must precede any control implementation.

As for the toluene tower, its RG analysis for the composition targets (Table 5, [5]) produced a 2.347 for the LV and 0.912 for the DV. Though it does not look like a sizeable difference, the biggest advantage of the material balance is that direct manipulation of the ratio D/F would result in a tighter and more robust option to minimizing the xylene losses. This advantage stems from the almost horizontal positioning of the material balance line showing that very tight control of the impurities at the top of the column is possible by keeping the D/F ratio constant. The inconveniently slow dynamic behaviour associated with the DV strategy is due to the lag imposed by the reflux drum. However, the use of the so-called Van Kampen decoupler [10] not only eliminates the problem but it can also convert the overhead accumulator into a lead, bringing the benefits of improved dynamics

Even though the original goal at the petrochemical facility was to bring stability to the operation of these towers that needed constant attention from the operators, the application ended up de-bottlenecking the operation and producing quantifiable benefits in excess of 1 MM$/year; for an ROI of 25 days. Also, it brought a noticeable increase in the much-sought stability of the operation.

CASE 3: Debutanizer – "The simulation shows that the conventional LV controls would work"

This tower was installed in an oil refinery. On one hand, its purpose is to control the volatility of the gasoline (Reid Vapour Pressure, RVP) during the warm months by limiting the amount of butanes in the bottom product. On the other hand, it had to ensure that the amount of pentanes in the distillate product were at a minimum.

The debutanizer was designed and commissioned for dual composition using the conventional energy balance configuration, LV. The decision to use LV as the control strategy was based on a computer simulation that determined this option would work. However, after several weeks of fruitless attempts to stop the continuous 1.5 hours period oscillation in both compositions, as depicted on the left-hand side of Figure 4, the refinery decided to implement a material balance control strategy.

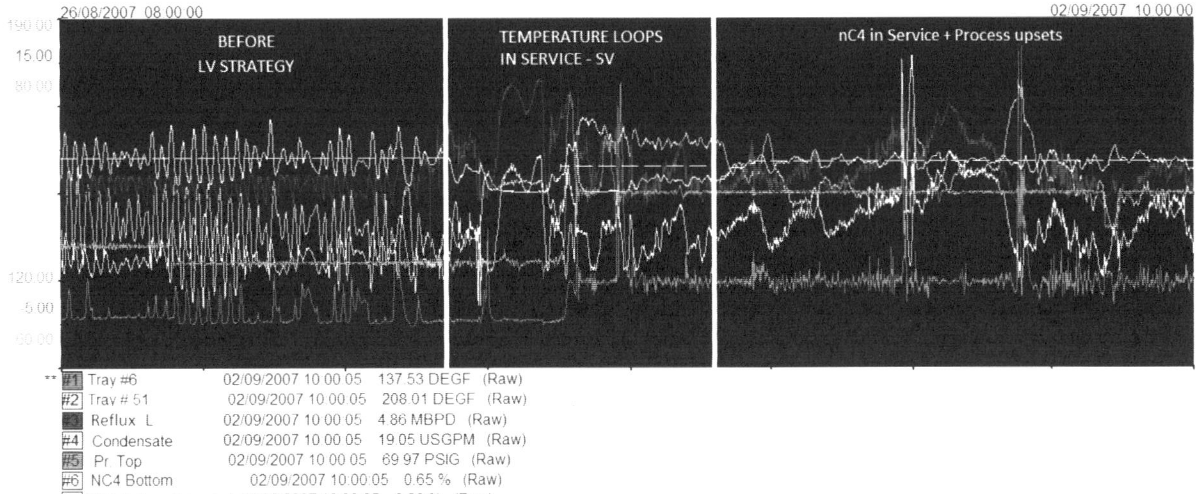

Figure 4 – Process Variables Trend – Before and After

As all the information required by the Shinskey's four-component model was available, it was possible to produce the relative gains and the corresponding operating curves. The following data was used for the specifications table: LLK is propane, LK is normal plus iso-butane, HK is normal plus iso-pentane, HHK is the balance.

The proximity of the L and V curves, shown in Figure 5, explains the strong interaction between the temperature control loops, confirmed by the value of 13.37 in the relative gain calculation. The chosen strategy is the SV as indicated by the grey cell in the RG table. The S-VR could have worked out as well; however, that would require controlling the tower bottom level with the heat input, which has the tendency to have an inverse response.

The control strategy, as implemented, can be seen in Figure 7, was proposed in 1981 by Ryskamp [6], and it features explicit and internal decoupling. The explicit decoupling between the reflux drum level control and the top composition control is performed by the summing block; when the top temperature controller changes the distillate flow (D), the reflux flow (L) will change by the same amount in the opposite direction. This way, the inventory remains constant. There is a partial internal decoupling between the controlled temperatures with the multiplier in the top temperature composition control circuit, as the RR stays constant when the controller output remains constant, thus preventing the top product composition from changing whenever there is a change in the heat input caused by the actions of the bottom temperature control.

Regardless of the PID optimal tuning parameters, the boundary will always be the proximity to the stability limit; once a controller has been brought up, the best practical performance limit can be pushed further by using feedforward control. Following the "adding only the required complexity provided that it outweighs the benefits" philosophy, the feed rate was added to the bottom temperature control loop as a feedforward through a dynamic compensator to pace the steam following a change in the feed.

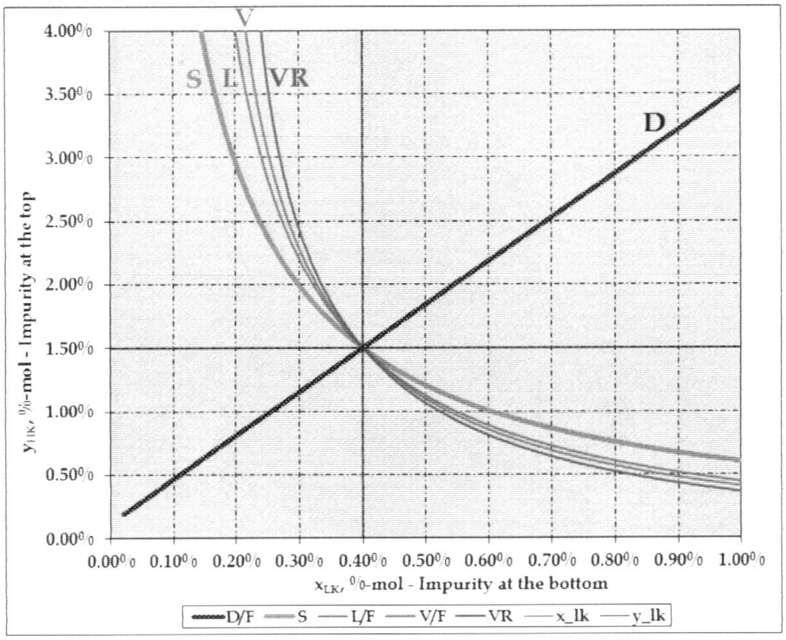

KEY		Feed	Distillate	Bottoms
		z%	y%	x%
Lighter	LLK	0.500	2.219	
Light	LK	22.000	96.281	0.400
Heavy	HK	50.000	1.500	64.103
Heavier	HHK	27.500		35.497
Theoretical Stages		35	D/F =	0.225
Reflux Ratio (RR)		3.7	α =	1.932

Relative Gains		MV2→y		
		D [D/F]	L [L/F]	S [L/D]
MV1→x	B [B/F]		0.412	0.481
	V [V/F]	0.607	13.371	3.323
	VR [V/B]	0.633	5.887	2.683

Figure 5 Debutanizer - Specifications, RG and Operating Curves

Although not explicitly illustrated in the column sketch, top and bottom controlled temperatures were Pressure Compensated by using the linear correction [8]:

$$T_c = T_{meas} - \left.\frac{\partial T}{\partial p}\right|_x \left(P_{meas} - P_{ref}\right) \quad (Eq. 4)$$

Where: T_c is the compensated temperature, T_{meas} is the measured temperature, P_{meas} is measured pressure, P_{ref} the reference pressure and the partial derivative is the slope of the vapour pressure curve as the normal product composition.

The tower propensity to flooding prompted the inclusion of a column differential pressure override (ΔPC-2 in the figure) in the strategy implemented in order to stop the heat input from increasing when the pressure drop across the column nears the incipient flood point.

The column came with three final control elements capable of controlling the tower pressure, which is not only an over-specification of the system but also the conflicting operation of variables that are not independent. This issue was highlighted at the time but was addressed only after the completion of this project.

Even if the end products are going to storage tanks, where a sustained oscillation in the product composition can average out, the cycling in the distillation column product compositions produces non-symmetrical oscillations in the energy input due to the nonlinear relation between energy and purity. Figure 6 depicts this situation as applied to this tower; assuming a uniform oscillation of 0.5% around the 1.0% impurity setpoint, it can be seen that much more energy is required to over-purify than to yield a lower column product concentration

.

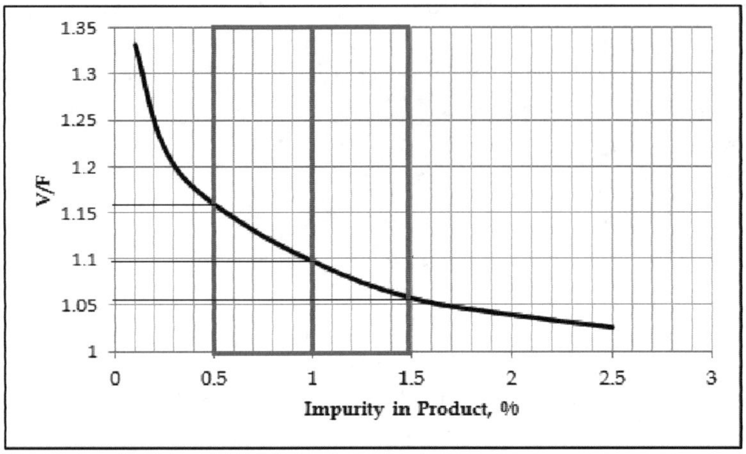

Figure 6 -Product Impurity VS. Energy Demand

The progressive improvement in the column captured in Figure 4 shows the transition toward the middle zone is probably the most impacting as the sustained oscillation in the column stopped immediately following the change of the top composition control strategy from reflux to reflux ratio. The right-hand side illustrates the steady state behaviour of the column operating under dual-composition control including instances of the ability of the control strategies to recover from upsets triggered by unmeasured disturbances.

As a result of this successful implementation of dual-composition controls, butane composition variability was reduced by almost 60%, making easier to maintain the RVP targets. The energy requirements were cut by 12%, or about $70000/yr. in a column that operates only 6 months per year. Lastly, because the tower was operated closer to the specifications, there has been a decrease of 14% in butanes lost to the bottom product.

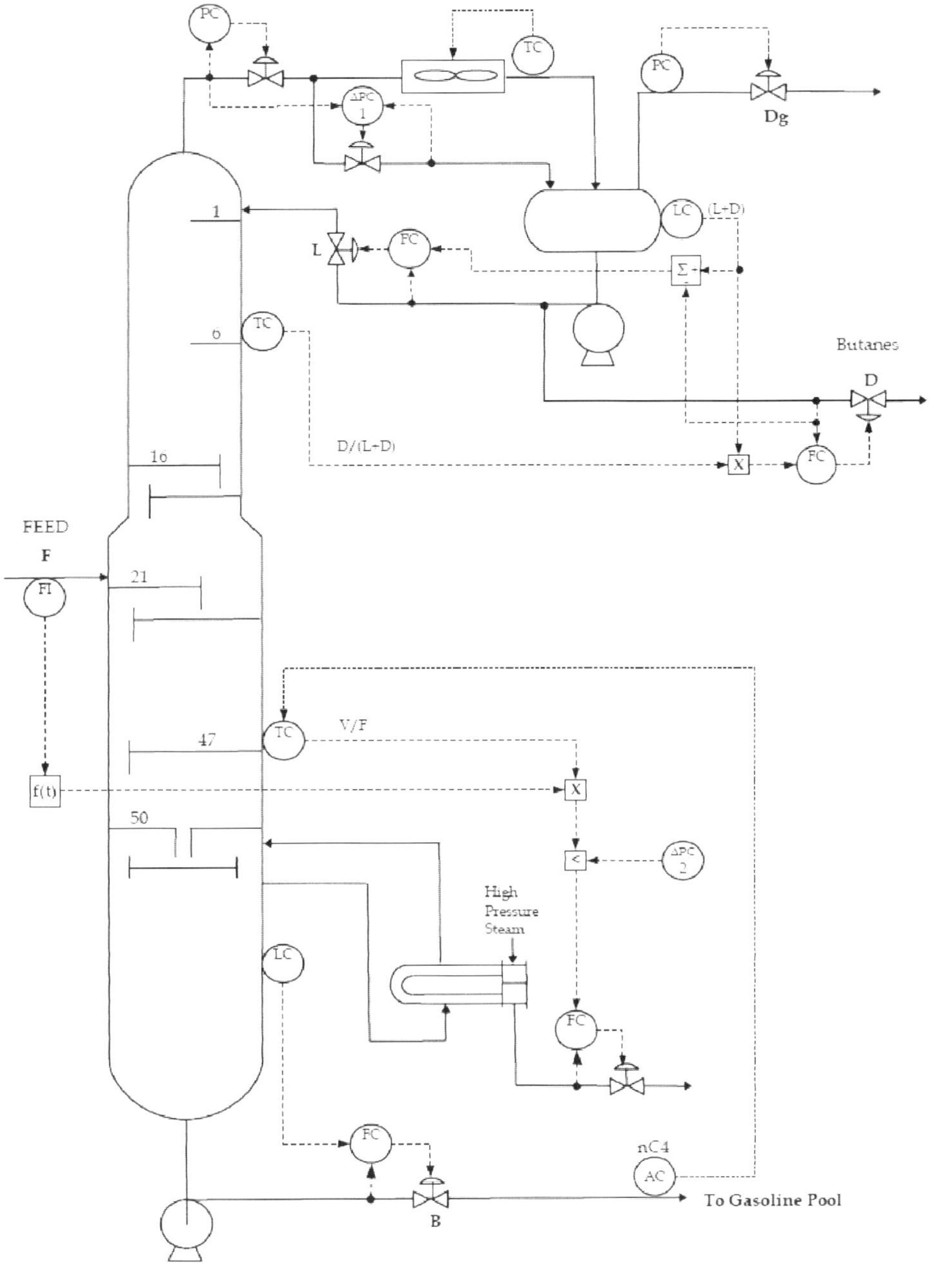

Figure 7 - Debutanizer - Control Strategies

FINAL REMARKS

Multiple applications of sound control practices applied to distillation columns, like those presented in this paper, have proven time after time that benefits can be accrued from improved stability and optimization.

Process control optimization can be the most lucrative investment, in terms of ROI, that can be done given that investment size is not as much as the level required for capital projects. An eye-opening Benefits/Investment ratio exceeding 10:1, achievable through process control engineering, should certainly be in the interest of any operation manager looking forward to maximizing the profitability of a process unit.

Optimization can be accomplished in single-composition controlled distillation column by minimizing the cost associated with its utilities and maximizing product recovery per unit feed.

Even if sometimes and oscillating PV can average out, the product composition-to-energy consumption nonlinear relation in a distillation column, causes a net excess in the heat input to the tower. Therefore, the elimination of oscillatory behaviour can result in immediate economic benefits because the cycling will cause the energy usage to be higher than it would be if the operation took place at the average values and compositions in steady state.

Distillation control is not beyond linear Advanced Regulatory Control (ARC). However, the added complexity in a control strategy is only justifiable if the benefits outweigh it. Simple controls work better and last forever - or until any of the field devices fail!

The estimation of the benefits of an optimized stable operation is difficult. However, in this author's experience, the optimization task is always profitable for the companies. Sound engineering based control applications have shown that once stability is attained, the opportunities for improvement will appear. In most cases, the benefits are well beyond expectations.

Once engaged in a plant project, the process control expert will have to find the way to work with the primary and final control elements that are already in place, setting aside the nice-to-haves. An initial survey can help, but unexpected situations will always arise.

MPC certainly has a place in process control; not as a rival to ARC but as a complement. The alleged ability of the MPC to work better than ARC, when there are complex dynamics involved, has always been baffling to the author; simply put, there are transfer functions that are uninvertible.

ACKNOWLEDGEMENT

The author wishes to dedicate this paper to F. G. Shinskey in recognition of his mentorship and support for over two decades, with the hope that his contribution to process control engineering will never be forgotten.

BIBLIOGRAPHY AND REFERENCES

[1] Bristol E. H. (1966) "On a New Measure of Interaction for Multivariable Process Control" IEEE Trans. Auto. Control, AC-11(1), p.133

[2] Douglas J. M., A. Jafarey, and T. J. McAvoy (1979):"Short-Cut Techniques for Distillation Column Design and Control. 1. Column Design", Ind. Eng. Chem. Process Des. Dev., V.18, No.2, pp.197-202

[3] Gordon, L. M. (1987) "Using Relative Gains to Design Dual Composition Controls". In Handbook of Process Control Systems and Instrumentation, Les Kane (Editor) Gulf Publishing

[4] McMillan, G. and S. Weiner (2015) "Is PID the algorithm of the future?" Control, June 2015, pp. 55-56

[5] Nino, S. (2011) "Improve material balance in high-purity distillation control". HPI, November

[6] Ryskamp, C.J. (1980) "New strategy improves dual composition column control". HPI, June.

[7] Shinskey, F.G. (1977) "Distillation Control" Second Edition, McGraw-Hill, New York, pp. 41

[8] Shinskey, F.G. (1984) "Distillation Control: For Productivity and Energy Conservation" Second Edition, McGraw-Hill, New York

[9] Shinskey, F.G. (2009) "Multivariable Control of Distillation", Parts 1, 2 and 3. Control Global, May, June and July

[10] Shinskey, F. G. (2012) "Control Systems: Investment and Return". Expertune website

[11] Van Kampen J.A., (1965) "Automatic Control by Chromatographs of the Product Quality of a Distillation Column". Proc. Instn. Mech. Engrs. 1964-65, Vol. 179 Pt. 3H

Details Matter When Designing Hot Vapor Bypasses

Daryl Hanson

2017 AIChE Spring Distillation Conference
March 26 – 30, 2017 – San Antonio, TX

Background

- **Hot Vapor Bypasses (HVB) are used extensively in the process industries**
- **Three key variables – two are controlled (independent), the other dependent**
  - Pressure
  - Temperature
  - Composition
- **Distillation primary utilizes pressure & temperature control**
  - Once these two are controlled, relative volatility is fixed for absorption/stripping
- **HVB systems have been designed successfully many times and are the primary method for pressure control in totally condensed systems**
- **Keys to designing HVB systems are in public domain papers**
- **Present two case studies that resulted in instability**
  - Hydrocracker Fractionator
  - Naphtha Hydrotreater Depentanizer
- **Testing indicated problematic area of design**

Hot Vapor Bypass Basics

- **PC is used to backpressure the tower – Primary Pressure control**
- **PDI is used to --**
  - Set condensation mass rate
  - Provide heating to return liquid surface to bubble point
- **System works whether condenser is elevated or submerged**
- **Many rules of thumb that form basis for successful system**
  - Control stability
  - Determining capacity of PDI
  - How to insure no- to little-mixing occurs in the OH accumulator

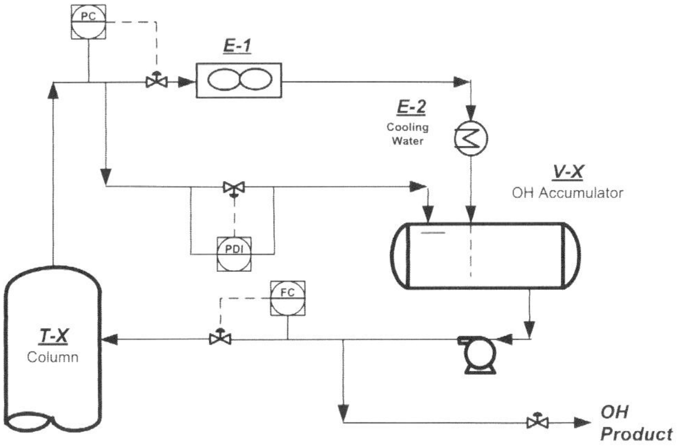

 3

Hot Vapor Bypass Basics

- PC is used to backpressure the tower – Primary Pressure control
- PDI is used to --
  - Set condensation mass rate
  - Provide heating to return liquid surface to bubble point
- System works whether condenser is elevated or submerged
- Many rules-of-thumb that form basis for successful system
  - Control stability
  - Determining capacity of PDI
  - How to insure no- to little-mixing occurs in the OH accumulator

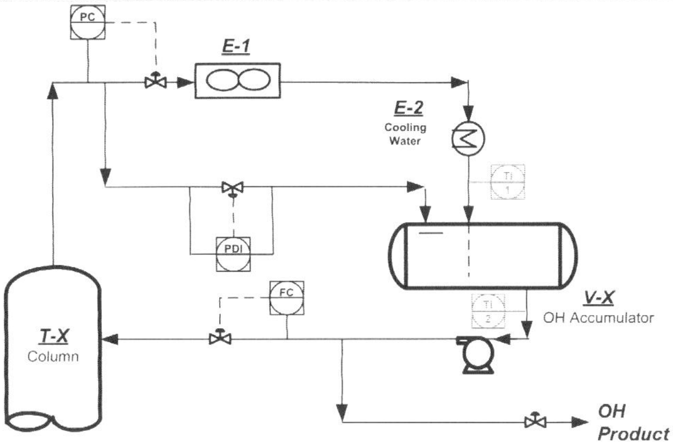

- **A HVB bypass may be stable, but is it working properly?**
- **Look at drum temperature stability**
  - Condenser effluent temperature (TI-1) should be the same as the OH Accumulator temperature (TI-2)
  - Does HVB bypass valve position seem to 'track' with TI-2 OH Accumulator temperature?
- **Is PDI control sluggish? It should be fast acting due to high energy content of vapor**
- **Some systems may work with HVB vapor and Condenser Effluent mixing, but our cases illustrate what can go wrong**

 4

Case 1: Hydrocracker Fractionator

- **In 2014, Valero revamped the Hydrocracker unit at the Meraux Refinery**
  - Feed rate increase
  - Conversion increase
  - Extensive reactor, separation, & fractionation scope
  - Details in 2016 AFPM paper AM-16-05 (*Moreland, et al., 2016*)
- **Fractionation scope involved tray capacity replacements, additional condensing duty for light products, new product rundown circuits**
  - Naphtha yield increased by 50%
  - Kerosene plus Diesel yield doubled
- **Scope for the Main Column was a targeted revamp**
  - All internals were retrofitted; structured packing and high capacity trays
  - Overhead air cooler capacity was increased
  - Reflux pump changes (swap of services)
- **No changes to the overhead accumulator or internals**

5

HCU Main Fractionator

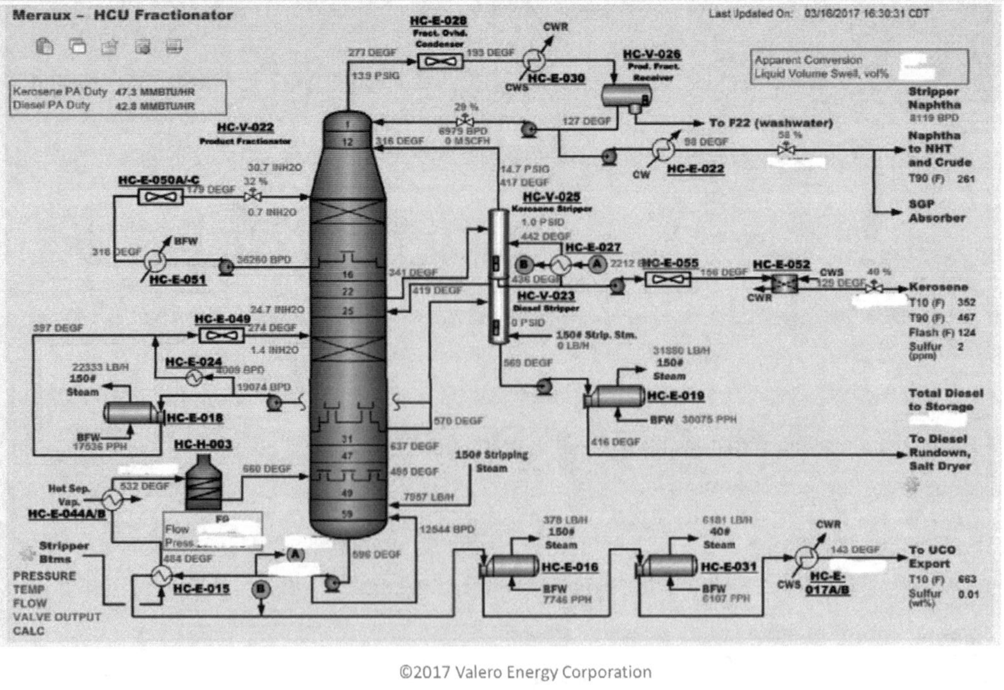

6

HCU Main Fractionator

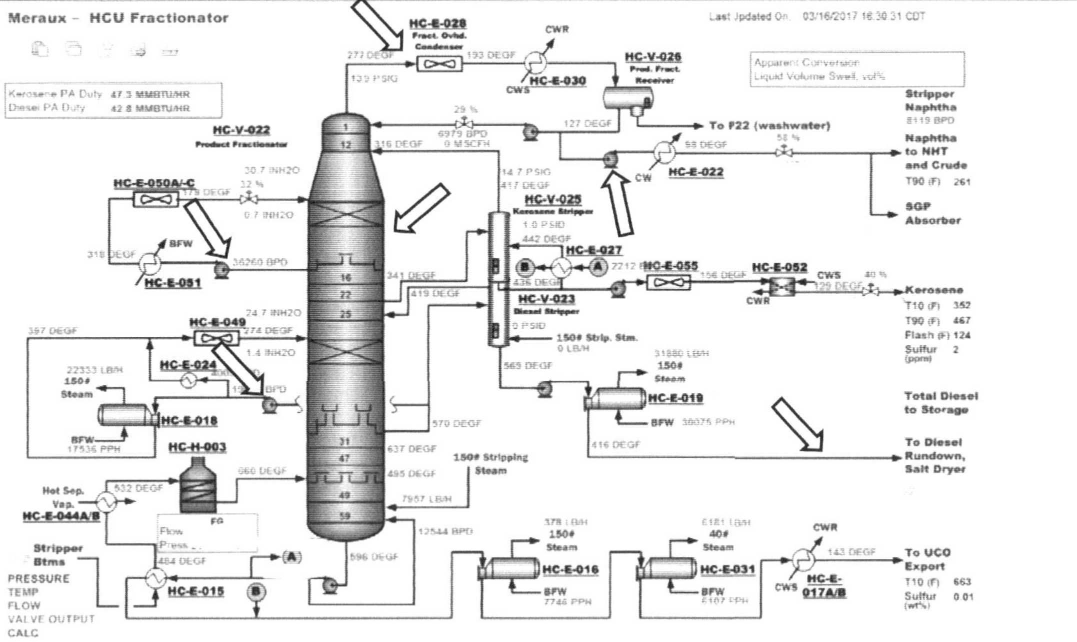

HCU Fractionator Overhead System

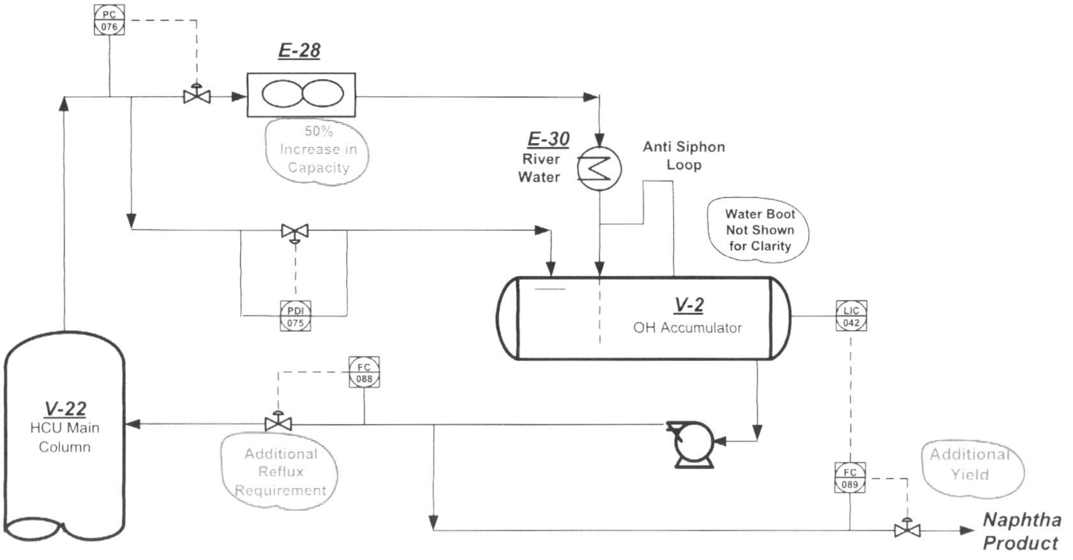

Design & Pains During Startup

DESIGN

OPERATION

- **Drum residence time was checked**
  - No changes were made to the drum, loading went up 35%
  - Detailed E&C company did not know about the slotted pipe issues
- **Hot vapor bypass & condensing branch hydraulics were confirmed**
- **Unit feed rate was increased and conversion was about half of design**
  - Operations reported 'shaking' on the 2nd deck
  - Severe instability was cited by the board operator in the overhead circuit
- **Startup progression was halted until problem was resolved**
- **Trip out to the 2nd deck showed**
  - Trim condenser was not stationary, entire 2nd deck would shake
  - Cyclic, severe 'banging noise' was being emitted from the trim condenser
  - Noise was so severe that double ear protection was required (>100 dB)
- **Severity was enough that discussion of unit shutdown was entertained**
- **Troubleshooting team was immediately activated**

HCU OH System Summary

- **Troubleshooting focused on the accumulator and condenser**
- **Drum had slotted pipe, capped end**
- **Standard design from the licensor**
- **OH liquid rate was approximately 8 MBD**
- **Liquid velocities**
  - OH effluent pipe: approx. 0.6 fps
  - Velocity out of slot (full height): ~0.4 fps
  - Velocity out of slot (50% uncovered): ~0.8 fps
- **System had run very close to bubble point in the past, with only 5-10 deg F subcool**
- **In theory, slot is to give you the flexibility handle vapor from the condenser**
  - Remainder of the system is not
- **Water Trim Condenser was not modified during outage (36" x 2000 sq.ft)**
  - Volume approx. 40 barrels

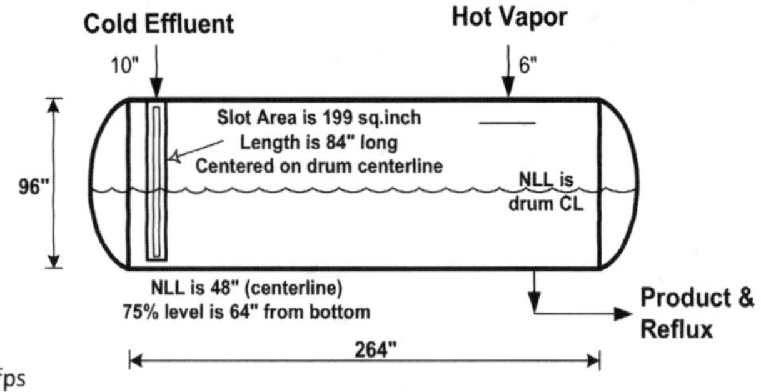

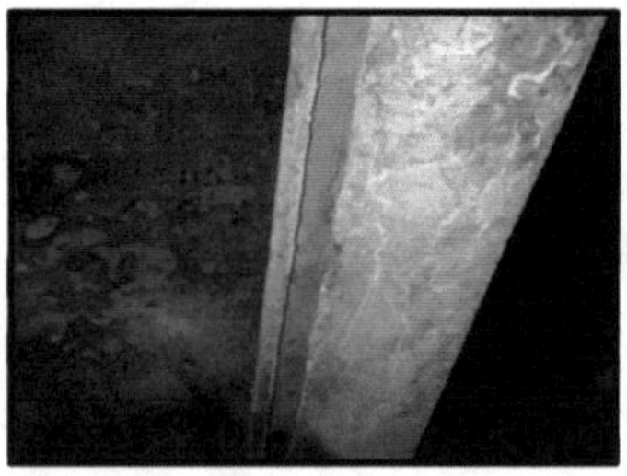

Troubleshooting Data

- **System pressure drop was not constant**
- **Hot vapor bypass was struggling to respond quick enough to very abrupt changes in the drum pressure**
- **Temperature changes in the OH drum, as HVB bypass was opening, signaled to us that vapor and cold liquid were mixing**
  - This also explained the loud banging noise
- **Rate of liquid drop (condensation & static) is staggering approx. 1000 bbl in 5 minutes**
  - Multiples of normal condensation rate
- **Key points**
  - Pre-12:07, vapor breakthrough is occurring, HVB dP indicates drum is dropping in Pressure
  - 12:10, after enough slot area is sealed up, the drum temperature immediately drops.
  - Drum pressures re-stabilize (pressure gain is 5 psi) which is about 20' liquid

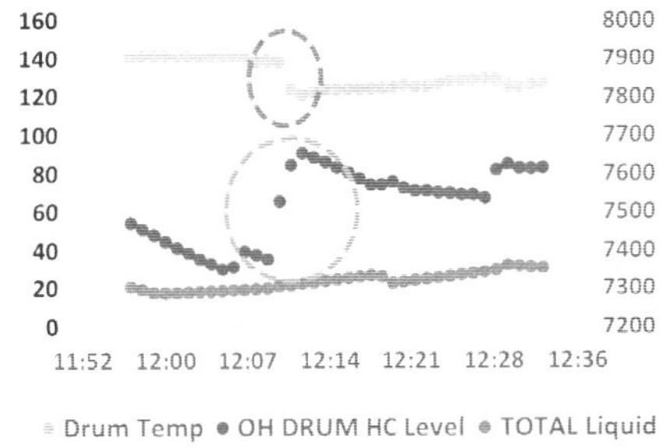

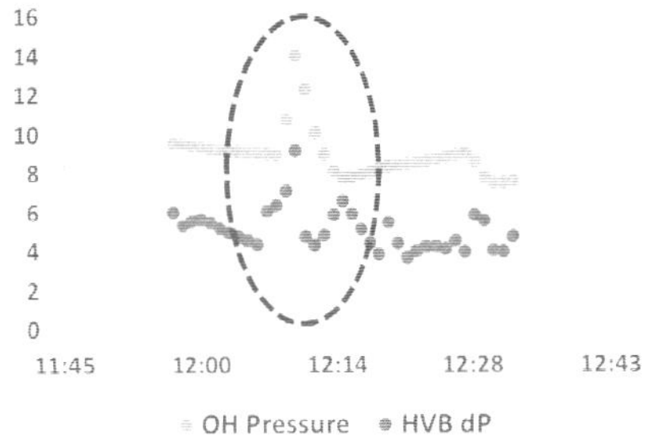

Covering Slot was not Sustainable

- **Vapor up-flow theory was realized by physically touching the effluent line during a cycle**
  - Temperature would increase and then decrease in concert with noises
- **During the drum filling, it was noticed that the PDI, level and temperature would become erratic**
- **Sealing up the slot would 'eliminate' or 'reduce' that from happening**
  - Slots were ~6" from the top of drum
- **Liquid level was increased, but became highly erratic as the pressure balance change was in a dynamic state; other events occurred simultaneously**
  - Hot vapor bypass dP became erratic (± 2 PSI)
  - Overhead condensing rate was erratic (± 5 %)
  - **Operation was not sustainable**
- **Temporary relief was gained by reducing cooling from trim condenser –closer to bubble point**
- **Problem was resolved by closing Hot Vapor Bypass and adding on an inert gas blanket into accumulator**
- **Valve downstream of condensers be a solution??**

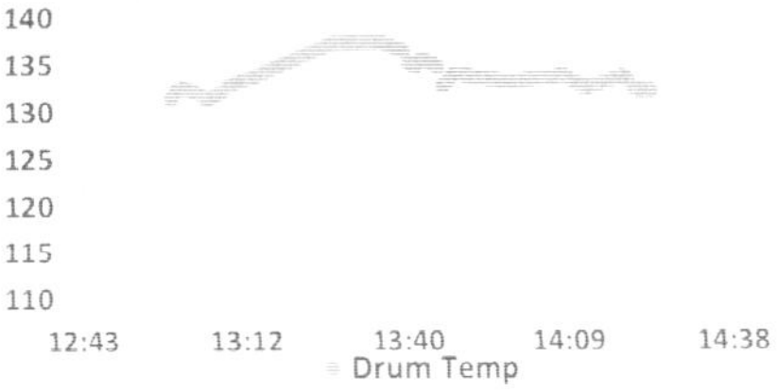

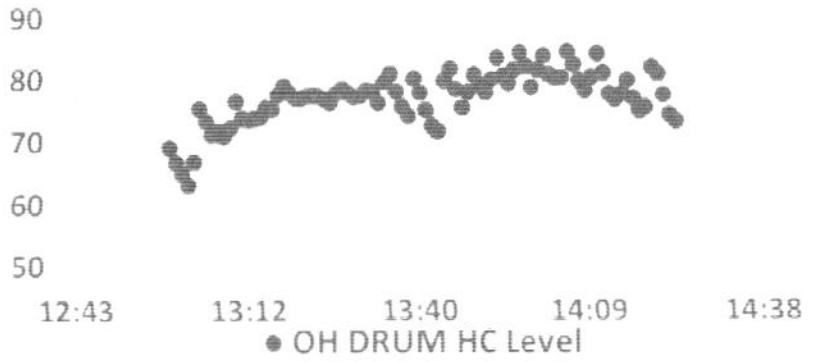

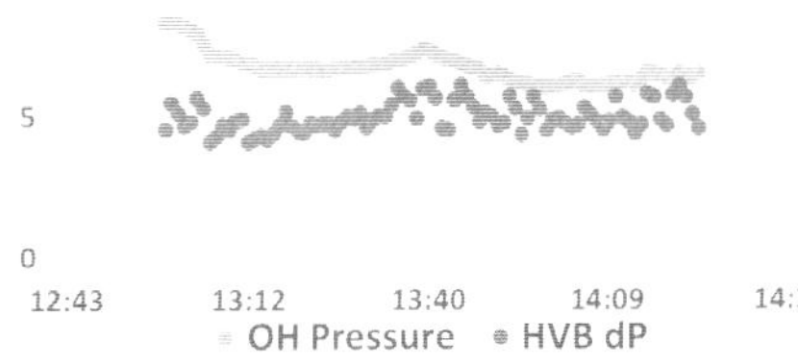

Case #2 Depentanizer Tower

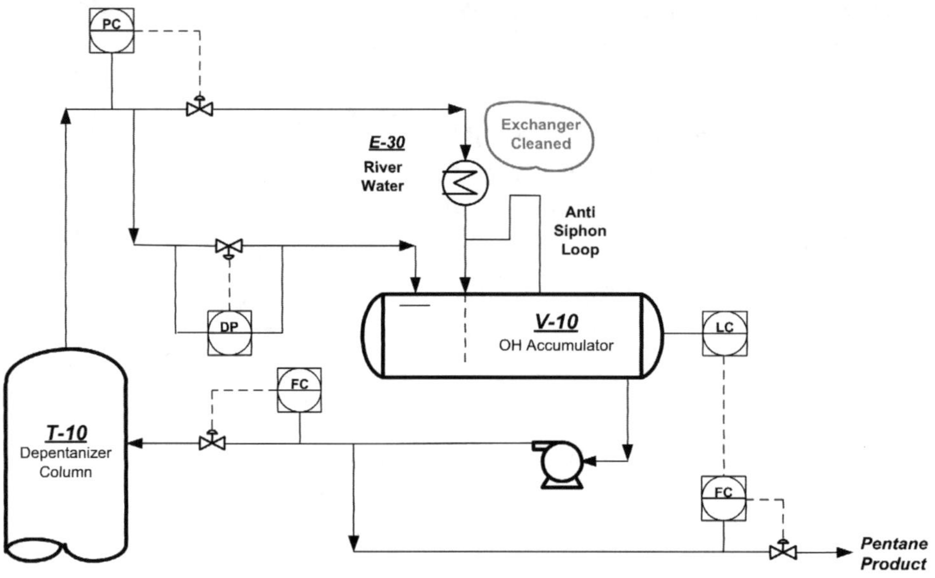

13

Case #2 Depentanizer Tower

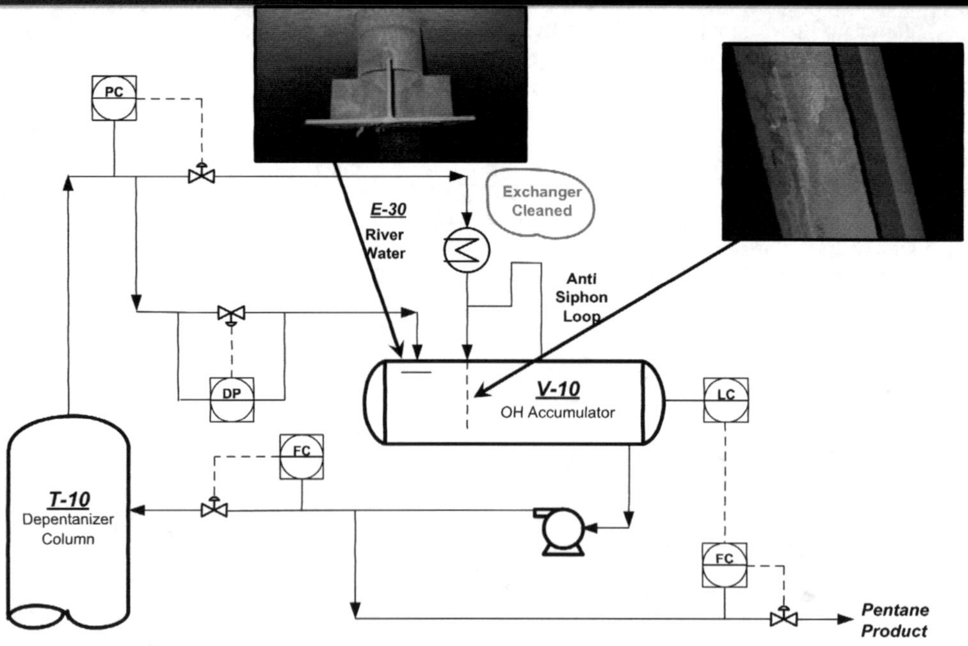

14

Case #2 Depentanizer Tower

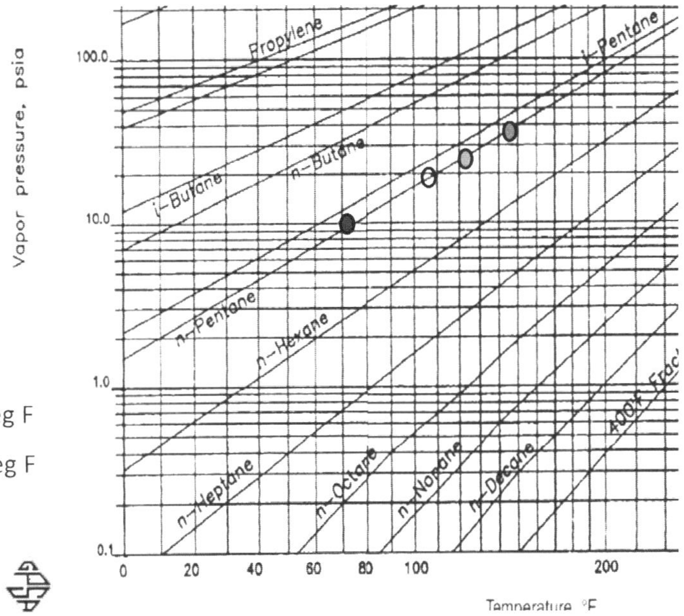

- Tower OH, 155 deg F/15 PSIG
- Drum, 10 PSIG
- New Condenser Outlet, 73 deg F
- Old Condenser Outlet, 105 deg F

 15

Case #2 Depentanizer Tower

- **Tower operating at 80% of pre-TAR operation**
- **Pre-14:37 timecycle**
  - Top temperature, accumulator temperature steady
  - Top pressure was steady
  - Hot vapor bypass dP was shaky
  - Material balance in accumulator was steady
  - But hot vapor bypass was trending up, signaling the drum was losing pressure (liquid head was building up in condenser circuit)
- **14:37 time-frame**
  - Condenser leg liquid buildup "DROPS"
  - Surge of cold liquid drops from condenser circuit and cools off the drum
  - Surge is equivalent to 3000 bbl in 15 minutes
  - Reflux and product are in manual, liquid builds up drum inventory
  - Cold reflux (drops 30 deg F) cools off tower top at 14:45
  - Hot vapor valve was closed, drum pressure drops to low pressure
- **Vapor up-flow scenario was proven in this case**

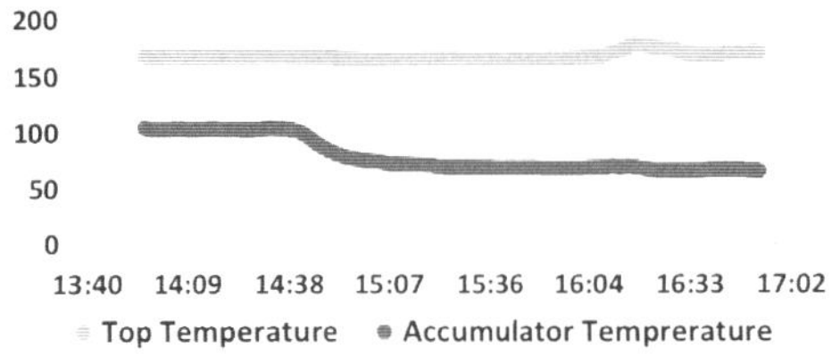

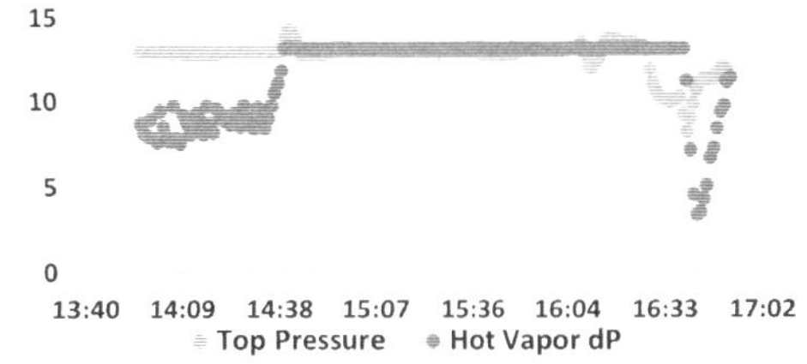

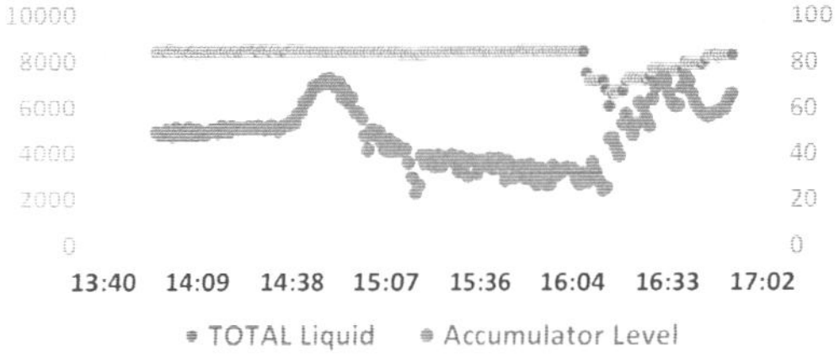

 16

Key Summary Points

- **When a malfunction happens, avoid getting "tunnel minded"**
  - Avoid rush to judgment
- **Slot design for Hot Vapor Bypass designs should be avoided**
- **Slot design is susceptible to**
  - Approach to bubble point for the effluent liquid
  - Ability to seal off the slot during all operating regimes to avoid vapor up-flow
  - Severe disturbances if operating conditions are not perfect
- **Inlet designs for two-phase drum feeds should not be used for totally condensed drums**
  - They require their own special design
  - If a slotted design is necessary, eliminate Hot Vapor bypass and go to blanket gas design
- **Inlet designs for subcooled drums should adequately separate the subcooled effluent from any interaction with the hot vapor**

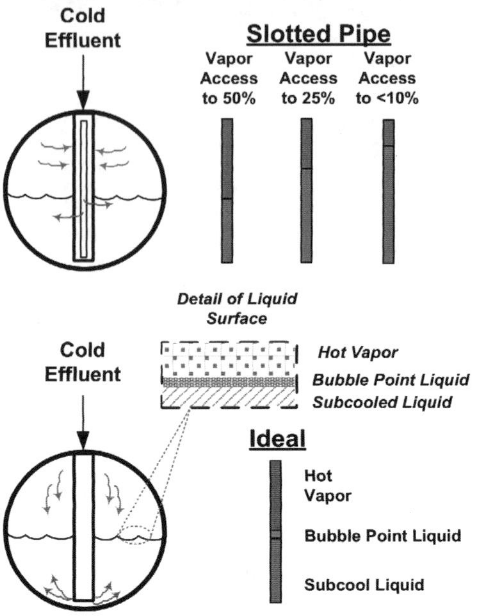

SUMMARY

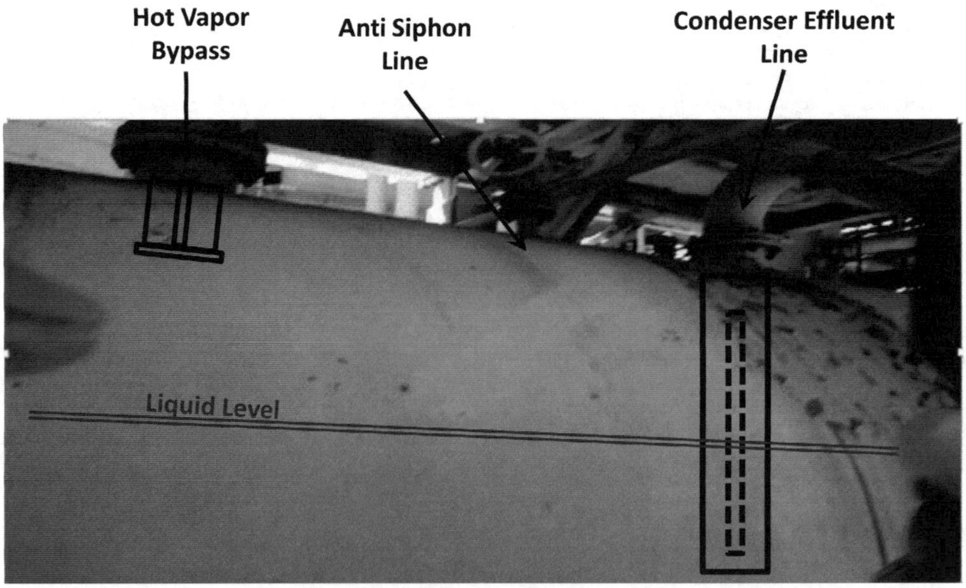

VALERO

SUMMARY

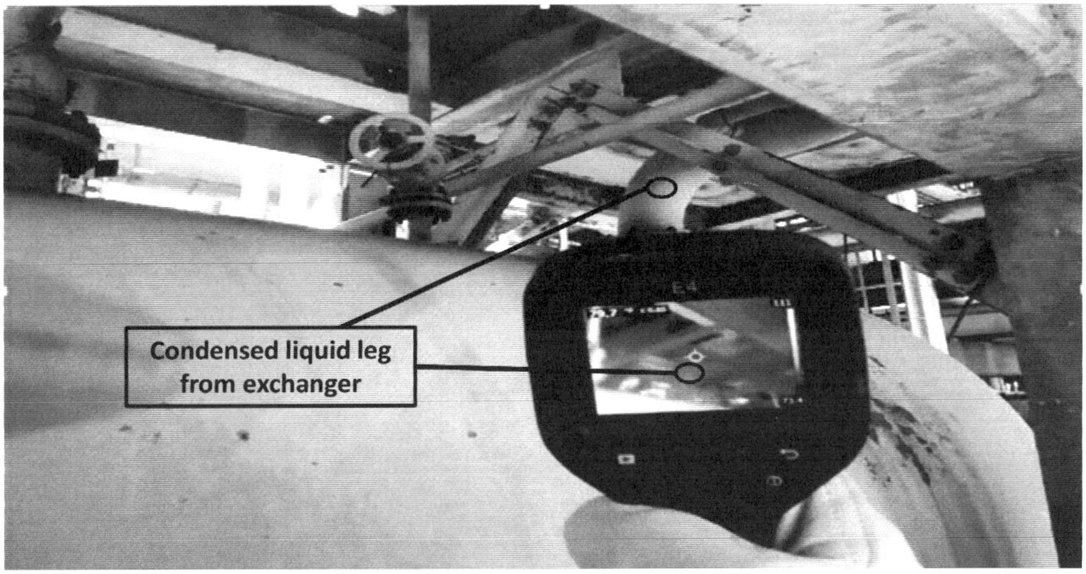

Courtesy of Jace Raney, VLO Meraux

VALERO

Condensate Leg Loss of Subcool

~60 Second Cycle

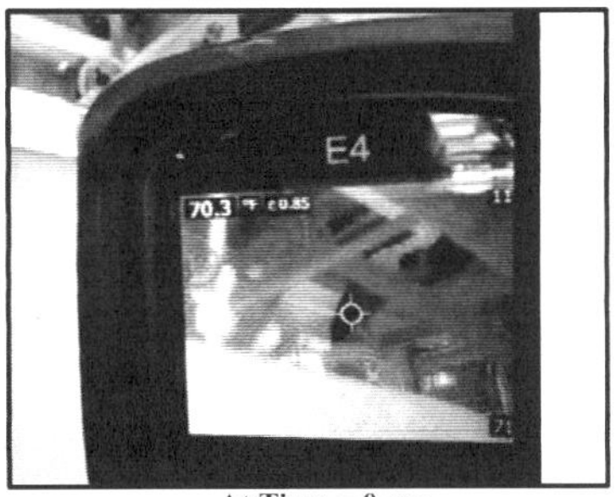

At Time = 0 sec

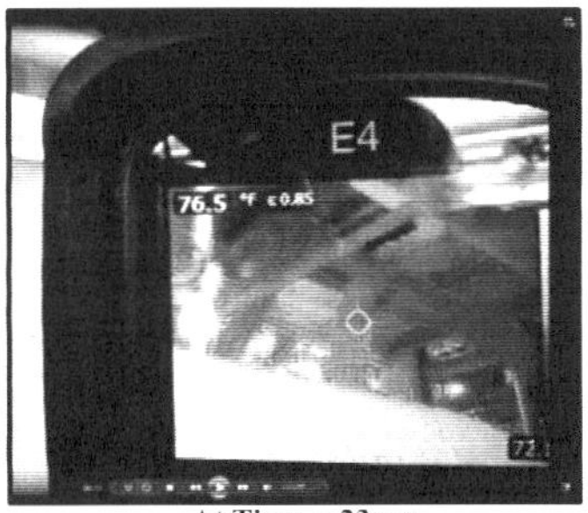

At Time = 23 sec

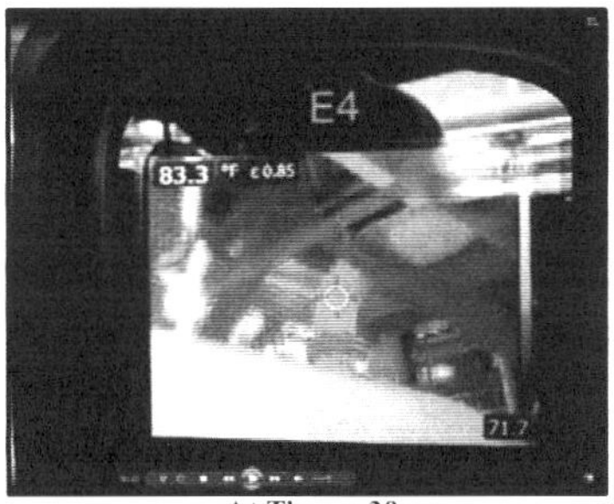

At Time = 30 sec

At Time = 46 sec

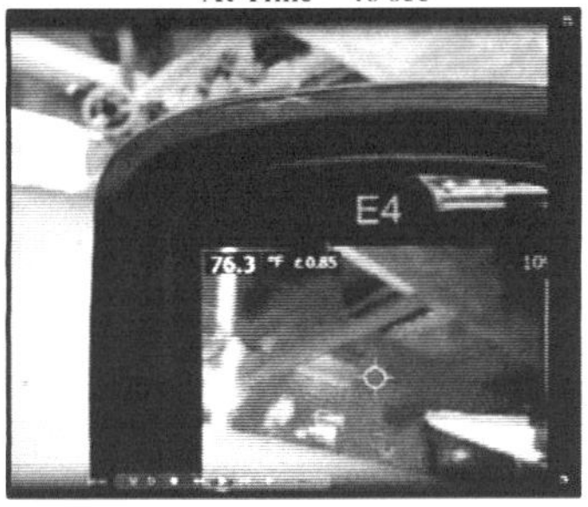

At Time = 57 sec

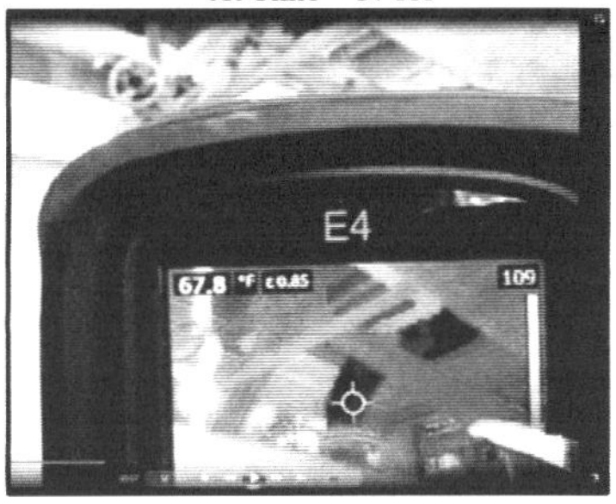

At Time = 72 sec

Questions?

Contact Information

Daryl Hanson
Valero Energy Corporation
Technical Advisor – Fractionation
(210) 345-5929
Daryl.Hanson@Valero.com

Silencing Hammering in a Flooded Condenser System

By

Henry Z. Kister, Fluor, Aliso Viejo, CA, USA

and

Casey Mueller and Matthew Gunn, HollyFrontier Refinery, Cheyenne, Wyoming

Presented at the Kister Distillation Symposium, AIChE Spring Meeting, San Antonio, Texas, March, 2017

SILENCING HAMMERING IN A FLOODED CONDENSER SYSTEM

Henry Kister, Fluor, Aliso Viejo, CA, Casey Mueller and Matthew Gunn, HollyFrontier, Cheyenne, Wyoming

SUMMARY

Flooded condensers are, and are likely to remain, the prime tower pressure control methods for total condensers generating liquid product only. The principles of these methods had been described in the literature more than 60 years ago, yet these methods continue to rank among the most troublesome distillation controls. Good understanding of the principles and learning from experience is the key for avoiding potential traps and improving operation for the industry. This paper describes a recent experience solving a problem that, until now, has been unknown and poorly understood in the industry.

In its first year in operation, a naphtha splitter experienced intermittent hammering in the line from the tower overhead flooded condenser to the reflux drum. The problem occurred at cold condenser outlet temperatures. In some of the hammering episodes, the hammering was severe, opening a flange and causing damage to a valve and a thermocouple.

Fluor, which was not involved in the original design, was requested to join the HollyFrontier task force investigating the hammering incidents. Our task force identified the large open slots at the top of the reflux drum dip pipe as a likely root cause. We also identified total closure of the tower overhead pressure control valve as another potential source of hammering.

Based on this diagnosis, we blanked the slots near the top of the dip pipe and added a DCS clamp on the overhead pressure control valve that prevents the valve from closing to less than 20%. HollyFrontier also emphasized in its operator training the importance of keeping tight control of the condenser outlet temperature at 180° F. The column was returned to service in March 2015, with no further hammering to date, and experiencing smoother operation.

This paper describes our investigation and our solution.

BACKGROUND

The HollyFrontier naphtha splitter had been installed to help the Cheyenne Refinery comply with the requirements of the MSATII rule for benzene in gasoline by removing benzene precursors in the feed to the naphtha reformer (a UOP Platformer ®). The tower was started up in February, 2014.

In its first year of operation, intermittent hammering occurred in the line from the tower overhead condenser to the reflux drum. The problem occurred at cold condenser outlet temperatures. In some of the hammering episodes, the hammering was severe, opening a flange and causing damage to a valve and a thermocouple. During the winter of 2014/2015, the naphtha splitter was shut down and HollyFrontier requested Fluor to join its task force to investigate the hammering incidents with the objective of implementing a solution prior to restart.

PROCESS DESCRIPTION

Figure 1 shows the HollyFrontier naphtha splitter. Process conditions shown are those at 10:30 pm on November 10th, 2014, just prior to the incident described below. Feed to the tower is the naphtha hydrotreater (NHT) debutanizer bottoms. This feed is preheated to 290° F by exchange with the tower bottoms. The preheated feed then enters the 9' ID, 70 tray naphtha splitter. The tower separates a light naphtha overhead product from a heavy naphtha bottom product. The tower is reboiled by a fired heater. Tower bottoms are pumped through the feed preheaters, an air cooler and a sulfur sorber to the Platformer. Tower overhead vapor at 217° F is condensed by an air condenser, which is a total condenser (no vapor product), and the condensate flows into the reflux drum. From the drum, most of the condensate is returned to the tower as reflux, and the rest is pumped to storage via an air cooler.

Tower pressure is controlled at about 29 pounds per square inch gauge (psig) using a control valve in the tower overhead line to the air condenser. There is a bypass around the condenser with a control valve that is manipulated by the differential pressure between the tower and the reflux drum, kept at 5-10 psi.

The condensate line leaving the condenser enters at the top of the reflux drum and is extended inside the drum by a dip pipe that takes the condensate to near the bottom of the drum. To raise tower pressure, the pressure control valve in the vapor line to the condenser is throttled. This reduces the pressure in the condenser, which in turn reduces the delta T (temperature difference between the condensing side and the cooling side) and also sucks liquid back from the drum. This liquid floods some of the condenser tubes, which reduces its condensation area, in turn reducing condensation, causing the tower pressure to rise. Conversely, opening the pressure valve lowers the condensate level, exposing more condenser tube area for condensation, and also raises the condenser pressure, which improves its delta T. Both increase the condensation rate and lower tower pressure.

The condenser bypass line helps keep up the pressure in the reflux drum. If the valve in the bypass line is closed, the pressure in the drum will be the vapor pressure of the condensate liquid. When the condensate is near its boiling point (no subcooling), the drum pressure will be much the same pressure as the pressure at the condenser outlet. In contrast, when the condensate is subcooled, its vapor pressure may be much lower, typically by about 0.4 psi per degree F of subcooling in naphtha splitters. The hot vapor bypass (HVB) line and valve prevent the drum pressure from falling too low. When condensate subcooling causes the pressure difference between the tower and the drum to rise, the HVB valve opens and diverts additional hot vapor to the drum. The hot vapor warms up the drum liquid surface, making the drum pressure equal to the vapor pressure of this warmed-up liquid surface (which is warmer than the subcooled bulk liquid). Process liquids are good thermal insulators, so, as long as the subcooled liquid does not reach the surface in excessive or fluctuating quantities, the surface temperature will remain steady and warm enough to sustain the differential pressure set by the dP controller. A detailed discussion is included in Reference 1.

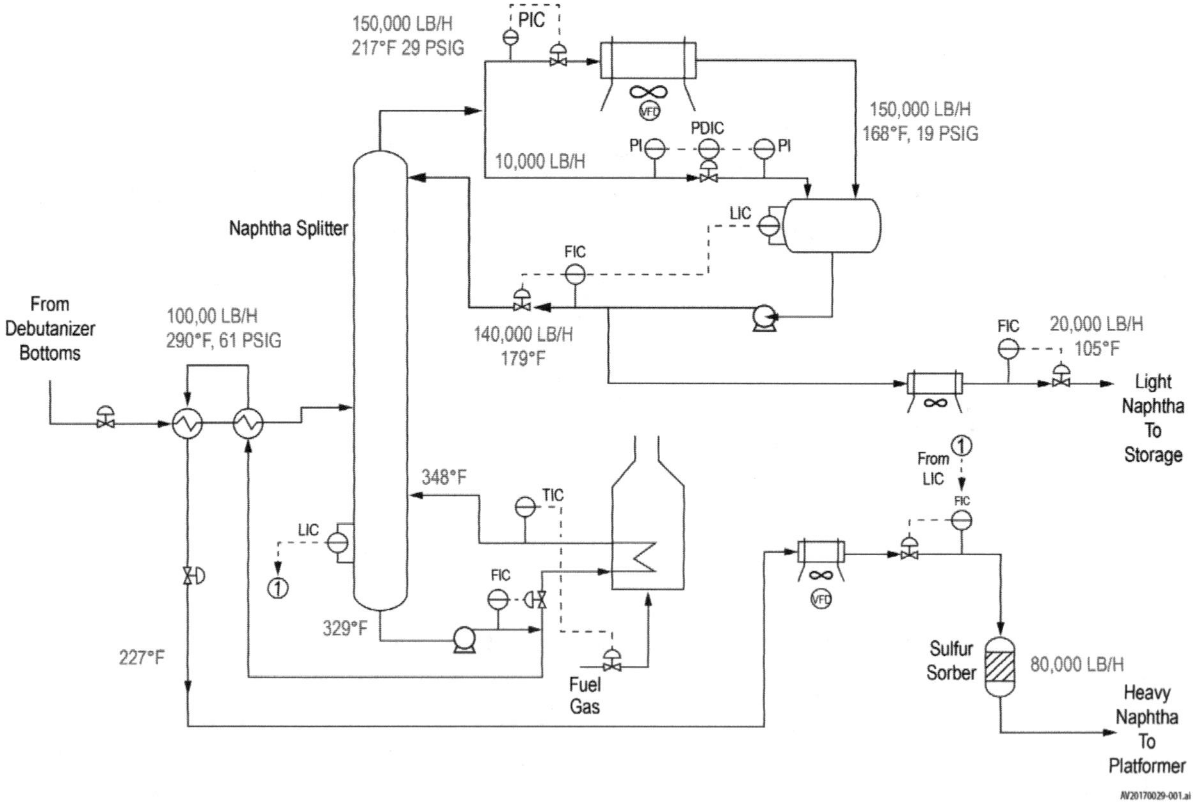

Figure 1. Naphtha splitter process scheme at beginning of incident (10:30 pm November 10, 2014)

FOCUS ON THE CONDENSER AND REFLUX DRUM

Because the hammering occurred between the condenser and the reflux drum, this area was the focus of our investigation. Figure 2 shows the physical layout of the piping from the air condensers to the reflux drum. Both the vapor and condensate lines descend to the drum with no pockets according to good design practices (1).

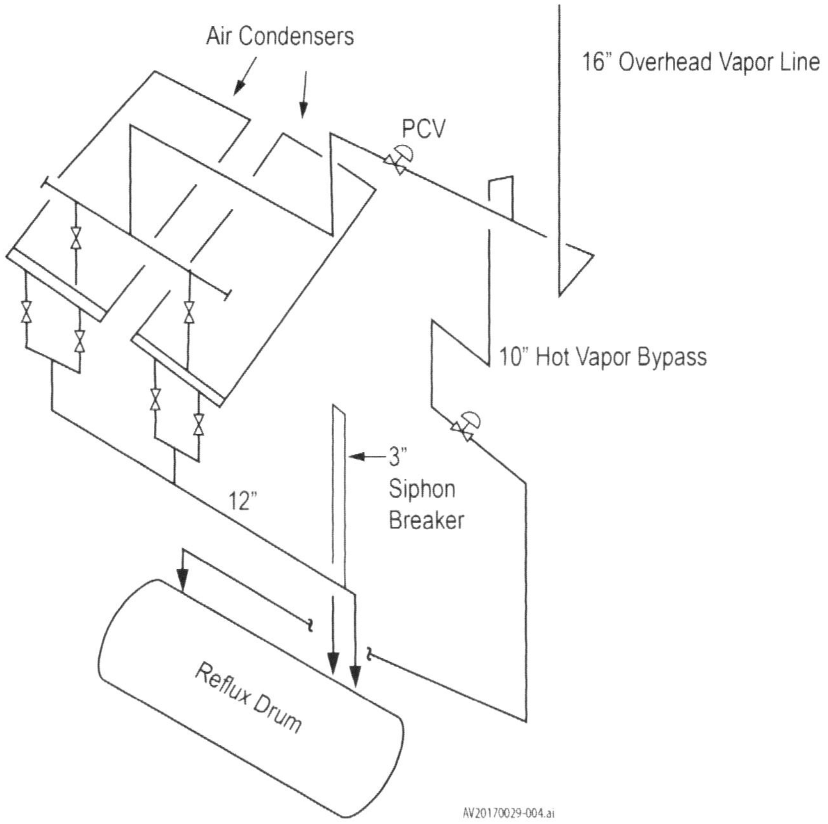

Figure 2. Layout of the naphtha splitter condenser and reflux drum piping

Figure 3 shows the reflux drum. The drum is 6.5 ft ID by 20 ft tangent to tangent horizontal drum. Condensate enters the drum via the 12 inch nozzle on the top right hand side of the drum in Figure 3. This nozzle is equipped with a dip pipe shown in detail in Figure 4. Condensate leaves the drum via the 8" nozzle at the bottom left of the drum in Figure 3. The vapor bypass around the drum enters via a 10" nozzle at the top left of the drum in Figure 3.

Figure 4 is a close up of the dip pipe. The pipe terminates 1 foot above the bottom of the drum. The bottom of the dip pipe is open. The dip pipe is equipped with four 2" wide by 12" long slots, starting 6" below the top of the drum, and continuing 12" to 18" below the top of the drum.

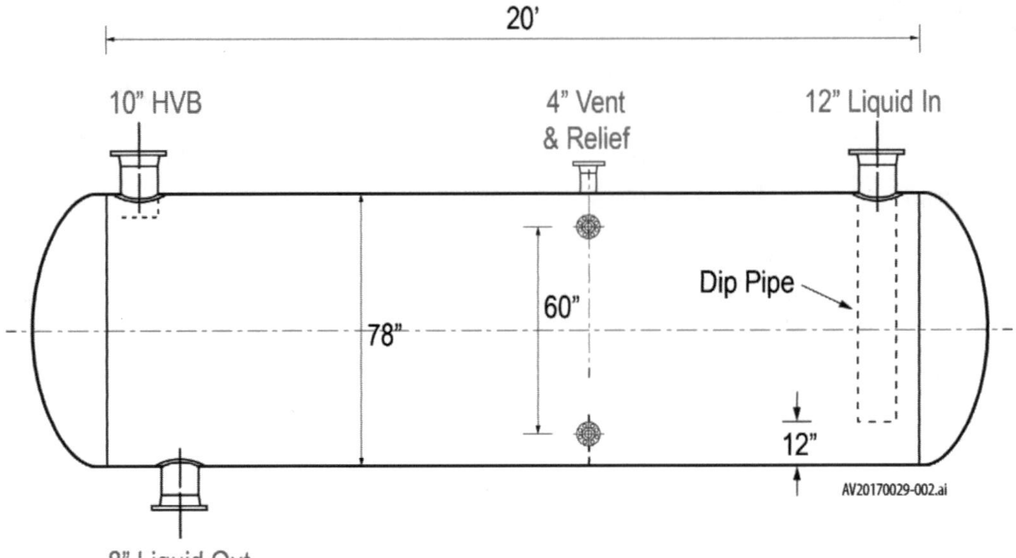

Figure 3. Naphtha splitter reflux drum

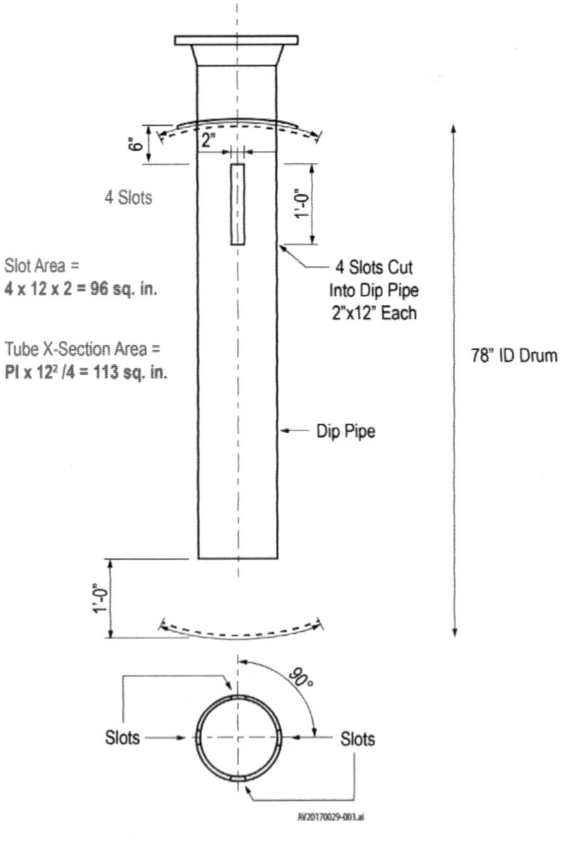

Figure 4. Naphtha splitter reflux drum dip pipe

During the design of the Naphtha Splitter, a concern was raised that non-condensables can be present in the column feed at times. The upstream Debutanizer may occasionally lose reboil, resulting in light ends (butanes and lighter) being present in the feed to the Naphtha Splitter. During such events it was envisioned that these light ends would need to be vented from the overhead accumulator; accordingly, slots were added to the dip tube to allow vapors to disengage from the liquid at the outlet of the condenser.

Emphasized in Figure 4 are the areas of the slots compared to the area of the opening at the bottom of the dip pipe. The slot area is 96 sq. inches, while that of the opening at the bottom of the pipe is only slightly higher, 113 sq. inches. The middle of the slots is 4.5 ft above the bottom of the pipe. When the drum level is about 50%, the additional head acting on the bottom of the pipe is about 2.25 ft higher than that acting on the slots. The total liquid head acting on the slots is therefore about 16 ft, compared to a head of about 18 ft acting on the bottom of the pipe. With the liquid heads being close, the liquid split between the slots and the pipe bottom will roughly follow the opening areas, 46% through the slots, 54% through the bottom of the pipe. That is without considering any vapor pressure effects.

OPERATING EXPERIENCE

The naphtha splitter was started up in February, 2014. It has been operating at rates close to those shown in Figure 1. These rates are below the future design rates of the system.

In its first year in operation, intermittent hammering occurred in the line from the overhead condenser to the reflux drum. The problem was first observed following the February 2014 startup, and occurred at cold condenser outlet temperatures. No hammering was observed when the condenser outlet temperature exceeded 170° F. The hammering was alleviated by tuning the condenser outlet temperature controller (which adjusts the speed of the fans motors) to avoid rapid changes in condenser outlet temperature.

There were two incidents where the hammering was particularly severe. During an outage in the summer of 2014, an effort was made to operate the tower at total reflux, however, some light naphtha product was still drawn. This resulted in heavier tower naphtha with overhead temperature reaching 300° F (normally 217° F). When the reboiler was shut down, the tower pressure appeared to collapse. The result was such strong hammering in the overhead line that flanges opened and the handle of one of the 6" valves at the condenser outlet came off.

A second strong hammering incident occurred during an outage on a cold night on November 10th, 2014. Again, the hammering followed the shutdown of the reboiler. In this incident, the hammering damaged the thermocouple used to control the air condenser fans. A decision was made to not restart the column until a solution to the hammering problem could be implemented.

In addition to these two strong hammering incidents, a number of several smaller hammering episodes occurred prior to the November, 2014 shutdown. In these incidents, the plant experienced condenser outlet temperature swings of 40-50° F, sometime falling from 170° F to

100° F in two minutes. Drum liquid temperature would drop by up to 20° F. At times the line from the condenser to the drum heated up.

Even at periods of no hammering, the tower overhead system appeared to experience frequent instability.

Following the November, 2014 naphtha splitter shutdown, HollyFrontier requested Fluor, who was not involved in the initial design but has wide distillation expertise, to join its task force investigating the hammering incidents, with the objective of implementing a solution prior to restart. Based on a review of the system, an event timing analysis, and conducting a comparative analysis using previous experiences with which Fluor has familiarity, the taskforce identified the problem and implemented a solution.

EVENT TIMING ANALYSIS

The event timing analysis was applied to an incident in which tower feed was lost between 10:30 pm and 11:20 pm on November 10, 2014. During this period, the tower feed rate dropped from 100,000 lb/h at 10:30 pm to 42,000 lb/h at 10:45 pm, reaching zero at 11:10 pm. Hammering in the aftermath of this incident later that night became so severe that it broke the condenser outlet temperature indicator.

The key parameter emerging from the event timing analysis was the tower overhead flow rate to the condenser, which was calculated from the pressure control valve opening. The variation of this flow rate is represented in Figure 5.

Between 10:30 pm and 10:45 pm, while the feed rate dropped from 100,000 to 42,000 lb/h, the bottom level held constant at 55%, and the bottom flow rate dropped from 80,000 lb/h to 61,000 lb/h. The reboiler duty was constant with the fuel gas rate to the reboiler at 40 MCFH.

The much larger reduction in feed rate compared to the bottom flow rate led to increased heating of the feed in the preheater, with the preheater outlet temperature rising from 290 F°to 304° F. This led to more vaporization in the feed and more tower overhead vapor. Based on the overhead control valve opening, the tower overhead flow rate rose from 150,000 lb/h at 10:30 pm to a peak of 157,000 lb/h at 10:45 pm.

Over the next 10 minutes the fuel gas flow to the reboiler dropped from 40 MSCFH to zero, leading to a decline of tower temperatures. The tower overhead vapor flowrate plunged. Based on the overhead control valve opening, the tower overhead flow rate declined to 123,000 lb/h by 11:00 pm and to 98,000 lb/h by 11:10 pm. Boil-up stopped and the bottom and distillate flow rates dropped to near zero. Instability set in about 11:10 pm. Tray temperatures dived, the condenser outlet temperature swung approximately 30° F peak to valley every 2-3 minutes, and with it the differential pressure between the tower overhead and the reflux drum swung 4-8 psi peak to valley. By 11:20 pm the tower overhead flow rate declined to 62,000 lb/h with the swings continuing.

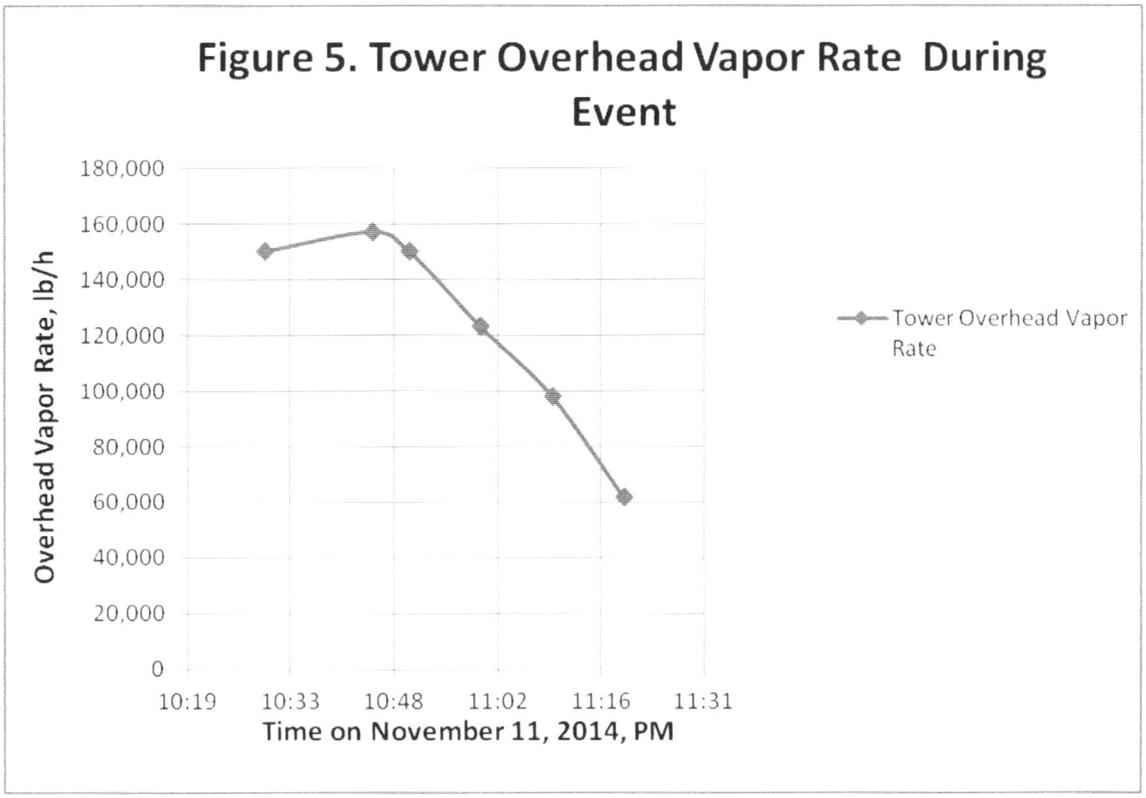

CONDENSER PERFORMANCE DURING THE EVENT

The overhead air condenser has two bays, each bay containing 238 34' long 1" OD finned 2-pass tubes arranged in four rows. The total bare tube area is 4240 ft^2. The tubes are sloped at 0.125" per foot, in accordance with good design practice. There are two fans per bay. The condenser outlet temperature controller manipulates the speed of the motor of one fan per bay. The other fan for each bay is on an on/off control mode. There are also louvers than can be manually adjusted to help maintain the desired condenser outlet temperature.

The condenser design case shown on the rating sheet has the splitter overhead temperature at 192° F, about 25° F colder than the splitter overhead temperature of about 217° F at the beginning of the November 10, 2014 event. Assuming the tower overhead is fully condensed at about 180° F (approximately the reflux temperature at the beginning of the November 10, 2014 event), the design case utilizes about 93% of the condenser area for condensing, and the remaining 7% for sub-cooling. The log mean temperature difference (LMTD) for the condenser design rating sheet was 55 F.

The condenser design rating sheet is based on a summer air inlet temperature of 95° F. On November 10th the ambient air temperature in Cheyenne was 10° F. With air temperature this cold, two of the four fans would normally be turned off, the others variable speed drives used for control. So the air flow would be approximately half the design, possibly even less. Even with two fans off, we estimated the LMTD to be about twice the design, around 100° F. This means that the condenser area utilized for condensation would be half or less the available condenser tubes area. Half the tubes would then be flooded. In the flooded area, the high delta T will give tremendous subcool.

Figure 6 shows our estimate of the number of unflooded tubes (i.e., those that perform condensation) during the event. When this number is 4, the condenser area is fully utilized for condensing, and the subcooling area is minimal. Figure 6 shows that at the beginning of the November 10 event (10:30 pm), until about 10:50 pm, there was little change in the number of unflooded tubes. After this, however, there was a rapid decline, with one row of tubes becoming submerged in liquid every 20 minutes or so. By 11:20 pm, the number of unflooded rows of tubes would decline to approximately 0.7.

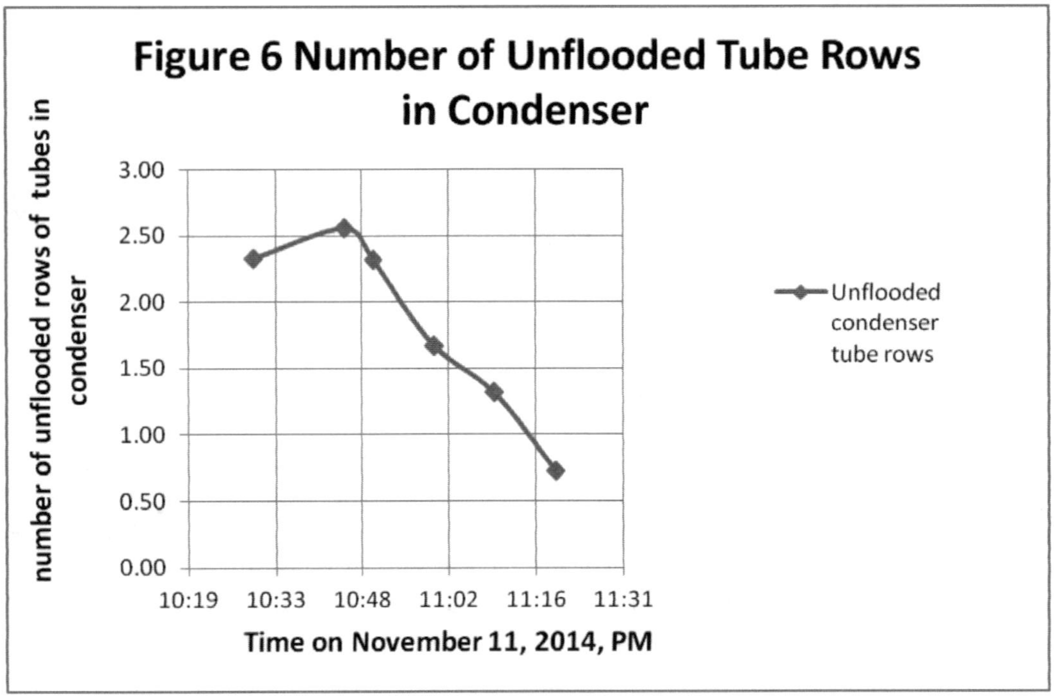

The scenario described in Figure 6 and in the two preceding paragraphs above is not realistic. It is based on an assumption that, on a cold night like November 10th, about half the normal air flow ascends through the air cooler. In reality, however, the condenser outlet temperature control and the operating crew will keep reducing the speed of the two operating air condenser fans to keep the subcool (as measured by the condenser outlet temperature) the same. Because subcooling increases as tubes flood, the control and operators will act to keep the liquid level at the air cooler constant. This also means keeping the number of unflooded tubes close to 4. It does not take much to subcool the liquid. We estimate that flooding less than 0.1 row of tubes will give a subcool of 10° F, even at the design (summer) conditions. With the cold air on the night of November 10th, flooding less than this will easily give 10° F subcool.

The only levers that the operators can pull are the fan speeds and louver openings. As the air gets colder and the duty declines, they would reduce fan speeds. The plant seldom manipulates the louvers to reduce air flow (rightly so) as this is an awkward way of controlling air condensers and can cause damage to the louvers. However, even if the operators turn off all fans, and even if they close all louvers, there would be a draft coming through the air condensers. If the louvers were left open, this draft will be quite strong. Figure 7 shows that before 11 pm on November 11th, there would have been quite a good condenser duty, and normal fan manipulation would have kept the subcool under control with condenser outlet temperature of about 170° F. At about 11:10 pm, the condenser duty dropped to just over half of the duty at 10:30 pm (Figure 7), and kept dropping as the reboil was stopped. With that rapid a decline in reboiler duty, it became difficult, if not impossible, to contain the subcooling, and the condenser outlet temperature dove precipitously as can be seen in Figure 8.

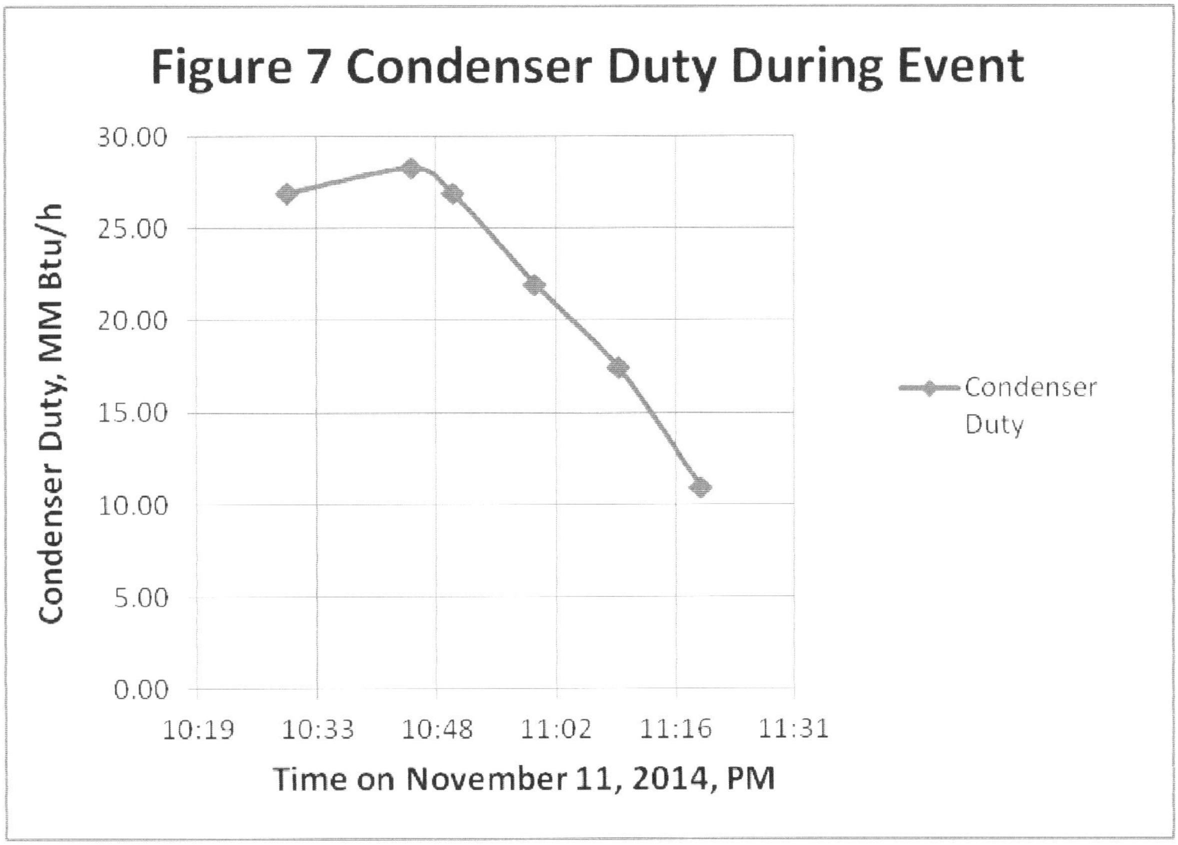

The liquid head between the condenser level and the middle of the slots in the reflux drum dip pipe is approximately 16 ft, equivalent to 4.2 psi. The naphtha vapor pressure changes about 0.4 psi per degree F. Accordingly, as long as the subcooling is less than about 11° F (= 4.2/0.4), the liquid head will exceed the difference in vapor pressures and there will be flow from the tube out. At larger subcool, the difference between the drum pressure and the vapor pressure of the liquid in the dip pipe will exceed the liquid head, generating a potential for vapor backflow from the drum into the dip pipe. Initially, the momentum of liquid exiting the slots as well as suction of some liquid from the bottom of the drum to the slots will prevent the back flow. However, as subcooling intensifies, the driving force (pressure difference between the drum pressure and the vapor pressure of the subcooled liquid) will increase. Eventually, a point will be reached when the driving force will be large enough to overcome the liquid head, and induce backflow of vapor from the drum into the dip pipe. This dew point vapor will collapse onto the subcooled liquid in the condensate pipe. When this vapor collapse occurs in the confines of the pipe, hammering will result. The vapor backflow may heat up the condensate pipe, as was observed during some of the hammering incidents.

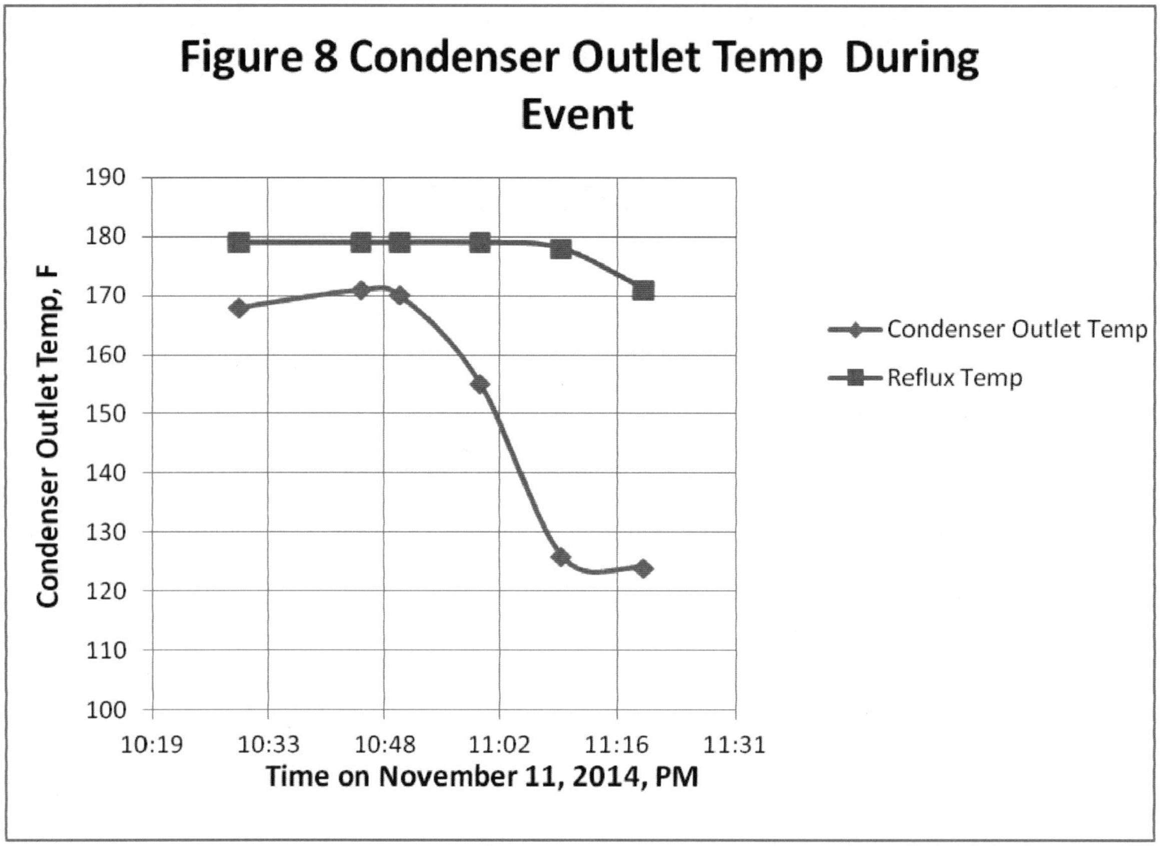

Figure 8 shows that until about 10:50 pm, there was a constant but low degree of subcool. Assuming the tower overhead is fully condensed at about 180° F (approximately the reflux temperature at the beginning of the November 10, 2014 event), this subcool was about 10° F. The hydraulic head exceeded the vapor pressure difference, so flow was from the slots in the dip pipe into the drum. Between 10:50 pm and 11:00 pm, there was a 15° F drop in the condenser outlet temperature, raising the subcool to 25° F. At this subcool, there was a 10 psi pressure difference between the drum pressure and the vapor pressure of the subcooled liquid, exceeding the head difference. However, from the plant charts it appears that this was not enough to initiate backflow of vapor from the drum to the dip pipe, as no hammering occurred.

From 11:00 pm to 11:10 pm, the condenser outlet temperature dropped by another 29° F, generating a subcooled liquid with a vapor pressure 22 psi less than the drum pressure. The drum pressure at 11:00 pm was steady at about 17° psig, in equilibrium with the liquid in the drum. By 11:10 pm, the vapor pressure of liquid in the pipe dropped to about 10 psia. This cold liquid exerted a strong suction on the drum vapor, much higher than the liquid head in the pipe. Therefore, instead of the liquid from the pipe pouring into the drum vapor space, this huge suction sucked vapor into the pipe. The dew point vapor from the drum rapidly collapsed onto the subcooled liquid, causing the hammering.

Had the slot area of the dip pipe been small compared to the open area at the bottom of the pipe, the resulting subcooled liquid that was sucked into the slots would have been less significant, resulting in minimal or no hammering. However, here the slot area was large, about 85% of the tube bottom area, so with high subcool, the suction could not match the vapor pressure difference.

Based on the above analysis, the key to prevention of further hammering incidents is blanking the slots in the dip pipe and/or limiting the subcool. Limiting the subcool can be readily achieved during normal operation, but may not be achievable during startups/shutdowns.

PRESSURE CONTROL VALVE CLOSURE

Complete closure of the tower overhead pressure control valve has been responsible for a number of severe hammering incidents such as Case 28.6 in Reference 2. If this valve fully closes, the cold air (especially during low rate operation and/or winter time) will quickly condense the small amount of vapor present downstream of the valve. The condenser pressure will immediately drop, and may go into vacuum, causing liquid to be rapidly sucked from the accumulator drum. The rushing liquid will bang against any objects in its path, producing a severe liquid hammer.

Figure 9 is a plot of the naphtha splitter tower overhead pressure controller output throughout the event. The diagram shows that at the lowest signal, the valve output was 38%, which coincides with an overhead vapor flow rate of 62,000 lb/h. A review of the operating charts during the hammering incident, throughout the night (until 4:30 am on November 11$^{th)}$ shows that the smallest valve opening recorded during this event was again 35%. We therefore conclude that during the analyzed event and the severe hammering that followed and damaged the thermocouple, the control valve did not get anywhere near total closure, so this mechanism did not contribute to the observed hammering on November 10th and 11th.

The observation that the overhead valve was kept away from closure probably resulted from good operation of the overhead condenser, where operators switch off fans and minimize fan speed during the low rates and cold weather. However, although valve closure did not occur in these cases, it does not preclude the possibility of an event in which the overhead valve will attempt to close fully during cold weather. We therefore added a DCS clamp that will prohibit the valve from reaching full closure.

POST-REVAMP OPERATION

Based on our joint task force review, we blanked the slots near the top of the dip pipe and added a DCS clamp on the overhead pressure control valve that does not allow the valve to close to less than 20%. HollyFrontier also emphasized in its operator training the importance of keeping tight control of the condenser outlet temperature at 180° F.

The column was returned to service on March 12, 2015 and operated continuously until May 13, when it was again shut down due to a refinery maintenance outage. It has been restarted and shut down on a number of occasions since then. No further hammering has occurred during normal operation nor during any of the startup/shutdown events, even though condenser outlet temperatures were as low as 100° F, and even less during startups and shutdowns. During normal operation, the condenser outlet temperature is always kept close to 180° F.

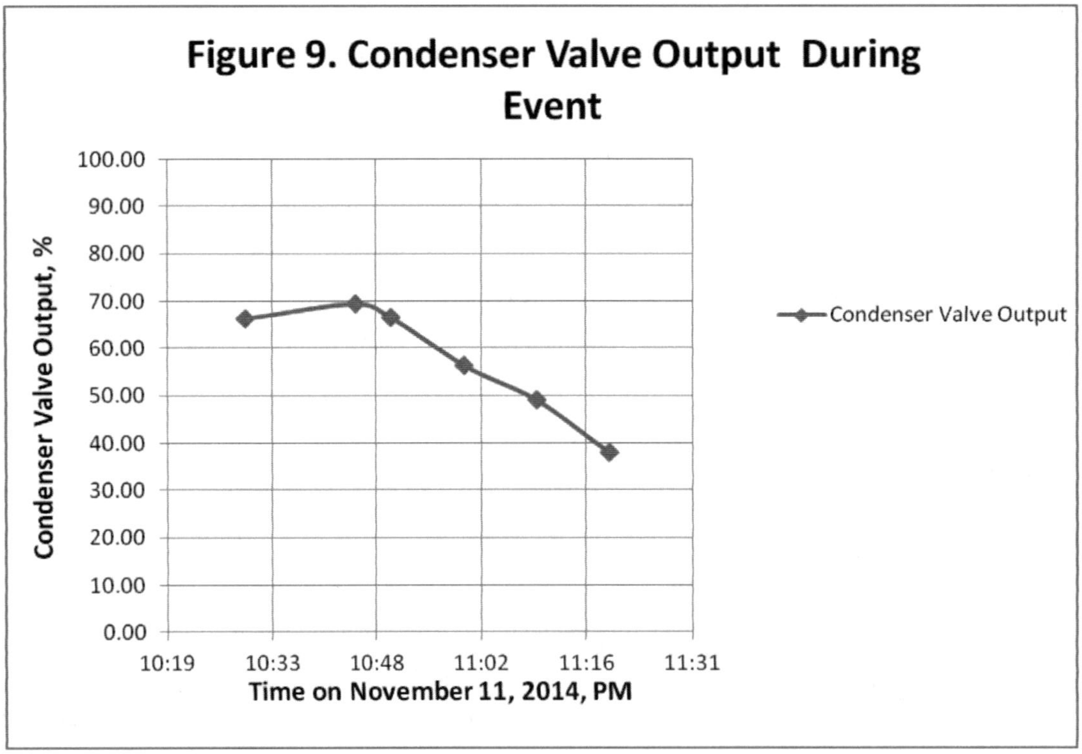

The operators noticed that since the restart, the system has operated much more smoothly. This too resulted from blanking the slots in the dip pipe. Prior to their blanking, the slots in the dip pipe poured subcooled liquid onto the liquid surface in the drum. This cold liquid reduced the vapor pressure in the drum. The amount of liquid pouring out of the slots was variable, changing with the pressure controller actions. The degree of subcool, therefore, would also vary. Varying amounts of subcooled liquid hitting the drum hot liquid surface would cause fluctuations to the drum pressure and therefore destabilize the system. Closing the slots stabilized the drum surface temperature and vapor pressure and therefore the drum pressure.

CONCLUDING REMARKS

Our experience teaches that when it comes to flooded condensers, "the devil is in the details." Adhering to good design practices, such as those described in reference 1, is a must. However, there are new things that we learn even in this well-established science, and experience is the best teacher. Event-timing analysis is a powerful tool for analyzing and acquiring the understanding that would pave the way for diagnosing flooded condenser control problems. With a correct diagnosis, minor equipment changes at negligible cost are often sufficient to provide total solutions and to reinstate trouble-free operation.

REFERENCES

1. Kister, H. Z. "Flooded Condenser Controls – Principles and Troubleshooting" , Chemical Engineering, January, p..37, 2016.
2. Kister, H. Z., "Distillation Troubleshooting", Wiley – Interscience, Hoboken, New Jersey, 2006.

TRICKY LIQUID DISTRIBUTORS

Petr Lenfeld

GTI Solutions
Brno, Czech Republic

David Lin

GTI Solutions
Flower Mound, Texas, USA

Prepared for Presentation

at the

Kister Distillation Symposium 2017

Topical Conference

at the

2017 Spring Meeting and 13th Global Congress on Process Safety

of the

American Institute of Chemical Engineers

San Antonio, Texas

March 26 – 30, 2017

Introduction

The design criteria of liquid distributor are well established for packed columns. However, liquid distributor selections and design are still challenging even for reputable vendors. Overlooking the basics often result in costly operational problems.

Case Study #1: Fire in Ammonia Absorber Column

An intermediate petrochemical product is manufactured using ammonia as the main feedstock. Unconverted ammonia is removed from the gaseous product by sulfuric acid absorption in the absorber column. Ammonia absorber column is a two stage absorber equipped with two beds of structured packing with high quality channel trough type distributors and two cooled pumparound circuits. Sulfuric acid is injected automatically to both sections, controlling pH in the pumparound loops.

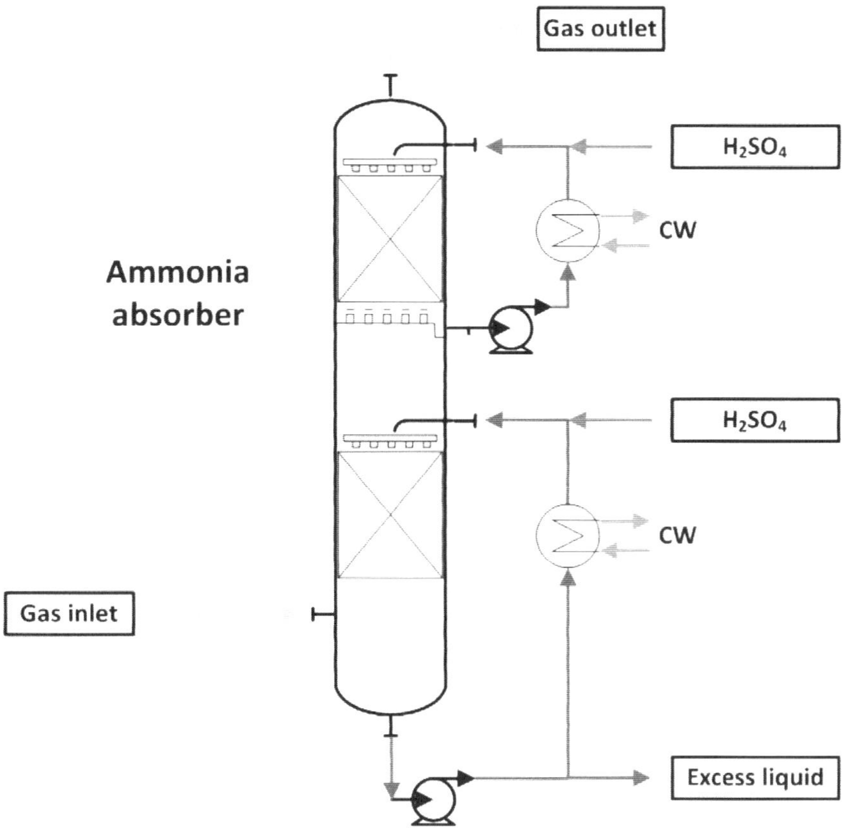

Figure 1: Ammonia absorber

Problem Description

Shortly after the shutdown for cleaning, the absorber column performance deteriorated. Sulfuric acid consumption in the bottom loop dropped, while the acid consumption rose in the top loop. In order to prevent ammonia break-through, operators manually increased acid dosing to the bottom loop and started the investigation to identify the root cause.

Column behavior indicated a problem with bottom section absorption efficiency. Tower internals vendor was consulted but the company excluded such problem could be caused by the new tower internals (namely liquid distributor). Other potential causes were investigated as well:

- sulfuric acid quality (low concentration)
- wrong control system readings (pH probes, acid consumption measurements)

Various troubleshooting actions were performed by the client, including cross-check of lab analyses for various streams, pH probes replacement and acid material balances; however, no abnormality was identified. The client ultimately decided to call in an independent troubleshooter to identify the root cause and to develop possible solutions.

Problem analysis

Field survey and observation indicated liquid distributor as a probable cause with possible plugging, damage and/or tower upset that may have tilted the liquid distributor. As no improvement was expected unless the liquid distributor problem was resolved, a short shutdown was planned for column inspection and repair of the liquid distributor.

When the absorber column was opened we discovered:

- bottom section liquid distributor plugged and tilted (as expected)
- massive damage of bottom section structured packing (a surprise)

To make matters even worse, after removal of damaged packing, it was found that the column shell was deformed in the bottom section. When thermal insulation was removed it was clear that the damage was caused by internal fire in the bottom structured packing bed, a consequence of pyrophoric compounds self-ignition during the shutdown operations.

What started as a small shutdown became as major one with serious safety issues.

Finding the Root Cause

Investigation revealed the following disastrous chain of events:

- Plugged distributor, leading to lower bottom section efficiency.
- Operators' wrong interpretation of column performance and subsequently overdosing of sulfuric acid.
- Packing damaged by high acid concentration.
- By "design", distributor was not properly secured. As a cost saving protocol, the vendor by default omitted the packing hold-down grid in the design. The liquid distributor was simply supported by the structured packing underneath it. For this particular case, liquid distribution quality deteriorated once the distributor shifted/tilted due to dissolved packing underneath it. Out of level liquid distributor led to localized dry spots across the packed bed causing further deterioration in column performance.
- Pyrophoric compounds accumulated in dry packing bed.
- Fire during shutdown.

Cause(s) of Distributor Plugging

Absorber column has been revamped some years back with structured packing and new high quality channel trough type distributors. Original column design had random packing and ladder type distributors. Revamped absorber column worked well for several years.

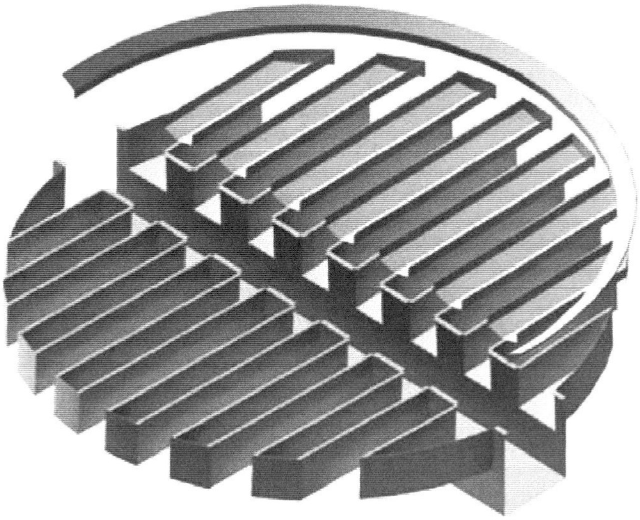

Figure 2: Typical channel through type distributor

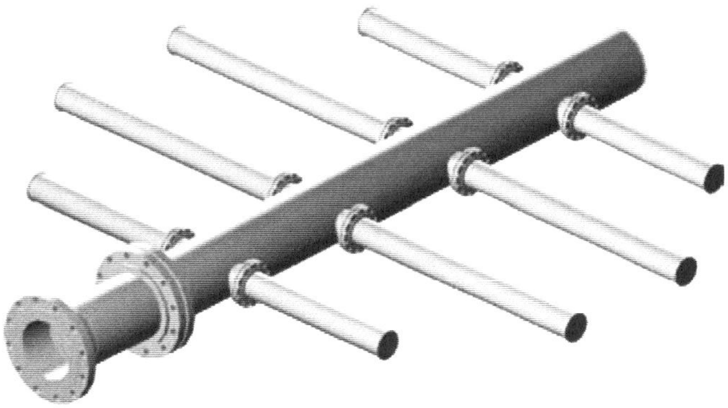

Figure 3: Typical ladder type distributor

Problem started shortly after the shutdown for cleaning. During the shutdown, pumparound coolers were cleaned; column packing and distributors were washed but piping was not washed nor flushed. It was postulated that liquid distributor was plugged by scale deposits released from piping and column shell walls after unit startup.

Rust and scale deposits formation is commonly seen and reported in column operations. This specific problem and design criteria were discussed with the tower internals vendor before the revamp took place. For unknown reasons, a distributor prone to plugging was selected but no preventive measures were applied (no filter requested). The design approach made plugging problem practically inevitable.

Previously, operators had never encountered distributor plugging because of different distributor type: ladder distributor did not suffer from plugging as its open construction allowed the packing underneath to be wetted or washed. The replacement with a "new and better" channel type distributor, though may be an optimum choice from process point of view but ignoring possible plugging, resulted in a serious problem.

Cause of Fire

The shutdown time schedule was very tight so the absorber column was blown with air before all the packing was properly cooled. Pyrophoric compounds formation was underestimated; this column has never experienced problems associated with pyrophoric material.

The overlooked moment was the wrong distributor selection that had led to fouled and dry packed bed. As long as the distributor was fine, packed bed below was always wetted and pyrophoric compounds did not form. The column could also be water washed and cooled properly before opening of tower manhole. Plugged liquid distributor allowed for pyrophoric compounds formation and prohibited proper water washing during shutdown.

Modifications

- Column shell was repaired in the vessel manufacturer's shop.
- New tower internals were provided on an emergency basis. Channel type distributor was replaced with a proven ladder type.
- Pressure transmitter was installed at distributor inlet piping as an indicator of distributor plugging.
- No filter was installed based on previous successful experience with ladder type distributor.

Results

Unit was successfully put into operation after a three weeks repair period. Absorber column is operating well since then.

Morals learned from the Case Study

1) Newer is not always better. High distribution quality does not always mean the best performance in real sense. When selecting packed tower internals, particularly liquid distributor, it is important to understand the application and the operational philosophy.
2) The case study demonstrated the importance of communication with clients, not only during design stage but also when clients consult for advice on shutdown procedures.
3) Special procedures and precautions shall be implemented for columns cleaning where pyrophoric components may form, especially if water washing efficiency is compromised by the damaged distributor.

Case Study #2: Atmospheric column vibrations

Atmospheric distillation of crude oil is a core process of any oil refineries. The technology is well known for many decades and equipment design had been standardized, apparently leaving no room for surprises.

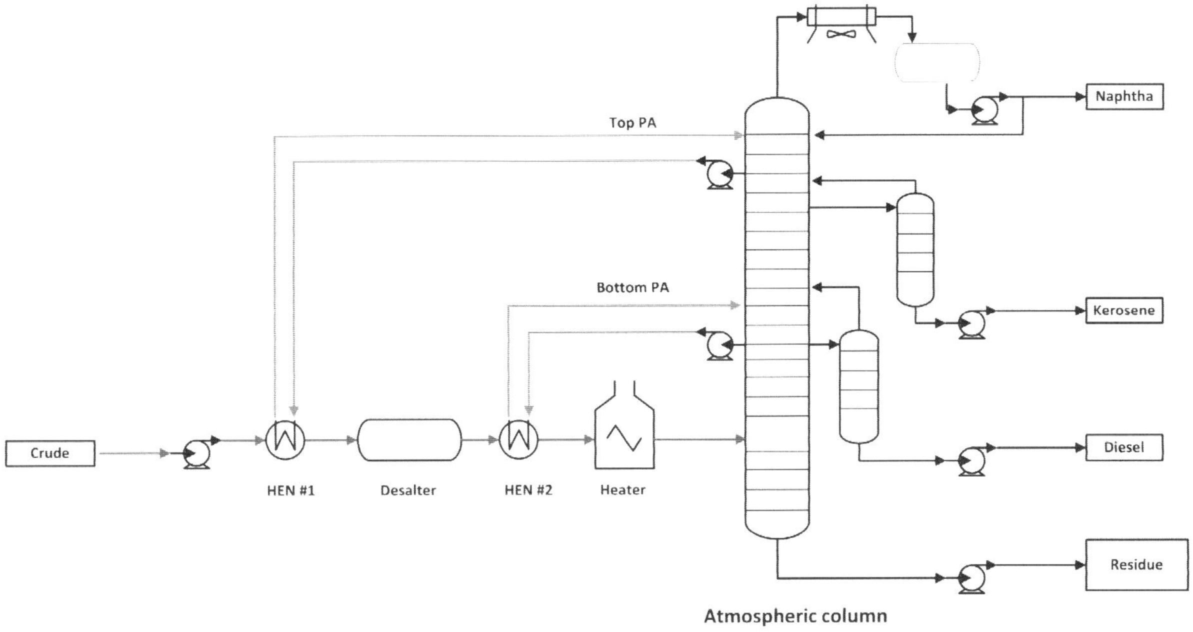

Figure 4: Crude oil distillation unit

Problem description

After the startup of a grass-root atmospheric distillation unit, a main fractionator column could not be operated at design conditions. When feed heater outlet temperature was gradually increased to design value, column vibration was observed because of frequent hydraulic shocks in pumparound section, and risk of flange gasket pop-out (what happened at the first startup). Hydraulic shocks could only be prevented at the expense of downgrading valuable diesel fraction to atmospheric resids (AR) by keeping heater outlet temperature lower than designed value.

Hydraulic shocks were most audible in column pumparound section thus a pumparound return distributor was suspected as a possible root cause. Tower internals supplier strictly rejected this hypothesis, and refused to visit the plant site claiming that "this behavior is not possible". Based on vendor statement, other potential causes were investigated:

- pumparound pumps cavitation
- vibration of heat exchangers and air coolers in the pumparound loops
- vibration / cavitation of check-valve at pumparound return nozzle
- liquid flashing at pumparound return nozzle / distributor

All the above possibilities were thoroughly investigated but both observation and calculations ruled out these possibilities. An independent troubleshooting effort was in pursuit to help resolve the problems.

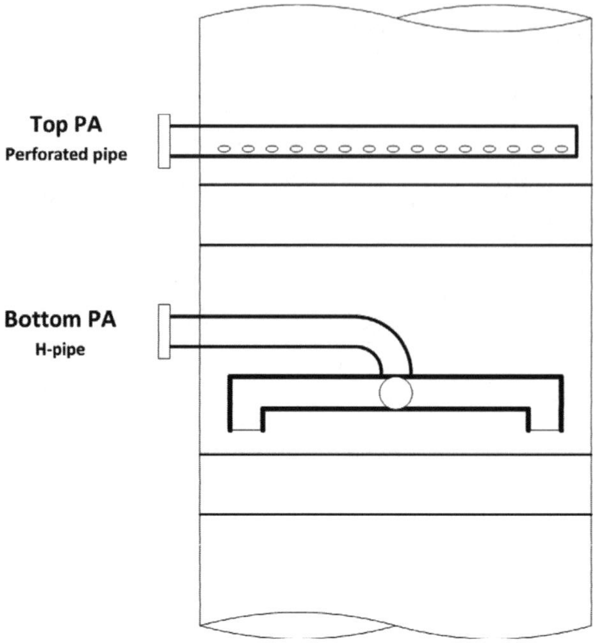

Figure 5: Pumparound inlet pipes before modification

Problem Analysis

Atmospheric column was started up in the following sequence:

- While crude oil is circulated through the unit, fired heater is lit on and atmospheric column feed is heated to its boiling point and above.
- Vapors from the column flash zone were condensed in the overhead system and all liquid is returned back as total top reflux.
- When collector trays are filled with liquid from the top reflux, top (TPA) and bottom (BPA) pumparound pumps are switched on. At the first moment liquid in the pumparound loops is subcooled.
- Column is slowly heated up and normal operating conditions are established.

The list below was gathered via interview sessions with plant operators:

- Operator did not observe column vibration when the column is operated only with top reflux.
- Hydraulic shocks started only when pumparound pumps are switched on (liquid is cold).
- Hydraulic shocks can be prevented only if P/A return temperature is kept high. *High P/A return temperature has limited pumparound cooling duty and consequently the feed heater outlet temperature. Maintaining a lower fired heater outlet temperature has resulted in loss of diesel to AR.*
- If P/A return temperature drops below a certain limit, hydraulic shocks started again.
- It was difficult to switch from top reflux to pumparound reflux.
- Attempts to keep P/A return temperature as high as possible has resulted in longer column startup and stabilization, up to12 hours.

Hydraulic shocks are usually induced by unsteady two phase flow, either by liquid flashing or vapor condensation. Based on field observations, high P/A return temperature prevented shocks while low temperature induces shocks thus liquid flashing at P/A return nozzle was excluded from the list. No matter how strange it sounded, the real cause of the problem is an instant vapor condensation on cold liquid surface inside of the P/A return distributor.

Based on field observations and troubleshooters' experiences, following sequence for hydraulic shocks was established:

Hydraulic shock (water hammer) mechanism:

- Hot vapor from column enters the distributor pipe and condenses on cold liquid surface.
- Instant vapor condensation created local vacuum (draft) inside the P/A return pipe, causing more vapor to be sucked into the pipe.
- High velocity vapor flow reversed the liquid flow direction, forming a slug inside of the pipe.
- The resultant slug flowed in the opposite direction against the main liquid stream. Hydraulic shock resulted when the slug struck the main liquid stream.
- Once the liquid exited the pipe the vicious cycle repeated itself.

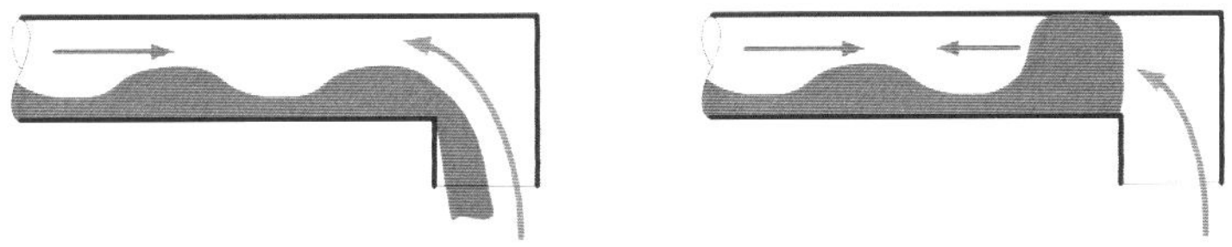

Figure 6: Vapor backflow causing water hammer

Modifications

Column internals modifications were performed during a short shutdown. Both top and bottom pumparound return distributors were equipped with hydraulic seals to prevent vapor backflow.

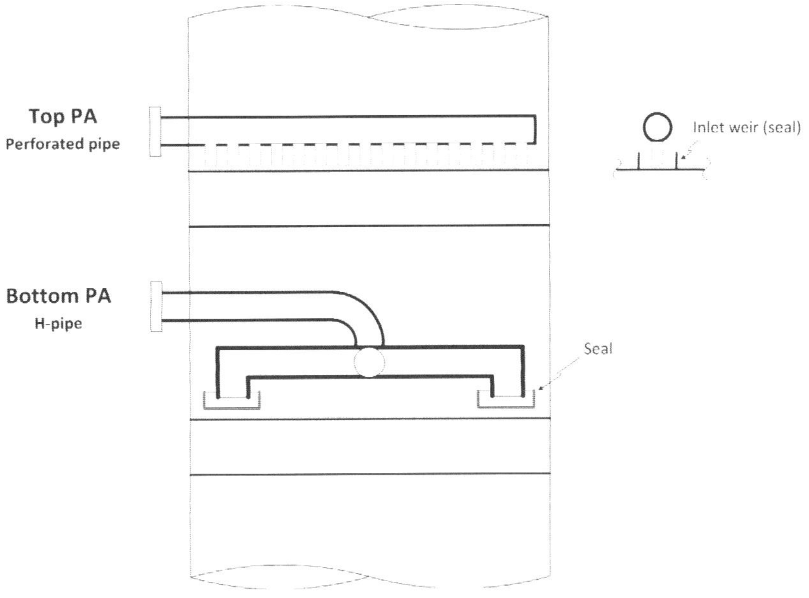

Figure 7: Pumparound inlet pipes after modification

Results

Crude oil distillation unit was successfully started after the modifications and reached design capacity and product quality without any sign of hydraulic shocks. Atmospheric column startup took only one hour, instead of 12 hours as previously experienced.

Morals Learned From the Case Study

1) Successful troubleshooting always starts from plant survey.
2) Overlooking the basics in column internal design often resulted in costly operational problems.

Distillation Tower Inspection
for
Turnarounds and Revamps

185 Kister Distillation Symposium 2017

Doug Bouck
Technical Consultant
Separation Solutions, Ltd

Safety

- Personal
  - Basic Refinery/Chemical Plant Safety Training
  - Confined Space Safety Training
  - Hole Watch Communication
    - Voice or Radio (Compatibility Issue)
  - Know Your Own Limits
    - Heights
    - Closed In Spaces
    - Darkness
    - Flexibility
    - Conditioning

2

Safety

- Hazards
  - Climbing
    - Outside: Cluttered Platforms and Ladders
    - Inside: Cords, Hoses, Damage
  - Encumbrances & Hang-ups: Harness, Meters, Radio, Tools, Coveralls
  - Falling Objects: Tray Parts & Tools
  - Trays & Other Internals: Loose, Corroded, Displaced, Sharp Edges
  - Partner or Other Workers/Inspectors
  - Tight Spaces – Exit Plan

3

Safety

- Hazards
  - Dust, Scale
  - Old Equipment - Asbestos Gaskets
  - Pyrophoric Dust & Scale Producing SO2/H2SO3 Over 104F
  - Packing Fire Prevention:
    - "Causes and Prevention of Packing Fires", Chemical Engineering, p. 34 (July 2007).
- YOU are Responsible for YOUR Safety
- YOU are Responsible for OTHER'S Safety

4

5

Typical Tower Internal Manways

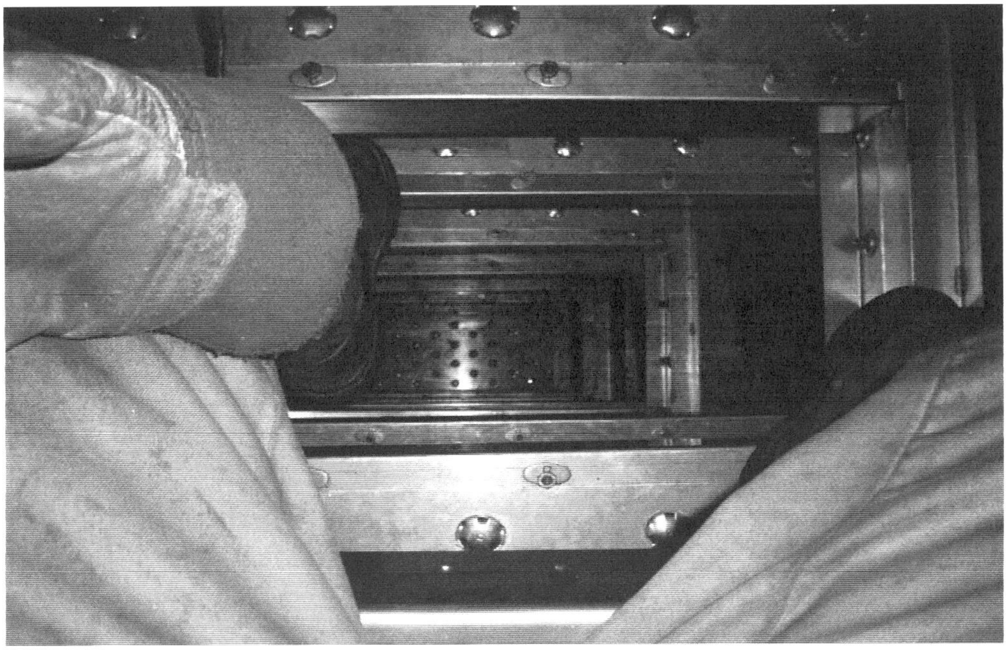

6

Tight Spaces

7

Multiple Personnel

8

Tight with Tray Damage

9

Cluttered Manway & Internals

10

Scaffold Inside Tower

11

Workplan Responsibilities

- Process Inspection
- Maintenance/Mechanical Inspection
- Vendor Reps./Inspectors
- Contractor Inspectors
- Operations Inspection
- Training Inspection Tours
- Punch Lists and Change Orders
- Follow-up Inspections
- Signoff Responsibilities

12

Inspection Equipment

- Typical in Waist Pack
  - Lock-out Lock and Tag
  - Zip Lock Bags & Magnet
  - Soapstone + Yellow Crayon + Black Marker
  - 10-25 ft Tape Measure
  - 9/16 Wrench & Plyers
- Other
  - Shin Guards and/or Knee Pads
  - Note Pad, Drawings & Inspection Plan
  - 2 Light Sources with Fresh Batteries
  - 2+ Pens/Pencils
  - Camera (Hot Work Permit)

13

Waist Pack Contents

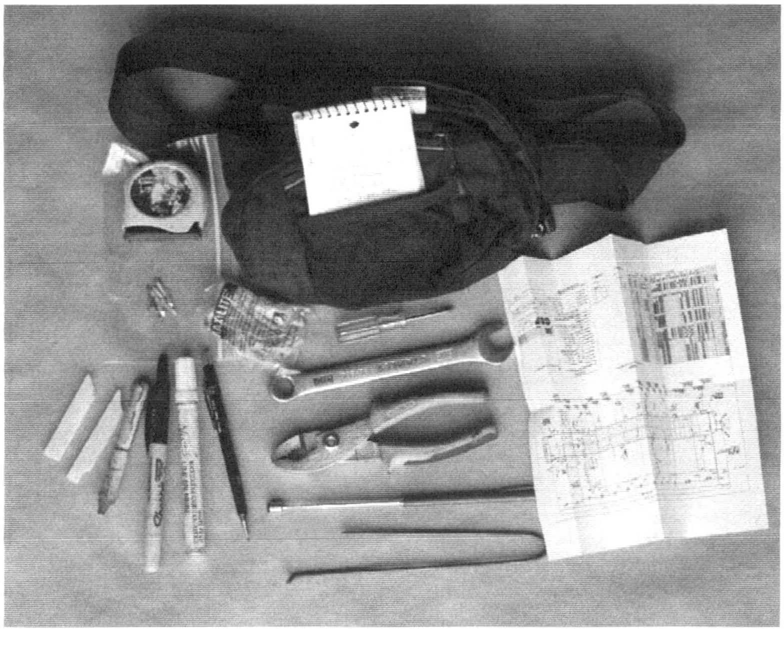

14

Other Equipment

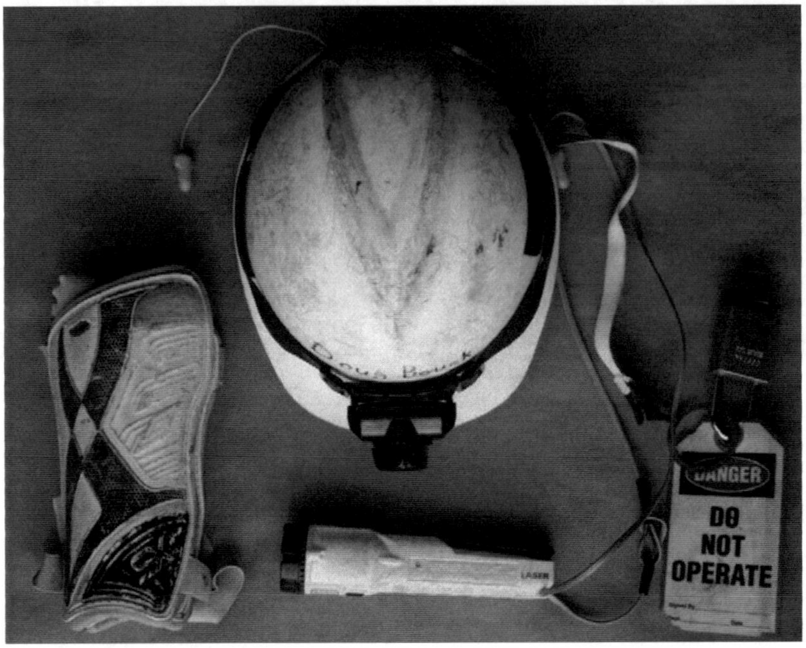

15

Inspection Equipment

- Special
  - Goggles, Dust Mask or Face Mask
  - Weir/Downcomer Dimension Template
  - 6 Inch Steel Rule
  - Dimension & Tolerance Table
  - Detailed Tray/Internals Drawings
  - Punch List and/or Drawings
  - Level Device: Bubble, Water or Laser
    - Device Accuracy Test
  - Distributor Test Equipment
  - Water Tightness Test Equipment

16

17

Keys to Process Inspection

- Understand How the Process Works
- Understand How the Parts Fit Together
- Understand How the Parts Work Together
- Know Why the Specs are Important
- Know How Deviations Affect the Process
- Be Able to Explain Operating and Economic Effects of Deviations
- Consider the Next Fix Chance May be 5 Years
  - Know When to Hold'em and When to Fold'em

18

Basic Preliminary Tray Inspection

- Number Trays for Reference
- Note Cleaning Issues – Take Photos
  - Does It Need Cleaning Before Full Prelim.?
  - Loose Deposits or Hard Scale?
  - Tray Decks/Downcomers and On/Under Valves
  - Any Special Cleaning Techniques Required?
- Note & Mark Missing/Stuck Valves
- Note & Mark Loose or Missing Hardware
- Note Cracked/Damaged/Corroded Trays
- Note Condition of Internal Manway Hardware
- Note Condition of Feed Pipes & Vortex Breakers
- Prepare Recommendations & Punch List

19

TAR Tray Inspection 1

- Objectives:
  - Return Equip. to Efficient Operation
  - Remember the Next Fix Opportunity May Be Years

- Cleanliness
  - Loose Scale/Deposits on Downcomers & Walls
    - Any Loose Scale Can Move to Downstream Equipment
  - Deposits on Decks
    - Plugging or Restriction of Valves, Holes or Caps
    - Bubble/Tunnel Caps Annular Space Must Be Clear
  - Accumulation in Downcomers, Sealpans & Drawpans

20

TAR Tray Inspection 2

- Missing/Worn/Stuck/Plugged Valves
  - One (1) Percent Acceptable Exception
  - Moving Valves Must Move Full Travel
    - Remove Deposits on Top, Legs or Around Holes of Valves
  - Check for Worn Valves & Tray Decks
  - Replacement Valves: Legs Bent for Retention
  - Replacement Valves with Cages: Crimping of Leg Tabs

21

Valve Cage Installation Problem

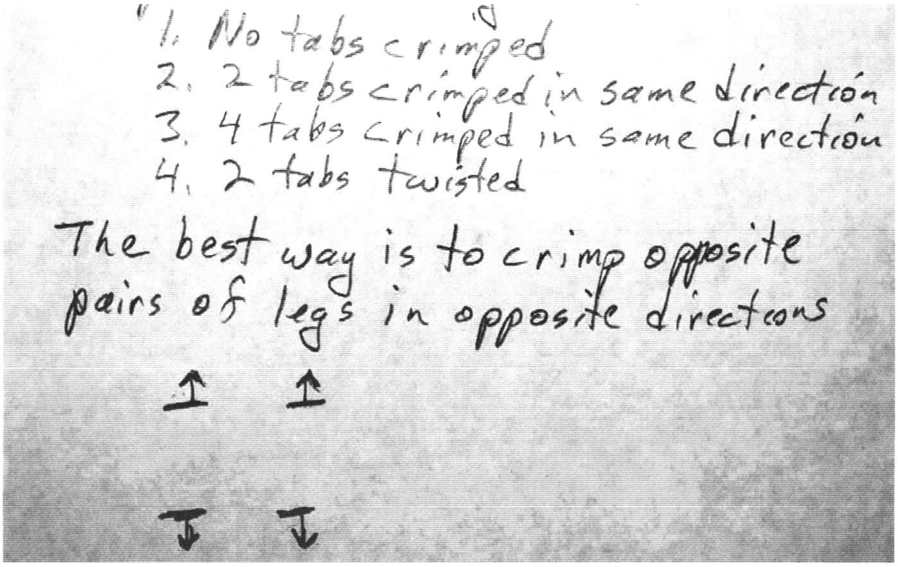

22

TAR Tray Inspection 3

- **Most Frequent Problem:**
  - Frozen/Damaged Internal Manway Hardware
    - Result of Hammer Release and/or Corrosion
    - Frozen Hardware Guarantees Poor Re-Closure
    - Poor Closure Causes Vibration Cracking of Trays
    - Poor Closure Causes Less Security Against Upset
    - Poor Closure May Result In Operating Deficiency
  - **Hardware Problems Must be Corrected Before Approval for Internal Closure**

23

TAR Tray Inspection 4

- Nutter MW Hardware Problems:
  - Subtle Correct Orientation
  - Rectangular Slots and Bolts
  - Boilermaker Hammer Release Damage
  - Same Poor Closure Issues
- Blind Fastening
  - Unusual
  - Difficult
  - Two Different Types of Hardware
  - Two Different Techniques
  - Boilermakers May Need Instruction

24

Nutter Internal Manway Hardware Orientation

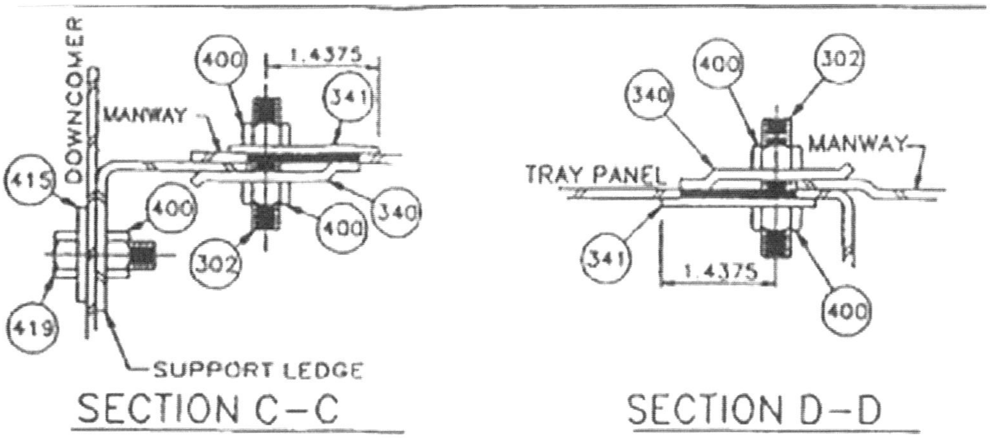

25

Blind Manway Bolting

26

Blind Manway Bolting

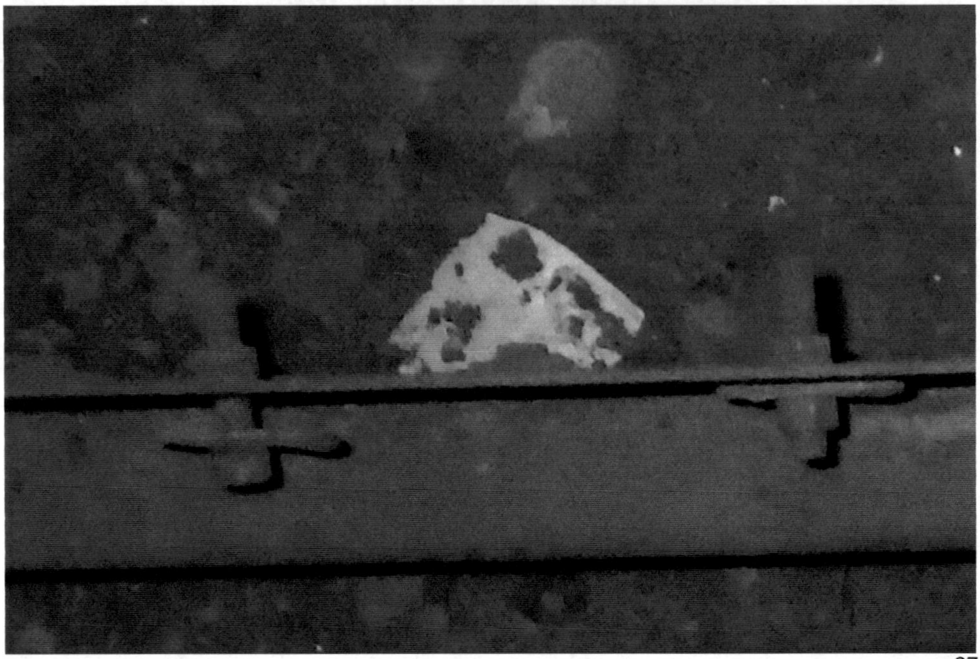

27

TAR Tray Inspection 5

- Other MW Hardware Problems
  - Light Wt. Football Washer Bends Causing Poor 2nd Closure
  - 3/16 in. Thick Heavy Duty Washers (Footballs) Recommended

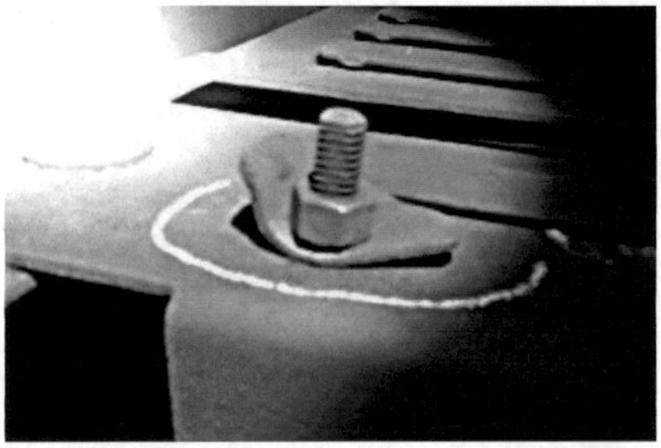

28

Holes in Downcomers

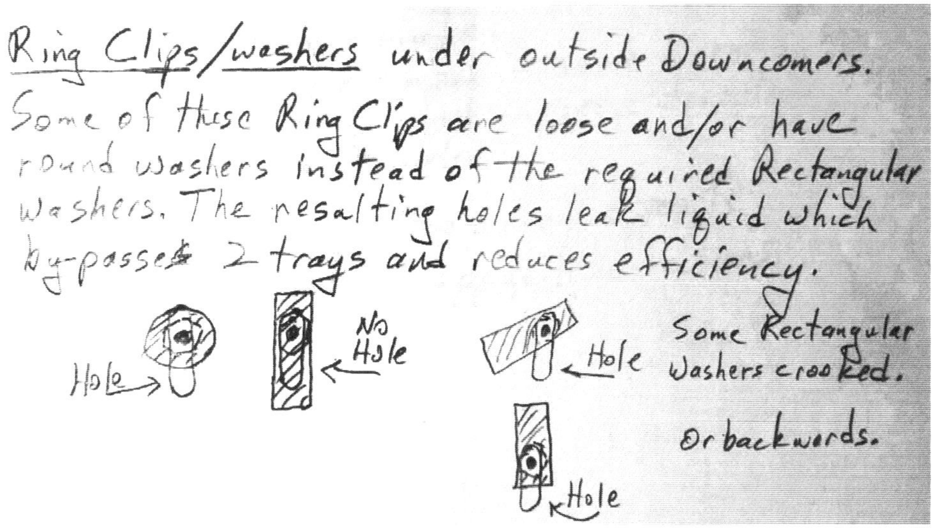

29

Internal Downcomer Loose Scale

30

Downcomer Scale

31

Dirty Round Valve Tray

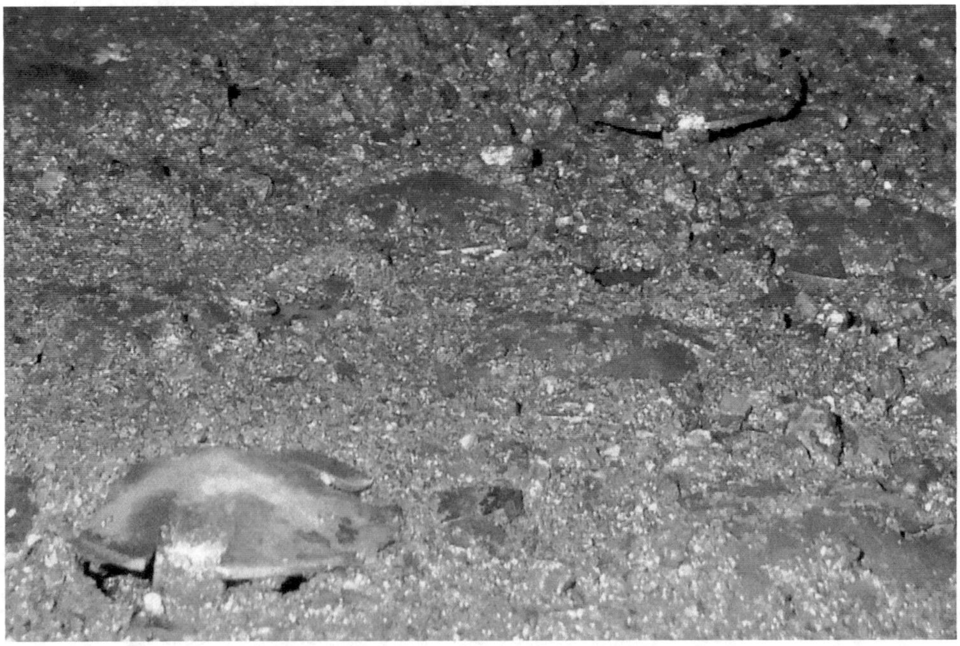

32

1- 1.5 Inch Deposits

33

Corroded and Plugged Bubble Caps

34

Displaced Weir

35

Bent/Loose Downcomer Clip

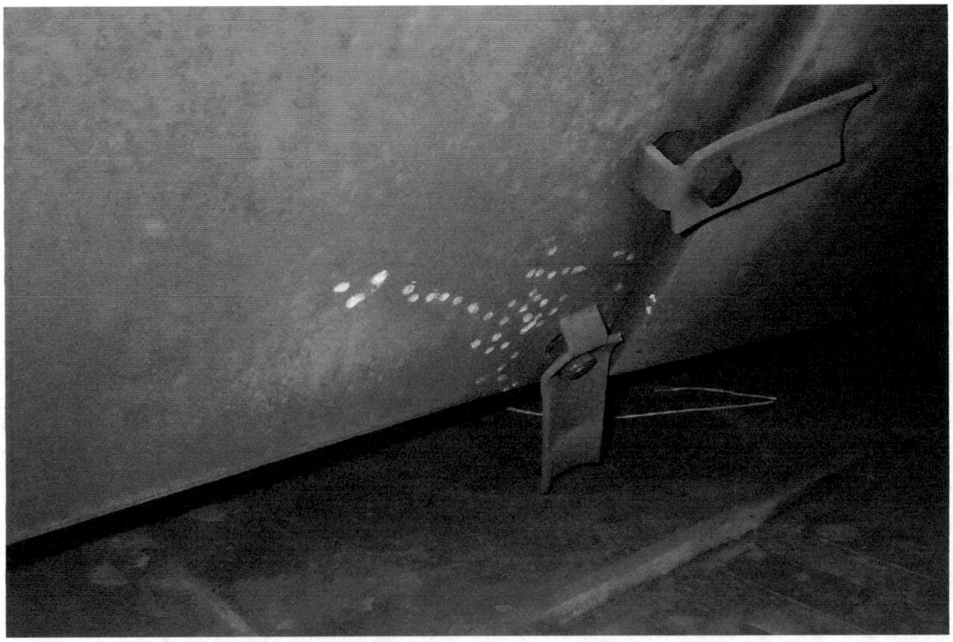

36

Bent Tray Downcomer & Clip

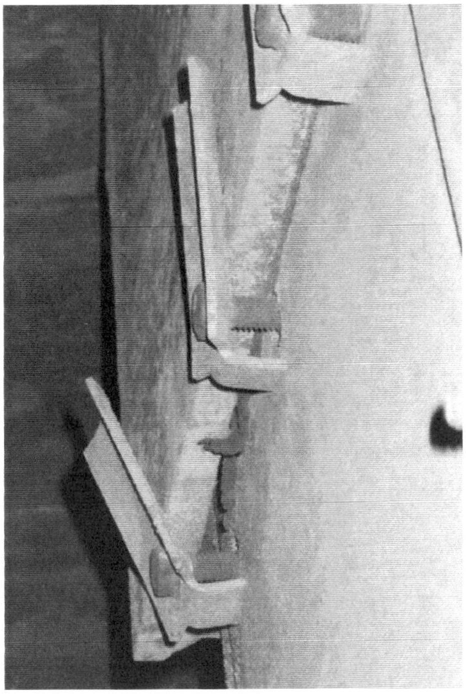

37

Ring Clip Askew

38

Displaced Valves

39

Displaced Valves

- Check the Legs Condition
  - Valve Legs Worn?
  - Improper Installation?
    - Legs not Spread Properly?
- Check the Hole Condition
  - Worn By Erosion-Corrosion?
- Blown Loose by Process Problem?
- Deck Thinning from Corrosion?

40

Missing Bubble Cap

41

Missing Swept Back Weir Piece

42

Missing Deck Bolts

43

Tray Damage Analysis

- Trays Bent or Cracked
  - Fatigued by Age and/or Weight
  - Overtightening
  - Level Loss or Explosion
- Trays Pushed Up
  - High Level + Percolating Vapor
  - Water Explosion Below
- Trays Pushed Down
  - High Level + Rapid Level Loss
  - Water Explosion Above

44

Boss says:

DOUGBERT, THE CEO SAYS THAT THE PROBLEM DIB COLUMN IS OPEN FOR INSPECTION AND YOU ARE THE MAN!

Dougbert says:

WHY ME? THAT TOWER IS OVER 200 FEET TALL!

Boss says:

HE SAYS YOU ARE THE ONLY ONE WHO KNOWS HOW THE INSIDES ARE SUPPOSED TO WORK.

45

Bent Tray Panel at Support Ring

46

Results of Loss of Level Control with Vapor Pushing Liquid Up Through the Trays

47

Uplifted Downcomer Panel

48

Lift Damage

49

Lift Damage

50

Lift Damage

51

Push Down Damage

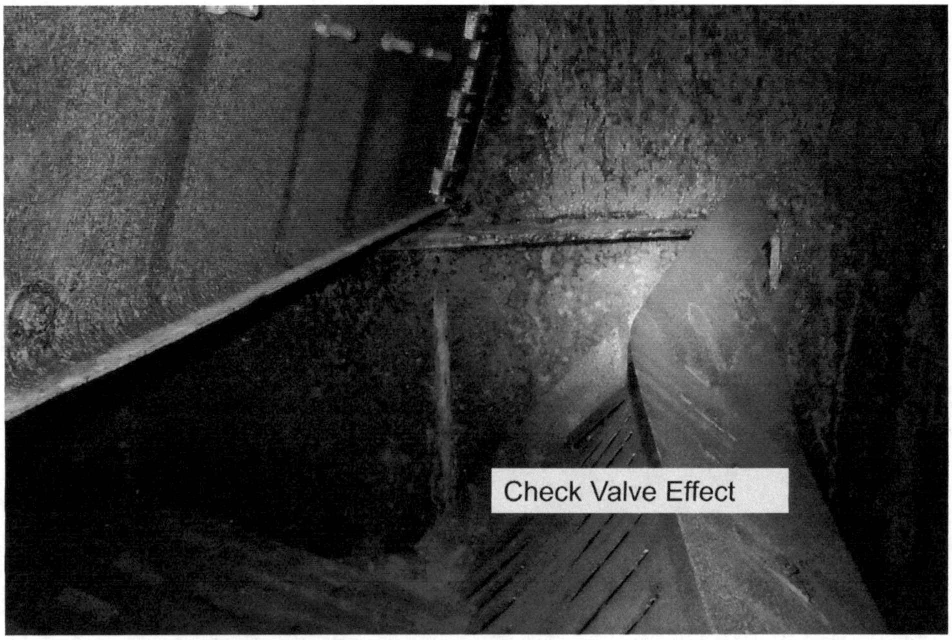

52

Push Down Damage

53

Brittle Failure

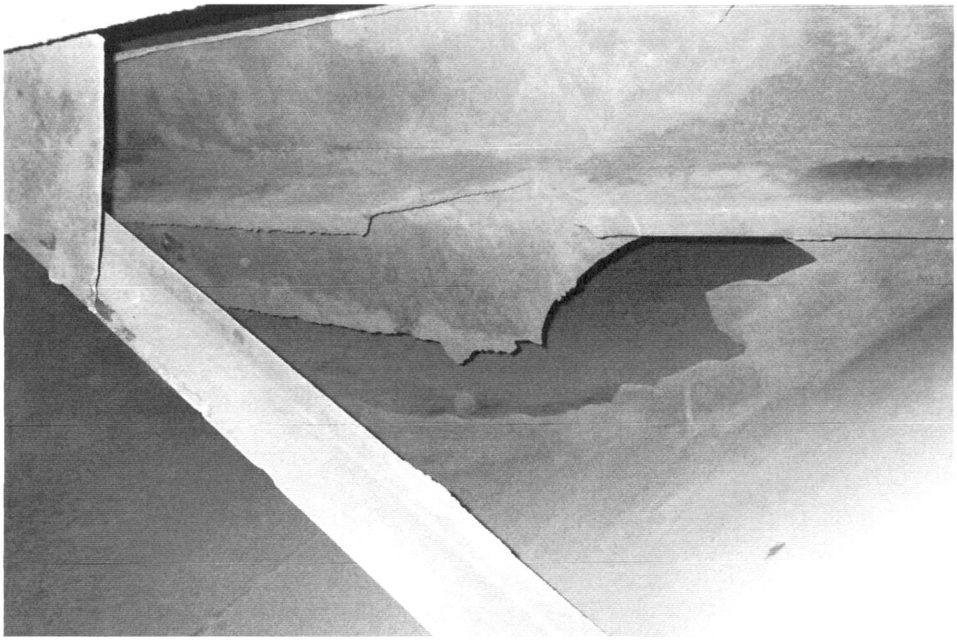

54

Process Revamp Review

- Pre-TAR Revamp Process Review
  - What is Different
  - Objective of Change
  - Impact of Change on the Process
  - Will the Changes Meet the Objectives
  - Are the Changes Consistent and Valid Engineering?
- Pre-TAR Mechanical Review
  - Mechanical Design meet Process Design?
  - Integration with Surrounding Systems: High Capacity Trays + Reboiler + Condenser?

55

Revamp Inspection

- Installed Mechanical & Process Check:
  - Installed per Drawings
  - DC Installation Blocks (Record DC Escapes & Any DC Changes)
  - Alignment of Tray Sections (Centering & Panel to Panel) Out of Round?
  - Correct Hardware Installation
  - Hardware Orientation & Proper Tightening
  - Punch List & Final Process Check

56

Revamp Field Inspection Examples

- Parts Verification Check: Example Valve Lift
- Small Parts Check and Security: Examples Twr. Closure Hold-up & Missing Nozzles
- Field Process Check: Example Missing Downcomers
- Field Dimension Check: Example Stripper Dbl. Flood
- Field Dimension Check: Support Ring Thickness
- Inspection Timing: Example Redesigned Feed Pipe
- Field Layout: Example Chimney Tray Water Draw

57

Example: Chimney Tray Water Draw

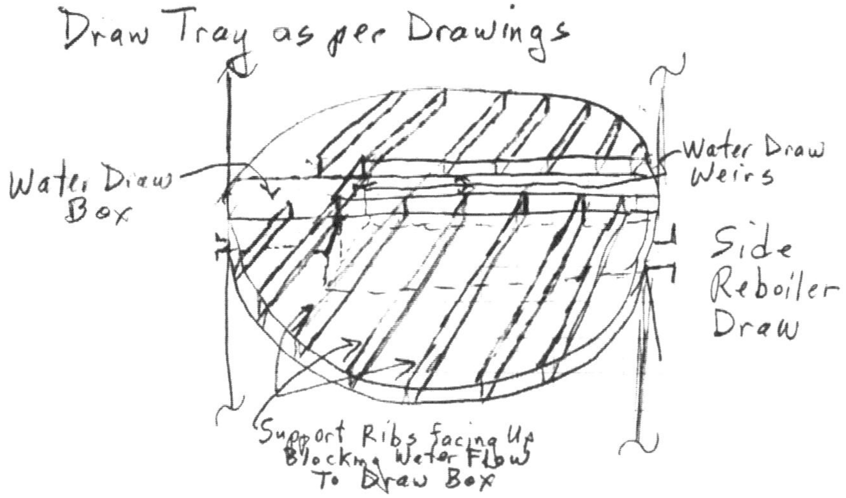

Bouck, Doug, Distillation Revamp Pitfalls to Avoid, Chemical Engineering Progress, p. 32-38, Feb. 2014

58

Special Inspections 1

- Feed and Return Pipe Distributors
  - Holes/Slots/Nozzles
    - Clear?
    - Corroded?
  - Flange Bolts Tight (Wrench Check)
  - Distributor Support Bolting Tight (Dbl. Nutting Use)
- False Downcomers & Distributor Impact Panels
  - Double Check for Loose Hardware
  - Need for extra security from vibration?
  - Vibration initiated cracking?

59

Distributor Hole Corrosion

60

Special Inspections 2

- Material Balance or Draw Problems?
- Flow or Level Control Problems?
- Suspected Leaks?
- Chimney Trays & Draw Boxes
  - Corrosion & Deposits in Draw and/or Piping
  - Thickness Checks in Suspect Corrosion Areas
  - Potential Leak Areas at Welds or Gasketing
  - Potential Leak Areas at Bottom (Under-deposit Corrosion)
  - Loose or Bent Hats, Chimneys or Decks

61

Chimney Draw Tray

62

Draw Box Corroded Through at Bottom

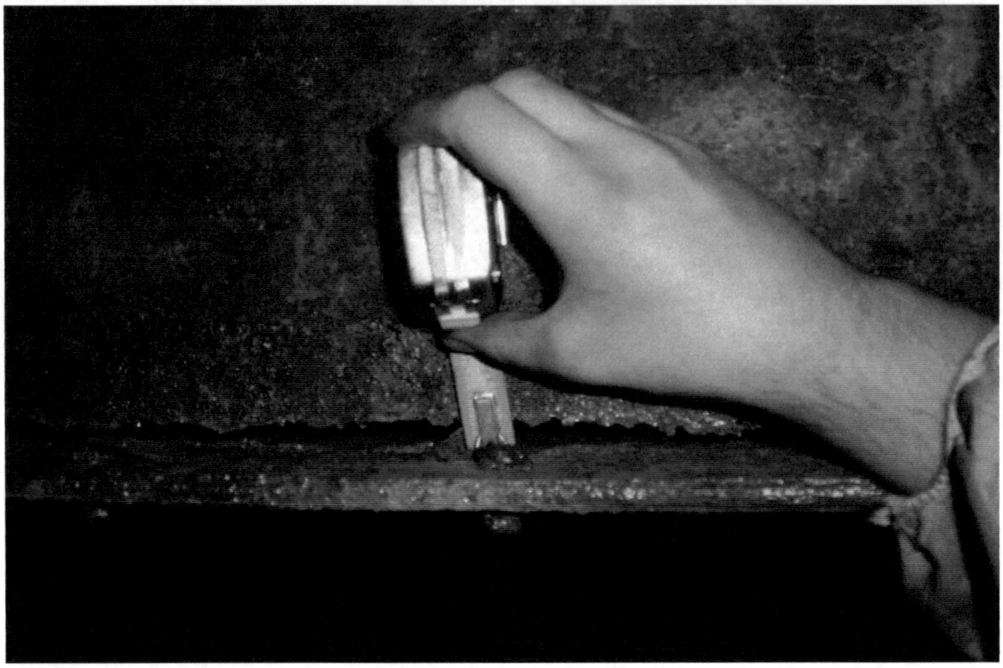

63

Draw Box Corroded Through at Welds

64

Special Inspections 3

- New or Revamped Chimney Trays
  - Gasketed or Welded
  - Leak Testing
    - Water Quality (Chlorides for Certain 300 Series Stainless and/or Solids in Firewater)
    - Quantity - Multiples Greater Than Leak Rate
    - Pressure - Pump Head Requirements (Water)
    - Test Depth - Approx. Operating Level
- Distributor Testing
  - On Site Spray Distributor Example
  - Off Site Packing Distributor Vendor Testing

65

Special Inspections 4

- Final Inspection Before Closure
  - All Internal Repairs & Cleaning Complete
  - Internal Manway Hardware Fixed or Replaced
  - Account For All Internal Manways – Directional?
  - Tools & Misc. Picked Up

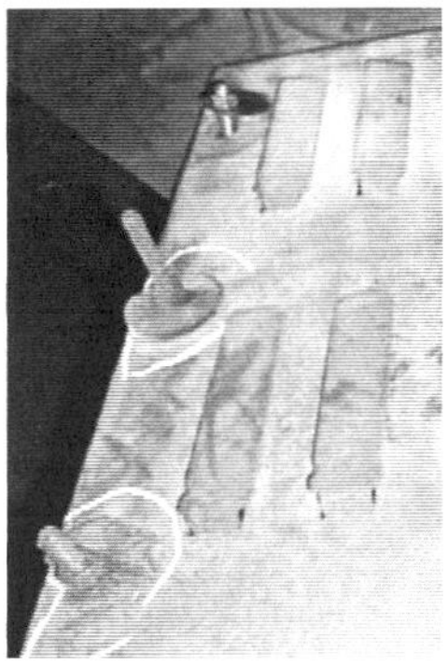

66

New Tower Inspection 1

- Tolerances
- Level Checks
  - Trays – Dependent on ring levelness
  - Weirs (If Adjustable)
- Downcomer Width – Top & Bottom
- Downcomer Escape
- Weir Heights (If Adjustable)
- Alignment/Centering of Tray Sections
- Valve Rise & Valve/Hole Counts

67

New Tower Inspection 2

- Hardware
  - Per Drawing
  - Correct Metallurgy
  - Bolting Snug Tight (~15 ft-lbs for 3/8 inch)
  - NOT Overtight
    - Bend Tray
    - Bend Clamps
    - Bend Washers
    - Gall Bolts

68

Pyrophoric Fire Prevention

- Pyrophoric Deposits
  - Iron Sulfide Can Be Most Anywhere Sulfur Is
  - Conversion Treatments Not Perfect
  - Prefer One-Step Processes Like ZymeFlow
  - Preventive Measures + Monitoring
  - FeS + O2, Dry and 104+ Degrees F. = Exothermic
  - Examples: Amine, Main Fractionators
- Carbon, Coke & Other Organics
- Hot Work
  - Read Chemical Engineering July 2007 Article on "Causes and Prevention of Packing Fires"

69

Packed System Inspection with NO Packing Removal

- No Specific Operating Problems
- Pre-TAR Gamma Scan
- Distributor Piping – Clean & Bolting Tight
- Pre-Distributor - Clean & Functional
- Distributor Pan, Troughs or Spray
  - Clean
  - Holes or Nozzles Clear
- Visual Inspection Top of Packing
- Visual Inspection Bottom of Packing

70

Structured Packing Installation

Feed Inlet Pipe
Predistrubutor
Distributor
Optional Holddown
Top Layer of Packing
90 Deg. Oriented Next Layer

Alternating Orientation of Layers

Bottom Layer of Packing
Packing Support
Liquid Collector

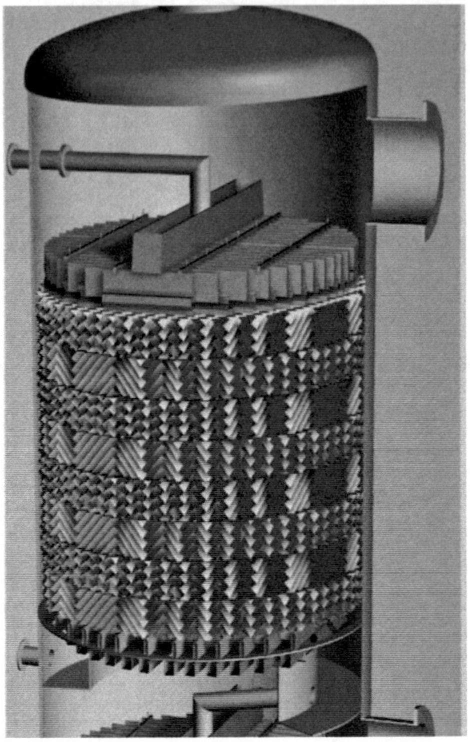

Sulzer Chemtech Brochure

71

Packed System Inspection With Packing Removal

- Before Re-Installation of Packing Check:
  - Draw Integrity – Check for Corrosion & Cleanliness
  - Support Pan/Grid – Clean/Unplugged
  - Check Support Orientation
  - Check Connections and Support Bolting Tightness
- Structured Packing Installation:
  - Verify Packing Size, Type & Metallurgy
  - Starting Orientation is Important
  - 90 Degree Alternate Layers (45 or 60 Degree Rotation for Grid)

72

Protect the Packing

- Walk on Planks NOT the Packing
- WHY for Structured Packing??
  - Structured Packing is Very Thin
  - Boots Smash the Top Crimps
  - Smashed Crimps Reduce Capacity
  - Smashed Crimps Reduce Efficiency
  - Hidden Damage is Untraceable
- Make Sure the Contractor is Complying

73

Protect the Packing

Stand on Planks
not the Packing

74

Protect the Hands/Arms

Koch-Glitsch Brochure

75

Installation

- Personal Safety
  - Gloves for Sharp Edges
- Orientation to Support
- Orientation Layer to Layer
- Shoehorns for Multi-piece Installation
- Distribute Excess Space Between Pieces
- Wall Wipers of Various Designs

76

Structured Packing Layers

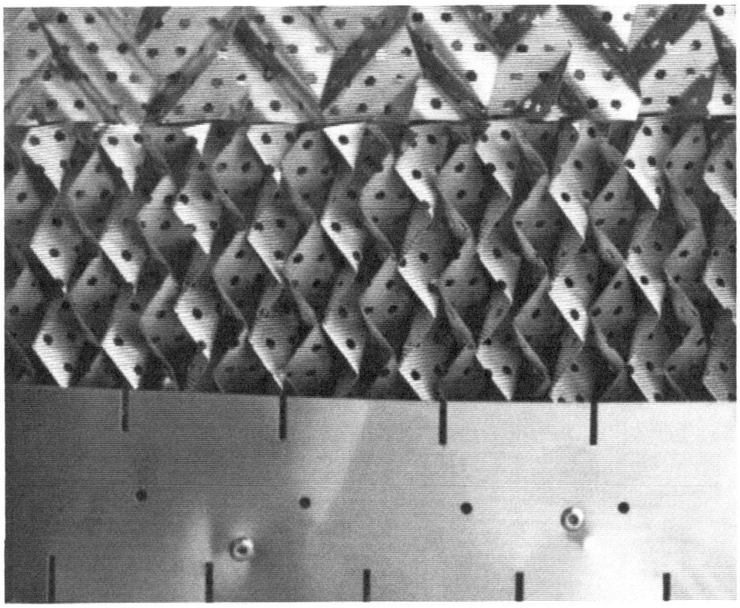

FLEXIPAC® Structured Packing with perforated and textured surface shown with attached wall wiper band

77

Other Wall Wipers

Sulzer Mellapac Structured Packing

78

Holddown Grid – Miss-installed

79

Distributor Inspections

- Holddown – Bolting Tight
- Level Distributors & Pre-Dist. 1/8 in. High to Low
- Installation Hold Point for Level Check/ Adjustments
- All Distributor and Predistributor Bolting Tight
- Distributor Clean
- Predistributor Clean
- Feed Piping Re-installed & Bolted Tight

80

Spray Distributors

- Check all Spray Nozzles and Headers
  - Remove Nozzles (Have Spares on Hand)
  - View Nozzles Internals – Clean or Replace
  - Save Nozzles in Safe Place
  - Check and Clean Headers as Necessary
  - Wrench Check all Nozzles for Tightness
  - Wrench Check all Flange Bolting
  - Double Nut and Check all Header Support Bolting

81

Detached Spray Headers

82

Upset Structured Packing (No Holddown Grid)

83

Redistributor Between Packed Beds

84

Random Packing Installation

Feed or Transfer Pipe
Predistributor
Distributor
Retaining Grid

Random Packing

Support Plate
Support Ring

Vapor Inlet

Liquid Level Collector/Sump

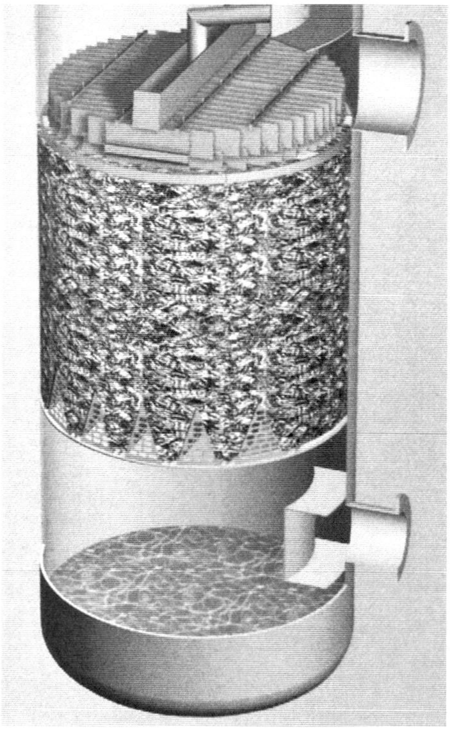

Sulzer Chemtech Brochure

85

Packing - Random

- Removal for Replacement
  - Remove Piping, Distributors & Retaining Grid
    - Safe Storage of All Parts
    - Record Orientations for Re-Installation
    - Clean All Parts
  - Remove Rings by Vacuum
  - Clean Any Collector or Draw System
  - Clean Packing Support
  - Check Support Integrity & Bolting

86

Packing - Random

- Installation
  - Verify Packing Size, Type and Metallurgy
  - Dump New Packing
  - Rake Top
  - **Do Not Overfill!**
  - If Necessary to Enter, Stand on Planks not Packing
  - Re-install Retainer Grid
  - Level Distributors & Pre-Dist. 1/8 in. High to Low
    - Installation Hold Point for Level Check/Adjustments
  - Install Feed Piping
  - Check Tightness of All Bolting

87

Protect the Packing

- Walk on Planks NOT the Packing
- WHY for Random Packing??
  - Random Packing is Thin
  - Boots Crush or Compress the Rings
  - Hidden Damage is Untraceable
  - Crushed Rings Reduce Capacity
  - Crushed Rings Reduce Efficiency
- Make Sure the Contractor is Complying
- Stay Totally Off, If Possible

88

The Process Engineer
(A Bit of Philosophy)

- Little Authority
- Much Responsibility
- Bridge Operations, Maintenance & Management
- Optimization Economics - Beyond Politics
- Relationships Create Influence
- Influence Creates Accepted Recommendations
- Recommendations Create Improvements
- Improvements Create More Profits
- More Profits Create Stable Company and Jobs

89

Take Ownership of the Future. The Role of Process Engineers in Inspection **Is** Important!

2 Peter 1:15

And I will make every effort to see that after my departure you will always be able to remember these things.

YOU ARE the FUTURE!

90

Quantification of the Effect of Viscosity on Point Efficiency of Sieve Trays in Distillation Columns

Raj Manivannan[1], Mason Dupré[1], Ken McCarley[2], Tony Cai[2], Sayeed Mohammad[1], and Clint P. Aichele[1]

(1) Oklahoma State University

(2) Fractionation Research Incorporated

Outline

- Overview
- Motivation
- Background
- Experimental Set Up
- Low and High Viscosity Data
- Model Comparison
- Future Work

Overview

- Methods for calculating number of theoretical trays can be considered as well developed
- Overall tray efficiency = $\frac{N_{theoretical\ trays}}{N_{actual\ trays}}$
- Prediction of distillation efficiency does not always give satisfactory results

Overview

- Under prediction of efficiency ⟶ Over design of the column ⟶ Increase in capital cost
- Over prediction of efficiency ⟶ Under design of the column ⟶ Desired capacity and product purity not met
- Efficiency depends on
  - Properties of the liquid mixture and
  - Mechanical design of the internal (tray or packing)

Importance of Viscosity

- One of the most important properties that affects distillation efficiency is viscosity
- Viscosity strongly influences diffusivity
  - Wilke Chang correlation

$$D_{AB} \propto \frac{1}{\mu_{Liquid}}$$

- Viscosity influences column hydrodynamics

Motivation

- Nearly all of the distillation efficiency correlations were developed based on data less than 1 cP
  - O'Connell's correlation(1946) was developed based on data from 0.07 to 0.5 cP
- Need for high viscosity data in order to
  - Quantify the effect of viscosity on distillation efficiency above 1 cP
  - Develop improved efficiency correlations

Background

- Drickamer and Bradford (1943)
  - Used viscosity to calculate efficiency
- O'Connell (1946)
  - Used viscosity and relative volatility to calculate efficiency
- Mahiout and Vogelpohl (1984)
  - Impact of viscosity in sieve trays for absorption efficiency
  - Conclusion: Increasing viscosity leads to a large reduction in mass transfer

Background

- Bradtmöller and Scholl (2015)
  - Proposed a new test mixture (isobutanol/tert-amyl alcohol) for distillation efficiency experiments at high viscosities

    (μ=0.5-3.5cP, with P=0.02 bara to 0.95 bara)
  - Demonstrated the inability of models to predict the influence of high-viscosity on packed column efficiency
- Insufficient distillation efficiency data available for tray columns at high viscosities (more than 1 cP)

Background

- Fair et al. (1983) – Oldershaw Columns
  - Fair investigated 5 different test systems at different operating conditions
  - Converted commercial column efficiency data into commercial point efficiency data
  - Plotted Efficiency Vs %Flood
  - Agreement of commercial point efficiency data with Oldershaw data in the region of practical interest (50-85% flood) is satisfactory

Fair, J.R., H.R. Null, and W.L. Bolles, *Scale-up of Plate Efficiency from Laboratory Oldershaw Data.* Industrial & Engineering Chemistry Process Design and Development, 1983. **22**(1): p. 53-58

Background

- What is an Oldershaw column?
  - Small diameter columns- available in 25mm, 50 mm and 75 mm ID
  - Sieve trays with central down comer
  - Due to the very small diameter of the column, point efficiencies can be directly measured by taking top & bottom samples and using Fenske equation

    (Oldershaw column efficiency = Point efficiency)

Oldershaw Trays

- Tray diameter=50 mm
- Hole diameter = 0.89mm
- No. of holes = 294
- % hole area $(\frac{hole}{bubbling} * 100)$ = 10%
- Downcomer diameter = 10 mm
- Weir height = 3 mm

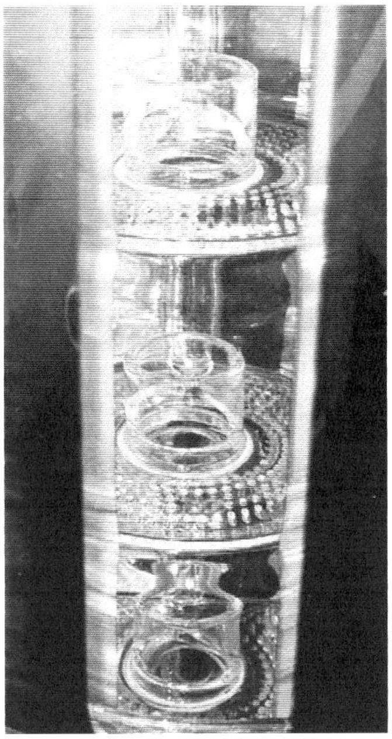

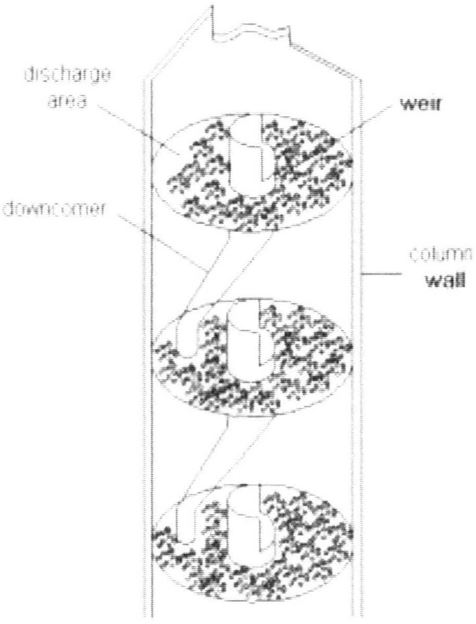

Experimental Set Up

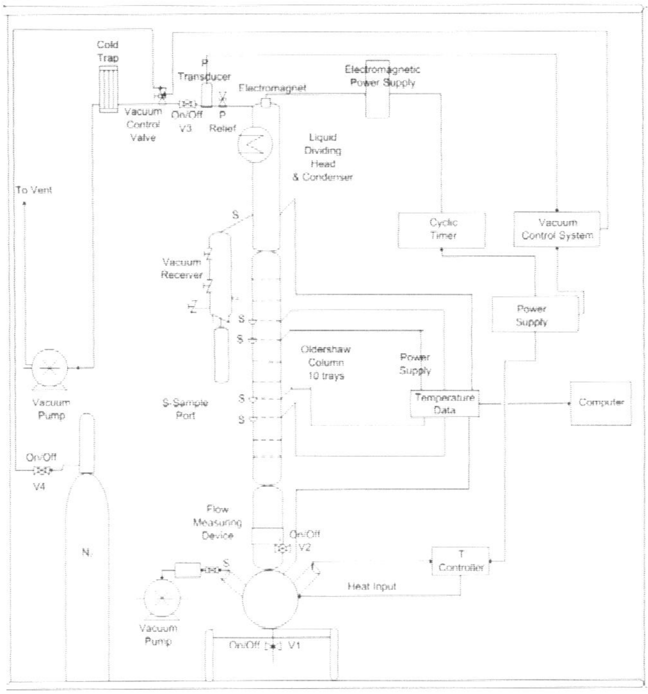

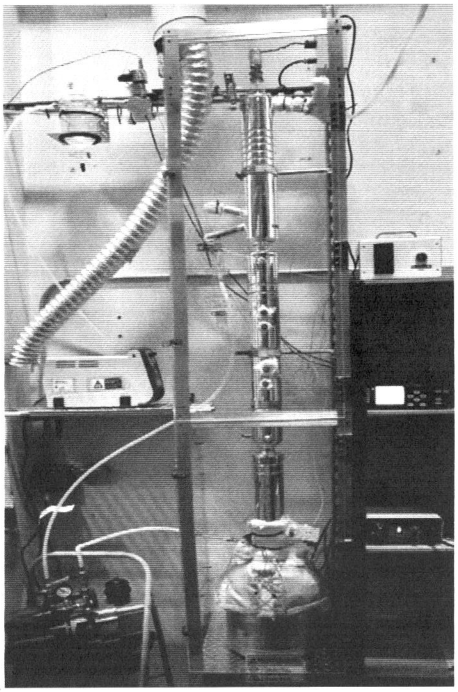

Column Validation

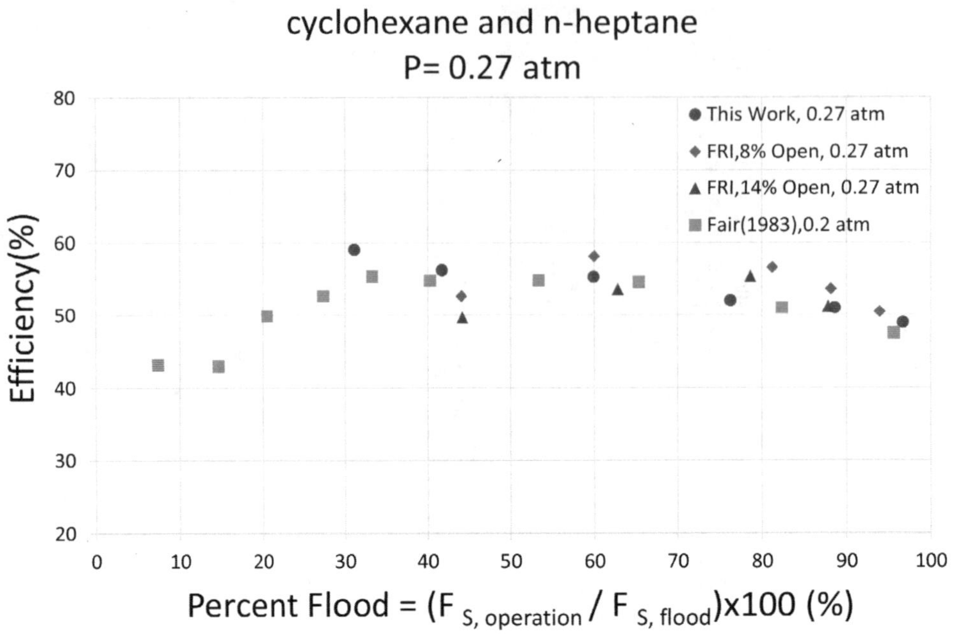

Average absolute difference in efficiency values between Fair and this work data = 1.56%

13

Data Consistency

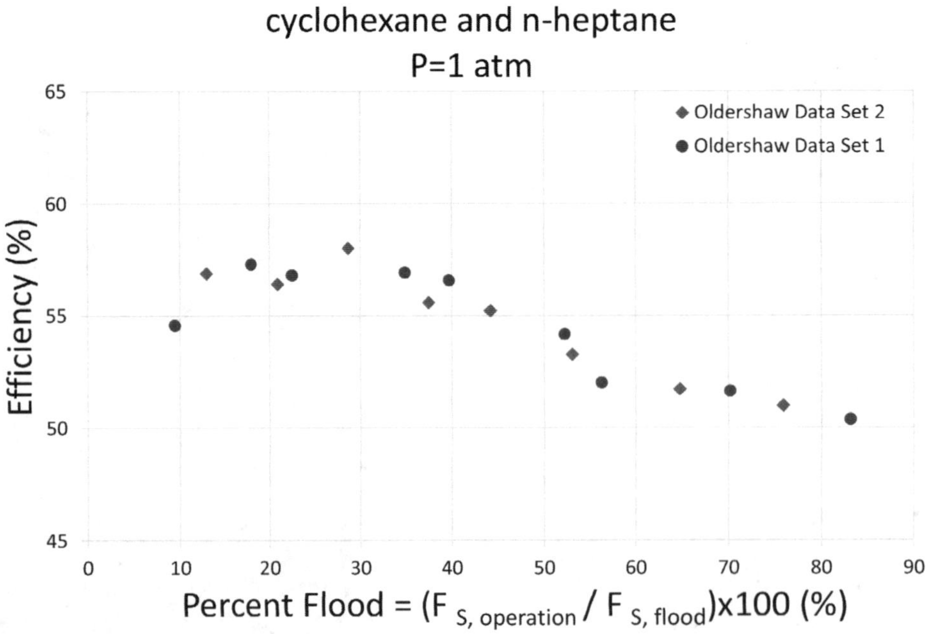

Average absolute difference in efficiency values between data set 1 and 2 = 0.763%

14

O'Connell's Correlation

- Based on test data from 29 commercial columns , 5 laboratory columns and 3 unpublished tests on commercial columns
- Overall tray efficiency, $E_0 = \frac{9.06}{(\mu_L \alpha)^{0.245}}$
  - μ_L= Liquid viscosity at average column temperature
  - α = Relative volatility at average column temperature and pressure

Low Viscosity Data cyclohexane/n-heptane

P=0.27 atm, μ_{avg}=0.41 cP

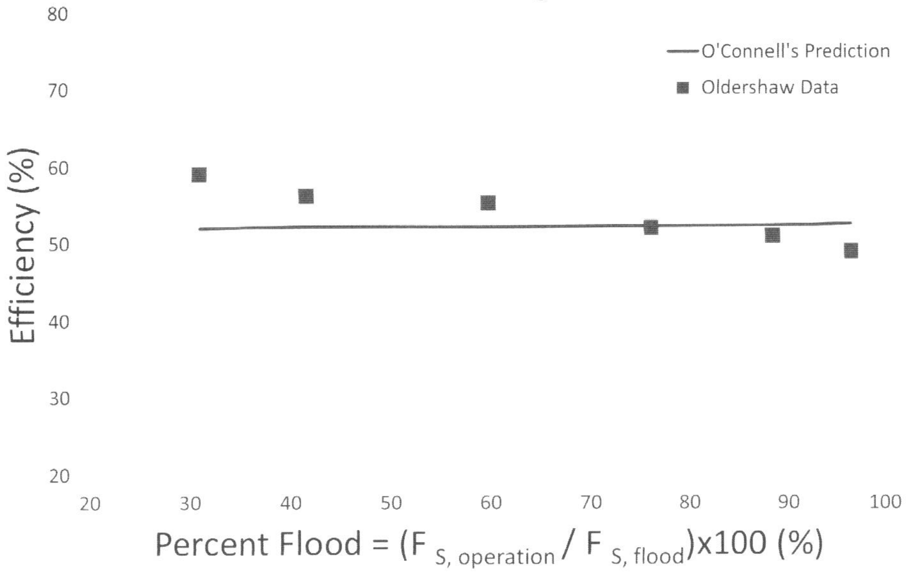

Average Relative Difference(50-85% Flood) = 1.62 %

High Viscosity Mixture

- Bradtmöller and Scholl (2015) test mixture
  - isobutanol/tertiary amyl alcohol
- VLE and viscosity data available at low pressures (0 to 1 bara)
- Operating at deep vacuum levels (less than 0.1 bara) helps to achieve high viscosities (1 cP - 3.5 cP)
  - Viscosity is inversely proportional to temperature

Low Viscosity Data isobutanol/tert amyl alcohol (IB/TAA)

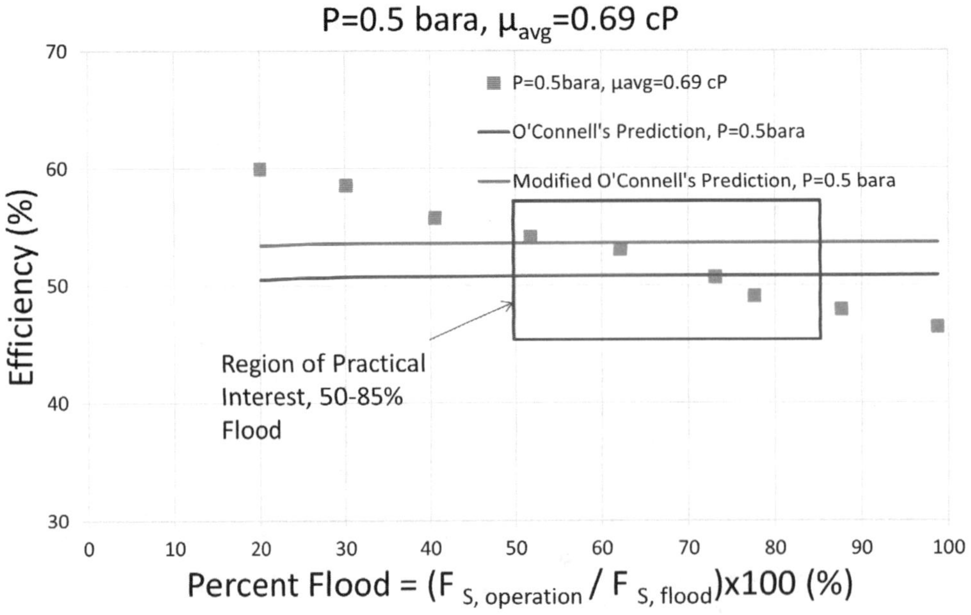

Average Relative Difference(50-85% Flood) = 3.57 %, 4.31 %

High Viscosity Data isobutanol/tert amyl alcohol (IB/TAA)

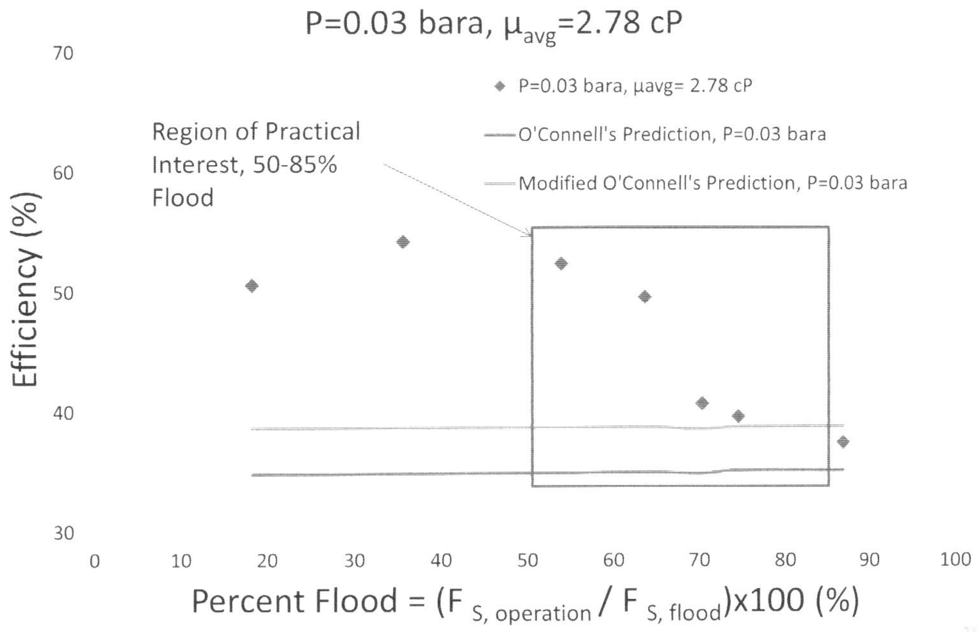

Average Relative Difference(50-85% Flood) = 13.83%, 22.13 %

Future Work

- P=0.02 bara (≈3.5 cP) experiment with IB/TAA
- Identify more high viscosity systems and collect high viscosity data
- Compare the collected new high viscosity data with predictions using point efficiency correlations
- Develop new or extend existing correlations for high viscosities

Acknowledgements

- Fractionation Research Inc. for funding this research
- Dr. Frank Seibert (UT Austin)
- Barbara Todd (Phillips 66 Research)

Questions?

Investigating the Impact of Feed-Air Saturation in Air-Water Spray Entrainment

Mason Dupré[1], Raj Manivannan[1], Tony Cai[2], Ken McCarley[2], and Clint Aichele[1]

Oklahoma State University[1]
Fractionation Research Incorporated[2]

Outline

- Motivation
- What is PDI?
- Equipment Overview
- Experiment Overview
- Results
- Discussion
- Conclusions
- Future Work

Motivation

- Build on prior FRI-OSU entrainment research
- Feed gas relative humidity varied due to:
  - Weather conditions
  - Building HVAC
  - Duration of operation
  - Time of day
- Attempt to validate prior work while validating future work
- Water volatility

What is PDI?

- PDI system measures, without any interaction or interference with the process:
  - Droplet diameter
  - Droplet velocity
  - Volume flux

What is PDI?

Forward-Scatter Diameter Analysis and Backscatter Diameter Analysis

	Forward-Scatter Results	*Backscatter Results*	*Glass-Bead Reference*
Mean Diameter (μm)	97.4	97.1	97.5
Standard Deviation(μm)	8.56	10.4	3.4
$P(Z \leq z)$ two-tail (Matching Glass Bead Data Sheet)	0.99	0.97	

What is PDI?

Air-Water Saturation System

- Air-Water Saturation Unit (<20% RH to ≥80% RH)
  - Requires adding 0.0083 lb water/lb feed air
  - 80% RH was determined to be an optimum target
  - Output humidity is controlled using water temperature and flow rate

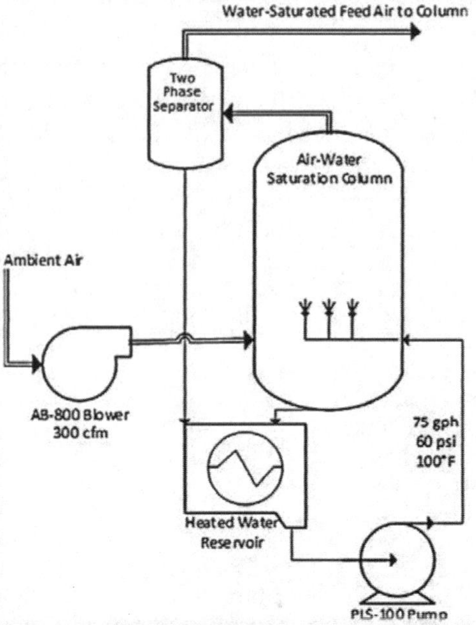

8-inch Test Column

- C-factor from 0 to 0.6 ft/s
- BETE 0.30 MPL spray nozzle
- Height: 13 feet
- Diameter: 8 inches

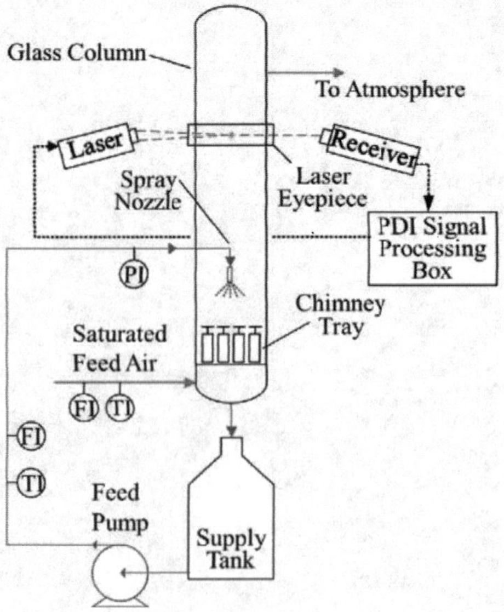

Experiment Overview

- PDI system was configured the same way for unsaturated and saturated feed measurements
- PDI measurements only taken after system hits steady-state
- All unsaturated data sets taken at once, then saturated data is taken after saturation hits steady-state

Experiment Overview

- All variables recorded at beginning and end of data sets to confirm steady-state
- PDI measurements will be made at radial positions:
  - R = 0.00 in.
  - R = 0.50 in.
  - R = 1.00 in.
  - R = 1.50 in.

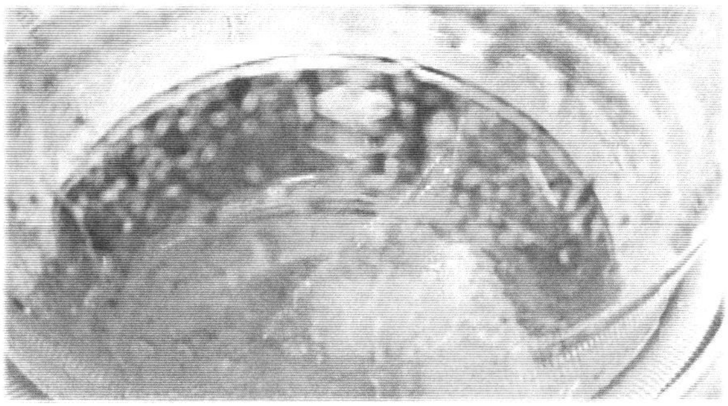

Experiment Overview

- Feed air velocity and temperature measured using hot-wire anemometer in two-phase separator effluent and using PDI
- Feed air humidity and temperature measured using hygrometer in two-phase separator effluent
- Two-phase separator removes all entrained droplets from saturation process prior to introduction to column

Motivation – What is PDI – **Overview** – Results – Discussion - Conclusions – Future Work 11 EFRI

Experiment Overview

- Definitions:
  - D_{10} = Arithmetic mean droplet diameter (µm)
  - D_{32} = Sauter mean diameter, or droplet with a surface area to volume ratio that matches mean surface area, volume (µm)
  - $D_{V0.5}$ = Median volume diameter, or diameter for which half of all volume flux occurred below this diameter (µm)
  - v_{95} = 95th percentile velocity – representative of superficial vapor velocity (m/s)
  - $D_{0.5}$ = Median droplet diameter

Motivation – What is PDI – **Overview** – Results – Discussion - Conclusions – Future Work 12 EFRI

Results

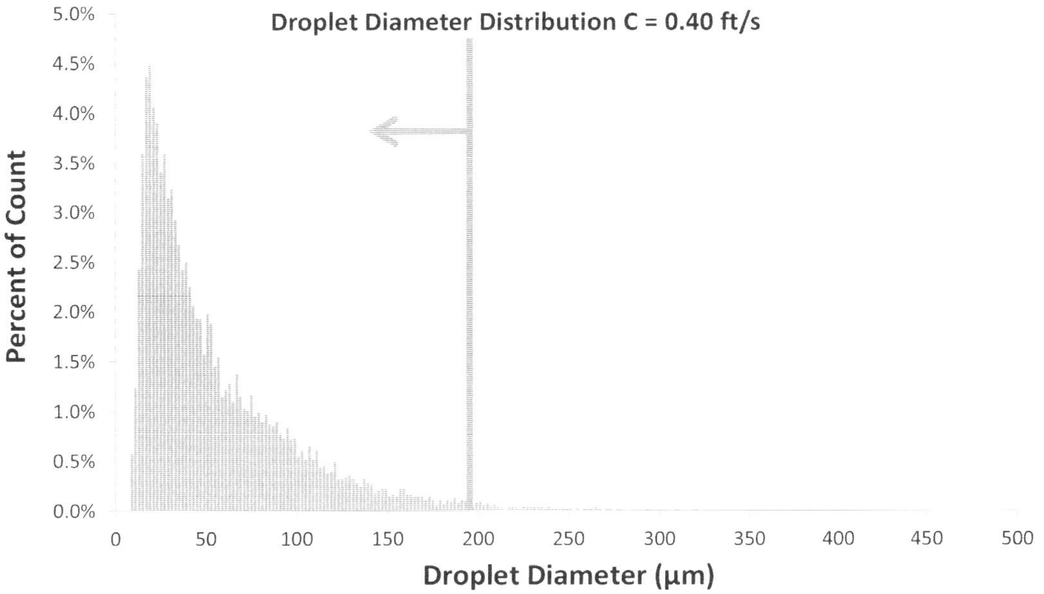

Results

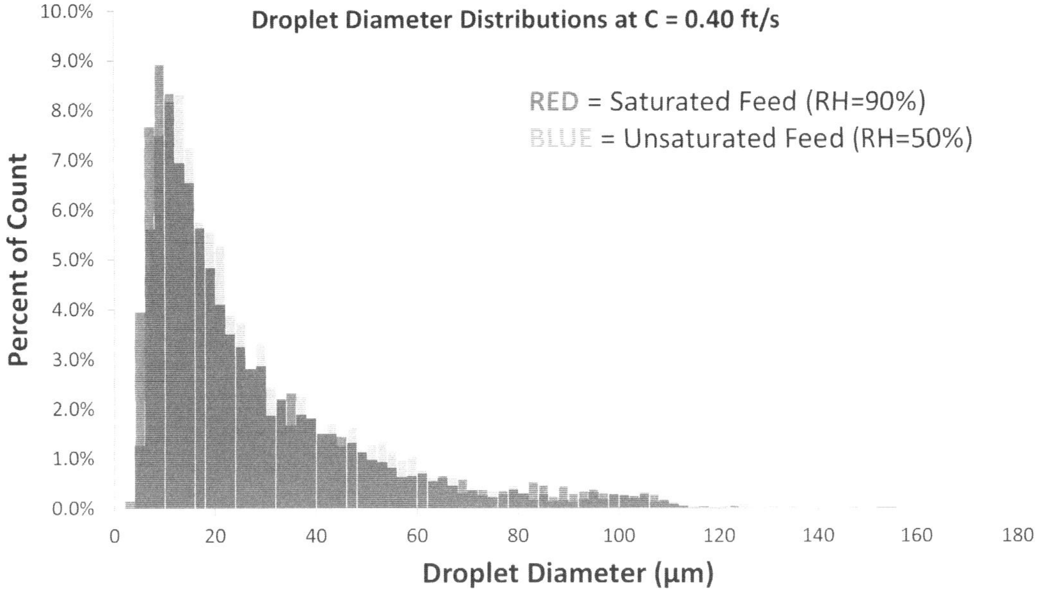

Results

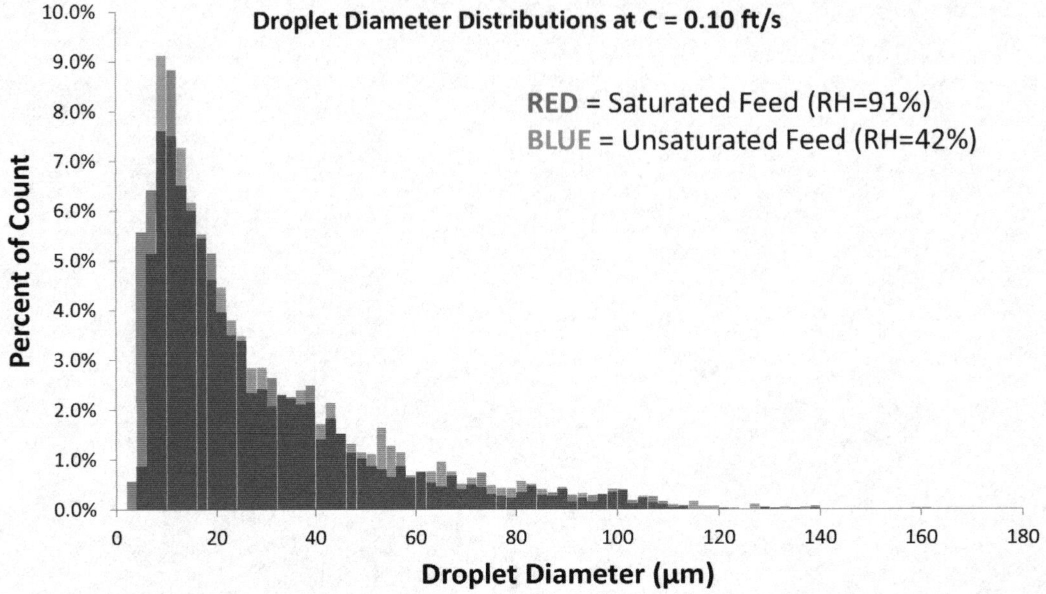

Motivation – What is PDI – Overview – **Results** – Discussion - Conclusions – Future Work

15 EFRI

Results

- Saturation of feed gas showed:
  - Decrease in Arithmetic Mean Diameter (D_{10})
    - For C = 0.10, 8.8% Decrease
  - Decrease in D_{32}
    - For C = 0.10, 5.4% decrease
  - Decrease in $D_{V0.5}$
    - For C = 0.10, 6.4% decrease
  - Decrease in $D_{0.5}$
    - For C = 0.10, 10.5% decrease
  - Increase in Liquid Water Content
    - For C = 0.10, 27.8% increase

Motivation – What is PDI – Overview – **Results** – Discussion - Conclusions – Future Work

16 EFRI

Discussion

$F_{Drag} \propto \frac{1}{2} \times \rho_{vapor} \times C_D \times V_t^2 \times A$

$\Delta v = v_{air} - v_{drop}$

Since $F_{Gravity}$ has a higher scaling power, as the radius increases so does Δv

$F_{Gravity} \propto \rho_L \times \frac{4}{3} \pi r_{Drop}^3 \times g$

Discussion

- For a given set of conditions, a maximum mass flux exists
  - $\frac{\dot{m}_{Total}}{A} = \frac{\dot{m}_{convection}}{A} + \frac{\dot{m}_{diffusion}}{A}$
- As $r \to r_{\min}$, $\Delta v \to 0 \therefore \dot{m} \to \dot{m}_{\min}$
  - Convective mass transfer diminishes
  - Small changes in mass lead to large changes in radius
- As $r \to r_{\max}$, Δv $increases \therefore \dot{m} \to \dot{m}_{\max}$
  - Convective mass transfer increases
  - Radius sensitivity to mass loss decreases

Discussion

- Largest entrained droplets experience the most mass transfer, smallest relative diameter change
- Smallest entrained droplets see least mass transfer, but diameter sensitivity is highest
- There must be some "sweet spot"
  - "Sweet spot" diameter range is affected by experimental conditions

Discussion

- Maximum drop that may be entrained is f(v)
- Drop size distributions matter!
  - If all droplets see the same velocity, collisions would be minimal post "turnaround" zone
  - Small droplets turn around quickly
  - Large droplets turn slowly, and further increases can lead to shedding or "eating" smaller droplets
  - Complex interactions increase with size of diameter range, increasing measurement variance

Discussion

C = 0.10 ft/s		
	Average	Rel. Std. Dev
D_{min} (μm)	2.6	17.5%
D_{10} (μm)	25.0	4.2%
D_{32} (μm)	64.4	1.7%
D_{max} (μm)	139.0	5.5%
D_{σ} (μm)	22.4	2.2%
v_{95} (m/s)	0.794	1.0%

C = 0.40 ft/s		
	Average	Rel. Std. Dev
D_{min} (μm)	8.2	55.6%
D_{10} (μm)	55.2	11.8%
D_{32} (μm)	155.1	8.3%
D_{max} (μm)	474.3	11.7%
D_{σ} (μm)	48.5	8.1%
v_{95} (m/s)	3.031	2.8%

Conclusions

- Saturation of feed does affect measurements
  - Arithmetic mean diameter (D_{10})
  - Sauter mean diameter (D_{32})
  - Median volume diameter ($D_{V0.5}$)
  - Liquid water content
  - Median droplet diameter ($D_{0.5}$)
- Small droplets experience the least mass transfer
- Largest droplets experience the most mass transfer
- Depending on process conditions, a "sweet spot" exists
- Increasing diameter distribution in entrainment increases measurement variance

Future Work

- Increase radial measurements
- Re-run trials at higher C factors
- Optimize experimental conditions to match equipment setup
- Further investigate the relationship between the "sweet spot" and process conditions to improve applicability
- Quantify entrainment rates in-column using only PDI
- Determine the effect of surface tension on entrainment

Motivation – What is PDI – Overview – Results – Discussion - Conclusions – **Future Work**

23

Acknowledgements

- Fractionation Research, Inc. (FRI)
  - Tony Cai, Ken McCarley
- Artium Technologies Inc.

24

Questions?

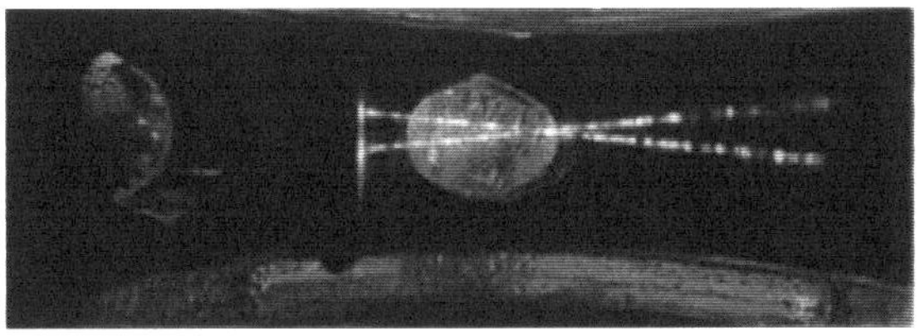

Effect of Liquid Viscosity on Mass Transfer for Packings

Di Song[a,b], A. Frank Seibert[b], Gary T. Rochelle[a,]*

[a]*McKetta Department of Chemical Engineering, The University of Texas at Austin, 200 E Dean Keeton St., C0400, Austin, TX 78712, USA*

[b]*Separations Research Program, Pickle Research Center, The University of Texas at Austin, 10100 Burnet Rd., Austin, TX 78758, USA*

Abstract

It is important to understand the effect of liquid viscosity (μ_L) on both the effective mass transfer area (a_e) and liquid film mass transfer coefficient (k_L) for packings to correctly predict the separation rate and properly design the column. Existing correlations are mostly based on water system and the application of them to viscous media could be problematic.

In this work, a_e and k_L were measured in a semi-works scale packed column (16.8 in. I.D.) with 10 ft packings installed as the internal. Seven structured packings (M 125Y, M 250Y, M 250X, GTO 250Y, B1 250MN, GTP 350Y, GTP 500Y), one random packing (RSR 1.5), and one hybrid (RSP 250Y) were tested. Liquid viscosity was varied from 0.8 to 70 cP by adding glycerol to water. The a_e was determined by reactive absorption of ambient CO_2 into dilute NaOH, and the k_L was determined by air stripping of toluene from the liquid. Liquid viscosity showed insignificant effect on a_e over the range of μ_L investigated. The total dependence of k_L on μ_L is -0.72, of which -0.35 is from the indirect influence through diffusivity, and -0.37 is from the direct influence of μ_L on k_L through liquid turbulence.

Models of a_e (Eq. 1) and k_L (Eq. 2) were developed based on the nine measured packings together with other 18 packings in the database of the Separations Research Program (SRP). The effect of secondary mass transfer (wall and end effect) has been included in the models.

$$\frac{a_{e,packing}}{a_P} = 1.26 \cdot \eta \cdot \left(We \cdot Fr^{-\frac{1}{3}}\right)^{0.117} = 1.26 \cdot \eta \cdot \left[\left(\frac{\rho_L}{\sigma}\right) g^{1/3} \left(\frac{u_L}{a_P}\right)^{4/3}\right]^{0.117} \quad (1.a)$$

$$a_{e,total} = a_{e,packing} + a_{secondary} \approx a_{e,packing} + a_{wall} \quad (1.b)$$

$$Sh = 0.124 \cdot Sc^{0.5} \cdot \left(Re \cdot Ca^{1/5} \cdot Fr^{-1/3}\right)^{1.08} \quad (2.a)$$

$$k_L = 0.124 \cdot u_L^{0.576} \cdot \mu_L^{-0.364} \cdot D^{0.5} \cdot a_P^{-0.44} \cdot g^{0.36} \cdot \rho^{0.58} \cdot \sigma^{-0.216} \quad (2.b)$$

With a relatively large equipment size and packing database, this work has the greatly expanded the data of mass transfer in viscous liquids available in the open literature.

Introduction

Many industrial applications encounter mass transfer in viscous liquid, such as electrolyte solutions, organic/polymer solutions, crude oil, and ionic liquids, etc. The increase in μ_L can significantly decrease k_L. Take post-combustion CO_2 capture with aqueous amine for example, the high solvent viscosity can cause slower diffusion of CO_2 to the bulk liquid, slower diffusion of free amine from the bulk liquid to the gas-liquid interface (surface depletion of alkalinity), slower diffusion of loaded amine back to the bulk liquid (which affects equilibrium partial pressure of CO_2 at the interface), and less liquid turbulence on the packing surface. Most of the existing k_L (k_La) correlations investigate only water and thus provide unsatisfactory predictions for the influence of μ_L [1]. Therefore, the application of them to amine solutions could lead to either inefficient or insufficient column design.

The SRP database for the air-water column contains valuable hydraulic and mass transfer data for various structured, random, and hybrid packings. Based on the database, models of a_e and k_L have been developed previously by Tsai [2] and Wang [3], respectively. The Tsai area model was developed seven years ago, and new data with totally different packing geometry have been generated continuously. For the k_L model developed by Wang, though it is relatively new, a large amount of new data with water and aqueous glycerol was generated during the past two years. Therefore, it is necessary to update both models with the expanded database.

Methods and Equipment

A systematic method for packing characterization with water has been developed at the Separation Research Program in the University of Texas at Austin. The a_e has been measured with chemical absorption of ambient CO_2 into aqueous NaOH because of the well-established kinetics [4]. The k_L has been measured with air stripping of toluene from water.

Glycerol was chosen as the liquid viscosity enhancer for its complete and easy dissolution in water, relatively cheap price, and the Newtonian behavior of its aqueous solution. Liquid viscosities of 65, 15, 4, and 1 cP (approx.) were obtained by adding 85, 65, 44, and 0 wt % glycerol to water. The reaction kinetics for CO_2/NaOH/H_2O/glycerol was calculated from a model developed based on the wetted-wall column data [5]. The glycerol will not change the controlling mechanism for mass transfer in the existing water systems. For toluene air stripping, though the infinite dilution activity coefficient of toluene decreases rapidly with increasing glycerol, the value will still be well above 100 at the highest proposed glycerol, which ensures the liquid film domination. For the reactive system, the effect of physical liquid film mass transfer (i.e. surface depletion of alkalinity) has been corrected in the kinetic model and is believed to be not an issue for the packed column experiments considering the relatively low CO_2 concentration and flux.

Packed column

The PVC packed column used in this work is sketched in Fig. 1. The inside diameter of the column was 16.8 in. The maximum packing height was 10 ft. The gas and liquid had countercurrent contact inside the column with air supplied by a 30-kW blower with variable speed drive from a duct (8 in outside diameter) below the packing support. The liquid was discharged by a centrifugal pump regulated with a variable speed drive with a maximum capacity of 150 gpm from a 350-gallon storage tank. The liquid could either flow to the column top and distributed by an F40 distributor (40 drip points/ft^2) or flow back to the column sump and storage tank via a bag filter for the pump to operate at a more favorable drive-speed region. Above the distributor was a Trutna collector to knock out any liquid reaching the column exhaust. Typical liquid inventory size was 250-300 gallon. Typical gas and liquid rates were 150-750 acfm and 1-30 gpm/ft^2 respectively.

Approximate 10 ft packing was installed for a_e measurement and k_L measurement with aqueous glycerol. A shorter 6 ft bed was used for k_L measurement with water because of the difficulty of GC analysis for the low outlet toluene concentration caused by high mass transfer efficiency. Each element was installed with 90° rotation with adjacent elements to facilitate liquid distribution as a routine practice.

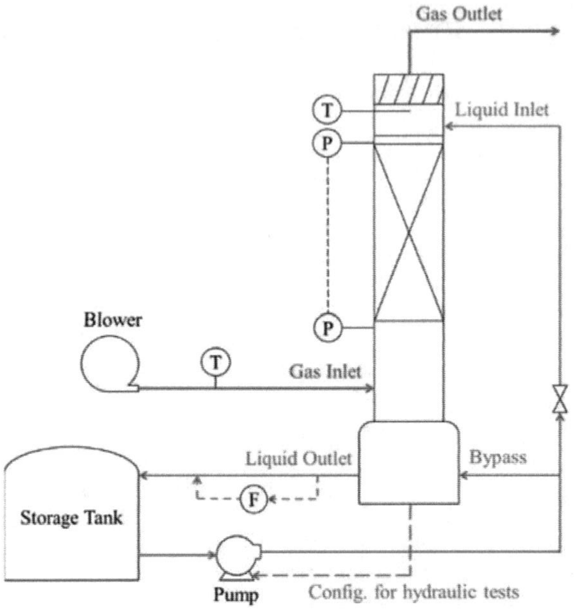

Fig. 1: Packed column experimental system.

Auxiliary Equipment

The CO_2 concentration in both the inlet and outlet gas was measured online by a XStream® CO_2 analyzer (Rosemount). Inlet CO_2 was ambient 370-400 ppm$_V$, and outlet CO_2 was 100-300 ppm$_V$ depending on the packing and gas flow rate. Toluene concentrations of the liquid samples were determined by gas chromatograph (Hewlett Packard 6890) analysis with 1-bromo-4-fluorobenzene used as the internal standard. Typical inlet toluene was 100-200 ppm$_W$, and outlet being 0.5-50 ppm$_W$ depending on the liquid flow rate and viscosity. A viscometer (TA Instrument AR-2000ex) was used for liquid viscosity measurement. A densitometer (Mettler Toledo DE40) was used for density measurement to determine glycerol concentration for reaction kinetic calculations.

Calculations

The total mass transfer area in the column, $a_{e,total}$, was determined experimentally by Equation 3. The k_g' for aqueous NaOH was calculated using the model of Pohorecki and Moniuk [4]. The k_g' for caustic glycerol solutions was calculated from a model developed based on the wetted-wall column data [5]. The k_La_e was determined experimentally by Equation 4.

$$a_{e,total} = \frac{K_{OG}a_e}{K_{OG}} \approx \frac{K_{OG}a_e}{k_g'} = \frac{u_L}{k_g' Z}\ln\left(\frac{C_{CO_2,\ in}}{C_{CO_2,\ out}}\right) \quad (3)$$

$$k_L a_{e,total} = \frac{u_L}{Z}\ln\left(\frac{C_{toluene,\ in}}{C_{toluene,\ out}}\right) \quad (4)$$

Physical properties (density, viscosity, surface tension, diffusivity, Henry's constant of CO_2) of the aqueous glycerol solutions were calculated using models developed previously [1].

The secondary effect (end and wall effect) on mass transfer area ($a_{secondary}$) has been corrected to give effective mass transfer area of packing surface ($a_{e,\text{packing}}$). To calculate the wall area, it has been assumed that the column wall was totally wetted not only in the packing section, but also in the auxiliary section (for this case with only one packing section, it's only the section between packing bottom and liquid sump):

$$a_{wall} = \pi \cdot D_{column} \cdot \left(Z_{packing} + Z_{auxiliary}\right) \quad (5)$$

The top area includes area generated by the liquid streams from distributor to packing top (both stream peripheral and impact area). Similarly, the bottom area includes area generated by the liquid streams from packing bottom to sump. It has been demonstrated in later section that wall area is always dominant in $a_{secondary}$.

Model of Mass Transfer Area (a_e)

The $a_{secondary}$ has been deducted from $a_{e,total}$ to give $a_{e,\text{packing}}$ (Eq. 5). With 16.8 in. I.D. column, the secondary area is not significant for normal to fine packings, but it starts to play an important role for course packings (Fig. 2). Since the wall area is dominant in the secondary area for the whole spectrum of packing a_P, top and bottom area have been omitted in the future correction of secondary effect for simplification. The column diameter will greatly affect the $a_{secondary}$ as expected (Fig. 3). Therefore, this correction will enable wider application of the area model to column with different sizes. Different regimes of packing area for MG 64Y have been shown in Fig. 4 for illustration.

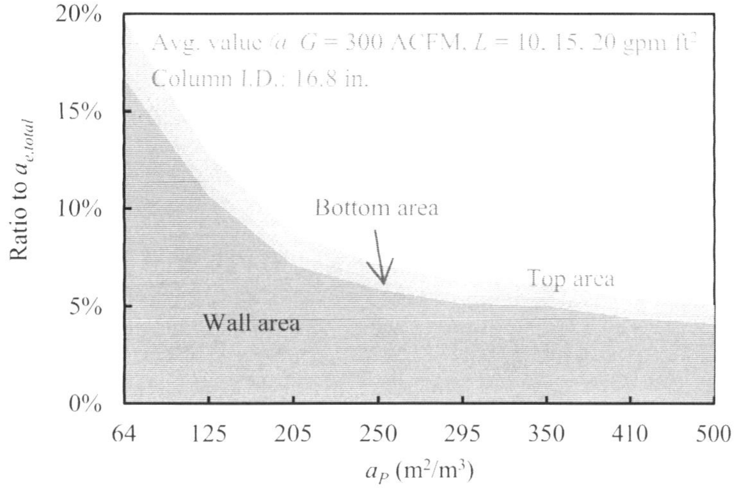

Fig. 2: Effect of a_P on $a_{secondary}$

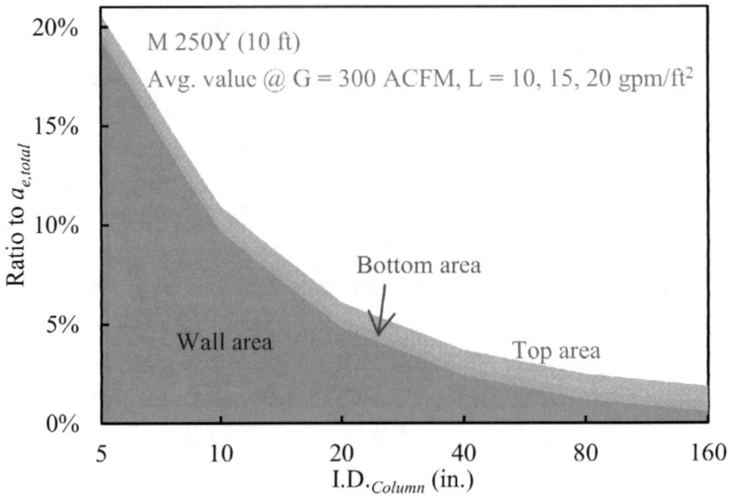

Fig. 3: Effect of column size on $a_{secondary}$

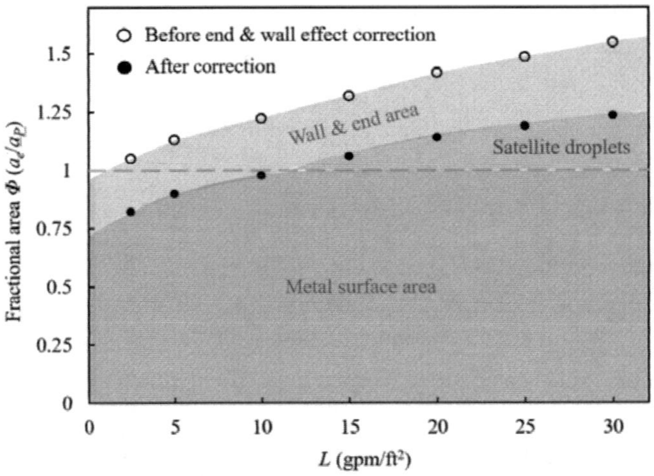

Fig. 4: Packing area regimes of MG 64Y

Figure 5 compares the measured and calculated area with water. An empirical correction factor has been included in the model to account for deviation of random/hybrid packings and plastic packings (Eq. 6 and Tab. 1). For 27 packings with drastically different geometry, the model shows an average absolute deviation (AAD) of only 6.3 percent (Fig. 6).

$$\eta = \eta_{type} \cdot \eta_{material} \tag{6}$$

Tab. 1: Packing correction factors for area model

η_{type}	Structured	Random/Hybrid
	1.0	$-2\times10^{-5}a_P^2+7\times10^{-3}a_P+0.64$
$\eta_{material}$	Stainless Steel	Plastic
	1.0	0.7

Per the model, surface tension affects $a_{e,packing}$ but not viscosity. This is confirmed by Fig. 7 & 8 where no bias of error was observed as a function of the two physical properties. For viscosity, both solutions of polyethylene glycol (PEG) and glycerol have shown no difference in the $a_{e,packing}$ compared to water.

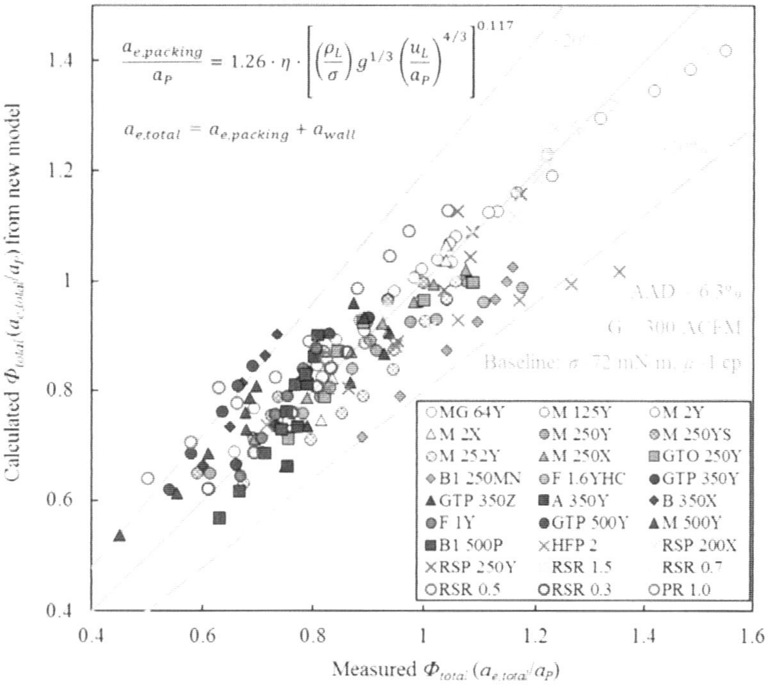

Fig. 5: Comparison of measured and calculated area from Eq. 1 for water

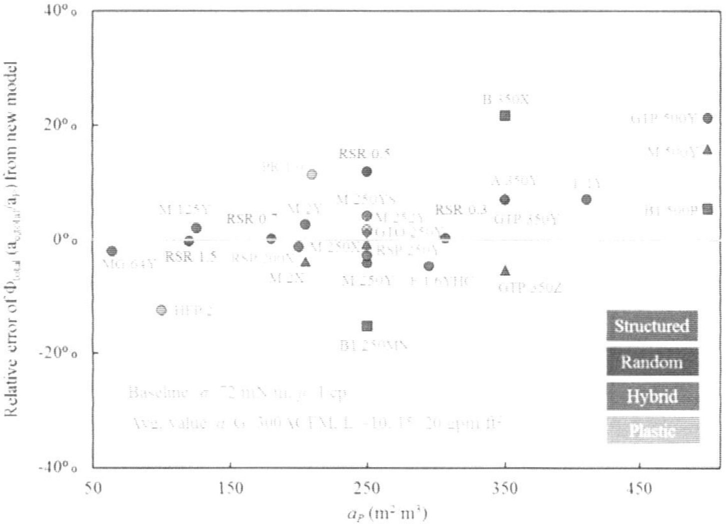

Fig. 6: Relative error of Eq. 1 for different packings

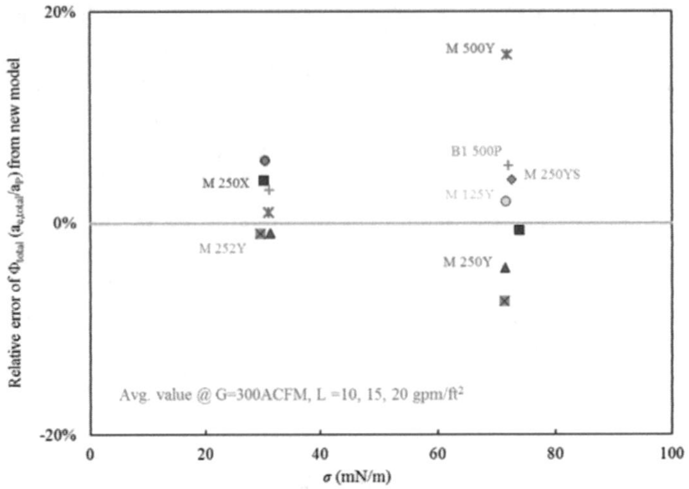

Fig. 7: Relative error of Eq. 1 for different surface tension

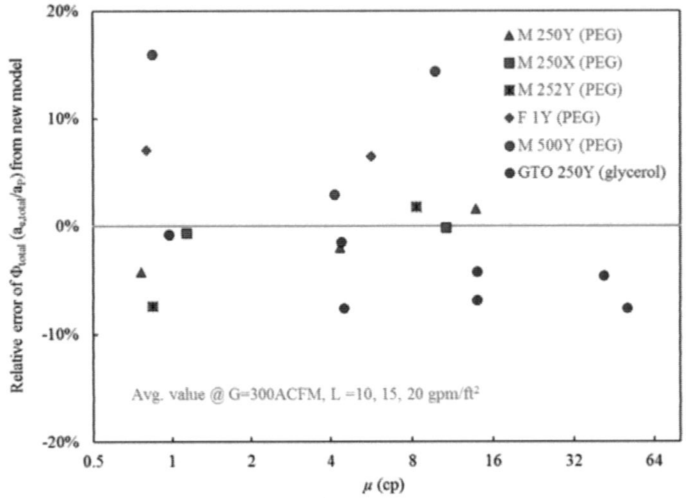

Fig. 8: Relative error of Eq. 1 for different viscosity

Model of Liquid Film Mass Transfer Coefficient (k_L)

The experimental protocol for k_L measurement of SRP has been changed over time. A total of 10 10 ft of packing was initially used. The height was then reduced to 6 ft due to the low outlet toluene concentration that is close to the detection limit of the gas chromatograph (GC). Theoretically, k_L should not be affected by packing height when there is no issue with liquid distribution and GC detection. However, greater k_L has been observed with the 6 ft bed than the 10 ft bed, regardless of packing type and experiment operators (Fig. 9).

The difference of k_L with different bed height may be caused by both the poorer liquid distribution and by issue with the GC detection limit at greater packing bed height. In the development of k_L model, k_L measured with 10 ft bed will be multiplied by 1.32 to be able to compare with those measured with 6 ft bed.

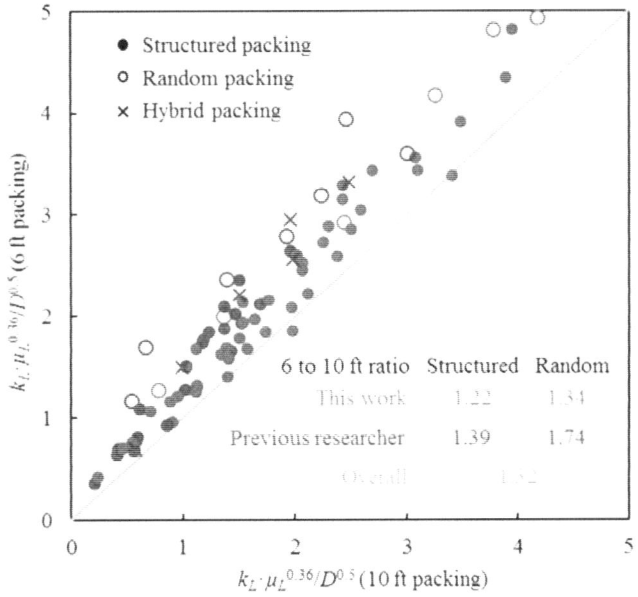

Figure 9: Comparison of k_L measured with 10 ft and 6 ft packing

The average k_L of structured, random, and hybrid packings at 300 ACFM gas rate and 10, 15, and 20 gpm/ft^2 liquid load have been plotted against a_P (Fig. 10). For structured packings, there is a counter-intuitive trend that the k_L decreases with increasing a_P. This might be caused by the longer flow path (and thus greater turbulence) of the liquid on the surface of course packings. The corrugation angle does not have a clear effect on k_L.

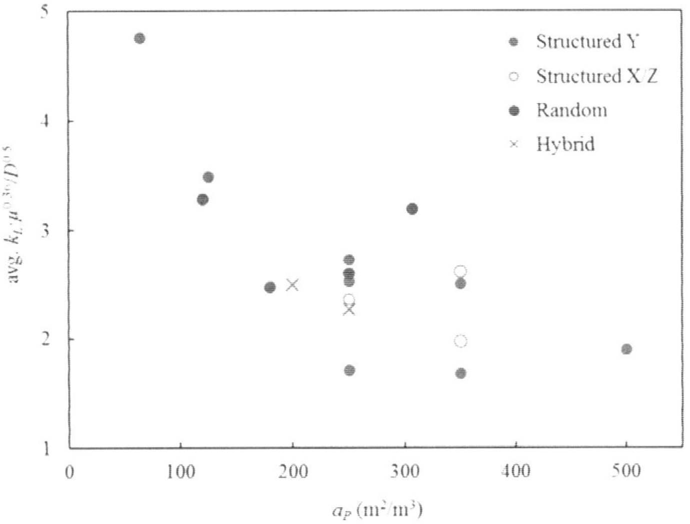

Fig. 10: Average k_L as a function of a_P

Figure 11 compares the measured k_L and calculated k_L from Eq. 2 for 21 packing. Glycerol and water tests have been performed on nine of the packings, and water tests only on the rest of 12 packings. The new model shows an AAD of 20% for all 21 packings. Unlike the area model, the k_L model does not require a correction for random or hybrid packings.

Figure 12 shows the total dependence on μ_L of k_L. The slope (-0.72) is the same for structured, random, and hybrid packings, which means the impeding effect of μ_L on k_L is the same for all three types of packings. Of the total -0.72 dependence, -0.35 is from the indirect influence of μ_L through diffusivity, and -0.37 is from the direct influence of μ_L on k_L through liquid turbulence. The indirect part is believed to be system dependent, and requires knowledge on the D-μ relationship. The direct part is universally applicable to all Newtonian liquids.

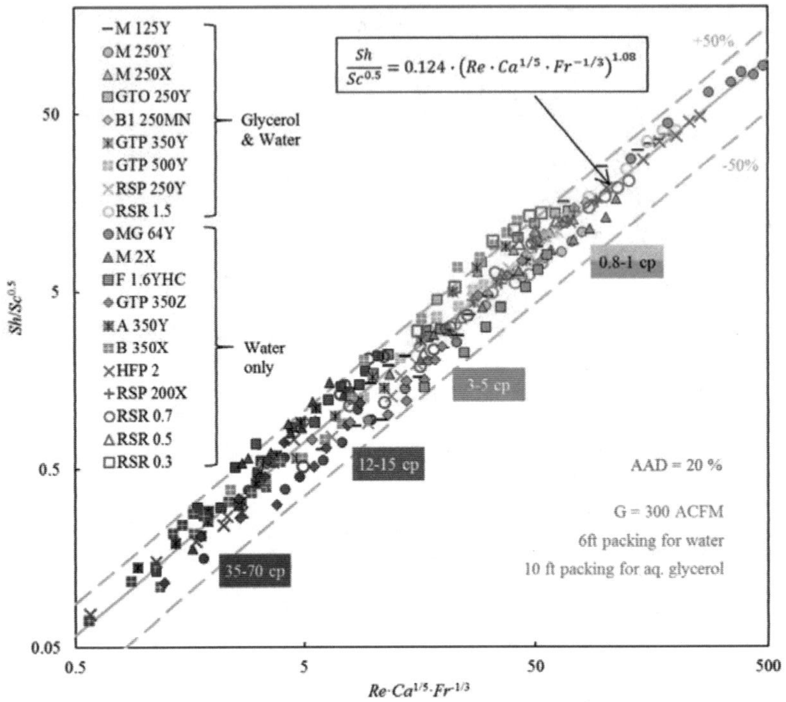

Fig. 11: Comparison of measured and calculated k_L from Eq. 2

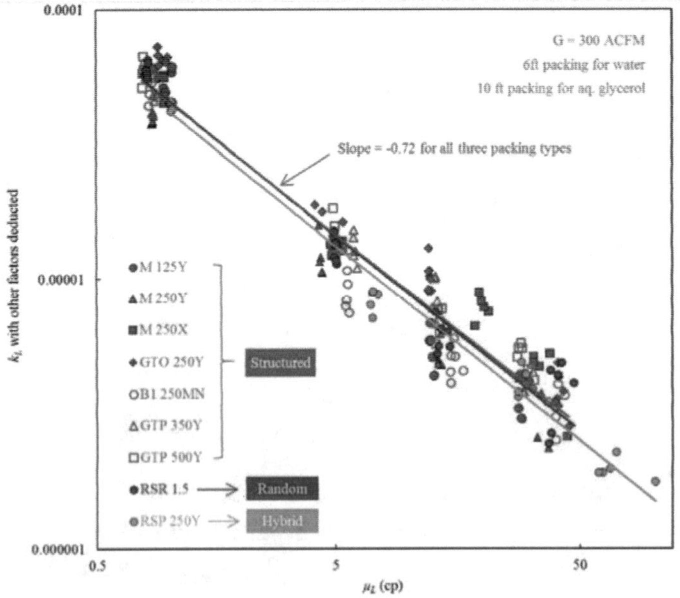

Fig. 12: Total dependence on μ_L of k_L

Figures 13 & 14 shows the relative error of Eq. 2 as function of a_P, and μ_L, respectively. No bias of error was observed as a function of a_P for water in Fig. 13. The model underpredicts k_L for water slightly. In Fig. 14, no bias of error was observed as a function of μ_L.

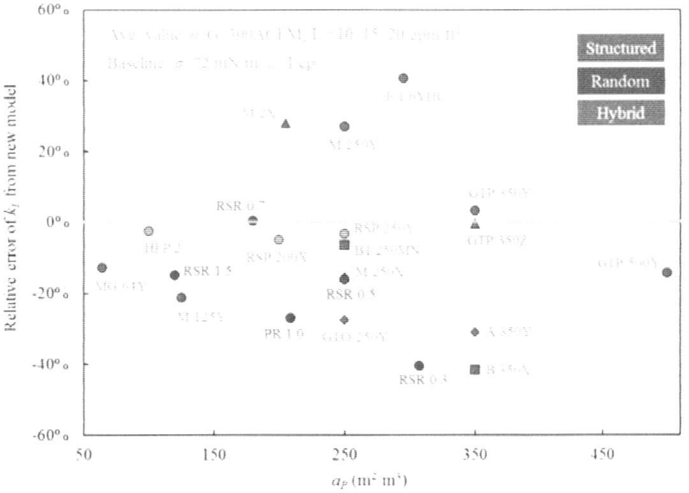

Fig. 13: Relative error of Eq. 2 for different a_P

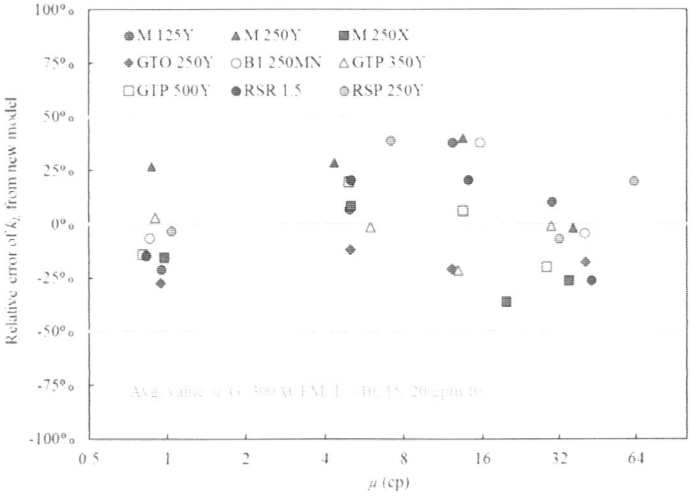

Fig. 14: Relative error of Eq. 2 for different μ_L

Conclusions

Models of a_e and k_L (Eq.s 1 & 2) have been developed based on the SRP air-water column database with 27 packings of various type, a_P, and materials. Liquid viscosity has been varied from 0.8 to 70 cP by the addition of glycerol. The μ_L does not affect $a_{e,packing}$, but it affects k_L with a total dependence of -0.72, of which -0.35 is from the indirect influence through diffusivity, and -0.37 is from the direct influence of μ_L on k_L through liquid turbulence.

The wall effect is dominant in the secondary effect on a_e, and it has been included in the a_e model. The k_L decreases with increasing a_P for structured packings, and shorter packing bed consistently shows greater k_L than taller bed with the same packing internal. With a relatively large equipment size and packing database, this work has the greatly expanded the data of mass transfer in viscous liquids available in the open literature.

Nomenclature

a_e – effective area of mass transfer per volume of column, m^2/m^3
a_P – specific area of packing per volume of column, m^2/m^3
C – concentration, ppmv or ppmw
D – diffusivity, m^2/s
d – characteristic length in dimensionless groups ($1/a_P$), m
g – gravitational acceleration, m/s^2
k_g' – liquid film mass transfer coefficient with chemical reactions expressed in gas unit, $mol/m^2 \cdot Pa \cdot s$
k_L – liquid film mass transfer coefficient, m/s
K_{OG} – overall mass transfer coefficient expressed in gas unit, $mol/m^2 \cdot Pa \cdot s$
u – characteristic velocity in dimensionless groups (liquid superficial velocity), m/s
Z – packing height

Greek letters

σ – surface tension, N/m
μ – viscosity, cP
ρ – density, kg/m^3
η – correction factor for area model

Dimensionless groups

Ca – Capillary number, $\mu \cdot u/\sigma$
Fr – Froude number, $u^2/g \cdot d$
Re – Reynolds number, $d \cdot u \cdot \rho/ \mu$
Sc – Schmidt number, $\mu/\rho \cdot D$
Sh – Sherwood number, $k_L \cdot d/D$
We – Webber number, $\rho \cdot u^2 \cdot d/\sigma$

Packing names

B1 – MONTZ-Pak Type B1TM
F – Flexipac®
GTO – GT-OPTIMPAKTM
GTP – GT-PAKTM
HFP – Hiflow® Plus
M – MellaPakTM
MG – MellaGridTM
PR – Pall Ring
RSP – Raschig Super-Pak®
RSR – Raschig Super-Ring®

References

1. Song, D., A.F. Seibert, and G.T. Rochelle, *Effect of Liquid Viscosity on the Liquid Phase Mass Transfer Coefficient of Packing.* Energy Procedia, 2014. **63**(0): p. 1268-1286.
2. Tsai, R.E., *Mass Transfer Area of Structured Packing*, in *Department of Chemical Engineering*. 2010, The University of Texas at Austin: USA.
3. Wang, C., *Mass Transfer Coefficients and Effective Area of Packing*, in *Department of Chemical Engineering*. 2015, The University of Texas at Austin: USA.
4. Pohorecki, R. and W. Moniuk, *Kinetics of reaction between carbon dioxide and hydroxyl ions in aqueous electrolyte solutions.* chemical Engineering Science, 1988. **43**(7): p. 1677–1684.
5. Song, D. and G.T. Rochelle, *Reaction kinetics of carbon dioxide and hydroxide in aqueous glycerol.* Chemical Engineering Science, 2017. **161**: p. 151-158.

AIChE Paper 7d

MOLECULE TRACKING - A NEW WAY TO ANALYZE DISTILLATION COLUMNS

Clifford Alan Maat

Senior Staff Engineer

TPC Group

Prepared for Presentation at the 2017 Spring National Meeting

Distillation Symposium

San Antonio, Texas, March 27 -30, 2016

Abstract: Molecule Tracking, tracing the paths of individual molecules through the trays of a distillation column, is proposed as a method of analysis to supplement conventional shortcut and rigorous tray calculations. The benefits of this method, in its current form, include better understanding of the dynamics of distillation operation and an objective measurement of expected performance variation around the design product purities.

Background

Distillation Column analysis and design work always treats the chemicals as continuous quantities rather than individual molecules. As a practical matter, this is perfectly correct. However, considering the individual molecules' behaviors can be useful for improving understanding of complex distillation problems.

Molecule Tracking

Consider the hypothetical demethanizer column in Figure 1. Methane and Ethylene are the light and heavy keys.

Figure 1

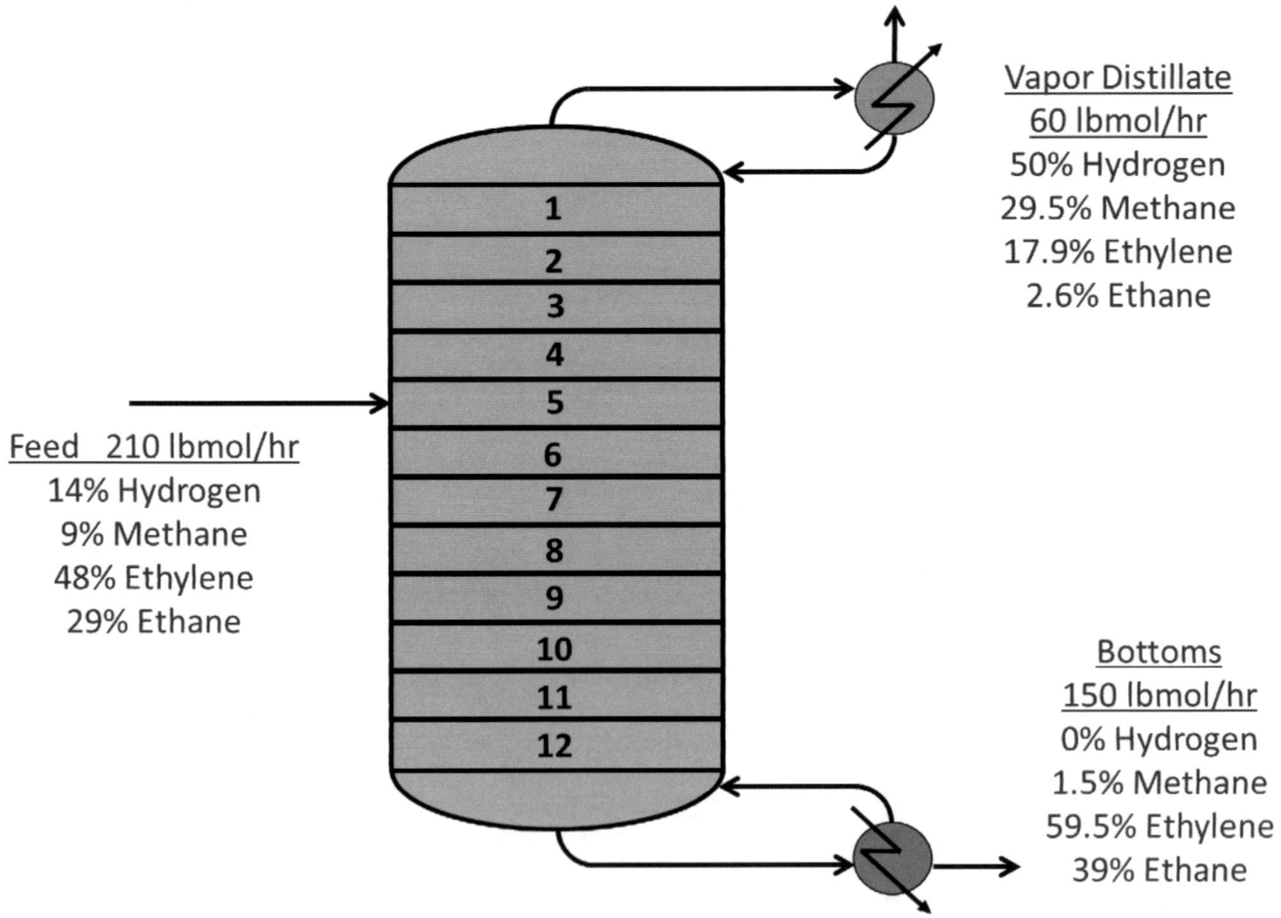

Each molecule in the feed enters the feed tray and then, sooner or later, must go up to the next higher tray or down to the next lower tray. Then it must, sooner or later, go back to the feed tray or away to the next higher/lower tray. The steps continue, either up or down each time, until the molecule exits to either the distillate or bottoms product.

A possible path for a methane molecule is shown in Figure 2. Paths for hydrogen and ethylene molecules are shown in Figure 3.

Figure 2 – A Possible Path for a Methane molecule

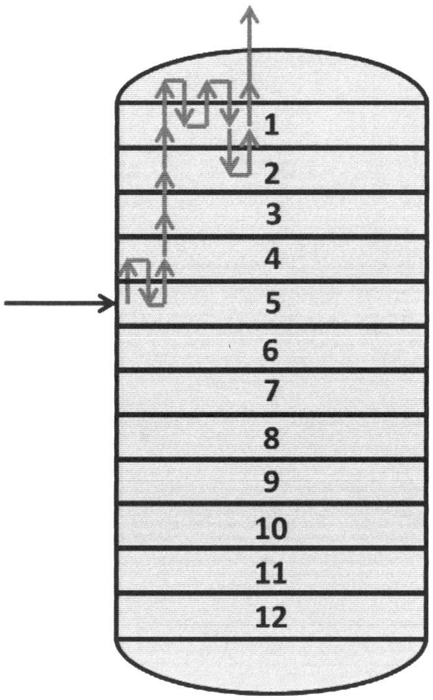

Figure 3 – Possible Paths for Hydrogen and Ethylene

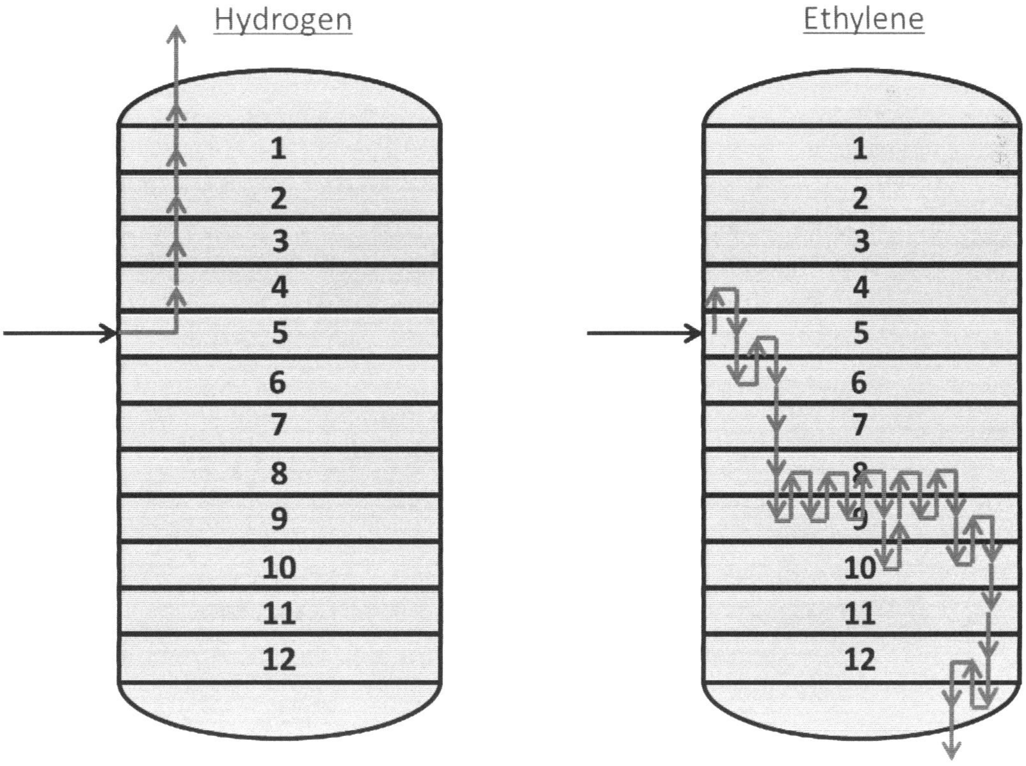

There is a random element to each step in the paths; however, clearly the more volatile molecules will tend to go upward while the less volatile molecules go down.

Mathematical Description of the Steps

A diagram of a theoretical tray is shown in Figure 4.

Figure 4 – Theoretical Tray

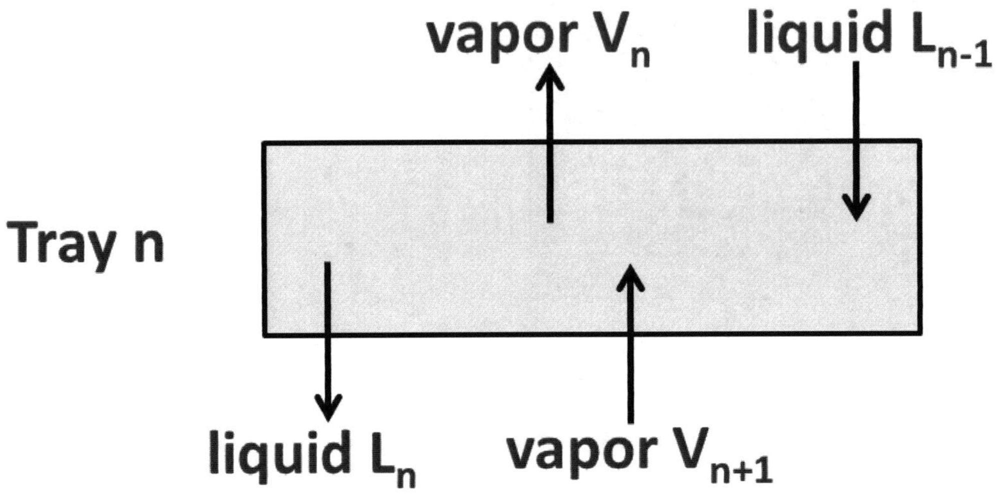

Entering the tray is liquid L_{n-1} from the tray above and vapor V_{n+1} from the tray below. Leaving the tray is vapor V_n to tray n-1 and liquid L_n to tray n+1, each with chemical composition in equilibrium with the other. At steady state, the amounts of each chemical entering equal the amounts leaving.

Consider a single molecule of component 1 in the tray. It must leave to either the tray above or the tray below. The probability that it will leave to the tray above is

$P_1(n) = y_1(n)*V(n) / (y_1(n)*V(n) + x_1(n)*L(n))$

There is a corresponding probability for each component on each tray. At each tray with a product stream (condenser, reboiler, and any sidedraw tray), there is another probability that the molecule leaves as a product.

Once all the probabilities are determined, the path of a molecule may be traced from the feed to the eventual exit.

Example: Optimization of a Sidedraw Column

A diagram of an operating column is shown in Figure 5.

Figure 5 – Baseline Column Operation

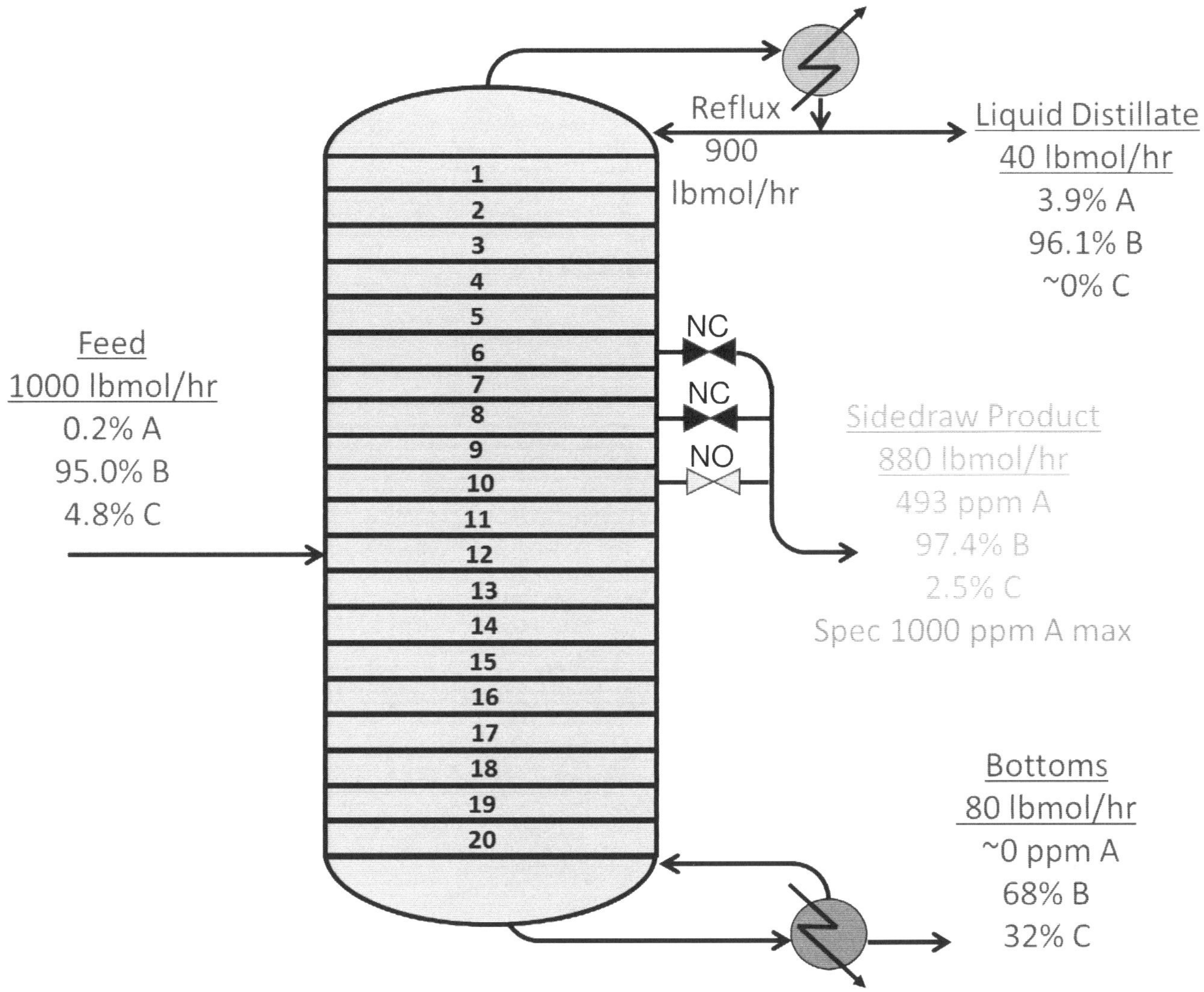

Column Purpose and Arrangement

This arrangement has been described as "one prominent product" [1]. The feed contains components A, B, and C, with A the most volatile. The valuable product is B. The contaminant most detrimental to product quality is A; the specification is 1000 ppm, but product with concentrations lower than the specification receive higher prices. The heavy impurity C also has a specification which is easily met and does not much affect the product value. The feed goes to Tray 12. The product sidedraw may be taken from Tray 6, 8, or 10. The trays have spare capacity and good turndown capability.

Normal Operation

The sidedraw was being taken from Tray 10 in order to minimize the amount of A in the product.

Optimization Study

Simulation of the column indicated that the sidedraw purity could be improved by increasing reflux; this was not surprising. However, it also indicated that the choice of sidedraw tray made no difference – drawing from tray 6, 8, or 10 all showed essentially the same product purity. Why?

A Molecule Tracking analysis of component A shows why the choice of sidedraw tray was insignificant, and also gives better insight to why the increase in reflux was effective. Look at the following paths for ten randomly chosen A molecules in Figure 6:

Figure 6

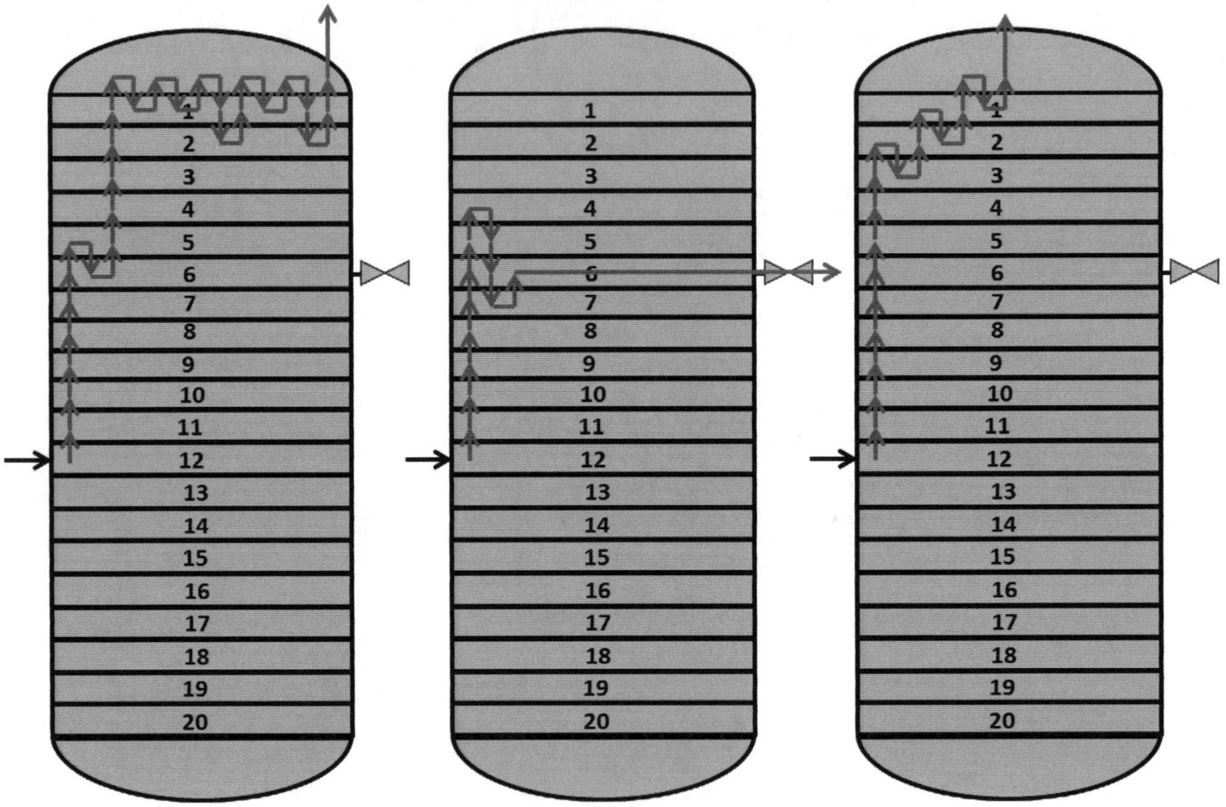

Figure 6, cont.

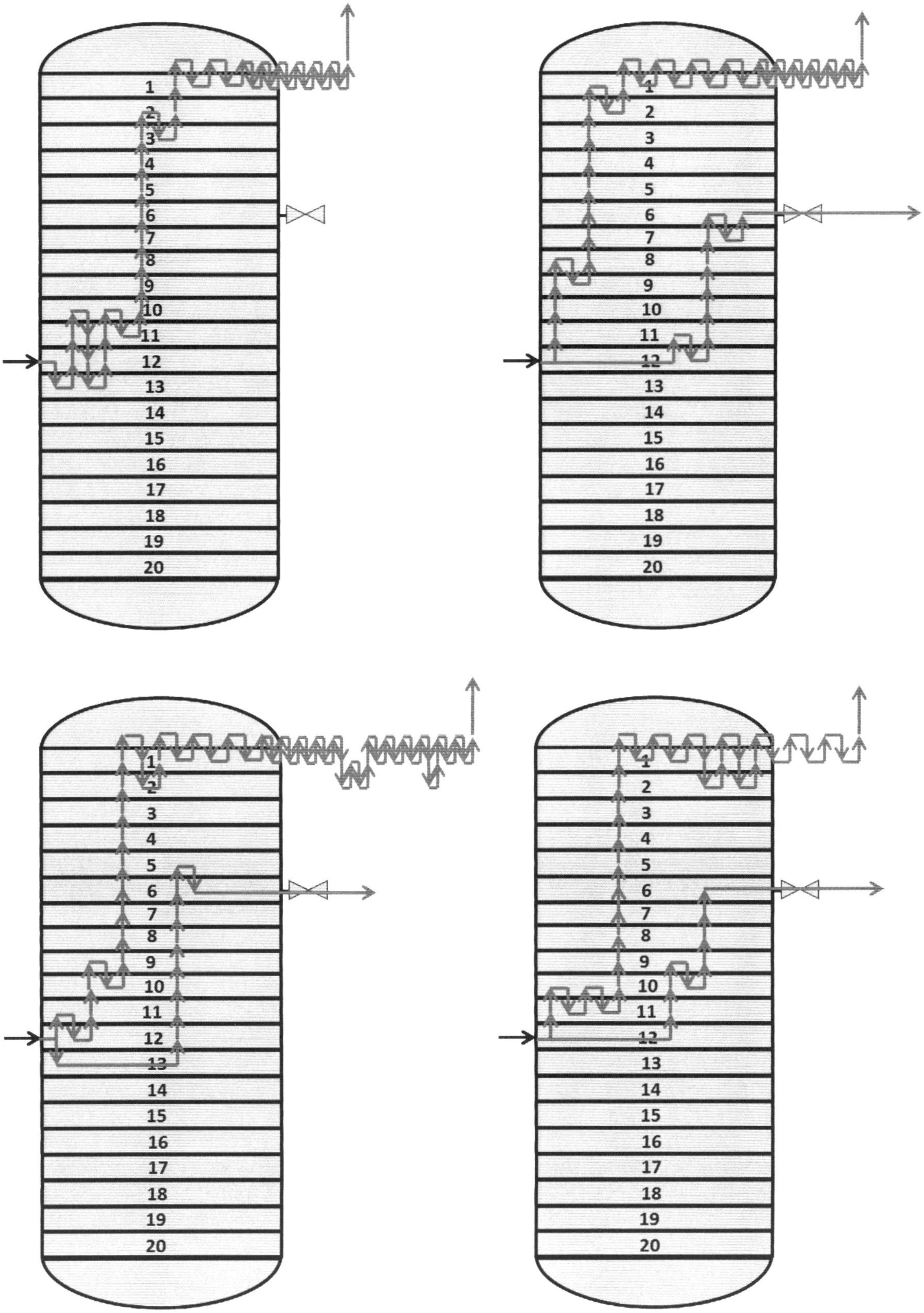

All of the A molecules that exited to the sidedraw came straight from the feed, or nearly so. Once an A molecule reached the overhead condenser, it had almost no chance of getting back down to tray 6, or 8, or 10. The flowrate of A entering the tray was the same for 6, 8, and 10, and the probability of leaving the tray as a vapor was the same, so the concentration of A in the sidedraw product was the same.

Increasing reflux led to a corresponding increase in vapor flow through all the trays. Since the total vapor leaving the sidedraw tray increased, while the total sidedraw flow remained the same, *the probability of an A molecule leaving with the vapor increased.* Therefore the sidedraw product purity improved. This is in contrast to most distillation separations; normally increases in reflux are effective because they increase the separation per tray, or the concentration gradient along the column. In this sidedraw column, the separation per tray was irrelevant and there was no concentration gradient between 6 and 10 (the concentration of A was the same at trays 6, 8, and 10, whichever reflux rate was chosen).

Thus raising the sidedraw point from tray 10 to tray 6 had no effect on the concentration of A. It did, however, decrease the concentration of C. This enabled a reduction in the bottoms flowrate, which increased the amount of B in the sidedraw, increasing total product volume and fractional recovery of B and further reducing the concentration of contaminant A in the product.

Final recommendations were to

1. Raise the sidedraw point to tray 6.
2. Increase reflux to the highest practical flow.
3. Increase bottoms temperature to reduce loss of B to the bottoms.

Figure 7 shows the improved column operation.

Figure 7 – Improved Column Operation

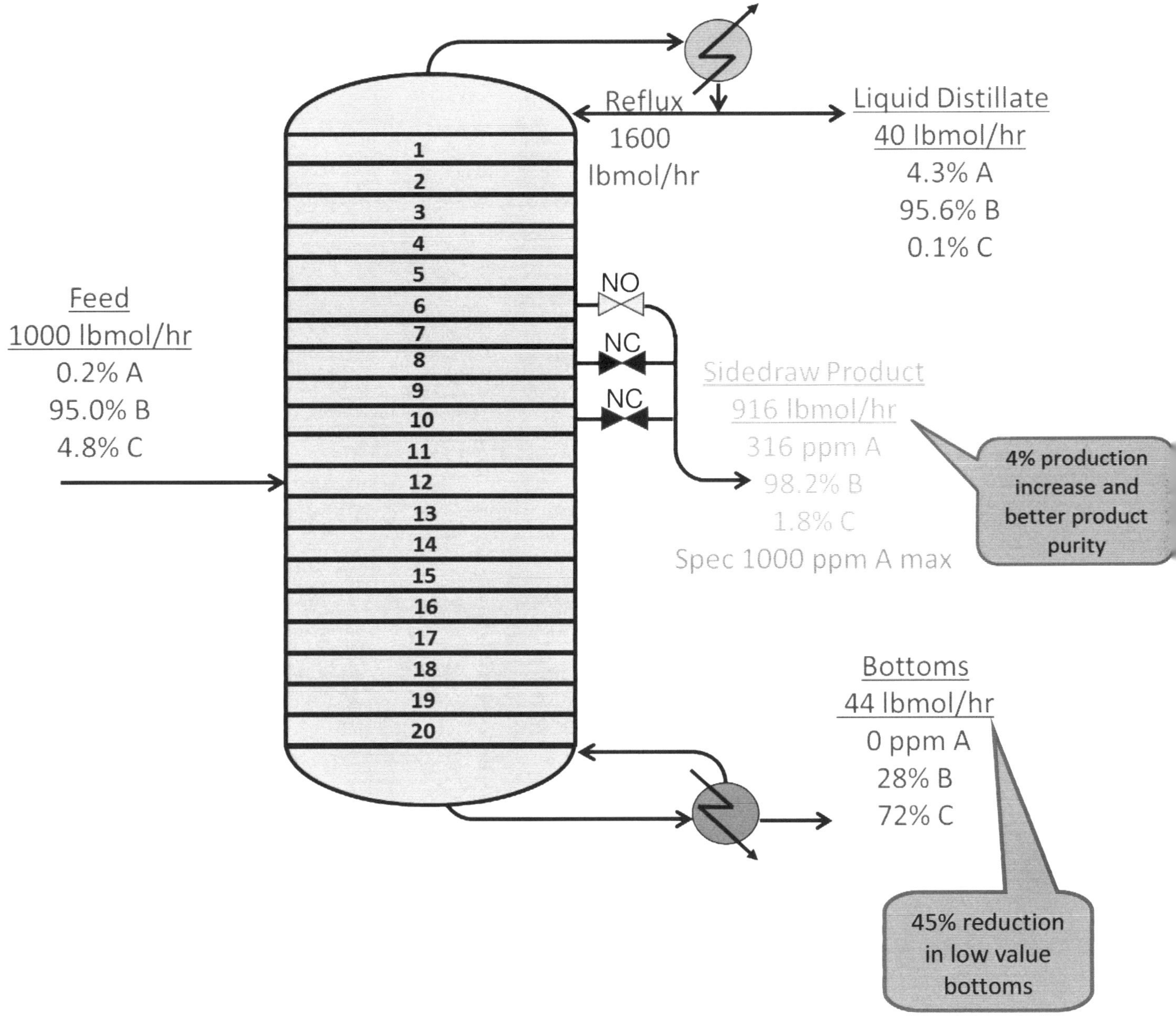

Measuring the Variability of Column Performance

A process simulation of a distillation column gives a single set of precise numbers for the composition of each product. However, as every operator knows, the compositions of the products from a real column vary somewhat with time even if all operating variables are held as constant as possible.

The Molecule Tracking method, repeated for a large number of molecules, also gives the composition of each product from a particular distillation column, but when done a second time it gives a slightly different solution. If the method is re-run a number of times, the composition of each component in each product forms a distribution around an average. An example distribution is shown in Figure 8 for the ethylene concentration in the distillate of the hypothetical demethanizer column:

Figure 8

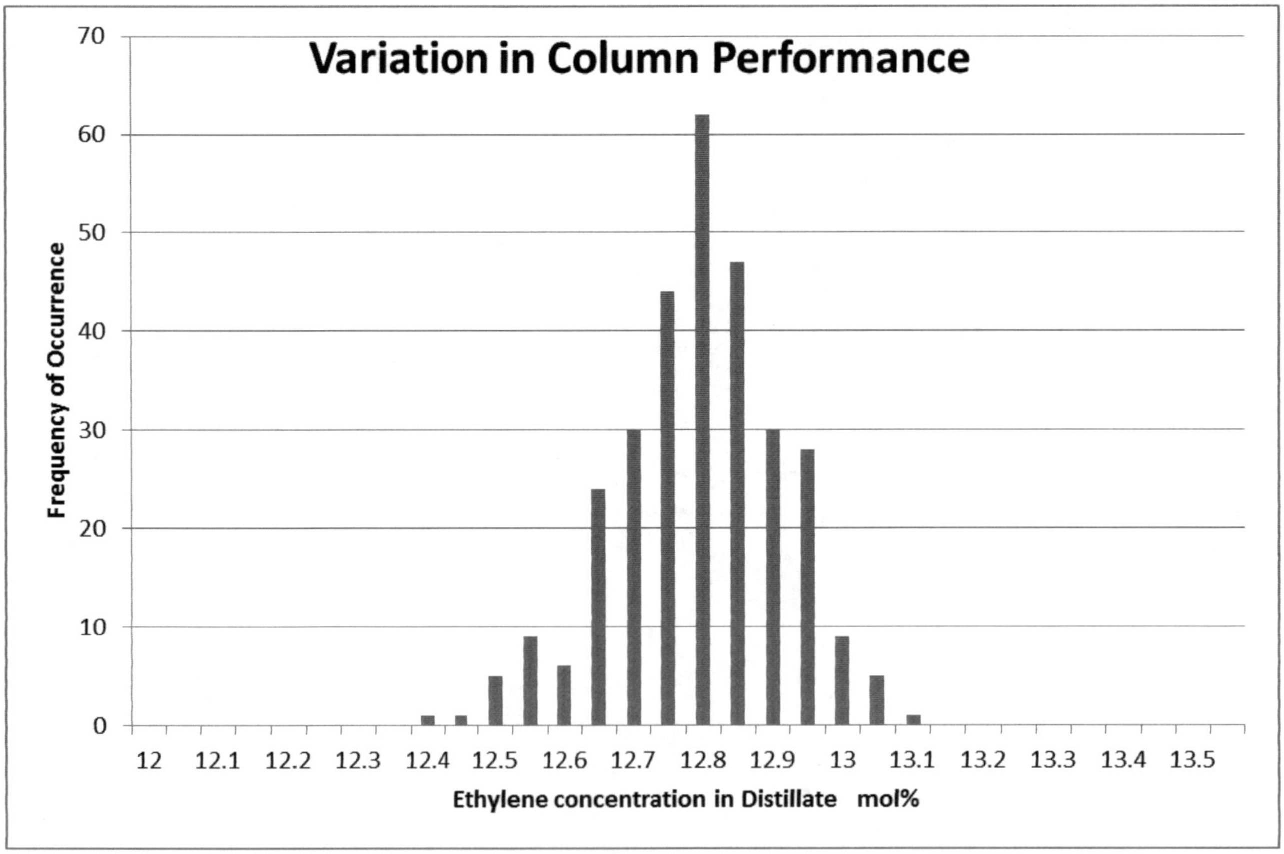

The average is 12.77 mole%, with a standard deviation of 0.12 mole%. It appears to be not quite a normal bell curve, skewed with a longer tail on the low side. This curve may provide insight to how the real column is expected to behave. For example, if it is desired to keep the ethylene concentration below 13% at all times, the bell curve indicates that controlling at 12.8% would not work, but controlling at 12.6% might. Process control strategies could then be adjusted accordingly.

Limitations of the Method in its Current Form

1. The Molecule Tracking method cannot simulate the column; a converged simulation is required in order to create the probability matrix.
2. The product composition does not exactly match the original simulation in the cases tried so far; for example, in the hypothetical demethanizer, the distillate ethylene concentration is 12.77 mol%, but the simulation shows 17.9%. Flashing feed has been observed to change the results. In general, it is not known whether the discrepancies are a problem with the program, the simulation output, or the Molecule Tracking method itself.

Summary

The new column analysis method of Molecule Tracking, tracing the paths of individual molecules through the trays of a distillation column, can be used for

- Training of operators or engineers new to distillation,
- Troubleshooting and optimization, when the operating condition or results of a column seem counterintuitive or confusing,
- Control strategy, when the random variation in column performance has a bearing on desired control points.

Further development of the Molecule Tracking method may improve its accuracy or reveal other profitable uses.

References

1. Kister, H. Z. (1990), "Distillation Operation," p. 593.

AUTHOR INDEX

AIChE
The Global Home of Chemical Engineers

AIChE
120 Wall Street, FL 23
New York, NY 10005-4020

ISBN 978-1-5108-4280-9

CURRAN ASSOCIATES INC.
proceedings.com